INTERNATIONAL SERIES OF MONOGRAPHS IN
CHEMICAL ENGINEERING
GENERAL EDITOR: P. V. DANCKWERTS

VOLUME 11

PREDICTION OF TRANSPORT AND OTHER PHYSICAL PROPERTIES OF FLUIDS

PREDICTION OF TRANSPORT AND OTHER PHYSICAL PROPERTIES OF FLUIDS

BY

S. BRETSZNAJDER

TRANSLATED FROM THE POLISH BY
Dr. J. BANDROWSKI

PERGAMON PRESS
OXFORD · NEW YORK · TORONTO
SYDNEY · BRAUNSCHWEIG

WYDAWNICTWA NAUKOWO-TECHNICZNE
WARSAW

Pergamon Press Ltd., Headington Hill Hall, Oxford
Pergamon Press Inc., Maxwell House, Fairview Park, Elmsford, New York 10523
Pergamon of Canada Ltd., 207 Queen's Quay West, Toronto 1
Pergamon Press (Aust.) Pty. Ltd., 19a Boundary Street,
Rushcutters Bay, N.S.W. 2011, Australia
Vieweg & Sohn GmbH, Burgplatz 1, Braunschweig

First English edition 1971

Originally published under the title
Własności gazów i cieczy
by Wydawnictwa Naukowo–Techniczne, Warsaw

Library of Congress Catalog Card No. 75–138678

Printed in Poland

08 013412 2

CONTENTS

NOTATION

A	cross-sectional area
a	constant in van der Waals' equation
a	activity
a	thermal diffusivity
b	constant in van der Waals' equation
C	condenser capacity
C	similarity constant
C	Sutherland constant
C_p	molal heat capacity at constant pressure
C_v	molal heat capacity at constant volume
c	concentration
c	heat capacity
c_p	specific heat capacity at constant pressure
c_v	specific heat capacity at constant volume
D	diffusivity
E°	standard energy of vaporization
E_d	activation energy for viscous flow
E_r	potential energy
E_v	activation energy of liquid flow
ΔE_v	internal energy of vaporization
e	electrical charge
F	electric field intensity
F	diffusion factor
F	Faraday's constant
ΔH	heat of activation
h	Planck's constant
I	viscosity constitutional constant
k	Boltzmann constant
L	molal heat of vaporization
L_b	molal heat of vaporization at normal boiling temperature
L_m	molal heat of fusion (melting)
l	distance between the poles of a dipole
l	mobility of ion
$\bar{l}$	mean free path of molecule
l_A	equivalent conductivity of anion
l_C	equivalent conductivity of cation
M	molecular weight
m	molality of solution
m	mass of single molecule
m	number of side chains in molecule
m_i	weight fraction of component i
N	Avogadro's number
N, n	number of moles
n	number of carbon atoms in the molecule
n	number of degrees of freedom
n	number of molecules in 1 cm^3
n	refractive index
O_r	orthochor
P	polarization

P_{ch} parachor
P, p pressure
p_c critical pressure
p_n pressure of saturated vapour
p_{pc} pseudocritical pressure
p_r reduced pressure
p_{rm} reduced pressure of mixture
Q heat
R gas constant
R molar refraction
R_{ch} rheochor
R_D molecular refraction
r distance between the molecules
r radius of diffusing molecule
r resistivity
S cross-sectional area
S entropy
ΔS entropy of activation
T absolute temperature
T_b normal boiling temperature
T_c critical temperature
T_f freezing temperature
T_m melting temperature
T_{pc} pseudocritical temperature
T_r reduced temperature
T_{rm} reduced temperature of a mixture
t independent variable
t temperature, °C
U_u molar sound velocity
u mobility of ion
u sound velocity
V molal volume
V_b molal volume at boiling temperature
V_c critical molal volume
V_m molal volume of solid at melting temperature
V_0 molal volume at absolute zero
V_{pc} pseudocritical volume
V_r reduced volume
v specific volume
v_c specific volume at critical point
v_M volume per molecule
v_m volume of single molecule
w velocity
w Wiener number
$\bar{w}$ mean linear velocity of molecule
x mole fraction (in liquid phase)
x length
y mole fraction (in gaseous phase)
z compressibility factor
z valency of ion
z_c critical compressibility factor
α polarizability of the molecule
α_c critical parameter
α_i contribution of ion i to thermal conductivity
β molar friction coefficient
β volumetric coefficient of expansion
β_s adiabatic compressibility
γ activity coefficient

γ specific weight
ε dielectric constant
ε force constant
θ angle between the vectors of the dipoles
$\varkappa$ ratio c_p/c_v
Λ^* characteristic parameter
λ thermal conductivity
λ_m thermal conductivity of a mixture
λ_r reduced thermal conductivity
μ viscosity (sometimes called "dynamic" viscosity)
μ_b viscosity of liquid at normal boiling temperature
μ_c viscosity at critical point
μ_m viscosity of liquid at melting temperature
μ_{mix} viscosity of mixture
$\bar{\mu}$ dipole moment
ν kinematic viscosity
ρ density
ρ_b density at normal boiling temperature
ρ_c density at critical point
ρ_f density at freezing temperature
ρ_l liquid density
ρ_m density at melting temperature
ρ_n density of saturated vapour
ρ_s density of saturated vapour
σ diameter of molecule
σ force constant
σ surface tension
σ_b surface tension at normal boiling temperature
σ_M molar surface tension
σ_m surface tension at melting temperature
σ_p surface tension at pressure p
τ time
ϕ shape factor of the molecule
φ fluidity
φ volumetric concentration of solid phase in liquid
ω acentric factor
ω expansion factor

CHAPTER 1

GENERAL METHODS FOR ESTIMATING PHYSICO-CHEMICAL PROPERTIES OF GASES AND LIQUIDS

1. INTRODUCTION

A knowledge of the transport properties of fluids, i.e. their viscosity, thermal conductivity and diffusivity, is frequently required for designing new technological processes and also in research work. This is particularly the case when operations involving heat or mass transfer are considered. Fluid densities and surface tensions are also frequently required. The values of such quantities may sometimes be obtained from tables but it is usually found that even the most extensive physico-chemical tables do not contain all the data necessary for designing a technological process. For example, the values of the physico-chemical quantities may be tabulated for conditions differing from those at which the process is to be carried out so that it is necessary to extrapolate or interpolate from the values given in the tables. Frequently also the physico-chemical properties of fluid mixtures are required. It is usually found in such cases that the properties have only been studied for the pure components from which the mixture is constituted and some method is required for estimating the properties of the mixtures from those of the pure substances. Such predictions usually entail considerable difficulties.

Finally, the most difficult case occurs when the properties of the compound or mixture in question have not been measured at all. In this case the only information available may be the structural formula of the given compound.

Although the values of the necessary quantities can sometimes be estimated, it would clearly be preferable to perform the appropriate measurements and to determine the values of the properties under consideration experimentally. Quite often this is not practicable since the determination of the values of some physical properties under the conditions at which a technological process is carried out, e.g. at a high pressure or temperature, requires the use of expensive special equipment which is not readily obtainable. For this reason a knowledge of the methods of computing the values of these physico-chemical quantities is of great importance for the technologist. It should theoretically be possible to compute the values of the necessary quantities from a knowledge of the structure of the molecules and the character of the forces joining the atoms. Unfortunately calculations of this kind, performed by means of the methods of statistical mechanics, are only applicable in the most simple cases and then not until after the introduction of some simplifications. Often therefore these methods are of no use for solving technical problems and resort must be made to semi-empirical methods which, while relatively quick, are usually rather approximate. In situations of this kind the following cases can be distinguished:

1. The required value of the physical quantity in question is unknown under the conditions at which the process is carried out, but we have one or more (experimentally determined) values of this quantity under other conditions and also the law according to which the given quantity varies (this relationship may be known in the form of a mathematical equation or of a graph).

2. Some values of the physical property in question are known for conditions differing from the process conditions, but we do not know the laws according to which this quantity varies.

3. No experimentally determined value of the physical quantity under consideration is available.

The solution of the problem in the first case requires the determination of the constants in the mathematical relationship by substitution of the known data; afterwards the required value may be easily computed. The general principles of such a procedure have been described by the present author in another book.[35]

The solution of the second case (when two experimentally determined values of the physical quantity in question are available) consists in performing an interpolation or extrapolation, the form of the function, i.e. of the relationship between the dependent variable (the physical quantity of interest) and independent parameters (temperature, pressure, etc.), being unknown. One general method frequently employed in this case is based on the assumption that a given physical quantity of the compound under consideration varies in much the same way as it does for some other compound, which is well known and accepted as a reference. The problem is most often solved graphically or, more rarely, analytically. The use of methods based upon the analogy in the behaviour of the investigated compound with that of some suitably chosen reference compound is discussed in section 7.

The third case, i.e. the computation of the value of a physico-chemical quantity if such a value has never been determined experimentally for a given compound, is the most difficult, and such calculations may be neither reliable nor accurate. Frequently, one of the following two procedures is employed. The first procedure consists in computing the necessary value by summing up the contributions of the atoms, bonds or atomic groups, which make up the molecule of the compound in question: this is applicable to a constitutive or additive property. Alternatively use may be made of the similarity of different compounds, e.g. of the theory of corresponding states and general relationships, on the assumption that when the corresponding reduced parameters are equal the reduced values of the physico-chemical quantities under consideration are the same. The reader will find a review of the general procedures available for computing the physico-chemical properties of compounds in the work of Karapet'yants and Chen-Huang Iuc.[191]

2. BASIC INFORMATION ON GASES AND LIQUIDS

Before proceeding to a discussion of practical methods for computing the values of physico-chemical quantities necessary for designing technological processes, it is desirable to recall some fundamental concepts and definitions and to discuss briefly certain topics concerning the theory of fluids and solids.

a. States of Matter

We shall begin our discussion of the basic concepts by giving definitions of the states of matter and advancing remarks concerning the transition from one state of matter into another. These transitions may take place under the influence of temperature or pressure changes. The conditions of coexistence of the phases of a single pure substance are presented in Fig. 1.1, where temperature T is plotted vs. pressure P. The region above the limiting curves SOT corresponds to the existence of the solid phase, the region above TOP to the liquid phase and the region below SOP to the vapour phase. The point O

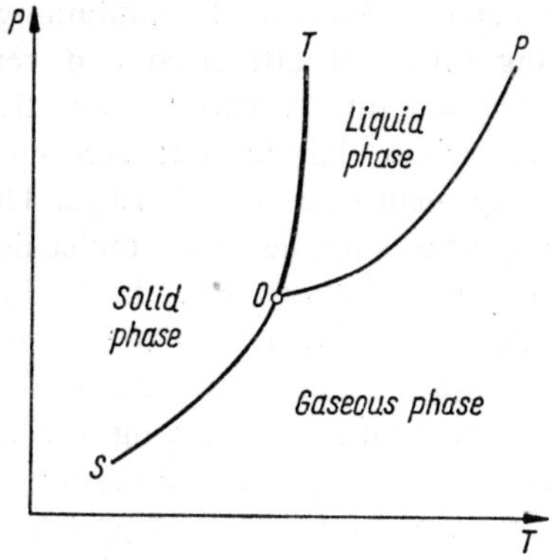

FIG. 1.1. Coexistence of the solid, liquid and gaseous phases (of a single pure substance). Temperature–pressure relation.

denoting the coexistence of all the three phases is, according to the requirements of the Gibbs phase rule, an invariant point. The coexistence of two phases in univariant systems is represented by the following limiting curves:

(1) The *sublimation curve* SO, corresponding to the equilibrium states existing between the solid and gas phases.

(2) The *melting curve* TO, corresponding to the equilibrium states existing between the solid and liquid phases.

(3) The *vaporization curve* OP, corresponding to the equilibrium states existing between the liquid and gas phases.

Elementary definitions of the solid state are based on the observation that solids retain their volume and shape without any external support. We know however that such a description is inadequate, since supercooled liquids in the vitreous or plastic state comply with this definition too. It is the regularity of arrangement of the elements of which the substance is constituted (e.g. ions, atoms, atomic groups or molecules in the points of the crystal lattice) that should be considered characteristic of the solid state of matter. In the crystals of real solids deviations from this ideal picture are always observed, e.g. in the structure of a lattice there may occur faults and defects, either affecting many elementary cells of the lattice (macrodefects) or consisting in the absence of an ion, molecule or atom at the lattice point, or—vice versa—in the inclusion of a superfluous ion, molecule or atom into the lattice (microdefects). The elements situated at the lattice points undergo an oscillating motion around the equilibrium positions corresponding to the lattice points. The amplitude of the oscillation depends on the temperature of the solid and increases with rise in temperature. At a certain temperature, called the *melting temperature* T_m, the lattice disintegrates, the molecules abandon their fixed positions and the substance changes its state from solid to liquid.

Little is known about the structure of liquids. According to the elementary definition of the liquid state, a given mass of liquid occupies a volume which is determined by its temperature and pressure but is independent of the volume of the vessel; it does not maintain any fixed shape but assumes the shape of the confining vessel. The incorrectness of such a definition has been discussed above. The mobility of the liquid provides evidence that its molecules may move quite easily; it may, however, be expected that the mean free path of the molecules is small and of the same order of magnitude as the molecular size. From the property of liquids of maintaining a definite volume we can conclude that counterbalancing forces of attraction and repulsion are acting between the liquid molecules. More detailed investigations show that the arrangement of the molecules in a liquid is not entirely a chaotic one: over short distances it resembles to some extent the molecular arrangement in a crystal lattice. The arrangement is, however, less regular and there are many spaces unfilled with molecules, a fact which makes their motion possible. Some properties of liquids, e.g. the boiling temperature, depend on the size of the space in the liquid not filled with the molecules.

The vapour phase above a liquid contains molecules of the same type as are present in the liquid. These molecules are continually leaving and reentering the liquid surface and their equilibrium concentration in the gas phase depends on the temperature. A measure of this concentration is provided by the vapour pressure p of the liquid. The higher the temperature, the higher is the vapour pressure. If the vapour pressure becomes equal to the pressure of the surrounding atmosphere above the liquid surface, the liquid in its whole mass begins to change into vapour—it boils. The temperature at which the liquid boils under a pressure of one atmosphere is often quoted as being one of the properties which characterize its behaviour. This temperature is called the *normal boiling temperature*, T_b.

If the pressure above a boiling liquid is increased, boiling ceases and it is necessary to increase the temperature to start it again. The phenomena of boiling only take place below a certain temperature, called the *critical temperature*, T_c; at this temperature the vapour pressure over the liquid is equal to p_c, the critical pressure, and 1 mole of the liquid occupies the critical volume, V_c. The quantities T_b, T_c, p_c and V_c will play an important part in our further considerations.

The conditions for the coexistence of the liquid and gaseous (vapour) phases are shown in the phase diagram in Fig. 1.2. Any point on the surface *ADRSEB* corresponds to a possible equilibrium condition for the system. Curves have been drawn on this surface corresponding to processes at a constant temperature (isotherms), e.g. *DABE, KCN* or *RS*. Points in the region *ADKC* correspond to the liquid state, while those in the region *BCNE* denote the vapour.† The boiling and condensation curve *AC* delimits the region of the liquid, and the saturated-vapour curve *CB* the region of vapour, from the region *ACB* where liquid and vapour coexist. The point *C* is the critical point and the critical isotherm *KCN* is the limiting isotherm beyond which (in the area *KCNSR*) there is no liquid phase. The higher the temperature, i.e. the more distant we are from the critical isotherm *KCN* in the direction *RS*, the more nearly do the shapes of the isotherms approach the shape of a perfect-gas isotherm. The curve *ACB* is called the *saturation curve*; on the branch *AC* we have the *bubble points* and on the branch *CB* the *dew points*.

† The gas below the critical isotherm.

It follows from the graph that the vapour can be converted into a liquid or vice versa, in a continuous way, i.e. without observing the formation of any separation boundary between the liquid and the gas phase. This can be accomplished by heating the liquid (initially in a state represented by a point in the region *ADKC*) and simultaneously increasing the pressure so that the critical isotherm is intersected between points *K* and *C* under a pressure which is higher than the critical pressure p_c. The system is then in the region *KCNRS* which corresponds to the gaseous state. Andrews (1889) observed that the possibility of producing the liquid $\rightleftharpoons$ vapour transition in a continuous way proves that both states of matter—the liquid and the gas—constitute no more than very distant stages in a sequence of physical transformations of the fluid state. (By *fluids* we understand *liquids*, *vapours* and *gases*).

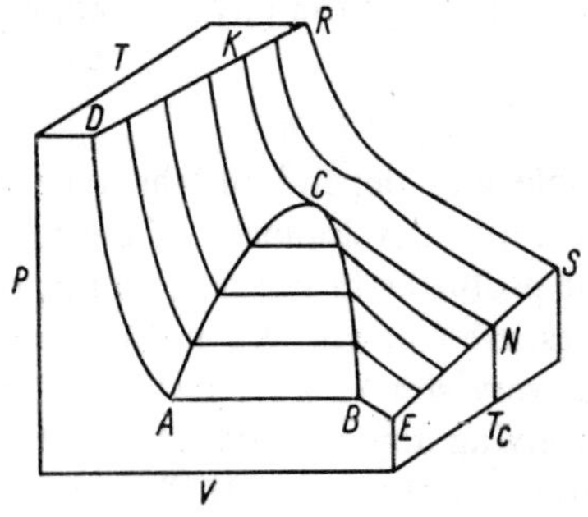

FIG. 1.2. PVT relation for a liquid–vapour system (after Partington[282]).

The picture presented above of the equilibrium between the vapour and liquid phases is a very simplified one. Some investigators have suggested that critical phenomena may not be confined to a point but that they may extend over a range of temperatures and pressures. (However, more recent experimental evidence seems to suggest that the simple classical picture is correct for a fluid in a negligible gravitational field (Wentorf, *J. Chem. Phys.* **24**, 607 (1956)—*Ed.*).

According to the elementary definition, the characteristic quality of the gaseous state consists in the fact that a gas does not maintain either a constant volume or a constant shape, but fills homogeneously the vessel in which it is contained. The first stage in the development of the theory of *the gaseous state* consists in considering the properties of a fictitious 'ideal' gas consisting of molecules moving chaotically in all directions and having the properties of perfectly elastic spheres. The size of the molecules is infinitely small compared with their mean free path (i.e. the path between two successive collisions) and no account is taken in this simple theory of the forces of interaction between the molecules. The properties of real gases are found to approach those of this hypothetical 'ideal' gas as the pressure tends to zero, the perfect gas law

$$PV = RT$$

being then obeyed. As the pressure is raised it is found that the volumetric behaviour of real gases deviates from this law. Under these conditions the molecular centres become closer together and the molecules spend a correspondingly larger fraction of their time in close contact with each other: it is believed that the molecules may then combine into assemblies, this association process being a result of the intermolecular attraction forces. Natanson (1883) first called attention to this phenomenon which has subsequently been taken into account in the theory of real gases developed in the works of Boltzmann, Ursell, Mayer, Band, Swietoslawski, Bogolubov, Vukalovich and others. A knowledge of the properties of real gases is required for the solution of many of the problems which arise in the design of technological processes in the chemical industry.

b. Properties of Molecules

The physical and thermodynamic properties of a substance depend on the mass, volume, and shape of the molecules of which the substance is composed, on the character of the bonds in the molecules, and on the character and magnitude of the intermolecular forces. The above-mentioned factors will be discussed below, though only in sufficient detail to permit their effect upon some of the physico-chemical properties to be described.

Types of Chemical Bonds

The properties of the molecules depend on the type of the bonds which link the atoms together in the molecule. The development of a rigorous theory of chemical bonding is one of the most difficult problems in physics; so far, only approximate quantitative solutions of the simplest cases are known. We must content ourselves, therefore, with representing the problem in a very simple way, just sufficient for distinguishing the four basic types of bonds, without giving any essential idea as to the character of the phenomena involved in the formation of these bonds.[368]

1. The *ionic bond,* called also the polar, electrovalent or heteropolar bond, arises from the coulombic forces between the positive and negative ions which are formed when one atom (+) easily gives off one or more electrons and the other atom (—) readily acquires them so that the electron passes from the (+) to the (—) atom. Such bonds, consisting in a mutual saturation of electrostatic charges, do not display any directional action and are characterized by large coordination numbers.

Examples of compounds with ionic bonds are CsF, NaCl. Such bonds appear particularly in the solid and in the liquid state.

2. The *covalent bond* is also called the homopolar bond. The explanation of the nature of this bond is based on considerations involving wave mechanics; simplifying the problem considerably we may imagine the covalent bond between two atoms as arising from a pair of electrons, common to both atoms and often, according to Lewis' suggestion, it is represented in the form of two dots between the symbols of the atoms. Covalent bonds appear in the molecules of gases and liquids such as F_2, H_2, Cl_2 and the organic compounds, and also in the crystals of molecular and atomic lattices, in diamond for example. This linkage is characterized by a bond energy of the order of 100 kcal/mole, by directional actions and by small coordination numbers which are explicable in terms of wave mechanics.

3. *Metallic bonds* appear only in the solid and liquid states. The metals in the solid phase consist of free atoms with loosely linked electrons (their ionization energy and energy of excitation being smaller than 5 eV). In the solid metal phase there exists a 'gas' composed of free electrons which are not connected with any single atom of the metal. The metallic bonds are characterized by a lack of directional action and by a large coordination number (i.e. each atom is surrounded by a large number of nearest neighbours).

4. *Van der Waals' bonds* are a result of the interaction of permanent dipoles (Keesom), permanent and induced dipoles (Debye) and of forces the appearance of which is justified by considerations of wave mechanics (London). Bonds of this type are generally weak and do not display any directional action. Any compounds which are formed as a result of these bonds are usually unstable. Such compounds may be formed between molecules, as in the case of hydrated salts or additive compounds, or between atoms, as in the case of HgKr. Van der Waals' bonds will be discussed in detail later.

Usually, a given bond cannot be simply classified into any definite bond type. For example it is found that in the molecules of the halides of the alkali metals in the gaseous state the character of the linkage is such that the contribution of the ionic bond is approximately 65 to 83%, and the contribution of the covalent bond 17 to 35% (in the crystalline state the contribution of the ionic bond is in all these cases higher).

Beside these four fundamental bond types, one more practically important type of linkage between the atoms, called the hydrogen bond, is known. The *hydrogen bond* often exists between the hydrogen atom of one molecule and the atoms of oxygen, nitrogen, fluorine, chlorine or sulphur in another (or in the same) molecule. The hydrogen atom has only one electron and may participate in one covalent bond only, but in certain groups, as O—H or C—O, the bonds maintain the character of partly covalent and partly ionic bonds. The hydrogen bond can form in such cases and give rise to association or solvation of the molecules; the forces of the hydrogen bond (bond energy 3·5 to 10 kcal/mole) are slightly higher than the van der Waals' forces (bond energy, 3 kcal/mole). In chemical formulae the hydrogen bonds are represented by a series of dots.

The character of the bonds occurring in a given molecule may be predicted to some extent, since the kind of interaction between the atoms depends on their electronic structure and on the diameter of the atoms or ions, and thus it is associated with the position of the elements in the periodic system.[201]

c. Shape and Size of Molecules

Molecules in the Gaseous State

Most of the molecules of the elements are, in the gaseous state, composed of two atoms, linked together by covalent bonds. Exceptionally, the molecule of ozone consists of three atoms of oxygen forming a triangle with an obtuse angle of about 117°. Phosphorus at its boiling point can be represented by the formula P_4 (the atoms form a regular tetrahedron); it is not until a temperature of about 1000°C is reached that this tetrahedron splits into two diatomic molecules. In the vapour of sulphur and selenium, molecules with the formulae S_8 and Se_8 occur; in the molecule of sulphur, the eight atoms form a ring with obtuse angles of 103° between the atoms. At higher temperatures the rings split, and above a temperature of 800° the sulphur molecules are diatomic.

In order to describe a molecule it is necessary to know the *interatomic distances* (calculated, for example by summation of the covalent radii, Table 1.1), and the molecular symmetry or the angles contained between the directions of the bonds. A rule, formulated by Poltorak,[291] may be helpful in the determination of these angles and of the configuration of the atoms in a molecule. According to this rule, in the molecule of the type $A(X_i)_n$, the highest symmetry is observed when the valency of atom A (which is the central atom of a molecule or of part of a group in a large molecule) is equal to the number of the group of the periodic system to which the element A belongs. If this condition is not fulfilled, lower symmetry will appear. The following groupings correspond to the higher symmetry: a linear grouping for AX_2, a planar triangle for AX_3, and a tetrahedron for AX_4. In the case of a "lower" symmetry we have: an angular grouping for AX_2, a pyramid for AX_3, and a square pyramid for AX_4. If there are various different radicals or atoms attached to the atom A, the general system of the symmetry is maintained, but the angles between the directions of the bonds (within the limits of a given symmetry type) may differ.

However, it is necessary to know not only the distance between the nuclei of the atoms in a molecule, equal to the sum Σr of the covalent radii, but also the diameter of the molecule's sphere of activity, i.e. the diameter of a molecule. This quantity may be characterized by three different values.[11, 25, 356]

1. The *collision diameter* σ_T, temperature-dependent, and computed according to the kinetic theory of gases.
2. The *distance* σ_0, the minimum distance to which a pair of molecules can approach during a collision when their kinetic energy at infinite separation is zero, i.e. that intermolecular separation at which the potential energy is zero (Fig. 1.4).
3. The *distance* σ_{min} *between two molecules* corresponding to the minimum potential energy of the system (σ_{min} is 10–25% higher than σ_0).

TABLE 1.1. DISTANCES Σr BETWEEN THE ATOMIC NUCLEI IN THE MOLECULE AND THE MOLECULAR DIAMETERS σ_T, σ AND σ_{min} (Å)

Gas	Σr	σ_T	σ_0	σ_{min}	Gas	Σr	σ_T	σ_0	σ_{min}
H_2	0·74	2·47	2·84	3·24	N_2	1·10	3·18	3·5	3·8
F_2	1·26	—	3·4	(4·0)	C_2	1·31	—	—	—
Cl_2	1·99	3·7	4·0	(4·5)	Li_2	2·67	—	—	—
Br_2	2·28	3·04	4·15	(4·7)	Na_2	3·08	—	—	—
I_2	2·66	4·46	—	(5·1)	K_2	3·92	—	—	—
O_2	1·21	2·98	3·2	3·5	CO	1·15	3·2	—	(4·1)
O_3	1·27	—	—	—	NO	1·15	3·0	—	(3·7)
P_2	1·89	—	—	—	HF	0·92	—	—	—
S_2	1·92	—	—	—	HCl	1·28	3·0	—	3·89
Se_2	2·19	—	—	—	HBr	1·41	3·12	—	4·17
					HI	1·6	(3·4)	—	4·5

In Table 1.1 the distances between the atomic nuclei in the molecule and the values of the diameter of the sphere of action are given for several elements.

It should be noted that the considerations given here refer to approximately spherical molecules.

Shape and Size of Organic Molecules

In technological practice we are concerned with only a few thousands of inorganic compounds and, in the vast majority of cases, the physico-chemical constants of these compounds are well known. The number of organic compounds is many times as large, it reaches several hundreds of thousands, and the properties of these compounds are frequently unknown. In the present section we shall, therefore, consider in the first place the factors influencing the geometric configuration of the atoms in the molecules and the size and shape of organic molecules.

The size and shape of the molecules have a great effect on certain properties of the compounds; e.g. if we assume that the surface tension of a liquid constituted of open-chain molecules is equal to $\sigma = 1$, we find (maintaining the remaining parameters such as the volume of the molecule, polarizability, and dipole moment, unchanged) the surface tension to be:

0·93 for molecules with branched chains,
0·87 for molecules of tetrahedral shape,
1·31 for molecules in the shape of rings.[231]

Modern electron diffraction, spectrographic and X-ray research methods enable us to determine the size, the interatomic distances and the configuration of the atoms in many of the molecules of organic compounds. As we know, the bonds linking the atoms in organic compounds are mainly covalent and, in distinction with bonds of the ionic type, are characterized by directional action. Omitting further discussion of the nature and kind of bonds in organic compounds, we shall confine ourselves to a study of the data for the intermolecular distances and bond angles in the molecules of some typical substances.

Measurements show that the distance between two atoms in a molecule depends not only on the kind of the atoms, but also on their configuration; hence, it also depends on the other atoms or radicals adjacent to the pair of atoms in question.

It follows from the data tabulated by Tat'evskii[359] for a large number of measurements of interatomic distances in molecules of different hydrocarbons, that for a certain definite configuration the distances between the two atoms in different compounds are almost the same (Tables 1.2 and 1.3).

TABLE 1.2. DISTANCES BETWEEN THE ATOM OF CARBON AND THE ATOM OF HYDROGEN IN MOLECULES OF HYDROCARBONS

Compound	Configuration	Distance $2r$, Å
CH_4	>C—H (tetrahedral, three single bonds)	1·0936
C_2H_4	>C—H (one double and one single bond)	1·071
C_2H_2	≡C—H	1·059

Comparing the distances between the carbon atoms in molecules of different hydrocarbons, it may be noticed that these distances become smaller on passing from the bond C—C (tetrahedral arrangement of the bond directions) to the bond C=C (plane system) and, finally, to the bond C≡C (linear system).

For a certain definite configuration of bonds these distances are almost the same for different molecules, irrespective of the state of aggregation of the compound investigated.

In order to estimate the distances between two different atoms, use is made of the quantity called the *covalent radius* of the atom, which in the case of two similar atoms is assumed to be equal to half of the distance between the centres of the atoms. It is assumed, for example, that the covalent radius of a carbon atom in the bond C—C is $\frac{1}{2}\times 1{\cdot}54 = 0{\cdot}77$ Å; for C=C the covalent radius equals $\frac{1}{2}\times 1{\cdot}20 = 0{\cdot}60$ Å, etc. Interatomic distances may be estimated by summing the appropriate covalent radii, though

TABLE 1.3. AVERAGE VALUES OF THE DISTANCES BETWEEN THE CARBON ATOMS IN VARIOUS HYDROCARBONS

Configuration	2r, Å	Configuration	2r, Å
→C—C←	1·54	≡C—C≡	1·37
→C—C⫽	1·52	>C═C<	1·34
→C—C⫽⫽	1·52	>C═C═	1·33
→C—C≡	1·46	═C═C═	1·29
⫽C—C⫽	1·47	—C≡C—	1·20
⫽C—C⫽⫽	1·47	>C═C<	1·39
⫽⫽C—C⫽⫽	1·47	>C═C<	1·42
⫽C—C≡	1·42	>C═C<	1·41 (?)
⫽⫽C—C≡	1·42		

this procedure is somewhat approximate since various factors may alter these radii somewhat. Errors may arise, for example, if the bond is partially ionic in character; the proximity of neighbouring bonds or groups may also alter the values of the covalent radii. This may be illustrated by considering the distance between the atoms of hydrogen and carbon. Taking the covalent radii to be equal to 0·30 Å for H and 0·77 Å for C the interatomic distance is 0·30+0·77 = 1·07 Å. The distances C—H, e.g. in the molecules of methane, ethylene and acetylene, computed by this method, should thus be the same, being equal to 0·77+0·30 = 1·07 Å, whereas according to Table 1.2 they are seen to amount to 1·094, 1·071 and 1·059 Å, respectively.

Average values of the covalent radii of several elements, as compiled by Syrkin and Dyatkina,[356] are listed in Table 1.4.

TABLE 1.4. COVALENT RADII, Å[356]

Element	Radius, Å	Element	Radius, Å
Li	1·34	Sb	1·41
Na	1·54	Se	1·17
K	1·96	Te	1·37
Si	1·17	F	0·72
Ge	1·22	Cl	0·99
Sn	1·40	Br	1·14
P	1·10	I	1·33
As	1·21		

In order to construct the model of a molecule, it is necessary to know, in addition to the internuclear distances, the angles between the directions of the bonds. Values of these angles in some inorganic and organic compounds are presented in Tables 1.5 and 1.6.[356] It is seen from the examples listed in these tables that the angles between the

TABLE 1.5. ANGLES BETWEEN THE BONDS IN THE MOLECULES OF SOME INORGANIC AND ORGANIC COMPOUNDS

Compound	Bonds	Angle	Compound	Bonds	Angle
H_2O	H—O—H	105°	SCl_2	Cl—S—Cl	103°
Cl_2O	Cl—O—Cl	115°	H_2Se	H—Se—H	~ 90°
CH=CH╲ │ O CH=CH╱	C—O—C	107°	NH_3	H—N—H	108°
$COCl_2$	O=C—Cl	117°	$N(CH_3)_3$	C—N—C	108°
H_2S	H—S—H	92°	PH_3	H—P—H	93°
S_8 (gas)	S—S—S	100°	$P(CH_3)_3$	C—P—C	100°
CH=CH╲ │ S CH=CH╱	C—S—C	91°	PCl_3	Cl—P—Cl	101°
$CSCl_2$	S=C—Cl	116°	$As(CH_3)_3$	C—As—C	96°
			$AsCl_3$	Cl—As—Cl	108°
			$SbCl_3$	Cl—Sb—Cl	104°
			$SiHCl_3$	Cl—Si—Cl	110°
			$SiHBr_3$	Br—Si—Br	110°

TABLE 1.6. ANGLES BETWEEN THE BONDS IN THE MOLECULES OF SOME CARBON COMPOUNDS

Compound	Bonds	Angle	Compound	Bonds	Angle
CH_4	H—C—H	109°28′	$(CH_3)_2C{=}C(CH_3)_2$	C=C—C	124°
CH_3—CH_3	H—C—H	109°28′	$(CH_3)_2C{=}CH_2$	C=C—C	124°
CH_3—CH_2—CH_3	C—C—C	111°30′	CH_3CHO	C—C=O	122°
$(CH_3)_2$—CH—CH_3	C—C—C	111°30′	HCOOH	O—C=O	125°
$(CH_3)_3$—C—Cl	C—C—C	111°30′	HCOONa	O—C—O	124°
C_2H_5—Cl	C—C—Cl	111°30′	NH_2CH_2COOH	O—C=O	122°
C_2H_5Br	C—C—Br	109°	COOH │ $\cdot 2H_2O$ COOH	O—C=O	124°
$(CH_3)_2CHCl$	C—C—Cl	109°	$(CH_3)_2O$	C—O—C	111°
CH_2F_2	F—C—F	110°	CH_3OH	C—O—H	111°
CH_2Cl_2	Cl—C—Cl	112°			
$CHCl_3$	Cl—C—Cl	112°			
$C_6H_{12}N_4$	N—C—N	110°			

bonds in various molecules are very similar for analogically situated groups and atoms, as also are the values of the contributions to the internuclear distances. This fact has enabled Hougen and Watson,[160] who made use of the data of Hirschfelder and Pauling, to present the average values of the covalent radii and the values of the angles between the bonds with sufficient accuracy for performing approximate calculations. These values are listed in Table 1.7.

It is seen from the table that when an atom is linked with two adjacent atoms in different ways, e.g. with one atom by means of a single bond and with the other one by a double bond, then, for the purpose of calculating the interatomic distances, two

TABLE 1.7. CONTRIBUTIONS OF THE INTERATOMIC DISTANCES (COVALENT RADII), AND THE BOND ANGLES IN THE MOLECULES OF ORGANIC COMPOUNDS (APPROXIMATE AVERAGE VALUES)

Atom	Contribution, Å	Bond	Arrangement of bonds and angles between them
Hydrogen	0·30		
Carbon			Regular tetrahedron 108°
single-bond carbon	0·77		
double-bond carbon	0·67	double bond	Planar
	0·77	single bond	112° C 124°, 124°
triple-bond carbon	0·60	triple bond	Linear
	0·77	single bond	≡C—
benzene carbon	0·695	each of the two C⩦C bonds	Planar
	0·77	bond extending outward	120° C 120°, 120°
Oxygen			
single-bond oxygen	0·66		111° O
double-bond oxygen	0·57		=O
Nitrogen			Flattened pyramid
amino nitrogen	0·70		108° 108° N 108°
nitrate nitrogen	0·65	double bond	Planar
	0·70	single bond	120° 120° N 120°
isonitrile nitrogen	0·55	double bond	Linear
	0·70	single bond	—N=
cyanide nitrogen	0·55		≡N
Sulphur			
single-bond sulphur	1·04		105° S
double-bond sulphur	0·95		=S
sulphate sulphur	0·95	double bond	Tetrahedron 108°
	1·04	single bond	108° 108° S 108° 108°
Sodium	1·81		
Chlorine	0·99		
Bromine	1·14		
Iodine	1·33		

different radii are assumed for the same atom. The "radii" presented above depend, therefore, not only on the nature of the atom but also on the nature of the bond, and the term "*radius*" is, possibly, not altogether apt.

In the case of hydrogen bonds (section 2b), the distances between the atoms differ from those involved in the covalent bonds considered above. Examples of interatomic distances corresponding to hydrogen bonds[356] are presented in Table 1.8. It should be noted that hydrogen bonds may form when interatomic distances are rather small (below 3 Å). In the case of the association of two molecules of formic acid, the distance O—H ... O is equal to about 2·73 Å. In the case of atomic bonds the distance between

TABLE 1.8. INTERATOMIC DISTANCES IN HYDROGEN BONDS

Bond	Distance between extreme atoms, Å	Compound
O—H...O	2·76	ice
	2·70	o-$C_6H_4(OH)_2$
	2·73	$(HCOOH)_2$
	2·55	$NaHCO_3$
F—H...F	2·25	KHF_2
N—H...N	2·96	NH_4N_3
N—H...O	2·98, 3·03	$CO(NH_2)_2$
N—H...F	2·63	NH_4F
O—D...O	2·76	$(CH_3COOD)_2$

the hydrogen and oxygen atoms amounts to about 1·07 Å, hence, the distance between the hydrogen and oxygen atoms linked by the hydrogen bond equals about 1·64 Å.

Similar calculations may be also carried out for other cases, e.g. the distance corresponding to the hydrogen bond H ... F in the associated molecules of hydrogen fluoride amounts to about 1·34 Å.

Knowing the interatomic distances and the angles between the straight lines joining the atomic centres, it is possible to reproduce in simple cases the model of an organic molecule and to calculate some quantities which are of interest, e.g. the moments of inertia of the molecule and the molal volume and density of the corresponding compound. Knowing these quantities it is sometimes possible to calculate the molal entropy and other properties of the compound.

When constructing a model, it should be remembered that the carbon chain of a molecule, e.g. of a hydrocarbon of the paraffin series, is not a straight but a zigzag line. According to what has been said above, the distance between the carbon atoms in the open chain equals $2r = 1{\cdot}54$ Å and the bond angle amounts to 109°28′. Investigations have shown that these quantities are almost the same for all gaseous, liquid and solid hydrocarbons, hence, the length of the molecule does not depend on the state of aggregation. Kurtz and Sankin[214] have computed the volume and cross-section of the part of the hydrocarbon chain corresponding to one $—CH_2—$ group. X-ray investigations of the structure of crystalline paraffins indicate that the distance L between the ends of the molecule of a normal paraffin with n carbon atoms may be calculated from the equation:

$$L = 1{\cdot}27n+2 \tag{1.1}$$

(since the carbon chain forms a zigzag line, the molecular length per carbon atom is equal to $1{\cdot}54 \sin(109°28'/2) = 1{\cdot}27$ Å). From the measurements of the molal volumes of liquid n-paraffins it follows that each g-mole of $—CH_2—$ occupies a volume of 16·28 cm^3, i.e. 27·03 $(Å)^3$ per group. Since the length of the group is 1·27 Å, this corresponds to a cross-sectional area for the chain of 21·3 $(Å)^2$. In crystalline solid n-paraffins the measured cross-sectional area is smaller and amounts only to 18·5 $(Å)^2$. The apparent increase in the cross-sectional area of the molecule during the melting process may be explained in part by the possibility of rotation in the molecule along the C—C bonds, i.e. by the possibility of a change in the shape of the molecule in the liquid. The above-mentioned phenomenon makes it difficult to calculate the physico-chemical constants for substances on the basis of the known molecular structure.

Compounds with Ionic Bonds

Knowing the values of the covalent radii of the atoms and of the angles between the bonds it is possible, in simple cases, to construct models for the molecules of organic compounds. In the case of compounds in which ionic bonds predominate, it is impossible to use the above-mentioned procedure for reproducing the structure of the substance. If it is assumed that the ion in a molecule or in a crystal is an incompressible non-deformable near-sphere (rigid but polarizable) with a definite radius and electric charge; the electrostatic forces forming the ionic bonds may be calculated from Coulomb's law. Since ionic bonds do not display any directional action, the ions in a crystal would be expected to adopt a close packed arrangement. In general there are three factors which determine the structure of a crystal lattice: (a) the ratios of the numbers of the various particles (molecules, atoms, ions) forming the lattice, (b) the size and shape of these particles, and (c) their interactions.[22, 196]

For predicting the structure of an ionic crystal lattice, it is helpful to introduce the concept of the ionic radius, which enables us to take into account to some extent the effect of the first two factors (the numbers of cations and anions, and their size) upon the crystal lattice. As electrically charged particles, ions may display a polarizing action on neighbouring ions. This action consists in the deformation and displacement of the external electronic shell of an ion with respect to its nucleus. Owing to such polarization, the bonds between the ions may not possess the character of strictly ionic bonds, and to some extent they may acquire the character of covalent bonds.

d. Dipole Moments †

It often happens that the different particles making up a molecule possess electrical charges opposite in sign. The molecule of hydrogen chloride may serve as a simple example; if we consider it as consisting of one proton and one chlorine ion, we have in the molecule two electric charges of opposite signs (charge: $4{\cdot}80\times10^{-10}$ esu or $g^{1/2}.cm^{3/2}.sec^{-1}$), placed at a distance of $1{\cdot}28\times10^{-8}$ cm. More complicated molecules may also be constructed of a few or many particles charged positively and negatively so that the algebraic sum of their charges equals zero.

Just as we may determine the centre of gravity of a molecule as that point about which $\Sigma ml = 0$, so we may determine by analogous geometrical addition of like charges the separate positions of the "centres of gravity" of the positive and negative charges in the molecule. If the molecule is asymmetric, the three centres are often found to occur in different places. Hence, such a molecule may be treated as a system in which two discreet electric charges of opposite sign are displaced from one another by a very small distance. Such a system is called a dipole. We assume that the positive charge $+e$ of the dipole is concentrated at the point called the *positive pole*, while the negative charge $-e$ is concentrated at the point called the *negative pole*; these points are connected by a straight line called the *dipole axis* (Fig. 1.3).

---

† Sections 2d and 2e as well as the numerical data and examples have been based on the refs. 11, 162, 356, 368 and 397.

The vector equal to the product of the absolute value of the charge e of one of the poles by the distance $AB = l$ between the poles of the dipole, directed along the dipole axis from the negative to the positive pole, is called the *dipole moment*

$$\bar{\mu} = e \times l \qquad (1.2)$$

The dipole moment of a molecule influences the interaction between that molecule and its neighbours and is of great importance for our further considerations. It is measured in units called Debyes (D)

$$1\ \text{D} = 10^{-18}\ \text{g}^{1/2}.\text{cm}^{5/2}.\text{sec}^{-1} = 10^{-18}\ \text{dyne}^{1/2}.\text{cm}^2$$

The coefficient 10^{-18} is introduced because the distances between the poles of the dipoles are of the order of 1 Å $= 10^{-8}$ cm, and the charges of the order of 10^{-10} esu. Dipole moments measured in D are therefore usually of the order of unity.

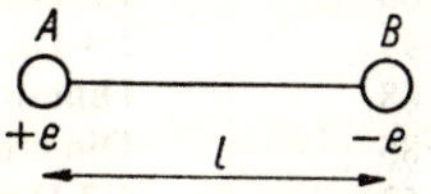

FIG. 1.3. Dipole: *A*—positive pole, *B*—negative pole, *l*—distance between the poles.

In Tables 1.9 and 1.10 the values of the dipole moments of the molecules of some common compounds are presented. Molecules which have permanent dipole moments are called *polar molecules*.

TABLE 1.9. DIPOLE MOMENTS OF SOME IMPORTANT INORGANIC COMPOUNDS[11, 356]

Compound	Dipole moment (Debye units)	Compound	Dipole moment (Debye units)	Compound	Dipole moment (Debye units)
H_2	< 0·015	N_2O	0·14	$COCl_2$	1·18
O_3	0·49	NO_2	0·4	COS	0·65
H_2O, D_2O	1·84	NO	0·13	HCN	2·93
H_2O_2	2·1	N_2O_4	< 0·4	C_2N_2	< 0·3
HCl	1·03	N_2O_5	1·39	$SiHBr_3$	0·79
HBr	0·79	PH_3	0·55	BCl_3	< 0·2
HI	0·38	PCl_5	0·8	$LiClO_4$	7·84
H_2S	0·93	PCl_3	1·1	NaI	4·9
SO_2	1·60	PBr_3	0·6	KCl	6·3
SCl_2	0·6	AsH_3	0·15	KBr	9
$SOCl_2$	1·58	As_2O_3	0·14	KI	9·2
$SO_2(NH_2)_2$	3·9	$AsCl_3$	2·1	$AlBr_3$	5·2
SO_2Cl_2	1·86	$SbCl_3$	3·2	AlI_3	2·5
NH_3	1·46	$SbCl_5$	< 1	$TaCl_5$	1·2
N_2H_4	1·84	CO	0·1	$Fe(CO)_5$	0·7

NOTE: $\bar{\mu} = 0$ for the following molecules:
O_2, N_2, Cl_2, Br_2, I_2, SF_6, PI_3, CO_2, CS_2, SiH_4, Si_2H_6, SiF_4, $SiCl_4$, $TiCl_4$, OsO_4, $Ni(CO)_4$, $HgCl_2$, $HgBr_2$, HgI_2, $SnCl_4$, $SnBr_4$, SnI_4.

TABLE 1.10. DIPOLE MOMENTS OF SOME ORGANIC COMPOUNDS[11, 356]

Compound	μ (Debye units)	Compound	μ (Debye units)
CH_4, C_2H_6, C_2H_4, C_2H_2, C_3H_8	0	*p*-Diaminobenzene	1·27
Acetaldehyde	2·55	1,1-Dibromoethane	2·12
Acetamide	3·6	*p*-Dichloroaniline	1·68
Acetic acid	1·73	*o*-Dichlorobenzene	2·26
Acetic anhydride	2·80	*m*-Dichlorobenzene	1·48
Acetone	2·80	*p*-Dichlorobenzene	0
Acetonitrile	3·4	1,1-Dichloroethane	1·99
Acetophenone	1·0	1,2-Dichloroethane	1·40
m-Aminophenol	1·83	1,1-Dichloroethylene	1·18
Amyl alcohol	1·63	1,2-*cis*-Dichloroethylene	1·82
Aniline	1·53	1,2-*trans*-Dichloroethylene	0
Benzaldehyde	2·75	1,2-Dichloronaphthalene	2·47
Benzene	0·08	1,3-Dichloronaphthalene	1·80
Benzophenone	2·99	1,4-Dichloronaphthalene	0
Benzyl alcohol	1·68	Difluoromonochloromethane	1·40
Benzyl chloride	1·85	Dimethyl ether	1·29
Bromobenzene	1·52	1,2-Dimethylbutadiene	0·5
Bromoform	1·1	1,3-Dimethylbutadiene	0·6
n-Butane, iso-Butane	0	1,1-Dimethylpropane	0
n-Butanol	1·72	*o*-Dinitrobenzene	6·00
iso-Butanol	1·58	Diphenyl	0
α-Butylene	0·4	Diphenylamine	1·3
β-Butylene	0	Diphenyl ether	1·14
iso-Butylene	0·49	Ethanol	1·70
Butyric acid	0·93	Ethyl acetate	1·81
Carbon tetrachloride	0	Ethyl bromide	2·00
o-Chloroaniline	1·80	Ethyl chloride	2·02
Chlorobenzene	1·69	Ethyl fluoride	1·92
Chloroform	1·10	Ethyl formate	1·92
1-Chloronaphthalene	1·54	Ethyl iodide	1·87
2-Chloronaphthalene	1·65	Ethyl mercaptan	1·39
o-Chloronitrobenzene	4·59	Ethylamine	1·2
m-Chloronitrobenzene	3·69	Ethylene chlorohydrin	1·88
p-Chloronitrobenzene	2·78	Ethylenediamine	1·94
o-Chlorophenol	1·3	1,2-Ethylene glycol	1·6...2·2
m-Chlorophenol	2·10	Ethylene oxide	1·88
p-Chlorophenol	2·22	Formamide	3·22
o-Chlorotoluene	1·35	Formic acid	1·5...2·1
m-Chlorotoluene	1·78	Furan	0·7
p-Chlorotoluene	1·90	Furfural	3·57
o-Cresol	1·41	n-Heptane	0
m-Cresol	1·54	Hexachlorobenzene	0
p-Cresol	1·57	n-Hexane	0
Cyanamide	3·8	Hydroquinone	2·47
Cyclohexane	0	Iodoform	0·92
Cyclohexanol	1·9	Methanol	1·68
Cyclohexanone	2·9	Methyl bromide	1·79
Cyclopentane	0·97	Methyl chloride	1·86
Decaline	0	Methyl fluoride	1·81
n-Decane	0	Methyl iodide	1·60
o-Diaminobenzene	1·44	Methylacetylene	0·77
m-Diaminobenzene	1·79	Methylamine	1·28

TABLE 1.10 (*cont.*)

Compound	μ (Debye units)	Compound	μ (Debye units)
α-Methylbutadiene	0·5	Propionic aldehyde	2·1
Methylene bromide	1·7	n-Propyl chloride	2·03
Methylene chloride	1·57	Propylene	0·35
Mononitrotrichloromethane	1·88	Quinoline	2·16
Naphthalene	0	iso-Quinoline	2·55
α-Naphthol	1·0	Quinone	0
β-Naphthol	1·3	Styrene	0
o-Nitroaniline	4·35	1,1,1,2-Tetrachloroethane	1·2
m-Nitroaniline	4·85	1,1,2,2-Tetrachloroethane	1·90
p-Nitroaniline	6·2	Tetraline	0·41
Nitrobenzene	3·95	Tiophene	0·63
Nitroethane	3·2	Tiophenol	1·33
Nitromethane	3·54	Toluene	0·37
1-Nitronaphthalene	3·7	*o*-Toluidine	1·57
o-Nitrophenol	3·10	*m*-Toluidine	1·43
m-Nitrophenol	3·90	*p*-Toluidine	1·27
p-Nitrophenol	5·02	1,2,4-Trichlorobenzene	1·25
o-Nitrotoluene	3·66	1,1,1-Trichloroethane	1·5
m-Nitrotoluene	4·17	1,1,2-Trichloroethane	1·55
p-Nitrotoluene	4·44	Trichloroethylene	0·94
n-Nonane	0	2,4,6-Trichlorophenol	1·62
n-Octane	0	2,2,3-Trimethylbutane	0
Paraldehyde	2·02	Trimethylethylene	0·47
n-Pentane	0	Triphenylmethane	0·6
Phenol	1·56	Vinyl chloride	1·66
n-Propanol	1·66	*o*-Xylene	0·52
iso-Propanol	1·64	*m*-Xylene	0·36
Propionic acid	1·69	*p*-Xylene	0

For the above-mentioned molecule of hydrogen chloride, if the bond between hydrogen and chlorine were purely ionic, the dipole moment would be

$$4{\cdot}8\times10^{-10}\ \mathrm{g^{1/2}.cm^{3/2}.sec^{-1}}\times1{\cdot}28\times10^{-8}\ \mathrm{cm} = 6{\cdot}14\times10^{-18}\ \mathrm{g^{1/2}.\,cm^{5/2}.sec^{-1}}$$

In fact, this moment is only one-sixth of this, viz. 1·04 D, probably because of the deformation of the electronic shell of the chlorine ion by the attractive forces of the proton, owing to which the bond loses its purely ionic character.

For molecules with a complicated but known geometrical structure, the values of individual dipoles in the molecule are computed and the vectors are added on the assumption that the principle of vectorial additivity of the dipole moments is valid.

As we know, the sum $\bar{\mu}_s$ of two vectors $\bar{\mu}_1$ and $\bar{\mu}_2$ making an angle θ with each other is equal to

$$\bar{\mu}_s = \sqrt{\bar{\mu}_1^2+\bar{\mu}_2^2 + 2\bar{\mu}_1\bar{\mu}_2\cos\theta} \tag{1.3}$$

If $\bar{\mu}_1 = \bar{\mu}_2 = \bar{\mu}$, the formula becomes simpler

$$\bar{\mu}_s = 2\bar{\mu}\cos\frac{\theta}{2} \tag{1.4}$$

If the groups, the dipole moments of which are added, are substituted into the benzene ring, we get for the position

$$\text{ortho:} \quad \sphericalangle\, \theta = 60°, \quad \bar{\mu}_o = \sqrt{\bar{\mu}_1^2+\bar{\mu}_2^2 \pm \bar{\mu}_1\bar{\mu}_2} \tag{1.5}$$

$$\text{meta:} \quad \sphericalangle\, \theta = 120°, \quad \bar{\mu}_m = \sqrt{\bar{\mu}_1^2+\bar{\mu}_2^2 \mp \bar{\mu}_1\bar{\mu}_2} \tag{1.6}$$

$$\text{para:} \quad \sphericalangle\, \theta = 180°, \quad \bar{\mu}_p = \bar{\mu}_1 \mp \bar{\mu}_2 \tag{1.7}$$

In these formulae, the upper signs correspond to the case when moments $\bar{\mu}_1$ and $\bar{\mu}_2$ are both directed towards, or both directed away, from the centre of the ring, and the lower signs apply when one is directed towards and one away from the centre of the ring.

Consider, for example, the dipole moment of the water molecule $\bar{\mu}_{H_2O} = 1{\cdot}84$ D (Table 1.9). The molecule is non-linear, and the bonds O—H make an angle equal to 105° (Table 1.5). If the principle of adding vectorially the dipole moments holds, then it is possible to calculate the dipole moment $\bar{\mu}_{OH}$ of the O—H bond from equation (1.4):

$$2\bar{\mu}_{OH} \cos \frac{105°}{2} = 1{\cdot}84$$

hence,

$$\bar{\mu}_{OH} = 1{\cdot}51 \text{ D}$$

The dipole moment of the molecule of methyl ether CH_3OCH_3 is equal to $\bar{\mu}_{(CH_3)_2O} = 1{\cdot}29$ D (Table 1.10), and the angle between the bonds O—CH_3 is 111° (Table 1.6). The dipole moment calculated on the basis of these data for the bond O—CH_3 is

$$\bar{\mu}_{OCH_3} = 1{\cdot}12 \text{ D}$$

If the principle of vector addition of the dipole moments is valid, knowing $\bar{\mu}_{OH}$ and $\bar{\mu}_{OCH_3}$, it is possible to estimate the dipole moment $\bar{\mu}_{CH_3OH}$ of the molecule of methanol. The angle between the bonds C—O—H in methanol is equal to 111° (Table 1.6). The sum of the two vectors $\bar{\mu}_{OH}$ and $\bar{\mu}_{OCH_3}$ is 1·71 D (Eq. (1.3)). The dipole moment of methanol, found experimentally, is $\bar{\mu}_{CH_3OH} = 1{\cdot}69$ D; as we see, this is very close to the computed value. The dipole moments of several bonds calculated by Syrkin and Dyatkina[356] are presented in Table 1.11.

The values of the dipole moments of the molecules calculated by the vector addition of the bond moments may differ considerably from the actual moments, owing to the interaction of the dipoles in the molecule. Therefore, attempts at determining the contributions (like those presented in Table 1.11) for the particular types of bonds occurring

TABLE 1.11. DIPOLE MOMENTS OF BONDS

Molecule	Dipole moment of the molecule (Debye units)	Angle between bonds	Bond	Distance in the bond, Å	Dipole moment of the bond (Debye units)
H_2O	1·84	105°	O—H	0·95	1·51
$(CH_3)_2O$	1·29	111°	O—CH_3	1·42	1·12
NH_3	1·46	109°	N—H	1·01	1·42
$N(CH_3)_3$	0·65	110°	N—CH_3	1·47	0·61
H_2S	0·93	92°	S—H	1·35	0·67

in organic molecules and at estimating the dipole moment as the vector sum of such contributions do not always lead to good results. The value of the bond contribution depends on the presence of other neighbouring groups in the molecule, which by induction change the dipole-moment contributions in their environment.

Application of the method of calculating molecular dipole moments by means of the vector addition of the contributions gives better results when the contributions are ascribed to some definite groups having a known environment.

TABLE 1.12. DIPOLE MOMENTS, $\bar{\mu}$, OF VARIOUS GROUPS IN DEBYES

Group	C_6H_5		CH_3		$C_nH_{2n+1}(n>1)$		Angle θ
	Gas	Solution	Gas	Solution	Gas	Solution	
$—CH_3$	0·36	0·4	0	—	0	—	180°
$—OCH_3$	1·35	1·25	1·30	—	—	—	55°
$—SCH_3$	—	1·27	—	1·40	—	—	52°
$—NH_2$	1·48	1·53	1·23	—	1·2	1·38	—
—I	1·6	1·30	1·64	1·5	1·87	1·8	0°
—Br	1·75	1·52	1·80	1·8	2·01	1·9	0°
—Cl	1·72	1·55	1·87	1·7	2·05	1·8	0°
—F	1·57	1·43	1·81	—	1·92	—	0°
—OH	1·4	1·6	1·69	1·66	1·69	1·7	62°
—COOH	—	1·7	1·73	1·6	1·73	1·7	74°
$—COOCH_3$	—	1·9	1·67	1·75	1·76	1·9	70°
—CHO	(3·1)	2·8	2·72	2·5	2·73	2·5	58°
$—COCH_3$	3·00	2·9	2·84	2·74	2·78	—	59°
—CN	4·39	4·0	3·94	3·4	4·00	3·57	0°
$—NO_2$	4·21	3·98	3·50	3·1	3·68	3·3	0°

The values of the group dipole-moment contributions in the gas molecules differ from those for the same molecules in solutions. Table 1.12 lists[397] both kinds of the contributions of various groups for three cases, viz., the group is linked with (a) C_6H_5, (b) CH_3, (c) C_2H_5 or longer chains, i.e. C_nH_{2n+1}, $n > 1$. Since the group moment is in the majority of cases a sum of several dipole moments, its direction may differ from that of the bond between this group and the rest of the molecule. Table 1.12 also lists the values of the angle θ made by the resultant dipole moment with the direction of the bond linking the group with the carbon atom in the remaining part of the molecule. The accuracy of the estimation of the dipole moment from the data given in Table 1.12 is rather low, as is seen from the following examples:[356]

1. The dipole moment of the molecule of *p*-chloronitrobenzene should be (Eq. (1.7))

$$4{\cdot}21-1{\cdot}72 = 2{\cdot}49 \text{ D}$$

The value of the dipole moment found experimentally is 2·78 D (Table 1.10), hence, the agreement with the estimate is fairly good.

2. The dipole moment of the molecule of *o*-dichlorobenzene should be, according to Eq. (1.4) and Table 1.12

$$2\times 1{\cdot}72\times \cos 30° = 2{\cdot}98 \text{ D}$$

The experimental value of the moment is, according to Table 1.10, much lower and amounts to 2·20 D; this fact is attributed to the interaction of the two dipoles C—Cl.

3. The dipole moment of the molecule of *m*-chloronitrobenzene, calculated from Eq. (1.6) as the vector sum of the dipole moments of chlorobenzene, $\bar{\mu}_1 = 1{\cdot}72$, and nitrobenzene, $\bar{\mu}_2 = 4{\cdot}21$ (Table 1.12), is $\bar{\mu}_{\text{calc}} = 3{\cdot}67$. The value measured experimentally (Table 1.10) is $\bar{\mu}_{\text{exp}} = 3{\cdot}69$; the agreement between the computed and measured values is excellent.

Analogous calculations yield for: *o*-chloronitrobenzene $\bar{\mu}_{\text{calc}} = 5{\cdot}26$, $\bar{\mu}_{\text{exp}} = 4{\cdot}59$; *p*-chloronitrobenzene $\bar{\mu}_{\text{calc}} = 2{\cdot}52$, $\bar{\mu}_{\text{exp}} = 2{\cdot}78$.

If the molecule has a complicated structure with several dipole groups, its dipole moment may be calculated by selecting a coordinate system *X*, *Y*, *Z* and calculating the values of the projections $\bar{\mu}_x$, $\bar{\mu}_y$, $\bar{\mu}_z$ of the individual moments on these axes. The sums of the projections on the individual axes are found and then the dipole moment of the molecule is obtained as the square root of the sum of squares of the sums of the projections

$$\bar{\mu}_s = \sqrt{(\bar{\mu}'_x+\bar{\mu}''_x+\ldots)^2+(\bar{\mu}'_y+\bar{\mu}''_y+\ldots)^2+(\bar{\mu}'_z+\bar{\mu}''_z+\ldots)^2} \tag{1.8}$$

The calculation is more complicated when the molecule contains several dipole groups the mutual positions of which may vary owing to their rotation around the bonds. In this case we calculate the mean value of the square of the resultant dipole moment for all possible positions of dipoles in the molecule.

e. Polarizability of Molecules

In molecules with permanent dipoles the "centres of gravity" of the positive charges (the nuclei) and the negative charges (the electronic shells) do not coincide. If, in the absence of external fields these "centres of gravity" do coincide the molecule has no permanent dipole moment. It may happen however that in an electric field such a molecule will acquire the properties of a dipole. If, for example, the molecule is situated between the plates of a charged capacitor the positive charges may shift towards the negative plate and the negative charges in the opposite direction, and the "gravity centres" of the positive and the negative charges will move apart, an induced dipole being thereby formed.

This phenomenon is known as *polarization* and is associated with the following processes:

1. The formation of an induced dipole owing to the displacement of electrons (*electron polarization*).
2. The shift or rotation of the atoms and polar groups of atoms in the molecules (*atom polarization*).
3. If the molecules present in the field have dipole moments, the orientation of the molecules to the position corresponding to the minimum energy of the system (*orientation polarization*).

A measure of the ease with which a substance may be polarized is provided by the so-called *molar polarization P*. This quantity may be defined by the equation

$$P = \frac{4\pi N\alpha}{3}$$

α is the polarizability of the molecules and is discussed further below. The three processes listed above gives rise to corresponding contributions P_{el}, P_{at} and P_{or} to the quantity P, so that

$$P = P_{el} + P_{at} + P_{or} \tag{1.9}$$

(All these quantities refer to 1 mole of the substance.) A relationship exists between the molar polarization P and the dielectric constant ε. (ε is equal to the ratio of the capacity C of a condenser containing the given compound between the plates to the capacity C_0 of the same condenser with vacuum between the plates.) For gases and dilute solutions of polar molecules in non polar solvents the relationship is

$$P = \frac{\varepsilon - 1}{\varepsilon + 2} \frac{M}{\rho} \quad \text{cm}^3/\text{mole} \tag{1.10}$$

where M is the molecular weight, and ρ is the density.

If the induced dipole is produced by a field of intensity F, then

$$\bar{\mu} = \alpha F \tag{1.11}$$

where α denotes the *polarizability* of the molecule and is equal to the magnitude of the dipole moment induced by a field of unit intensity. The polarizability α, like the polarization P, has the dimension of volume (cm^3).

In the general case the values of polarizability in the directions of the principal axes of the molecule are unequal so that

$$\alpha_1 \neq \alpha_2 \neq \alpha_3$$

Polarizabilities deduced from dielectric constant measurements are arithmetic mean values,

$$\alpha = \frac{\alpha_1 + \alpha_2 + \alpha_3}{3}$$

Table 1.13 lists illustrative values of the polarizabilities α_1, α_2, α_3 for several kinds of anisotropic molecules.

TABLE 1.13. POLARIZABILITIES $\alpha \times 10^{24}$ cm^3 IN THE DIRECTIONS OF THE PRINCIPAL AXES OF THE MOLECULE (DATA TAKEN FROM THE LITERATURE[11])

	Cl_2	HCl	H_2S	SO_2	NH_3	NO	CO_2	CS_2	C_2H_2	CH_3OH
α_1	6·6	3·13	4·21	5·49	2·42	5·32	4·10	15·14	5·12	4·00
α_2	3·62	2·39	3·21	2·72	2·18	1·83	1·93	5·54	2·43	2·56
α_3	3·62	2·39	3·93	3·49	2·18	1·83	1·93	5·54	2·43	3·14

The polarizability of large and complex molecules can be considered to result from the polarizability of the individual bonds in these molecules. Bond polarizabilities in the direction parallel to the bond ($\alpha \parallel$) usually differ considerably from those in the direction perpendicular to the bond ($\alpha \perp$).

The values of the polarizabilities of some bonds, according to Denbigh,[77] are given in Table 1.14.

TABLE 1.14. POLARIZABILITY OF BONDS IN THE DIRECTION PARALLEL ($\alpha \parallel \times 10^{24}$ cm³) AND PERPENDICULAR ($\alpha \perp \times 10^{24}$ cm³) TO THE BOND[77]

Bond	C—C	C_{arom}—C_{arom}	C=C	C≡C	C—Cl	C—Br	C=O	C=S
$\alpha\parallel \times 10^{24}$ cm³ $\alpha\perp \times 10^{24}$ cm³	1·88 0·2	2·25 0·48	2·86 1·06	3·54 1·27	3·67 2·08	5·40 2·88	1·99 0·75	7·57 2·77

Bond	C≡N	C—H	N—H	S—H	O_2	N_2	CO	HBr
$\alpha\parallel \times 10^{24}$ cm³ $\alpha\perp \times 10^{24}$ cm³	3·10 1·40	0·79 0·58	0·58 0·84	2·30 1·72	2·43 1·19	2·43 1·43	2·60 1·62	4·23 3·32

In simple cases the polarizability of a molecule may be calculated directly from the bond polarizabilities. For the molecule of acetylene we find

$$(C{\equiv}C)+2(C{-}H) = 3{\cdot}54+2\times 0{\cdot}79 = 5{\cdot}12$$

The observed value of the polarizability of an acetylene molecule is also $5{\cdot}12\times 10^{-24}$ cm³.

For gases and dilute solutions of polar molecules in non polar solvents, the relationship between the polarization P, the dielectric constant ε, and the polarizability α is accurately represented by the *Clausius–Mossoti equation*

$$P = \frac{\varepsilon-1}{\varepsilon+2}\frac{M}{\rho} = \frac{4}{3}\pi N\alpha \tag{1.12}$$

where N is the Avogadro number.

Analogously to polarization (1.9), the following polarizability components may be distinguished

$$\alpha = \alpha_{el}+\alpha_{at}+\alpha_{or} \tag{1.13}$$

where α is the polarizability of a molecule,
α_{el} is the electron polarizability,
α_{at} is the atom polarizability,
and α_{or} is the orientation polarizability.

Hence,

$$P = \frac{4}{3}\pi N\alpha_{el}+\frac{4}{3}\pi N\alpha_{at}+\frac{4}{3}\pi N\alpha_{or} \tag{1.14}$$

The estimation of the value of polarizability α amounts to the calculation of the values of the addends in Eq. (1.14).

According to Debye, the orientation polarization is

$$P_{or} = \frac{4}{9}\pi N\frac{\overline{\mu^2}}{kT} \tag{1.15}$$

where N is the Avogadro number,
$\bar{\mu}$ the dipole moment,
k the Boltzmann constant,
and T is the absolute temperature, °K.

Hence,

$$\alpha_{or} = \frac{\bar{\mu}^2}{3kT} \tag{1.16}$$

The electron polarizability α_{el} is approximately equal to the cube of the radius $\sigma/2$ of the molecule:

$$\alpha_{el} \approx \frac{\sigma^3}{8} \tag{1.17}$$

Hence, in the case of some non-polar molecules ($\alpha_{at} = 0$ and $\alpha_{or} = 0$) we get

$$P = \frac{4}{3}\pi N\alpha_{el} = \frac{4}{3}\pi N\frac{\sigma^3}{8} = Nv_m = V_m \tag{1.18}$$

where: $v_m = \frac{4}{3}\pi(\sigma^3/8)$, the volume of a single molecule, and V_m is the actual volume of the molecules in 1 mole. If V_m can be calculated, Eq. (1.18) may be used to estimate the polarizability α_{el} of a molecule which possesses no permanent dipole (so that $\alpha_{or} = 0$) and also for which $\alpha_{at} = 0$. One very simple, though rather approximate, method is to make use of van der Waals' equation $(p+a/V^2)(V-b) = RT$. According to the crude theory upon which this equation is based, the constant b should be equal to four times the volume V_m. Hence, the polarization P is given by

$$P = \frac{b}{4} \tag{1.19}$$

and the polarizability by

$$\alpha = \frac{3b}{16\pi N} \tag{1.20}$$

It will be seen later that the quantity b is approximately additive and may be calculated by adding the contributions of the atoms making up a molecule (cf. section 5). This affords a simple, though somewhat approximate, method for estimating α for molecules which have no permanent dipole moment.

A more accurate calculation of α_{el} may be made using the following relationship between the dielectric constant ε and the refractive index n_∞ corresponding to light of infinite wavelength

$$\varepsilon = n_\infty^2 \tag{1.21}$$

Substituting this expression for ε into Eq. (1.12) it follows that

$$P = \left(\frac{n_\infty^2-1}{n_\infty^2+2}\right)\frac{M}{\rho}$$

But the molar refraction R_∞ corresponding to light of infinite wavelength is given by the Lorenz–Lorentz equation to be

$$R_\infty = \frac{n_\infty^2-1}{n_\infty^2+2}\,\frac{M}{\rho} \tag{1.22}$$

It is then a consequence of equation (1.9) that for those non-polar molecules for which $P_{at} = P_{or} = 0$,

$$P = P_{el} = R_{\infty} \tag{1.23}$$

The value of the atom polarization P_{at} is, on average, equal to 3–5% (only exceptionally it reaches 15%) of the value of the electron polarization P_{el}, and the values of the usually measured molar refraction R_D (for the sodium—D line) are about 3–4% higher than the values of R_{∞}. Therefore, for non-polar molecules, it can be assumed with accuracy sufficient for technological calculations that

$$P = P_{el}+P_{at} = R_D \tag{1.24}$$

The molar refraction R_D may be calculated with quite high accuracy by summing up the contributions of the atoms, groups and bonds in a molecule (cf. section 4). It is a consequence of Eqs. (1.14) and (1.22) that the polarizability of non-polar molecules is given by

$$\alpha = (\alpha_{el}+\alpha_{at}) = \frac{3R_D}{4\pi N} \tag{1.25}$$

Some polarizability contributions given by Ketelaar[196] are presented in Table 1.15. Approximate mean values of the electron polarizability can be calculated by summing up these contributions.

TABLE 1.15. ELECTRON POLARIZABILITY CONTRIBUTIONS, $\times 10^{24}$ cm³

Group	Contribution	Group	Contribution
F	0·38	N (primary)	0·87
Cl	2·28	N (secondary)	0·93
Br	3·34	N (tertiary)	1·03
I	5·11	H	0·42
CN	2·12	C	0·93
H	0·42	CH_2	1·77
O (hydroxyl)	0·59	Double bond	0·58
O (ether)	0·64	Triple bond	0·86
O (carbonyl)	0·84	—C_6H_5	9·38

EXAMPLE 1.1. Estimate the mean polarizability of the methanol molecule.

a. According to Table 1.13, the polarizabilities of the molecule of methanol ($\alpha_2 \perp$ to the COH surface) are

$$\alpha_1 = 4{\cdot}00\times 10^{-24}\ \text{cm}^3, \qquad \alpha_2 = 2{\cdot}56\times 10^{-24}\ \text{cm}^3, \qquad \alpha_3 = 3{\cdot}14\times 10^{-24}\ \text{cm}^3$$

The mean polarizability

$$\alpha = \frac{(4{\cdot}00+2{\cdot}56+3{\cdot}14)}{3}\times 10^{-24}\ \text{cm}^3 = 3{\cdot}23\times 10^{-24}\ \text{cm}^3$$

b. The molar refraction of methanol, $R_D = 8{\cdot}23$ cm³.
According to Eq. (1.25), the sum of the electron and atom polarizabilities is

$$\alpha = \frac{3\times 8{\cdot}23}{4\times 3{\cdot}14\times 6{\cdot}06\times 10^{23}}\ \text{cm}^3 = 3{\cdot}24\times 10^{-24}\ \text{cm}^3$$

c. Addition of the group contributions (Table 1.15) gives for H—CH_2—O—H:

$$\alpha = (0{\cdot}42+1{\cdot}77+0{\cdot}59+0{\cdot}42)\times 10^{-24}\ \text{cm}^3 = 3{\cdot}18\times 10^{-24}\ \text{cm}^3$$

3. INTERMOLECULAR FORCES

The idealized models of the gaseous, liquid and solid states, which are familiar from elementary physics, are insufficient for a quantitative treatment of such phenomena as viscosity or the thermal conductivity of fluids. For such a treatment to be made, it is necessary to know the character and magnitude of the intermolecular forces in more detail and to develop a picture of the structure of the phase of interest (gaseous, liquid or solid) which is closer to reality than that of a perfect structure.

In our considerations we shall use simplified relationships describing the forces† acting between molecules. We shall not discuss the methods by which these relationships have been deduced.

a. Energy of a System of Two Molecules

When discussing the forces acting between a pair of interacting molecules, it is convenient to consider the relationship between the intermolecular potential energy E and the distance r between the molecular centres. If E depends only on r, the magnitude of the interaction force between the two molecules is $-\mathrm{d}E/\mathrm{d}r$.

Let us consider, for instance, the forces acting between two molecules of a real gas, generally known as *van der Waals' forces*. The net force is the sum of a repulsion force, which is dominant at small distances (of the order of 10^{-8} cm), and forces of attraction which are dominant at larger distances. The latter forces can be due to: (a) electric forces when the molecules in question have dipole moments (Keesom); (b) direct formation of induced dipole moments (Debye); (c) attraction between the molecules (also between those which are not dipoles), justified by the wave theory of matter (London). At large separations the *total energy of a two molecule system* is the sum of the potential energies resulting from the above-mentioned three kinds of attractive forces, and may be expressed in the form

$$E_p = -\frac{c}{r^6} \tag{1.26}$$

where the quantity c depends, among other things, on the value of the dipole moment $\bar{\mu}$ and the polarizability α of the molecule.

When the intermolecular distances are small, forces of mutual repulsion appear. The potential energy E_{rep} of the repulsive forces in the system may be calculated according to the equation

$$E_{\mathrm{rep}} = \frac{d}{r^q} \tag{1.27}$$

The exponent q depends on the number N_A of the external electrons in the atoms of the molecule.[405] According to Altenburg[3]

$$q = 8+0{\cdot}288N_A \tag{1.28}$$

---

† In the following sections the author has made considerable use of material taken from the works of Bird, Hirschfelder and Curtiss,[25] and Novikov and Vukalovich.[260]

Lennard-Jones[25] has assumed that $q = 12$ is a fair approximation for the case of interaction between non-polar molecules, and proposed the equation

$$E(r) = 4\varepsilon\left[\left(\frac{\sigma}{r}\right)^{12} - \left(\frac{\sigma}{r}\right)^{6}\right] \tag{1.29}$$

where σ and ε are force constants dependent on the nature of the interacting molecules.

In the case of interaction between a pair of molecules of complicated structure, it is impossible to carry out exact calculations; if such is the case, we make use of *empirical functions* and calculate the values of ε and σ from such experimental data as the compressibility factor of a real gas, viscosity, thermal conductivity, etc., because the relationships between these quantities and ε and σ are known.

At present, the estimation of the energy $E(r)$ for pairs of non-polar spherical or almost spherical molecules, is usually carried out using the Lennard-Jones function (1.29). The dependence of the energy E on the distance r between two molecules is shown in Fig. 1.4. The constant σ, expressed in Å, represents the intermolecular distance correspond-

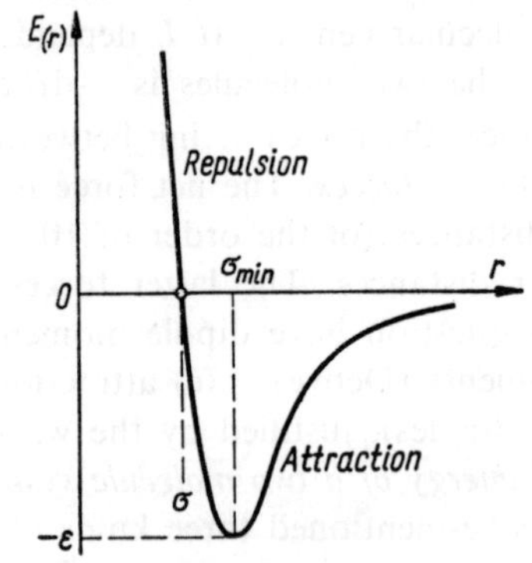

FIG. 1.4. Variation of the potential energy $E(r)$ with the distance r between two non-polar molecules (after Lennard-Jones[25]).

ing to $E(r) = 0$. The constant ε denotes the minimum energy of the system of two molecules, and the quantity ε/k (where k is Boltzmann's constant) is expressed in °K. The value ε of the potential energy corresponds to a certain distance between the molecules, equal to

$$\sigma_{\min} = \sigma\sqrt[6]{2} \tag{1.30}$$

If the distance r between the molecules is larger than $\sigma_{\min}$, the molecules attract themselves; if $r < \sigma_{\min}$, the molecules repel one another.

For calculating the potential energy of a pair of polar molecules use is often made of an empirical formula devised by Stockmayer, which is sufficiently accurate when simple polar molecules are involved

$$E(r, \theta_1, \theta_2, \phi_2-\phi_1) = 4\varepsilon\left[\left(\frac{\sigma}{r}\right)^{12} - \left(\frac{\sigma}{r}\right)^{6}\right] - \frac{\bar{\mu}^2}{r^3} f(\theta_1, \theta_2, \phi_2-\phi_1) \tag{1.31}$$

$$f(\theta_1, \theta_2, \phi_2-\phi_1) = 2\cos\theta_1\cos\theta_2 - \sin\theta_1\sin\theta_2\cos(\phi_2-\phi_1) \tag{1.32}$$

where $\bar{\mu}$ denotes the dipole moment, and θ_1, θ_2 and $(\phi_2-\phi_1)$ are the angles defining the orientation of dipoles in the system.

Equation (1.31) is cumbersome, and the accuracy of the calculations for polar gases is similar to that for non-polar gases when Eq. (1.29) is applied.[254] In the latter case Eq. (1.29) often yields results as accurate as those obtained with more complicated equations.[385]

b. Calculation of the Intermolecular Force Constants

If σ and ε are known, Eqs. (1.29) and (1.31) enable the interactions between all pairs of molecules in a system to be calculated and this in turn enables certain of the properties of gases to be estimated using statistical mechanical arguments. For example in later chapters formulae based on the methods used in statistical mechanics will be given which enable properties such as the viscosity, thermal conductivity and diffusivity of a gas to be estimated when the values of σ and ε are known. The most reliable way of finding the values of σ and ε is, however, the reverse procedure, viz., calculation according to the above formulae, provided some experimental values, e.g. of viscosity or thermal conductivity, are known. The whole problem amounts to solving a slightly troublesome system of implicit equations.

Often no experimental values of these properties are available. In such cases use may be made of some empirical, though not very accurate, relations between ε/k and the critical temperature T_c of the compound, its boiling temperature T_b or melting temperature T_m, and between σ and the molal volumes V_c, V_b, and V_m at the critical, boiling and melting points, respectively. The expressions for ε/k are:

$$\begin{aligned} \varepsilon/k &= 0{\cdot}77\,T_c \\ \varepsilon/k &= 1{\cdot}15\,T_b \\ \varepsilon/k &= 1{\cdot}92\,T_m \end{aligned} \tag{1.33}$$

The expressions for σ may be given in the form:

$$\begin{aligned} b_0 &= 0{\cdot}75\,V_c \\ b_0 &= 2{\cdot}0\,V_b \\ b_0 &= 2{\cdot}3\,V_m \end{aligned} \tag{1.34}$$

$$\text{and} \quad b_0 = 18{\cdot}4\,T_c/p_c \tag{1.35}$$

where p_c denotes the critical pressure of the compound.

In these expressions b_0 is given by

$$b_0 = \frac{2}{3}\pi N \sigma^3 \tag{1.36}$$

(N being the Avogadro number). When σ is expressed in Å and b_0 in cm³/mole

$$b_0 = 1{\cdot}2615\,\sigma^3 \tag{1.37}$$

The value of σ is calculated from Eqs. (1.36) and (1.37).

Stiel and Thodos[347] have found that the mean deviations of σ and ε/k calculated according to these equations from data based upon the measurements of their transport properties, are 4·6 and 20·4%, respectively. These authors recommend the use of the formulae:

$$\sigma = 0{\cdot}1866\, V_c^{1/3} z_c^{-6/5}$$

which shows an average deviation of 1·9%

and
$$\varepsilon/k = 65{\cdot}3\, T_c z_c^{18/5}$$

with an average deviation of 6·6%. In these formulae T_c is expressed in °K, V_c in cm^3/mole, σ in Å, and $z_c = \dfrac{p_c V_c}{RT_c}$.

In Table 1.16 the values σ and ε/k in the Lennard-Jones equation are given for a number of compounds. The second and third columns list the values of ε/k and σ evaluated from experimental viscosity data. The fourth and fifth columns give the values calculated from the second virial coefficient (i.e. from the measurements of the compressibility factors of real gases). The values listed in the table are taken from Hirschfelder *et al.*;[25, 154] for references the reader is referred to the original. The authors recommend the application of the values from the second and third column for the estimation of viscosity, thermal conductivity and diffusivity; for calculation of the compressibility factor as well as for thermodynamic calculations, the values contained in the fourth and fifth columns are more suitable.

The table also gives the values of ε/k and σ, calculated for substances with polar or non-spherical molecules, i.e. those for which Eq. (1.29) cannot, strictly speaking, be used. However, since so far no simple and accurate method for calculating the compressibility factor, viscosity, etc., has been developed for such cases, we can make use of the values ε/k and σ as presented in the table and consider the results obtained as an approximation.

Force constants calculated from the various transport properties for a given substance may differ considerably. For example investigations on the self-diffusivity of carbon dioxide[87] in the temperature range 1180–1680°K yielded $\sigma = 2{\cdot}70$ Å and $\varepsilon/k = 871$°K; viscosity studies in the range 200–1500°K gave: $\sigma = 3{\cdot}94$ Å and $\varepsilon/k = 200$°K. According to the authors, these differences arise because the molecules of carbon dioxide are not truly spherical.

Values of ε/k and σ for use with the Stockmayer equation (1.31), are known for a limited number of compounds. Table 1.17, compiled by Hirschfelder *et al.*,[25] shows values of ε/k, σ and t^*, calculated from second virial coefficient data. The quantity t^* is defined by the equation

$$t^* = \frac{\bar{\mu}^2}{\varepsilon\sigma^3\sqrt{8}} \tag{1.38}$$

where $\bar{\mu}$ is the dipole moment; t^* characterizes the difference in the behaviour of polar and non-polar substances. In Table 1.17, the dipole moments are expressed in Debyes, ε/k in °K, and σ in Å.

The constants ε and σ in the Lennard-Jones and Stockmayer equations, listed in Tables 1.16 and 1.17 or calculated according to Eqs. (1.33), (1.34), (1.35) and (1.36), refer to the interaction between two identical molecules. For mixtures containing two kinds of molecules interactions between molecules of different species must also be considered. The empirical estimation of the force constants for interaction of this type is considered below.

TABLE 1.16. FORCE CONSTANTS IN THE LENNARD-JONES EQUATION

Substance	Constants computed from viscosity data		Constants computed from compressibility factors	
	ε/k, °K	σ, Å	ε/k, °K	σ, Å
H_2	10·22*	2·576*	6·03	2·63
H_2	33·3*	2·918*	29·2	2·87
	38·0†	2·915†	—	—
D_2	39·3	2·948	31·1	2·87
Ne	27·5†	2·858†	34·9	2·78
A	116†	3·465†	122	3·40
Kr	190	3·61	171	3·60
Xe	229	4·055	221	4·100
Air	97·0*	3·617*	99·2	3·522
	84·0†	3·689†	102	3·62
N_2	91·5*	3·681*	95·05	3·698
	79·8†	3·749†	—	—
O_2	113*	3·433*	117·5	3·58
	88·0†	3·541†	—	—
CO	110*	3·590*	100·2	3·763
	88·0†	3·706†	—	—
CO_2	190*	3·996*	189	4·486
	213†	3·897†	—	—
NO	119*	3·470*	131	3·17
N_2O	220*	3·879*	189	4·59
CH_4	137*	3·882*	148·2	3·817
CCl_4	327	5·881	—	—
SO_2	252*	4·290*	—	—
F_2	112	3·653	—	—
Cl_2	357	4·115	—	—
Br_2	520	4·268	—	—
I_2	550	4·982	—	—
HCl	360	3·305	—	—
HI	324	4·123	—	—
AsH_3	281	4·06	—	—
$SnCl_4$	1550	4·540	—	—
Hg	851	2·898	—	—
C_2H_2	185	4·221	—	—
C_2H_4	205	4·232	199·2	4·523
C_2H_6	230	4·418	243	3·954
C_3H_8	254	5·061	242	5·637
n-C_4H_{10}	410	4·997	297	4·971
i-C_4H_{10}	313	5·341	—	—
n-C_5H_{12}	345	5·769	—	—
n-C_6H_{14}	413	5·909	—	—
n-C_7H_{16}	—	—	282	8·88
n-C_8H_{18}	320	7·451	—	—
n-C_9H_{20}	240	8·448	—	—
Cyclohexane	324	6·093	—	—
C_6H_6	440	5·270	—	—
CH_3OH	507	3·585	—	—
C_2H_5OH	391	4·455	—	—
CH_3Cl	855	3·375	—	—
CH_2Cl_2	406	4·759	—	—
$CHCl_3$	327	5·430	—	—
C_2N_2	339	4·38	—	—
COS	335	4·13	—	—
CS_2	488	4·438	—	—

NOTE: The values marked with an asterisk and a dagger have been calculated on the basis of measurements performed in the temperature range 80–300°K and 300–1000°K, respectively.

TABLE 1.17. FORCE CONSTANTS IN STOCKMAYER'S EQUATION

Substance	$\bar{\mu}$ (Debye units)	t^*	ε/k, °K	σ, Å
$CHCl_3$	1·05	0·1	1060	2·98
$CHCl_2F$	1·29	0·1	381	4·82
C_2H_5Cl	2·02	0·2	320	5·41
CH_3Cl	1·89	0·6	380	3·43
CH_3COCH_3	2·74	0·7	520	3·76
CH_3OH	1·66	0·8	630	2·40
NH_3	1·47	1·0	320	2·60
CH_3F	1·82	1·07	207	3·36
H_2O	1·83	1·2	380	2·65
CH_3CN	3·5	1·2	400	4·02
CH_3CHO	2·7	1·4	270	3·68

When both kinds of the substances contained in the mixture are composed of non-polar molecules, the potential energy of their interaction can be expressed by means of the Lennard-Jones equation using force constants $\sigma_{1,2}$ and $\varepsilon_{1,2}$ given by

$$\sigma_{1,2} = \frac{\sigma_1+\sigma_2}{2}, \qquad \varepsilon_{1,2} = \sqrt{\varepsilon_1\varepsilon_2} \tag{1.39}$$

In these empirical equations the subscripts 1 and 2 refer to interactions between like molecules (1 and 2 are the constituents of the mixture) and 1,2 to interactions between unlike molecules. σ_1, σ_2, ε_1 and ε_2 may be inferred from the properties of the pure components as described above.

Equation (1.39) is an approximation which would only be expected to be valid when $\sigma_1 = \sigma_2$. A rather more rigorous, and frequently more accurate equation has been proposed:[161]

$$\varepsilon_{1,2} = 2^6 \times \frac{\sqrt{\varepsilon_1\varepsilon_2}\ \sigma_1^3\sigma_2^3}{(\sigma_1+\sigma_2)^6} \tag{1.40}$$

The same relationship (1.40) may be employed for the estimation of the values of the constants $\sigma_{1,2}$ and $\varepsilon_{1,2}$ for mixtures of polar molecules, provided that the values σ_1, σ_2 and ε_1 and ε_2 have been calculated for the polar molecules (or taken, for example from Table 1.17); the interaction of two different polar molecules is then represented by the Stockmayer equation (1.31) in which we substitute the calculated values $\sigma_{1,2}$ and $\varepsilon_{1,2}$ as the force constants. In the case of the interaction of two different polar molecules, 1 and 2, the value of the parameter t^* is

$$t^*_{1,2} = \frac{\bar{\mu}_1\bar{\mu}_2}{\varepsilon_{1,2}\sigma^3_{1,2}\sqrt{8}} \quad \text{or} \quad t^*_{1,2} = \sqrt{t^*_1 t^*_2} \tag{1.41}$$

If a mixture is composed of two compounds, one of which contains polar (subscript p) and the other non-polar molecules (subscript n), it may be approximately assumed that the potential energy for interaction between the polar molecules and the non-polar molecules has the same form as that between non-polar molecules and is thus given by the Lennard-Jones equation when suitable "force constants" are inserted. The appropriate constants may be computed from the following equations:

$$\left.\begin{aligned}\sigma_{pn} &= \tfrac{1}{2}(\sigma_p+\sigma_n)\,\xi^{-1/6}\\ \varepsilon_{pn} &= \sqrt{\varepsilon_p\varepsilon_n}\,\xi^2\end{aligned}\right\} \quad (1.42)$$

where

$$\xi = \left[1+\tfrac{1}{4}\sqrt{2}\,\alpha_n^* t_p^* \sqrt{\varepsilon_p/\varepsilon_n}\right]$$

and

$$\alpha_n^* = \alpha_n/\sigma_n^3$$

α_n is the polarizability of the non-polar molecule. Values of σ_n and ε_n may be taken from Table 1.16, and those of σ_p, ε_p and t_p^* from Table 1.17. Alternatively these constants may be determined by one of the methods described above.

c. Application of the Lennard-Jones and Stockmayer Equations to the Estimation of some Physico-chemical Properties of Fluids

Knowing the values of the force constants σ and ε, and, in the case of polar fluids the value of t^* for the appropriate molecular interactions it should, in principle, be possible to estimate values of compressibility, viscosity, thermal conductivity, and diffusivity for the compound or mixture in question. Methods by which such estimations may be carried out are discussed below. The accuracy of estimations made in this way is not very high. One reason for this being that the effect of the shape of the molecule, often critical, is not taken into account; the introduction of the angles describing the orientation of dipoles into the Stockmayer equation (1.31) fails to allow adequately for the influence of various atomic groups in the molecule. The practical utility of estimates of physico-chemical properties based on intermolecular force constants is, therefore, rather limited.

The essential value of these considerations consists in the elucidation of the theoretical links existing between the properties of the individual molecules such as dipole moment, polarizability, molecular weight, etc. and the macroscopic properties of the system, i.e. compressibility, viscosity, thermal conductivity and diffusivity. A further benefit of these theoretical studies has been the development of relationships between the viscosity, the thermal conductivity and the diffusivity, which enable one of these properties to be calculated from a knowledge of the other two.

The above considerations also permit more intelligent use of relationships such as the corresponding states theorem and the method of reference compounds, which are discussed in detail in later sections.

4. CONSTITUTIVE AND ADDITIVE PROPERTIES OF CHEMICAL COMPOUNDS

The measurement and tabulation of the physico-chemical constants for many hundred thousands of compounds would be exceedingly costly and time-consuming. Special attention is, therefore, paid to the methods of calculating the molecular and thermodynamic quantities. The simplest case occurs when the dependence of the physico-chemical property on the molecular constitution (i.e. on the kind and number of atoms, atomic groups and the ways of their interconnection in the molecule) is known in a simple form. When the same contribution to the required physico-chemical property is made by an element of a molecule (atom, atomic group, bond, etc.), regardless of the nature of the rest of the molecule this property is called an *additive* one; a constitutive additive property may be easily and quickly estimated (provided the structural formula of the molecule is known) by summing up the contributions corresponding to all elements of the molecule. In such a case there is no need for compiling extensive tables listing the quantity for many different substances: for the calculation it is sufficient to use a short tabulation of the contributions corresponding to the individual atoms, atomic groups and bonds. We have already come across the method of additive calculation of the constitutive properties when estimating the polarizability (cf. Table 1.15).

Unfortunately, we know only a relatively small number of physico-chemical properties which may be calculated by making use of the principle of additivity, and in the majority of these few cases the rule of additivity holds only approximately.

Generally speaking, it is possible to compute additively quantities which depend either on the mass of the elements of which the molecule is constructed (e.g. molecular weight) or on their volume and shape (e.g. molal volume, constant b in van der Waals' equation) or, finally, on the magnitude of the intermolecular forces (e.g. constant a in van der Waals' equation). Some physico-chemical quantities depend on several of the factors mentioned above. This fact complicates the use of the principle of additivity in such cases.

According to Lutskii,[231] the macrophysical properties (y) of substances (e.g. surface tension, viscosity, thermal conductivity, etc.) and the quantities characterizing the relations between these properties (y') depend only indirectly on the atomic structure of the molecules of the substance, and directly on such properties of the molecule as the weight (m), volume (V), shape (ϕ), polarizability (α), dipole moment ($\bar{\mu}$) and electric charge (e).

The principle of additivity relates the sought properties y or y' only with the composition and structure of the molecules of which the substance in question is composed. Hence, the macroproperties y or y' of the compound in question can be theoretically predicted if we know the relationships existing between the composition and the structure of the molecules and their basic properties m, V, ϕ, α, $\bar{\mu}$, and e.

Usually, however, the dependence of the above-mentioned parameters on the molecular structure is unknown and the determination of the values of the contributions (the sum of which gives us the values of the sought additive quantity y or y') has to be carried out by trial and error.

In the general case the properties y or y' cannot strictly obey the law of additivity, since the changes of the properties of a molecule, due to some new substituent (e.g. a new atom or atomic group or else other changes in the molecular structure) do not depend merely on the character of the introduced substituent but also on the constitutive factors, i.e. atoms and environment with which the introduced new substituent has been connected in the molecule, on bonds by which it is linked, etc. The quantities V, ϕ, α, and $\bar{\mu}$, influencing the desired parameters y or y', depend on all these factors. For this reason any change expressed by ascribing to the new substituent only one value of the contribution will lead to an inaccurate scheme of additive calculations; while discrimination of the contributions and relating their values with the environment into which the new substituent is introduced, must give, on the other hand, a scheme with a very large number of contributions and which is therefore practically useless.

In practice, tables are used which contain from a dozen to several tens of contributions, and thus do not take into account all the complicating factors mentioned here, and usually do not give any possibility of obtaining high accuracy of estimation of the quantities required.

Molecular weight is a strictly additive quantity.† We calculate it by summing up the atomic weights of all the atoms contained in the molecule.

a. Molal Volume

From our considerations concerning the shape and the size of the molecules (section 2c) it may be concluded that since the interatomic distances and angles between bonds in the molecule are constant, the volume of the molecule should also be constant and dependent on the magnitude of these distances and angles. Consequently, it should be assumed that the volume of one molecule (and hence perhaps the volume of one mole of a substance composed of such molecules) may be calculated on the basis of contributions which correspond to the size of the atoms, atomic groups and bonds.‡

In simple cases, experiments confirm this presumption. It is found that whenever some definite atomic group is introduced into a molecule the molal volume V of the compound increases by approximately the same increment ΔV. The value of the increment depends first of all on the kind of the group introduced, but also, to some extent, on the presence of other atoms, atomic groups and bonds in the molecule. This may be illustrated by the dependence of the molal volume of hydrocarbons of the aliphatic series, C_nH_{2n+2}, on their molecular structure (Table 1.18).

TABLE 1.18. MOLAL VOLUMES AND CONTRIBUTIONS OF THE $—CH_2—$ AND $—CH_3$ GROUPS OF ALIPHATIC HYDROCARBONS

n	C_nH_{2n+2}		$(CH_3)_2—CH—C_{(n-3)}H_{(2n-5)}$			$(CH_3)_3—C—C_{(n-4)}H_{(2n-7)}$		
	V_{mol}	ΔV_{CH_2}	V_{mol}	ΔV_{CH_2}	$\Delta V'_{CH_3}$	V_{mol}	ΔV_{CH_2}	$\Delta V''_{CH_3}$
5	115·21		116·43			(117·7)		
		15·47		15·56	16·78		(15·03)	16·3
6	130·68		131·99			132·73		
		15·86		15·67	16·98		15·99	16·73
7	146·54		147·66			148·72		
		15·99		16·12	17·20		15·56	16·62
8	162·53		163·74			164·28		
		16·19		16·02	17·23		16·23	16·77
9	178·71		179·76			180·51		
		16·22		16·12	17·17		16·19	16·94
10	194·93		195·88			196·70		

NOTATION:[359]

n number of carbon atoms in the molecule,

$V_4^{20} = M/\rho_4^{20}$ (V_4^{20} is the molal volume in ml),

M molecular weight, g/mole,

ρ_4^{20} density, g/ml,

ΔV_{CH_2} contribution of the $—CH_2—$ group to the molal volume, ml/mole,

$\Delta V'_{CH_3}$ contribution of the first $—CH_3$ group in the 2-position,

$\Delta V''_{CH_3}$ contribution of the second $—CH_3$ group in the 2-position.

† The general theory of additivity of the constitutive properties of molecules was dealt with by Tat'evskii.[359]

‡ The molal volume is usually expressed in ml/mole; in order to find the volume per molecule, expressed in cubic angstroms, the volume given in ml/mole should be multiplied by 1·6604.

Comparing the figures given in Table 1.18, we notice that the chain extension of the molecule of an aliphatic hydrocarbon due to introducing the $—CH_2—$ group causes an increase in the molal volume ranging from 15·47 ml/mole (difference between the molal volumes of n-pentane and n-hexane) to 16·22 ml/mole (the corresponding value for n-nonane and n-decane), i.e. on the average 16 ml/mole. Similar increments of the molal volume are observed for aliphatic hydrocarbons with branched chains.

The introduction of the side $—CH_3$ group into the chain of an aliphatic hydrocarbon causes a somewhat larger increment of the molal volume, viz., ranging from 16·78 ml/mole (the difference between the molal volumes of 2-methylpentane and n-pentane) to 17·23 ml/mole (the corresponding value for 2-methyloctane and n-octane), on the average 17·15 ml/mole.

The addition of the second $—CH_3$ group to the same carbon atom causes an increase of the molal volume ranging from 16·3 ml/mole (difference between the molal volumes of 2,2-dimethylbutane and 2-methylbutane) to 16·94 ml/mole (the corresponding value for 2,2-dimethyloctane and 2-methyloctane), on the average by 16·8 ml/mole.

The magnitude of the contribution of the $—CH_2—$ group to the molal volume depends, therefore, on the length of the chain into which the group is introduced; the magnitude of the contribution of the $—CH_3$ group does not only depend on the chain length, but also on the presence of other groups bonded with the carbon atom to which the $—CH_3$ group is added.

Let us now calculate the contributions corresponding to the $—CH_3$ group at the ends of the straight chain of an aliphatic hydrocarbon. For this purpose we subtract the contributions of the $—CH_2—$ groups of the chain from the molal volume of the given compound and divide the difference by 2 (two $CH_3—$ groups at the ends of the chain):

$$\text{n-pentane:}\ \frac{115{\cdot}2-3\times 16}{2} = 33{\cdot}6\ \text{ml/mole}$$

$$\text{n-octane:}\ \frac{162{\cdot}53-6\times 16}{2} = 33{\cdot}27\ \text{ml/mole}$$

$$\text{n-decane:}\ \frac{194{\cdot}93-8\times 16}{2} = 33{\cdot}45\ \text{ml/mole}$$

It may be seen that the contribution of the terminal $CH_3—$ group amounts to about 33·5 ml/mole.

EXAMPLE 1.2. Estimate the molal volume of 2,2,4-trimethylpentane

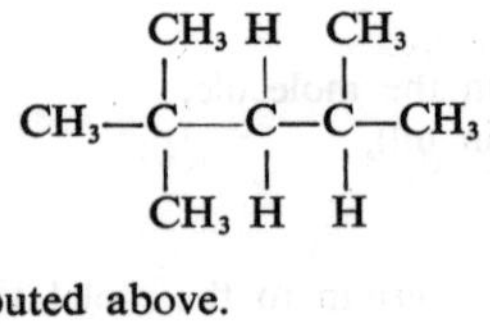

making use of the contributions computed above.

Solution

Three $—CH_2—$ groups	3×16·0	= 48·0
Two terminal $—CH_3$ groups	2×33·5	= 67·0
Two $—CH_3$ side groups in the 2- and 4-positions	2×17·15	= 34·3
One $—CH_3$ group in the 2-position		16·8
		166·1

$V_{calc} = 166{\cdot}1$ ml/mole
The value found experimentally: $V_{exp} = 165{\cdot}07$ ml/mole.
The error of the estimation is $+0{\cdot}66\%$.

The relations illustrated in this example are also valid when various groups are introduced into the molecule, containing beside the C and H atoms also O, S, P, Cl atoms, etc. We can tabulate the contributions of the groups and bonds which are often encountered in organic molecules and, after summing up these contributions, the molal volumes of these substances can be calculated. The higher the required accuracy of calculation the more extensive must be the table of contributions since the influence of environment must be considered in more detail. These contributions cannot be treated as actual volumes occupied by atomic groups, but merely as some average computational quantities best suited for estimation. This remark is dictated in part by the fact that the interatomic distances in a molecule may vary considerably under the influence of various factors which cause deformations and strains of the atomic groups in the molecules. For example, an important factor influencing the change of the molal volume of hydrocarbons may be the previously mentioned variation of the cross-sectional area of the molecule which arises from the rotation of groups in the molecule around the axis passing through the C—C bond of the chain.

Molal Volume of Liquids at the Normal Boiling Point

The estimation of the molal volume of a liquid on the basis of additivity is rendered more difficult because of the phenomenon of association, which causes considerable deviations from the rule of additivity. The effect of this phenomenon may be obviated to a certain extent by always considering the liquid under some standard conditions. Several rules exist which make it possible to estimate the volume of a liquid at its normal boiling point by summing up the appropriate contributions to the molal volume.

The oldest of these rules is Kopp's law which involves summing up the contributions of the atoms, without taking into consideration the way in which these atoms are linked in the molecule:

$$V_b = 11{\cdot}0(\mathrm{C})+7{\cdot}8(\mathrm{O}')+12{\cdot}2(\mathrm{O}'')+5{\cdot}5(\mathrm{H})+22{\cdot}8(\mathrm{Cl})+ +27{\cdot}8(\mathrm{Br})+37{\cdot}5(\mathrm{I})+22{\cdot}6(\mathrm{S}) \quad \text{ml/mole} \tag{1.43}$$

where (C) is the number of carbon atoms in the molecule,
(O') is the number of oxygen atoms linked with two carbon atoms 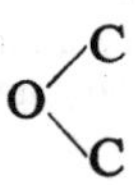,
(O'') is the number of oxygen atoms connected with one carbon atom O=C,
(H) is the number of hydrogen atoms,
(Cl) is the number of chlorine atoms,
(I) is the number of iodine atoms,
and (S) is the number of sulphur atoms.

Kopp's formula is not accurate, because it does not take into account the differences caused by the different grouping of the atoms in the molecule and by the differences between the bonds.

EXAMPLE 1.3. Estimate by Kopp's law the molal volume of furan at the boiling point

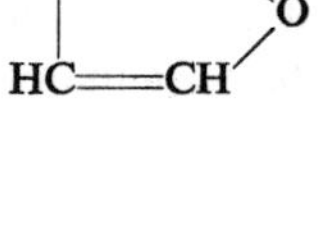

Solution

$(V_b)_{calc} = 4\times11{\cdot}0+4\times5{\cdot}5+7{\cdot}8 = 73{\cdot}8$ ml/mole

The molal volume of furan at the boiling point, found experimentally, is $(V_b)_{exp} = 76{\cdot}3$ ml/mole; the error is +3·3%.

According to Schroeder[282] the molal volume of liquids at the boiling point can be calculated more accurately by using the following contributions in cm³/mole: 7 for carbon, hydrogen, oxygen, nitrogen, or double bond C=C; 10·5 for fluorine; 24·5 for chlorine; 31·5 for bromine; 38·5 for iodine; 21 for sulphur; —7 for the ring, and 14 for the triple bond C≡C. As usual, when applying the rule of additivity the contributions are multiplied by the number of the corresponding atoms and bonds and then summed up (as in Eq. (1.43)). In this way quite satisfactory results are obtained; according to the data taken from the literature,[306] the average error for 35 cases was 2·8%.

EXAMPLE 1.4. Estimate by Schroeder's method the molal at the boiling point of:

a. Ethyl propyl ether, C_2H_5—O—C_3H_7.

$(V_b)_{calc} = 18\times7 = 126$ cm³/mole,

$(V_b)_{exp} = 129$ cm³/mole; the error amounts to −2·3%.

b. Bromobenzene, C_6H_5Br.

$(V_b)_{calc} = 11\times7+31{\cdot}5+3\times7-7 = 122{\cdot}5$ cm³/mole,

$(V_b)_{exp} = 120$ cm³/mole; the error +2·0%.

Frequently, the table of contributions compiled by Le Bas[282] (Table 1.19) is employed, which generally gives better results than Kopp's or Schroeder's formulae. The accuracy of these estimations is illustrated in Table 1.20.[306]

TABLE 1.19. CONTRIBUTIONS TO THE MOLAL VOLUME V_b (cm³/mole) OF A LIQUID AT ITS BOILING POINT, AFTER LE BAS

Carbon	14·8	Fluorine	8·7
Hydrogen	3·7	Chlorine	21·6
Oxygen		Bromine	27·0
double bond	7·4	Sulphur	25·6
methyl esters and ethers	9·1	Ring	
ethyl esters and ethers	9·9	three-membered	−6·0
higher esters and ethers	11·0	four-membered	−8·5
acids	12·0	five-membered	−11·5
bonded with S, P, N	8·3	six-membered	−15·0
Nitrogen		naphthalene	−30·0
primary amines	10·5	anthracene	−47·5
secondary amines	12·0		

MOLAL VOLUMES V_b (cm³/mole) OF SOME LIQUEFIED GASES

Hydrogen H_2	14·3	Ammonia NH_3	25·8
Oxygen O_2	25·6	Chlorine Cl_2	48·4
Nitrogen N_2	31·2	Bromine Br_2	53·2
Nitric oxide NO	23·6	Iodine I_2	71·5
Nitrous oxide N_2O	36·4	Water H_2O	18·9
Hydrogen sulphide H_2S	32·9	Carbon dioxide CO_2	34·0
Air	29·9	Sulphur dioxide SO_2	44·8
Carbon monoxide CO	30·7		

Molal Volumes of Hydrocarbons. Accurate methods for estimating the molal volumes of hydrocarbons are of great practical importance because they provide a basis for calculating the molal volumes of the vast majority of organic compounds. The molal

TABLE 1.20. COMPARISON OF CALCULATED AND EXPERIMENTAL MOLAL VOLUMES V_b OF LIQUIDS AT THE NORMAL BOILING POINT ON THE BASIS OF THE ADDITIVE CONTRIBUTIONS OF LE BAS

Compound	Volume at the boiling point, cm^3/mole		
	Experimental	Calculated	Error, %
Fluorobenzene	102	101	−1·0
Chlorobenzene	115	114	−0·9
Methanol	42·5	42	−1·2
n-Propanol	81·8	86	+5·1
Dimethyl ether	63·8	62	−2·8
Acetone	77·5	74	−4·5
Acetic acid	64·1	64	0
Methyl formate	62·8	61	−2·9
Diethylamine	109	112	+2·7
Carbon tetrachloride	102	101	−1·0
Phosgene	69·5	65	−6·5

volume of the hydrocarbon from which the given compound may be considered to be derived is first calculated and contributions are then added for the various groups which have been substituted for hydrogen atoms in forming the compound. Tables of contributions enable us to estimate the molal volume of hydrocarbons with a relatively high accuracy (e.g. according to Tat'evskii,[23, 359, 361] the error incurred is of the order of 0·5%), although in order to carry out such estimations a large number of contributions have to be known and the calculations become complicated.

Good results are obtained by employing the simple formula of Kurtz and Lipkin:[97]

$$V = 16{\cdot}28\,N_1 + 13{\cdot}15\,N_2 + 9{\cdot}7\,N_3 - 6{\cdot}2\,N_4 + C \tag{1.44}$$

where V is the molal volume, ml/mole, at 20°C under a pressure of 1 kg/cm^2,
N_1 is the number of CH_3, CH_2 and CH groups in open chains,
N_2 is the number of CH_2 and CH groups in rings, but not ring junctions,
N_3 is the number of CH groups at ring junctions,
N_4 is the number of double bonds in the molecule (refers also to benzene and other rings),
$C = 31{\cdot}2$ for all hydrocarbons, except n-alkanes,
and $C = 32{\cdot}6$ for normal aliphatic chains (the constant C contains contributions due to the hydrogen atoms at the chain ends of the hydrocarbon molecule).

EXAMPLE 1.5. Estimate the molal volume of the hydrocarbon with the formula

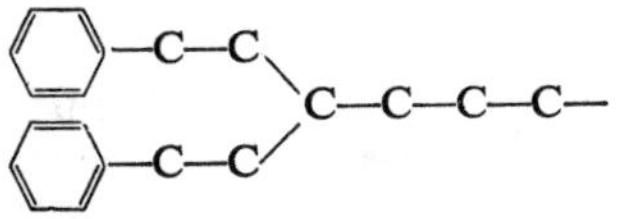

Solution

$N_1 = 8$; $N_2 = 17$; $N_3 = 0$; $N_4 = 6$

The molal volume computed from Eq. (1.44) is

$V_{calc} = 16{\cdot}28\times8+13{\cdot}15\times17-6{\cdot}2\times6+31{\cdot}2 = 347{\cdot}79$ ml/mole

The molal volume determined experimentally is $V_{exp} = 348{\cdot}8$ ml/mole; the error $+0{\cdot}29\%$.

The authors[97] have checked Eq. (1.44) on 380 hydrocarbons; the average error in the estimation does not exceed 1·5%; usually, however, it is lower.

Kurtz and Sankin[214] have also presented a more general formula for estimating the molal volume of hydrocarbons for the temperature range −253 to +200°C and for a pressure range up to 10,000 kg/cm²

$$V_{t \cdot p} = F_1K_1N_1+F_1K_2N_2+F_1K_3N_3-F_1K_4N_4+F_5K_5 \tag{1.45}$$

In this equation N_1, N_2, N_3 and N_4 have the same meaning as in Eq. (1.44). The values of the contributions K_1, K_2, K_3, K_4 and K_5 depend on temperature; they can be taken from Table 1.21. At a temperature of 200°C the values K_1, K_2, K_3, K_4, K_5 (ml/mole) are:

$$K_1 = 18{\cdot}11, \quad K_2 = 14{\cdot}63, \quad K_3 = 10{\cdot}79, \quad K_4 = 6{\cdot}90,$$

the contributions at 200°C are by about 12% higher than those at 20°C; the value K_5 at +200°C is 60·2.

TABLE 1.21. VALUES OF THE CONTRIBUTIONS TO THE MOLAL VOLUMES OF HYDROCARBONS IN KURTZ AND SANKIN'S EQUATION (ml/mole)

Temperature, °C	K_1	K_2	K_3	K_4	K_5
200	18·11	14·63	10·79	6·90	60·2
100	17·11	13·82	10·20	6·52	40·3
50	16·59	13·40	9·89	6·32	34·6
20	16·28	13·15	9·70	6·20	31·2
0	16·07	12·98	9·57	6·12	29·5
−50	15·54	12·55	9·26	5·82	26·0
−100	15·02	12·13	8·95	5·72	23·1
−253	13·15	11·13	8·90	4·0	

The coefficient F_1 depends only on the pressure; for a pressure of 1000 kg/cm², $F_1 =$ 0·965. The coefficient F_5 depends both on pressure and temperature. The values of F_1 and F_5 are given in Table 1.22; they have been calculated on the basis of measurements performed by Bridgman. Within the temperature range 0–100°C and the pressure range

TABLE 1.22. VALUES OF THE COEFFICIENTS F_1 AND F_5 IN KURTZ AND SANKIN'S EQUATION (EFFECT OF PRESSURE ON THE MOLAL VOLUME OF HYDROCARBONS)

Pressure kg/cm²	F_1	Values of F_5			
		Temperature, °C			
		0	20	50	100
0·99	1·0	1·0	1·0	1·0	1·0
250	0·990	0·900	0·900	0·890	0·845
500	0·982	0·840	0·835	0·840	0·745
1000	0·965	0·755	0·750	0·735	0·650
2000	0·938	0·655	0·660	0·640	0·550
3000	0·915	0·600	0·595	0·590	0·490
4000	0·895	0·580	0·575	0·560	0·455
5000	0·875	0·550	0·550	0·540	0·425
7500	0·839	0·530	0·530	0·515	0·410
10,000	0·807	0·510	0·510	0·490	0·390

of 1–10,000 kg/cm², the values computed for aliphatic hydrocarbons C_5—C_{10} do not differ by more than ±2% from the values of the molal volumes determined experimentally.

b. Molar Refraction

The ratio of the velocity of light W_0 in vacuum to the velocity of light W in a given compound, denoted by

$$n = \frac{W_0}{W} \tag{1.46}$$

is called the *refractive index* and can be defined as the ratio of the sine of the angle of incidence to the sine of the angle of refraction.

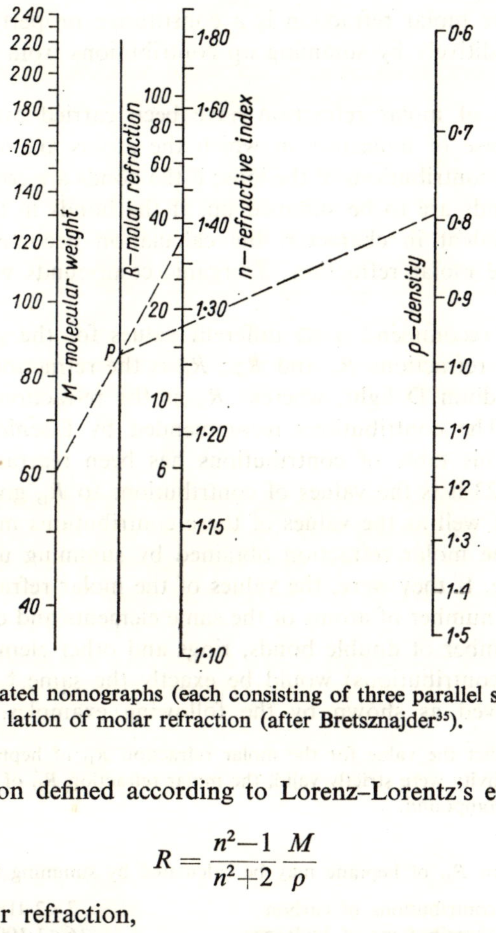

FIG. 1.5. Two conjugated nomographs (each consisting of three parallel scales) for the calculation of molar refraction (after Bretsznajder[35]).

The molar refraction defined according to Lorenz–Lorentz's equation (1.22) is

$$R = \frac{n^2-1}{n^2+2}\,\frac{M}{\rho} \tag{1.47}$$

where R is the molar refraction,
n is the refractive index,
M is the molecular weight,
and ρ is the density of the substance.

For rapid determination of the molar refraction R when the refractive index n, the density ρ and the molecular mass M are known, use can be made of the nomograph presented in Fig. 1.5.

EXAMPLE 1.6. The refractive index of isopropyl alcohol at 20°C is $n = 1{\cdot}377$, density $\rho = 0{\cdot}789$ g/cm^3, molecular weight $M = 60{\cdot}1$. Estimate the molar refraction R.

Solution

In order to find the value of the molar refraction, the point on the left axis of the nomograph, corresponding to the value of $M = 60{\cdot}1$, should be connected with the point of the right hand scale on the central axis, which corresponds to the value of $n = 1{\cdot}377$. This straight line intersects the reference axis (without scale). The intersection point obtained is to be connected with the point on the right axis, corresponding to the value of $\rho = 0{\cdot}789$. This second straight line intersects the left scale of the central axis at the point $R = 17{\cdot}5$, corresponding to the required value of the molar refraction.

According to the considerations presented in section 2e, the value of the molar refraction R_D (calculated on the basis of the refractive index n_D for a wavelength which corresponds to the line D of sodium light) provides an approximate measure of the electronic and atomic polarization of the molecules (1.24). Like the electronic polarizability (Table 1.15), the molar refraction is a constitutive property of the molecules and may be computed additively by summing up contributions from groups of atoms in the molecule.

Extensive studies of molar refraction have been carried out by Fajans and other authors.[97, 20] In the case of molecules in which the atoms are linked together by ionic bonds, we sum up the contributions of the ions; if the bonds are covalent, the contributions of the atoms and bonds are to be summed up. If the bonds in the molecules are partly ionic and partly covalent in character, the calculation becomes rather complicated.[20] The estimation of the molar refraction of organic compounds with covalent bonds will be considered first.

Various authors recommend quite different values for the group and bond contributions to the molar refractions R_D and R_∞; R_D is the refraction for a wavelength corresponding to the sodium D light, whereas R_∞ is the refraction calculated for light of infinite wavelength. The contributions recommended by Eisenlohr are employed most frequently. Recently his table of contributions has been rearranged and supplemented by Vogel.[382] Table 1.23 lists the values of contributions to R_D given by Eisenlohr, Vogel and other authors, as well as the values of these contributions measured for gases.[11, 20, 97]

The values of the molar refraction obtained by summing up the contributions are not precisely accurate. If they were, the values of the molar refractions of isomers (composed of an identical number of atoms of the same elements and containing in their molecules an identical number of double bonds, rings and other elements taken into account in summing up the contributions) would be exactly the same.† In fact, however, some deviations are observed as shown by the following examples.

EXAMPLE 1.7. Predict the value for the molar refraction R_D of heptane.

If the rules of additivity were strictly valid, the molar refraction R_D of heptane should be the same for all isomers of this compound.

Solution

The molar refraction R_D of heptane may be calculated by summing the Eisenlohr contributions:

7 contributions of carbon	$7 \times 2{\cdot}418 = 16{\cdot}926$
16 contributions of hydrogen	$16 \times 1{\cdot}100 = 17{\cdot}600$
The computed value of R_D for heptane is	$34{\cdot}526$

† A method for the prediction of the molar refractivities of hydrocarbon isomers which allows for this has been presented by Nutt (*Trans. Far. Soc.* **53**, 1538 (1957)). It is based on group contribution methods for the calculation of the numbers of optical dispersion electrons and their characteristic frequencies, in the molecules.

TABLE 1.23. VALUES OF CONTRIBUTIONS TO THE MOLAR REFRACTION OF ORGANIC COMPOUNDS

Atom, group or bond	R_D Eisenlohr[11]	R_D Vogel[382]	R_D Other authors	R_D for the element in gaseous state	R_∞ Vogel[382]	R_∞ for the element in gaseous state
—H	1·100	1·03	1·051, 1·025	1·03	1·03	1·02
—F	1·0–1·35	0·81	—	1·44	0·76	1·60
—Cl (linked with alkyl)	5·967	5·84	—	5·77	5·75	5·71
—Cl (linked with carbonyl)	6·336	—	—	—	—	—
—Br	8·865	8·74	—	8·66	8·54	8·09
—I	13·900	13·95	—	15·90	13·25	14·08
Oxygen						
—O—(H) (in hydroxyl)	1·525					
O (in ethers)	1·643	from 1·61 to 1·76			from 1·51 to 2·12	
=O (in carbonyl)	2·211			2·00		1·99
$—O_2—$ (peroxides)	4·035					
Sulphur						
C—S^{II}—C (sulphur S^{II})	7·80					
C—S^{IV}—C (sulphur S^{IV})	6·98	7·92	—	8·22	7·60	7·6
C—S^{VI}—C (sulphur S^{VI})	5·34					
Nitrogen						
O—N=C (hydroxy-imido compounds)	3·901					
C—N=C (carbo-imido compounds)	4·10	2·74	—	2·21	2·57	2·20
N—N=C (hydrazones)	3·46					
$>C<$	2·418	2·59	2·501, 2·519	—	2·46	—
Se	—	—	10·78	11·58	—	10·8
Te	—	—	17·0	18·45	15·3	14·4
P	—	—	9·14	9·0	8·59	8·6
As	—	—	11·55	11·5	11·03	10·3
Si	—	—	7·30	—	—	—
Ge	—	—	9·0	—	8·40	—
B	—	—	3·56	—	3·50	—
Zn	—	—	9·50	15·2	8·86	12·1
Cd	—	—	13·19	19·8	12·18	17·5
Hg	—	—	12·43	13·8	11·75	13·0
$—NH_2$ group	—	—	—	—	4·30	—
$>NH$ group	—	—	—	—	3·47	—
—OH group	—	—	—	—	2·49	—
—SH group	—	—	—	—	8·42	—
—CN group	5·415	—	—	—	5·29	—
—NC group	6·136	—	—	—	5·96	—
$>CH_2$ group	4·711	—	—	—	—	—
—SCN group	—	—	—	—	12·98	—
—NCS group	—	—	—	—	14·85	—
$—NO_2$ group	—	—	—	—	6·47	—

Continued overleaf

TABLE 1.23 (*cont.*)

Atom, group or bond	R_D Eisenlohr[11]	R_D Vogel[382]	R_D Other authors	R_D for the element in gaseous state	R_∞ Vogel[382]	R_∞ for the element in gaseous state
—ONO group	—	—	—	—	6·95	—
—NO_3 group	—	—	—	—	8·73	—
>SO_3 group	—	—	—	—	11·04	—
Double bond	1·773	—	1·707	—	—	—
Triple bond	2·336	—	2·10	—	—	—
5-membered ring	+0·04	—	—	—	—	—
6-membered ring = 3×1·733	5·199	—	—	—	—	—
8-membered ring	−0·47	—	—	—	—	—
15-membered ring	−0·62	—	—	—	—	—

Molar refractions R_D of heptanes, determined experimentally, are

n-heptane	34·51
2,2-dimethylpentane	34·61
2,3-dimethylpentane	34·31
2,4-dimethylpentane	34·61
3,3-dimethylpentane	34·32
2-methylhexane	34·57
3-methylhexane	34·45
2,2,3-trimethylbutane	34·39

It follows from this example that the deviations from the rule of additivity are, in this case, rather small, but larger deviations sometimes also occur.

EXAMPLE 1.8. Estimate the value of the molar refraction R_D of fluorobromobenzene BrC_6H_4F.

Solution

The molar refraction R_D of fluorobromobenzene may be calculated by summing the bond and atomic contributions

6 contributions of carbon	6×2·418 =	14·508
4 contributions of hydrogen	4×1·100 =	4·400
1 contribution of bromine		8·865
1 contribution of fluorine		0·810
3 contributions of double bonds	3×1·733 =	5·199
The calculated value of R_D for fluorobromobenzene is		33·782

The experimentally determined values of the refraction R_D for the fluorobromobenzenes are

m-fluorobromobenzene	30·82
o-fluorobromobenzene	31·81
p-fluorobromobenzene	31·21

The differences between the calculated and measured values for the individual isomers are considerable.

EXAMPLE 1.9. Estimate the value of the molar refraction R_D for trichloroethane $C_2H_3Cl_3$.

Solution

The molar refraction R_D for trichloroethane may be calculated by summing the contributions

2 contributions of carbon	2×2·418 =	4·836
3 contributions of hydrogen	3×1·100 =	3·300
3 contributions of chlorine	3×5·967 =	17·901
The calculated value of R_D for trichloroethane is		26·037

The experimental values of the molar refraction R_D for trichloroethane are

1,1,1-trichloroethane	26·22
1,1,2-trichloroethane	25·83

EXAMPLE 1.10. Estimate the molar refractions of benzyl alcohol and methylcyclohexanone on the basis of the contributions tabulated by Eisenlohr.

a. For benzyl alcohol $C_6H_5—CH_2—OH$

Solution

7 contributions of carbon	$7\times2·418 =$	16·926
8 contributions of hydrogen	$8\times1·100 =$	8·800
1 contribution of oxygen in (OH)		1·525
3 contributions of double bonds	$3\times1·733 =$	5·199
Molar refraction R_D for benzyl alcohol calculated	=	32·450
Molar refraction R_D for benzyl alcohol determined experimentally is		32·41

b. For methylcyclohexanone $CH_3—CH—(CH_2)_4{=}CO$

Solution

7 contributions of carbon	$7\times2·418 =$	16·926
12 contributions of hydrogen	$12\times1·100 =$	13·200
1 contribution of carbonyl oxygen		2·211
Molar refraction R_D for methylcyclohexanone calculated	=	32·337

Molar refractions R_D for methylcyclohexanones determined experimentally are

m-methylcyclohexanone	32·51
o-methylcyclohexanone	32·44
p-methylcyclohexanone	32·57

A study of these examples indicates that this method for estimating the molar refraction of organic compounds is relatively accurate and that the principle of additivity of the molar refraction holds, on the whole, rather well.

Owing to the relation which exists between the molar refraction and the polarizability of the molecules (the latter of these two quantities being of decisive consequence in the case of some technologically important properties of the substances) the additive calculation of the molar refraction is very useful and is frequently employed.

c. Parachor

The parachor is a constitutive quantity which characterizes a liquid compound and depends on surface tension.

The surface tension of a liquid depends on the magnitude of the forces acting on the molecules at the liquid surface. These molecules are under the influence of the attractive forces exerted by the molecules of deeper-situated layers as well as of the attractive forces exerted by these molecules in the gaseous phase. The magnitude of the interaction forces depends on the concentration of the molecules, their dipole moments, their polarizability, etc.

It might be expected that the surface tension of a liquid would depend on the difference in the concentrations of this compound in the liquid and the gaseous phase. In fact, McLeod found experimentally in 1923 the relationship

$$\sigma = C(\rho_l-\rho_v)^4 \tag{1.48}$$

where σ is the surface tension, dynes/cm,
C is a constant,
ρ_l is the density of the liquid, g/cm³,
and ρ_v is the density of the saturated vapour, g/cm³.

This relationship holds fairly well. Multiplying Eq. (1.48) by the molecular weight *M* and transforming it, Sugden[352] obtained a new constant, called the *parachor*.†

$$P_{ch} = \sigma^{1/4}\frac{M}{\rho_l - \rho_v} \quad g^{1/4}.cm^3.sec^{-1/2}.mole^{-1} \tag{1.49}$$

The parachor of organic compounds is a constitutive quantity; its value may be approximately calculated by summing up additively the contributions of atoms, bonds and groups; this quantity is almost independent of temperature.

Equation (1.49) can be simplified when the value of the parachor determined refers to a temperature considerably lower than the critical temperature of the liquid, i.e. when the vapour pressure over the liquid is rather small. Then $\rho_v \ll \rho_l$ and the density ρ_v of the saturated vapour can be disregarded in the equation. If we write for the molal volume of a liquid $V = M/\rho_l$, Eq. (1.49) can be transformed as follows:

$$P_{ch} = \sigma^{1/4} V \tag{1.50}$$

If the vapour pressure over a liquid is rather high, the simplified equation, (1.50), cannot be employed, and the complete Sugden equation, (1.49), must be used.

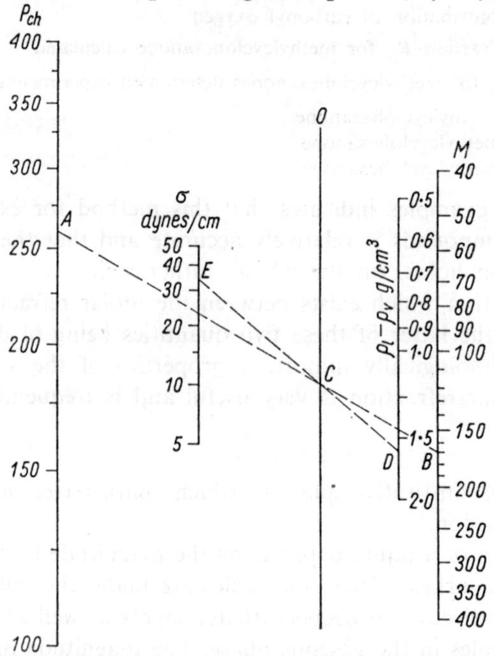

FIG. 1.6. Nomograph for calculating the parachor (after J. Johnson *et al.*, *Chemistry in Canada*, 28th August, 1934).

For a rapid determination of the value of the parachor according to the simplified equation, (1.50), the nomograph constructed by Johnson (Fig. 1.6) is applicable.

In order to find the value of the parachor, point *E* on the nomograph, which corresponds to a surface tension σ (dynes/cm) of the liquid, is connected with point *D* corresponding to the difference of its densities $\rho_l - \rho_v$ (g/cm³). Through the point *C* on the reference axis *O* (which is determined by the intersection with the line *ED*), and the point

† According to certain authors (Fergusson, Kennedy, Felang) the exponent in Eq. (1.48) is not 4 but *m*, where *m* is a constant characteristic for the compound, often (though not always) close to 4.

B (corresponding on the *M*-axis to the molecular weight of the compound), we construct another straight line. This line intersects the parachor axis at point *A* which corresponds to the required value of the parachor.

Within a limited range of temperatures, the value of the parachor varies but little. The dependence of the parachor on temperature for several normal hydrocarbons of the aliphatic series is illustrated below.[97]

t, °C	n-C_6H_{14}	n-C_7H_{16}	n-C_8H_{18}	n-C_9H_{20}	n-$C_{10}H_{22}$	n-$C_{11}H_{24}$
20	270·8	311·3	351·1	391·1	431·2	470·6
30	270·7	311·2	351·1	391·1	431·1	470·9
40	270·8	311·4	351·3	391·1	431·5	471·1

A closer examination of this problem shows that the temperature independence of the value of the parachor is the fortuitous result of compensation by several factors.[231]

The values of contributions used in the calculation of the parachor have been given by Sugden, and revised by several other authors. In Table 1.24 the contributions according to Sugden,[352] Mumford and Phillips,[97] and Quale[306] are given. A simpler, though less accurate system of contributions has been suggested by McGowan.[245]

The calculation consists in the summation of the atomic contributions presented in Table 1.25. From the sum of the contributions of each bond (no matter whether a single, double or triple) the number 19 is subtracted. The average error involved in this method checked on 36 inorganic and organic liquids, is ±1·6%.

TABLE 1.24. CONTRIBUTIONS TO THE PARACHOR FOR ORGANIC COMPOUNDS

Atom, group or bond		Sugden[11,352]	Mumford and Phillips[97]	Quale[306]
CH_2 in —$(CH_2)_n$	$n < 12$	—	—	40·0
	$n > 12$	—	—	40·3
C		4·8	9·2	9·0
H		17·1	15·4	15·5
H in (OH)		11·3	—	10·0
H in HN		—	—	12·5
O		20·0	20·0	19·8
O_2 in esters, acids		60·0	60·0	54·8
N		12·5	17·5	17·5
S		48·2	50·0	49·1
P		37·7	40·5	40·5
F		25·7	25·5	26·1
Cl		54·3	55·0	55·2
Br		68·0	69·0	68·0
I		91·0	90·0	90·3
Se		62·5	—	63
Si		25·0	—	31
Al		(38·6)	—	55
Sn		57·9	—	64·5
As		50·1	—	54
Double bond:	terminal			19·1
	2,3-position	23·2	19·0	17·7
	3,4-position			16·3
Triple bond		46·6	38·0	40·6

Continued overleaf

TABLE 1.24 (*cont.*)

Atom, group or bond		Sugden[11,352]	Mumford and Phillips[97]	Quale[306]
Carbonyl bond in ketones				
Total number of carbon atoms =	3			22·3
	4			20·0
	5			18·5
	6			17·3
	7			17·3
	8			15·1
	9			14·1
	10			13·0
	11			12·6
Hydrogen bridge				−14·4
Chain branching (per branch)				−3·7
Alkyl groups: 1-methylethyl				133·3
1-methylpropyl				171·9
1-methylbutyl				211·7
2-methylpropyl				173·3
1-ethylpropyl				209·5
1,1-dimethylethyl				170·4
1,1-dimethylpropyl				207·5
1,2-dimethylpropyl				207·9
1,1,2-trimethylpropyl				243·5
Two adjacent methyl groups at secondary carbons				−1·6
Two adjacent methyl groups, one at a secondary carbon, the other — at a tertiary carbon				−2·1
Two adjacent methyl groups at tertiary carbons				−4·5
Position differences in benzene:				
ortho-meta		—	—	1·8–3·4
meta-para		—	—	0·2–0·5
ortho-para		—	—	2·0–3·8
Rings				
3-membered		16·7	12·5	12·5
4-membered		11·6	6·0	6·0
5-membered		8·5	3·0	3·0
6-membered		6·1	0·8	0·8
7-membered		—	−4·0	−4·0

TABLE 1.25. ATOMIC CONTRIBUTIONS TO THE PARACHOR (AFTER MCGOWAN[245])

Element	Contribution	Element	Contribution	Element	Contribution
C	47·6	F	30·5	Sb	110·3
H	24·7	Ne	24·8	Te	104·6
O	36·2	P	73·5	I	98·9
N	41·9	Ar	56·3	Xe	93·2
S	67·7	Ge	93·3	Pb	113·8
Cl	62·0	As	87·6	Bi	108·1
Si	79·2	Se	81·9	Po	102·4
Be	59·1	Br	76·1	Rn	90·9
B	53·4	Kr	70·3	Al	96·7
He	19·0	Sn	116·0		

Comparison of the Methods for Estimating the Parachor

The accuracy of calculating the parachor using the contributions recommended by various authors, is illustrated by the following examples.

EXAMPLE 1.11. Estimation of the value of the parachor P_{ch} for 3-ethylpentane $(C_2H_5)_2$—CH—CH_2—CH_3 = C_7H_{16}. The value of P_{ch} determined experimentally is 305·4.

a. The value of P'_{ch} calculated after Sugden

7 contributions of carbon	7×4·8 =	33·6
16 contributions of hydrogen	16×17·1 =	274·6
The calculated value of P'_{ch} is		308·2
Error: 0·91%		

b. The value of P''_{ch} calculated after Quale

Contribution of the group $(C_2H_5)_2CH$		209·5
2 contributions of carbon	2×9·0 =	18·0
5 contributions of hydrogen	5×15·5 =	77·5
The calculated value of P''_{ch} is		305·0
Error: 0·13%.		

EXAMPLE 1.12. Estimation of the value of the parachor P_{ch} for acetone CH_3—CO—CH_3(=C_2H_6O). The value of P_{ch}, determined experimentally, is 161·7.

a. The value of P'_{ch} calculated after Sugden

3 contributions of carbon	3×4·8 =	14·4
6 contributions of hydrogen	6×17·1 =	102·6
1 contribution of oxygen	1×20·0 =	20·0
1 double bond	1×23·2 =	23·2
The calculated value of P'_{ch} is		160·2
Error: 0·93%.		

b. The value of P''_{ch} calculated after Quale

3 contributions of carbon	3×9·0 =	27·0
6 contributions of hydrogen	6×15·5 =	93·0
1 contribution of oxygen	1×19·8 =	19·8
1 double carbonyl bond (3C)	1×22·3 =	22·3
The calculated value of P''_{ch} is		162·1
Error: 0·25%.		

c. The value of P'''_{ch} calculated after McGowan

3 contributions of carbon	3×47·6 =	142·8
6 contributions of hydrogen	6×24·7 =	148·2
1 contribution of oxygen		36·2
		327·2
9 bonds	(−9)×19 =	−171·0
The calculated value of P'''_{ch} is		156·2
Error: 3·4%.		

EXAMPLE 1.13. Estimation of the value of the parachor (Sugden's contributions) for:

a. Benzophenone C_6H_5—CO—C_6H_5

13 contributions of carbon	13×4·8 =	62·4
10 contributions of hydrogen	10×17·1 =	171
1 contribution of oxygen	1×20 =	20
2 contributions of the ring	2×6·1 =	12·2
7 contributions of double bonds	7×23·2 =	162·4
The calculated value of P'_{ch} is		428·0
The value of P_{ch}, determined experimentally, is		428·2

b. Benzaldehyde C_6H_5—CHO

7 contributions of carbon	7×4·8 =	33·6
6 contributions of hydrogen	6×17·1 =	102·6
1 contribution of oxygen	1×20 =	20·0
4 contributions of double bonds	4×23·2 =	92·8
1 contribution of the ring	1×6·1 =	6·1
The calculated value of P'_{ch} is		255·1
The value of P'_{ch}, determined experimentally is		256·2

Extensive tests have shown that calculated values of the parachor correspond fairly accurately with those determined experimentally. Any discrepancies which have occurred have usually been found to be due to differences between the hypothetical structure of the molecule and the actual structure. In order to obtain a correct value for the parachor it is necessary for example to make allowances for the location of side chains, a factor which has been disregarded in the table of Sugden's contributions, but taken into account by Quale.

TABLE 1.26. COMPARISON OF THE VALUES OF THE PARACHOR P_{ch} DETERMINED EXPERIMENTALLY WITH THOSE CALCULATED FROM THE SUGDEN AND QUALE CONTRIBUTIONS[97]

Hydrocarbon	P_{ch} experim.	P_{ch} after Quale	Error, %	P_{ch} after Sugden	Error, %
2,5-dimethylhexane	347·0	346·6	0·11	351·0	1·1
2,4-dimethylhexane	345·3	345·2	0·03	351·0	1·7
3,4-dimethylhexane	342·7	342·2	0·14	351·0	2·4
2-methyl-3-ethylpentane	338·1	341·2	1·00	351·0	3·8
2,2,4-trimethylpentane	343·7	343·7	0·00	351·0	2·1
2,2,3-trimethylpentane	339·0	338·7	0·09	351·0	3·5
2,3,4-trimethylpentane	340·8	339·2	0·47	351·0	3·0
Average error of calculation			0·26		2·5

The influence of this factor is illustrated in Table 1.26,[97] in which experimentally determined parachors for various octanes are compared with values calculated from Sugden's tables and also with values calculated from Quale's tables.

d. Rheochor

The *rheochor* is a constitutive quantity that characterizes an organic liquid at its normal boiling point, and is dependent on viscosity. This quantity was introduced by Friend and Hargreaves in 1943.[109] It is defined as

$$R_{ch} = \frac{M(10\mu_b)^{1/8}}{(\rho_l+2\rho_v)} \tag{1.51}$$

where M is the molecular weight,
μ_b is the viscosity at the boiling point, cP,
ρ_l is the liquid density, g/cm³,
and ρ_v is the density of the saturated vapour, g/cm³.

The quantities μ_b, ρ_l, ρ_v are taken at the normal boiling point. Since at this temperature the density ρ_v of a saturated vapour is very small, it may be disregarded. The simplified equation takes the form

$$R_{ch} = \frac{M(10\mu_b)^{1/8}}{\rho_l} \quad \text{or} \quad R_{ch} = (10\mu_b)^{1/8} V_b \tag{1.52}$$

where V_b is the molal volume of the liquid at its normal boiling point.

The value of the rheochor may be calculated by summing the contributions which correspond to the atoms and atomic groups in the molecule. Such a calculation is necessary, for example, for the estimation of the viscosity μ_b of a liquid.

The contributions to the rheochor, used for estimating this quantity, are presented in Table 1.27. For many compounds (e.g. those containing sulphur, fluorine, iodine) the contributions have not been determined. Other, more complicated systems of contributions for calculating the value of the rheochor are known.[292]

TABLE 1.27. VALUES OF THE CONTRIBUTIONS TO THE RHEOCHOR FOR ORGANIC COMPOUNDS

Atom or atomic group	Contribution	Atom or atomic group	Contribution
C	12·8	NH_2	20·6
H (in C—H)	5·5	NH	13·6
H (in O—H)	10·0	ONO	39·3
O (in ethers)	10·0	NO_2	38·9
O (in alcohols)	13·0	CN	33·0
O (in aldehydes and ketones)	13·2		
Cl	27·3	COO (acids and esters)	36·0
Br	35·8	C_6H_5	101·7
N	6·6		

A relationship exists between the rheochor and the molal volume V_c of a compound at the critical point

$$R_{ch} = 0{\cdot}426\, V_c \tag{1.53}$$

In the case of non-associated liquids Eq. (1.53) is obeyed to within ±5%.[282]

Estimation of the Rheochor

Values of the liquid molal volume at the boiling point (V_b) which are needed for the calculation, are usually unknown. This quantity may be estimated by summing the bond and group contributions according to the method of Le Bas (Table 1.19).

EXAMPLE 1.14. Estimation of the value of the rheochor for:

a. *p*-Dichlorobenzene $C_6H_4Cl_2$

$M = 147{\cdot}0$, $\rho_l = 1{\cdot}11$ g/cm³, $\mu_b = 0{\cdot}304$ cP,

$$R_{ch} = \frac{147{\cdot}0 \times (10 \times 0{\cdot}304)^{1/8}}{1{\cdot}11} = 151{\cdot}3$$

The value of the rheochor R_{ch}, determined experimentally, is thus 151·3.

The value of the rheochor may be calculated by summing the contributions:

1 contribution of C_6H_5		101·7
2 contributions of Cl	$2\times 27·3 =$	54·6
		156·3
−1 contribution of hydrogen		−3·7
The calculated value of the rheochor R_{ch} is		152·6

Error: +0·86%.

b. Ethyl methyl ketone $C_2H_5—CO—CH_3$

$V_b = 96$ cm³/mole (calculated from the contributions after Le Bas)

$\mu_b = 0·244$ cP

$R_{ch} = (10\times 0·244)^{1/8}\times 96·2 = 108·2$

The value of the rheochor R_{ch}, determined experimentally, is thus 108·2.

The value of the rheochor may be calculated by summing the contributions:

4 contributions of carbon	$4\times 12·8 =$	51·2
8 contributions of hydrogen	$8\times 5·5 =$	44·0
1 contribution of oxygen (ketone)	$1\times 13·2 =$	13·2
The calculated value of the rheochor R_{ch} is		108·4

Error: +0·19%.

c. n-Butyl bromide n—C_4H_9Br

$V_b = 96·2$ cm³/mole (calculated from the contributions after Le Bas)

$\mu_b = 0·304$ cP

$R_{ch} = (10\times 0·304)^{1/8}\times 96·2 = 135·5$

The value of the rheochor R_{ch}, determined experimentally, is thus 135·5.

The value of the rheochor may be calculated by summing the contributions:

4 contributions of carbon	$4\times 12·8 =$	51·2
9 contributions of hydrogen	$9\times 5·5 =$	49·5
1 contribution of bromine	$1\times 35·8 =$	35·8
The calculated value of the rheochor R_{ch} is		136·5

Error: +0·74%.

It should be noted that owing to the presence of the power exponent 1/8 in Eq. (1.51), the accuracy of calculation of μ_b from R_{ch} and V_b may be rather low.

e. Orthochor

Discussing Bachinskii's equation,[17] which represents the relationship between the *fluidity* $1/\mu$ and the specific volume v of a liquid, Litvinov[227] has noticed that O_r, the smallest possible molal volume of a liquid that can be obtained by lowering the temperature (fictitious extrapolated value), should be

$$O_r = \frac{\rho_1\mu_1-\rho_2\mu_2}{\mu_1-\mu_2}M \tag{1.54}$$

where ρ_1 and μ_1 are the density and viscosity at temperature T_1,
ρ_2 and μ_2 are the density and viscosity at temperature T_2,
and M is the molecular weight.

Litvinov called this quantity the *orthochor*.

As a limiting volume, the orthochor should not depend on temperature; as an additive quantity it may be calculated by summing the bond contributions, which are listed in Table 1.28.

The additive calculation of the orthochor proved fairly accurate, and for half of the 300 compounds examined the author found that the deviation of the calculated from the measured value was smaller than 1%.

TABLE 1.28. VALUES OF THE CONTRIBUTIONS TO THE ORTHOCHOR (ACCORDING TO LITVINOV[227])

Bond	Fairly accurate values	Approximate values	Rough values
C—H	6·722		
C—C	3·872		
C=C	10·621		
C—O	5·722		
C=O	13·483		
C—N	3·725		
C=N		14·525	
C≡N	19·755		
C—F			14·074
C—Cl	21·461		
C—Br	26·018		
C—I	33·773		
C—S		10·835	
C=S		23·184	
O—H	10·677		
O—S			17·070
N—H	7·375		
N—O	13·186		
N=O	11·563		
N—N			8·641
N—Cl			29·075
Br—Br			45·999
		−6·101	
		−1·258	

EXAMPLE 1.15. The orthochor for acetaldehyde, CH_3—CH_2—CHO, may be estimated by summing the contributions:

6 C—H contributions	$6\times6{\cdot}722 = 40{\cdot}332$
2 C—C contributions	$2\times3{\cdot}872 = 7{\cdot}744$
1 C=O contribution	$13{\cdot}483$
The value of the orthochor of acetaldehyde	$O_r = 61{\cdot}559$

f. Molar Velocity of Sound[24]

Rao found in 1940 that the quantity

$$U_u = \frac{u^{1/3}M}{\rho} \tag{1.55}$$

(where u is the velocity of sound, M is the molecular weight, and ρ is the density), called the *molar velocity of sound*, is a constitutive and additive quantity. It may be calculated by summing the contributions listed in Table 1.29.

The molar velocity of sound, U_u, can be calculated more accurately by summing the group and not the bond contributions.[306, 314] It is generally assumed that each structure of a compound can be derived from one of the basic compounds (methane, benzene, cyclohexane, naphthalene) by substitution of the corresponding groups. The contributions of the basic compounds and those of the groups are presented in Table 1.30.†

† It should be noted that when calculating the sonic velocities from the contributions presented in Table 1.29, we get u in m/sec, whereas using the contributions from Table 1.30, u is given in cm/sec.

TABLE 1.29. BOND CONTRIBUTIONS FOR ESTIMATING THE MOLAR VELOCITY OF SOUND IN ORGANIC COMPOUNDS

Bond	Contribution	Bond	Contribution
C—H	95·2	C—Cl	230
C—C	4·25	C—Br	247
C—N	20·7	C—I	305
C—O	34·5	N—H	90·7
H—O	99·0	—CH_2OH	315
C=C	129·0	—CHO	282
C=N	145	—COOH	308
C=O	186	>C=N—OH (N bonded to HO)	285

TABLE 1.30. GROUP CONTRIBUTIONS FOR ESTIMATING THE MOLAR VELOCITY OF SOUND IN ORGANIC COMPOUNDS

	Contribution to U_u		Contribution to U_u
I. Basic structure		II. Substituted radicals (*cont.*)	
Methane	1850	—NH—	638
Benzene	4534	—NH_2	478
Cyclohexane	5363	—COOH	942
Naphthalene	6566	—C≡N	819
II. Substituted radicals		—O—	273
		—OH	137
		—Cl	610
		—Br	692
>C<, >CH—, —CH_2, —CH_3	872	—I	893
		—NO_2	893
—COO—	1220	—S	550
		=S	550
—C(=O)—H	449	III. Double bond	−254
		Triple bond	−507
		IV. Ortho-position	0
		Meta-position	59
—C(=O)—	872	Para-position	117

The average error of calculation of the molar velocity of sound using the contributions presented in Table 1.30 is ±2·6%, and the maximum error is about 8%.

Equation (1.55) can be presented in the form:

$$U_u = u^{1/3} V_m \tag{1.56}$$

where V_m is the molal volume of the compound.

The molar velocity of sound U_u, like the quantities discussed above (molar refraction R_D, parachor P_{ch}, rheochor R_{ch}), is related to the constitutive and additive quantity—the molal volume V_m of the compound, and within a rather wide range it is independent of temperature.

Calculated values of the molar sonic velocity U_u and hence of the sonic velocity u, may be used for the estimation of *adiabatic compressibility* β_s, the constant b in van der Waals' equation of a real gas and the thermal conductivity λ of liquids, etc.

The adiabatic compressibility is defined as

$$\beta_s = -\frac{1}{V}\left(\frac{\partial V}{\partial p}\right)_s \tag{1.57}$$

It can be calculated from the relationship

$$u^2 = \frac{1}{\beta_s \rho} \tag{1.58}$$

Relationships between the sound velocity in a liquid, the dipole moment of the molecule of the liquid and their polarizability are known to exist.

EXAMPLE 1.16. Estimate the sonic velocity u in propyl alcohol,

$$CH_3—CH_2—CH_2—OH.$$

$M = 60{\cdot}1$, $\rho_{24°} = 0{\cdot}8011$ g/cm³.

Solution

a. By using the bond contributions (Table 1.29)

5 C—H contributions	$5 \times 95{\cdot}2 = 476$
2 C—C contributions	$2 \times 4{\cdot}25 = 8{\cdot}5$
1 $—CH_2OH$ contribution	315
Molar sonic velocity	$U_u = 799{\cdot}5$

$$\text{Sonic velocity } u = \left(\frac{799{\cdot}5 \times 0{\cdot}8011}{60{\cdot}1}\right)^3 = 1200 \text{ m/sec.}$$

b. By using the group contributions (Table 1.30)

$$CH_4 + (—CH_2—) + (—CH_2) + (—OH)$$

Molar sonic velocity $U_u = 1850+872+872+137 = 3731.$

$$\text{Sonic velocity } u = \left(\frac{3731 \times 0{\cdot}8011}{60{\cdot}1}\right)^3 = 1{\cdot}234 \times 10^5 \text{ cm/sec.}$$

The experimentally determined sonic velocity in propyl alcohol is $u = 1234$ m/sec. The error in case (a) is equal to 2·76%, and in case (b) is zero.

5. METHODS FOR ESTIMATING THE CRITICAL CONSTANTS AS CONSTITUTIVE QUANTITIES

Knowledge of the critical constants is indispensable when applying certain methods for calculating transport properties. Since in most cases the critical constants are not known, the methods for their estimation are of great practical importance.

a. Relationship between the Structure and the Molecular Properties of a Compound and the Values of its Critical Parameters

The methods for calculating the critical parameters may be divided into two groups:

1. Methods based upon the constitutive properties of the molecules, consisting in summation of additive contributions.

2. Methods requiring the knowledge of certain properties of a compound which must be determined experimentally (e.g. density, boiling point, vapour pressure, etc.).

The critical parameters of some compounds are listed in Table 1.31.

TABLE 1.31. CRITICAL TEMPERATURES, PRESSURES AND DENSITIES OF SOME COMPOUNDS (Ref. 184, 260, 312)

Compound	t_c, °C	p_c, atm	ρ_c, g/cm³
Acetaldehyde	187·9	—	—
Acetic acid	321·6	57·2	0·351
Acetic anhydride	296·0	46·0	
Acetone	235·0	47·0	0·268
Acetonitrile	274·7	47·7	0·240
Acetylene	36·0	62·0	—
(Air)	(−140·7)	(37·2)	(0·35)
Aluminium chloride	356·3		
Ammonia	132·4	111·5	0·235
iso-Amyl alcohol	309·7	—	—
Aniline	427	52·4	—
Argon	−122·1	48·0	0·531
Benzene	288·5	47·7	0·304
Benzonitrile	426·0	41·6	—
Bromobenzene	397	44·6	0·486
1,3-Butadiene	152	42·7	
n-Butane	152·5	36·0	—
iso-Butane	133·5	36·5	—
iso-Butene	150·5	—	—
Butyl acetate	305·8	—	—
iso-Butyl acetate	288·2	31·4	—
n-Butyl alcohol	287	48·4	—
iso-Butyl alcohol	265·0	48·0	—
iso-Butyl formate	278·1	38·3	—
Butyl propionate	326·5	—	—
iso-Butyl propionate	315·9	—	—
n-Butyric acid	355·0	—	0·302
iso-Butyric acid	336·0	—	0·304
Carbon dioxide	31·1	73·0	0·460
Carbon disulphide	273·0	72·9	0·441
Carbon monoxide	−138·7	34·6	0·31
Carbon oxysulphide	31·3	(65·1)	—
Carbon tetrachloride	283·1	45·0	0·558
Chlorine	144·0	76·1	0·573
Chlorobenzene	359·0	44·6	0·365
Chloroform	260·9	54·9	0·516
m-Cresol	432·0	45·0	—
o-Cresol	422·0	49·4	—
p-Cresol	426·0	50·8	—
Cyanogen	128·2	59·75	—
Cyclohexane	281·0	40·4	0·27
Diethylamine	223·5	36·2	0·246
Dimethylamine	164·6	52·7	
Dimethylaniline	415·0	35·8	—
Dipropylamine	277	31·0	—
Ethane	32·1	48·8	0·21
Ethyl acetate	250·1	37·8	0·308
Ethyl alcohol	243·1	63·1	0·275
Ethylamine	183·2	55·5	—
Ethylbenzene	346·4	38·1	—
Ethyl bromide	230·8	61·5	0·513

TABLE 1.31 (*cont.*)

Compound	t_c, °C	p_c, atm	ρ_c, g/cm³
Ethyl n-butyrate	293·0	30·0	0·276
Ethyl chloride	187·2	51·7	0·33
Ethylene	9·7	50·9	0·22
Ethylene chloride	288·4	53	0·45
Ethylene fluoride	260	54·9	—
Ethyl ether	193·8	35·5	0·263
Ethyl formate	235·3	46·6	0·323
Ethyl mercaptan	315·0	—	—
Ethyl propionate	272·8	33·2	0·297
Fluorine	−101	55	—
Fluorobenzene	286·0	44·6	0·354
Formaldehyde	223·5	—	—
Helium	−267·9	2·3	0·069
n-Heptane	266·8	26·8	0·234
n-Hexane	234·8	29·5	0·234
Hexene	280	39·8	—
Hydrogen	−239·9	12·8	0·031
Hydrogen bromide	90	84·4	—
Hydrogen chloride	51·4	81·6	0·42
Hydrogen cyanide	183·5	59·9	0·20
Hydrogen fluoride	230·2	—	—
Hydrogen iodide	151·0	82·0	—
Hydrogen selenide	137·0	91·0	—
Hydrogen sulphide	100·4	88·9	—
Iodobenzene	448·0	44·6	0·581
Krypton	−63	54	0·78
Mercuric chloride	702·8	116	—
Mercury	1477	1150	—
Methane	−82·5	45·8	0·162
Methyl acetate	233·7	46·3	0·325
Methyl alcohol	240·0	78·7	0·272
Methylamine	156·9	73·6	—
Methyl bromide	194	51·6	—
Methyl n-butyrate	281·3	34·2	0·300
Methyl isobutyrate	267·4	33·9	—
Methylene chloride	245·1	—	—
Methyl ether	126·9	52·0	0·271
Methyl fluoride	44·9	62·0	—
Methyl formate	214·0	59·2	0·349
Methyl iodide	255	54·6	—
Methyl mercaptan	196·8	71·4	0·323
Methyl propionate	257·3	39·5	—
Monosilane	−3·8	47·8	—
Naphthalene	468·2	39·2	0·314
Neon	−228·7	25·9	0·484
Nitric oxide	−94·0	65·0	0·52
Nitrogen	−147·1	33·5	0·311
Nitrogen dioxide	−158·8	100·0	—
Nitrous oxide	36·5	71·7	0·45
Oxygen	−118·8	49·7	0·430
n-Pentane	197·2	33·0	0·232
iso-Pentane	187·7	—	—

Continued overleaf

TABLE 1.31 (*cont.*)

Compound	t_c, °C	p_c, atm	ρ_c, g/cm³
Pentene-1	202·5	40·4	—
iso-Pentene	191·5	33·9	—
Phenol	419	60·5	—
Phosgene	182·0	56·0	0·52
Phosphine	51·0	64·0	0·30
Phosphorus (P_4)	677	80	—
Pyridine	344·0	60·0	—
Propane	96·8	42·0	—
Propene	92·3	45·0	—
Propionic acid	339·5	53·0	0·515
Propionitrile	291·2	41·3	0·241
Propyl acetate	276·1	33·2	—
n-Propyl alcohol	263·7	50·0	0·273
iso-Propyl alcohol	235·0	53·0	—
Propylamine	223·8	46·3	—
Propyl butyrate	326·7	—	—
Propyl isobutyrate	316	—	—
n-Propyl chloride	230·0	45·2	—
Propyl formate	254·7	40·1	—
Propyl propionate	(304·7)	—	—
Radon	104·0	62·0	—
Silicium tetrachloride	230	(37)	—
Silicium tetrafluoride	−1·5	50·0	—
Sulphur	700	—	—
Sulphur dioxide	157·2	77·7	0·52
Sulphur trioxide	218·2	83·8	—
Tiophene	317·0	48·0	—
Toluene	320·6	41·6	0·292
Triethylamine	262·0	30·0	0·251
Trimethylamine	161·0	41·0	0·234
Water	374·15	218·4	0·323
Xenon	16·6	58·2	—

The values of the critical parameters of a substance must depend on the properties of the molecules of which it is composed. It can be assumed most generally that the following factors influence the values of T_c, p_c and V_c: the mass of the molecule, m, its volume, v_m, the shape factor, ϕ, and the intermolecular forces, hence the dipole moment $\bar{\mu}$ and the polarizability α of the molecule. The discussion of intermolecular forces in section 3 shows that the dependence of the critical parameters on the above molecular properties must be very complicated.

Lutskii[232] attempted to determine the molecular properties on which the value of the critical temperature T_c of a compound is dependent. For this purpose he compared the critical temperatures of substances differing significantly in one property only, e.g. molecular weight, dipole moment, polarizability, etc. From a comparison of the substances which differ only in molecular weight it followed that, most certainly, this factor does not influence the critical temperature, e.g. the weight of the molecules of NH_3 and ND_3 differ by 18% from each other, whereas the critical temperatures are almost equal. Similarly he concluded that the critical temperature of a compound does not depend on the volume v_m occupied by the molecules. For instance, the molecules of C_2H_5Br and n-C_3H_7Cl differ only in their weight m and volume v_m, but the critical temperatures of both compounds are almost equal (504°C, 503·7°C). Comparisons involving extensive experimental material have led to the conclusion that the value of the critical temperature of a given compound depends in the first place on the shape factor, ϕ, the dipole moment, $\bar{\mu}$, and the polarizability, α, of the molecules of the substance.

Lutskii has shown that the relationship between the critical temperature of an organic liquid and the polarizability, α, of the molecules of this compound (or molar refraction R_D, on account of the relation existing between α and R_D, cf. Eq. (1.24)) can be presented in the form

$$T_c = \text{constant} \times \alpha^x \quad \text{or} \quad T_c = \text{constant} \times R_D^x \tag{1.59}$$

For the majority of compounds the power exponent x is 0·5. This quantity, characterizing the effect of the polarizability of the molecules on the critical temperature of the compound, is the same for substances with similar dipole moments and molecular shape. On the other hand, compounds belonging to the same group may differ in the volume v_m and weight m of their molecules. For example, linear relationships between T_c and R_D^x have been found for the following groups of substances:

$x = 0·5$ (a) CH_3F, CH_3Cl, CH_3Br, CH_3I
(b) C_6H_5Cl, C_6H_5Br, C_6H_5I, CO_2, CS_2
(c) the series n-$C_nH_{2n+1}CH{=}CH_2$

$x = 0·4$ the series n-C_nH_{2n+2}
$x = 0·67$ tetrahedral molecules:
(d) SiF_4, CF_3Cl, CF_2Cl_2, $CFCl_3$, CCl_4
(e) CH_4, SiH_4, SiF_3Cl, SiF_2Cl_2, $SiFCl_3$, $SiCl_4$

$x = 0·75$ molecules of other shape
(f) F_2, Cl_2, Br_2, I_2
Ne, Ar, Kr, Xe

On the plots representing the relationship between T_c and R_D^x, the points for the same group, i.e. for equal values of x, but for different sub-groups (a), (b), (c), (d) ... fall on straight lines of different slope dependent on the value of the constant in Eq. (1.59).

The plot for $x = 0·5$ is presented in Fig. 1.7.

In this case the straight lines lie close to each other, because the values of the constants in Eq. (1.59) differ but slightly. In order to investigate the dependence of the critical temperature on the dipole moment of the molecule, Lutskii[232] compared the relationship between the quantity $T_c/R_D^{0·5}$ and the dipole moment $\bar{\mu}$ for a number of substances; by dividing T_c by $R_D^{0·5}$, he corrected for the influence of the polarizability of the molecules, which is approximately proportional to $R_D^{0·5}$.

For groups of compounds consisting of molecules similar in shape, and identical in polarizability, Lutskii found the relationship

$$T_c = A + B\bar{\mu} \tag{1.60}$$

where $\bar{\mu}$ denotes the dipole moment of the molecule. This relationship holds, for example, in isologous series: the constant A is in this case equal to the critical temperature of that compound of the series for which the value of the dipole moment is zero. In some cases, e.g. for aromatic compounds containing OH or NH_2 groups, Eq. (1.60) is inapplicable.

In the graphs representing the relationship between the quotient $T_c/R_D^{0.5}$ and the dipole moment, $\bar{\mu}$, the points which correspond to various compounds generally lie along

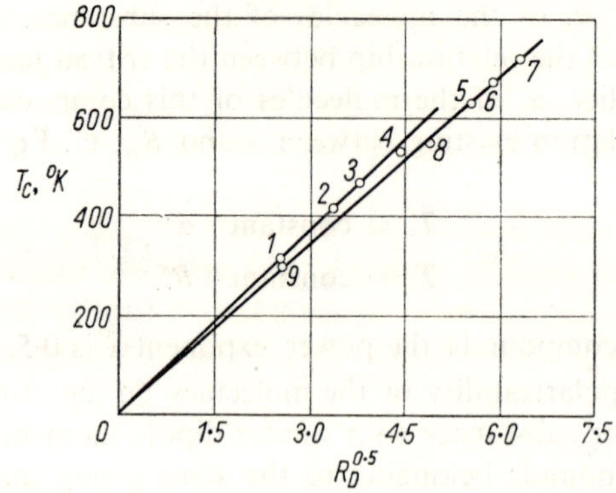

FIG. 1.7. Critical temperature (T_c) vs. square root of the molar refraction ($R_D^{0.5}$); *1*—CH_3F, *2*—CH_3Cl, *3*—CH_3Br, *4*—CH_3I, *5*—C_6H_5Cl, *6*—C_6H_5Br, *7*—C_6H_5I, *8*—CS_2, *9*—CO_2 (after Lutskii[232]).

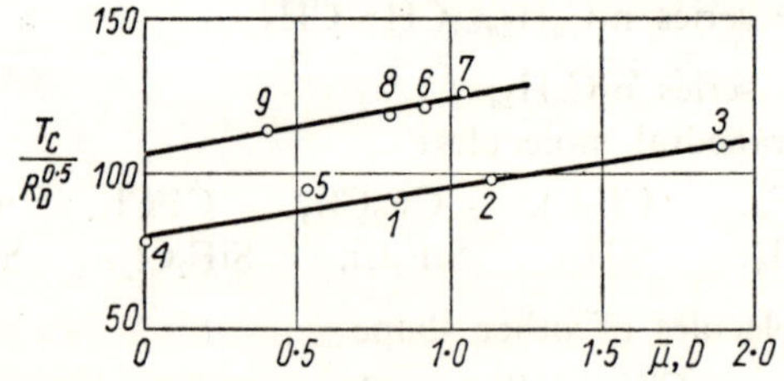

FIG. 1.8. Critical temperature (T_c) vs. dipole moment (the molecules of the substances differ in polarizability and dipole moment); *1*—$(C_2H_5)_3N$, *2*—$(C_2H_5)_2O$, *3*—C_2H_5F, *4*—SiH_4, *5*—PH_3, *6*—H_2S, *7*—HCl, *8*—HBr, *9*—HI (after Lutskii[232]).

straight lines (Fig. 1.8). It follows therefrom that for compounds with molecules similar in shape the dependence of the critical temperature, T_c, on the dipole moment, $\bar{\mu}$, and polarizability, α, (or molar refraction, R_D) can be represented by the equation

$$T_c = R_D^x(A' + B'\bar{\mu}) \tag{1.61}$$

which is derived by combining Eqs. (1.59) and (1.60); in most cases $x = 0.5$.

The influence of the shape of the molecule may be estimated by comparing the critical temperatures of compounds having molecules differing in shape but identical in polarizability and dipole moment.

According to Lutskii, the effect of the structure (and shape) of the molecules may be represented by the following relations:

$$\left[\frac{T_{c(i)}}{T_{c(n)}}\right]_{\bar{\mu},\alpha} = 0.98, \quad \left[\frac{T_{c(t)}}{T_{c(n)}}\right]_{\bar{\mu},\alpha} = 0.92, \quad \left[\frac{T_{c(c)}}{T_{c(n)}}\right]_{\bar{\mu},\alpha} = 1.075 \pm 0.05 \tag{1.62}$$

where the subscripts have the following meaning:

i refers to a molecule with a branched carbon chain,
n refers to a molecule with a normal chain,
t refers to a tetrahedral molecule, and
c refers to a cyclic molecule.

On the whole, metameric compounds with molecules of similar shape and similar polarizability and dipole moment, have similar critical temperatures. In the case of associated molecules, e.g. alcohols or acids, the critical temperatures can be several or even some score per cent higher than the critical temperatures of the corresponding ethers or esters.

Equations (1.59), (1.60) and (1.61) are used for compounds with non-associated molecules. In the case of alcohols, phenols, and aromatic amines, large deviations have been observed. If an associated molecule contains, on the average, Z simple molecules and its dipole moment is $\bar{\mu}'$, according to Lutskii the following relationship is valid:

$$(T_c)_{ass} = (ZR_D)^x(A-B\mu') \tag{1.63}$$

Equation (1.63) may be used to determine the average number Z of molecules forming the associated molecule by comparison, with the corresponding metameric non-associated compound.

b. Constants *a* and *b* in van der Waals' Equation

According to van der Waals' equation

$$\left(p+\frac{a}{V^2}\right)(V-b) = RT \tag{1.64}$$

where p is the pressure, V is the molal volume, T is the absolute temperature and R is the gas constant. This equation contains two constants, a and b, which characterize deviations in the behaviour of a given gas from the behaviour of a perfect gas. The constants a and b are usually evaluated from the critical constants, which have been determined experimentally, using the equations:

$$a = \frac{27}{8}RT_c b \qquad b = \frac{RT_c}{8p_c}$$
$$T_c = \frac{8a}{27Rb} \qquad p_c = \frac{a}{27b^2} \tag{1.65}$$

In these equations T_c is the critical temperature and p_c is the critical pressure. Van Laar[260] has shown that the constants a and b are constitutive and additive in character.

Constant a

The magnitude of the constant a depends on the magnitude of the forces of interaction between the molecules of the gas. According to van Laar, the square root of the constant a for a given compound can be calculated by summing the contributions corresponding to those atoms which occupy the external positions in the molecule. The contributions of the atoms which are inside the molecule and are "screened", i.e. surrounded by other external atoms, can be neglected. The contributions $\sqrt{a_i}$ employed for such calculations, are listed in Table 1.32.

TABLE 1.32. CONTRIBUTIONS FOR CALCULATING THE CONSTANT a IN VAN DER WAALS' EQUATION (THE TABLE GIVES THE VALUES OF $\sqrt{a}\times 10^2$)

H = 1·1(3·2, 1·6)				
C = 3·1	N = 2·9	O = 2·7	F = 2·9	
	Si = 5	P = 6·6	S = 6·3	Cl = 5·4
	Ge = 7	As = 7	Se = 7	Br = 7·0
	Sn = 9	Sb = 8·9	Te = 9	I = 8·9
	Pb = 11	Bi = 11		
Hg = 12·3				

It is worth noting that the values of the contributions in one horizontal series of the periodic system are approximately equal. The value of the contribution increases approximately by an increment of 2 when we pass to the next lower horizontal series:

Series	from	to	$\sqrt{a}\times 10^2$
I	H	He	≈ 1
II	C	Ne	≈ 3
III	Si	Ar	≈ 5
IV	Ge	Kr	≈ 7
V	Sn	Xe	≈ 9
VI	Pb	Rn	≈ 11

EXAMPLE 1.17. Estimate the value of $\sqrt{a}\times 10^2$ by summing up the contributions for:

a. Carbon monoxide, CO:

1 contribution for C	3·1
1 contribution for O	2·7
The calculated value of $\sqrt{a}\times 10^2$ =	5·8

The value of $\sqrt{a}\times 10^2$ determined experimentally is 5·5.

b. Carbon tetrachloride, CCl_4

Since the carbon atom is surrounded by four chlorine atoms, its contribution can be disregarded.

Contribution for C	0
4 contributions for Cl $4\times 5{\cdot}4$ =	21·6
The calculated value of $\sqrt{a}\times 10^2$ =	21·6

The value of $\sqrt{a}\times 10^2$ determined experimentally is 20.

c. n-Pentane, C_5H_{12}

Since all the carbon atoms in the chain are surrounded by hydrogen atoms, their contributions are disregarded.

Contributions for C	0
12 contributions for H $12\times 1{\cdot}6$ =	19·2
The calculated value of $\sqrt{a}\times 10^2$ =	19·2

The value of $\sqrt{a}\times 10^2$ determined experimentally is 19·7.

d. Propylamine, NH_2—C_3H_7:

The nitrogen atom and the carbon atoms are surrounded by hydrogen atoms

Contribution for N	0
2 contributions for H linked with nitrogen $2\times 3{\cdot}2$ =	6·4
7 contributions for H linked with carbon $7\times 1{\cdot}6$ =	11·2
The calculated value of $\sqrt{a}\times 10^2$ =	17·6

The value of $\sqrt{a}\times 10^2$ determined experimentally is 17·5.

The agreement between the calculated and experimental results is not very good, if we consider the fact that the value of the constant a is obtained by squaring the values of $\sqrt{a}$.

Constant b

The constant b in van der Waals' equation is approximately equal to four times the volume occupied by the molecules contained in 1 mole. If we refer to the considerations concerned with the molal volume of organic compounds (Section 4), it can be easily predicted that the constant b should have the properties of a constitutive quantity. The additive calculation of the constant b consists in the summation of the corresponding contributions of all the atoms contained in the molecule, making use of the data presented in Table 1.33.

TABLE 1.33. CONTRIBUTIONS FOR CALCULATING CONSTANT b IN VAN DER WAALS' EQUATION ($b \times 10^5$ cm^3)

H = 59 (34, 14)				
Li = 145	C = 100 (75)	N = 85 (60)	O = 70 (50)	F = 55
Na = 270	Si = 155	P = 140	S = 125	Cl = 100 (115)
K = 480	Ti = 180	As = 180		
Cu = 110	Ge = 210		Se = 180	Br = 165
Rb = 580	Zr = 235			
Ag = 150	Sn = 265	Sb = 250	Te = 235	I = 220
Cs = 710				
Au = 150	Pb = 320	Bi = 305		

In some compounds the atomic contributions of certain elements may have different values, e.g. for hydrogen (59, 34, 14), carbon (100, 75), etc. The values of contributions which are seldom employed are given in parenthesis.

EXAMPLE 1.18. Estimate the values of $b \times 10^5$ by summing the atomic contributions for:

a. Hydrogen chloride, HCl

1 contribution for H	59
1 contribution for Cl	115
The calculated value of $b \times 10^5$ =	174

The value of $b \times 10^5$ determined experimentally is 182.

b. Carbon dioxide, CO_2:

1 contribution for C	100
2 contributions for O 2×50 =	100
The calculated value of $b \times 10^5$ =	200

The value of $b \times 10^5$ determined experimentally is 191.

c. Naphthalene, $C_{10}H_8$

10 contributions for C 10×75 =	750
8 contributions for H 8×14 =	112
The calculated value of $b \times 10^5$ =	862

The value of $b \times 10^5$ determined experimentally is 866.

The agreement between the calculated and experimental values of the constant b is fairly good.

Constants a *and* b *for Aliphatic Hydrocarbons* (*Thodos' Method*)

Thodos[365] has elaborated a method which makes it possible to calculate the values of the constants a and b in van der Waals' equation for saturated and unsaturated aliphatic hydrocarbons with a fairly high accuracy.

The procedure for the calculation of the constants a and b is as follows. The saturated hydrocarbon corresponding to the actual hydrocarbon in question is considered first. Paraffinic molecules are considered to be derived from methane by successively replacing hydrogen atoms by methyl groups. Each such step introduces a well-defined contribution to the value of *a* and *b*. The actual value of this increment depends partly on the number of hydrogen atoms remaining unchanged on the carbon atom subjected to the substitution and partly on the number of hydrogen atoms on carbon atoms adjacent to it. For this the nature of the carbon atom can be characterized by a type number according to the following table:

Type number:	1	2	3	$\bar{3}$	4	$\bar{4}$
Group:	$-CH_3$	$-CH_2-$	$-\underset{\vert}{CH}-$	$=CH-$	$-\underset{\vert}{\overset{\vert}{C}}-$	$=C=$

TABLE 1.34. CONTRIBUTIONS FOR CALCULATING THE CONSTANTS *a* AND *b* IN VAN DER WAALS' EQUATION (AFTER THODOS[306,365])

Group	ϕ_a	ϕ_b	Formation of double and triple bonds	ϕ_a	ϕ_b
Methane (base value)	9509	17·371	First double bond between		
			1–1	−1941	−2·126
Primary methyl-group substitution	7065	6·497	2–1	−1518	−1·940
			2–2	−1204	−2·123
			3–1	−801	−1·692
Further methyl substitutions			3–2	−561*	−1·858*
1 (1)	6388	6·583	Second double bond between		
1 (2)	6417	6·686	3–1	−610*	−1·158*
1 (3)	5742	5·450	$\bar{3}$–1	−713*	−1·814*
1 (4)	5968	4·870	3–2	−428*	−1·811*
2 (1, 1)	5629	6·475	$\bar{3}$–2	−670*	−1·978*
2 (1, 2)	5261	5·879	$\bar{3}$–3	−637*	−2·142*
2 (1, 3)	5250	5·065	$\bar{3}$ ← 2–1	−792	−1·549
2 (1, 4)	5189	4·351	$\bar{3}$ ← 2–2	−610	−1·68*
			$\bar{4}$ ← 2–1	−750	−1·615
2 (2, 2)	4865	4·663	Triple bond between		
2 (2, 3)	4758	3·882	1–1	−2143	−3·876
2 (2, 4)	4750	3·354	1–2	−942	−3·260
2 (3, 3)	5137	3·481	2–2	−628*	−3·443*
3 (1, 1, 1)	4895	5·817			
3 (1, 1, 2)	4575	4·455			
3 (1, 1, 3)	4307	3·781			
3 (1, 1, 4)	4087	3·471			
3 (1, 2, 2)	4082	3·658			
3 (1, 2, 3)	3814	2·984			
3 (1, 3, 3)	3719	2·200			
3 (2, 2, 2)	3770	2·401			

NOTE: Estimated values are denoted by an asterisk. The arrow points away from the carbon atoms involved in the formation of the new double bond and towards the type of carbon adjacent.

Numerical values of the contributions are shown in the left hand half of Table 1.34, e.g. the contributions $\phi_a = 3814$ and $\phi_b = 2{\cdot}984$ result from replacement by methyl groups of a hydrogen atom of a —CH— group (with a vertical bond below the CH) to which are attached $-CH_3$, $-CH_2-$ and —CH— (with a vertical bond below the CH) groups, i.e. a type 3 (1, 2, 3) group increment.

The sums of the base value, primary methyl group substitution and further methyl substitution increments $\Sigma\phi_a$ and $\Sigma\phi_b$ are related to the van der Waals' constants by

$$\Sigma\phi_a = a^{0{\cdot}626}$$
$$\Sigma\phi_b = b^{0{\cdot}76} \tag{1.66}$$

The right hand half of Table 1.34 denotes the increments resulting when double and triple bonds are formed between carbon atoms, to convert the saturated paraffin to the compound under consideration (after Reid and Sherwood[306]).

Thodos recommends the following order for computing the contributions:

(a) By successive substitutions of methyl groups, the longest unbranched chain in the molecule is obtained.

(b) Single side methyl groups are added to the chain always in a clockwise direction.

(c) Methyl groups are successively substituted in the side chains until these chains reach the desired length.

(d) The contributions of double and triple bonds, obtained by removal of the corresponding hydrogen atoms are added. Values of a and b obtained by the above method may be used to calculate the critical parameters of aliphatic hydrocarbons using Eq. (1.65).

EXAMPLE 1.19. Use Thodos' method to estimate the constants a and b in van der Waals' equation, and the critical pressure, for 2,4,4-trimethylpentane-1.

CH₃ on C[2]; two CH₃ groups on C[4] (one above, one below):

$CH_2{=}C{-}CH_2{-}C{-}CH_3$
[1] [2] [3] [4] [5]

Solution

After the first substitution we obtain the formula: $CH_3{-}CH_3$ ([1] [2])

The next substitution, i.e. introduction of the third methyl group, is symbolized by the group of type 1 (1), since the group $-CH_3$ [2] of type 1, in which hydrogen is replaced by a methyl group, is also adjacent to the group of type 1, viz. to ($-CH_3$) [1].

After this substitution we get the formula

$CH_3{-}CH_2{-}CH_3$
[1] [2] [3]

The introduction of the next methyl groups is symbolized by the group of type 1 (2), since one hydrogen is substituted in the group $-CH_3$ [3], i.e. in the group of type 1, and this group is now adjacent to the group of type 2, viz. to $-CH_2-$ [2]

Further substitutions are accomplished in a similar way; the replacement of hydrogen at the carbon [4] by a methyl group, for example, when the formula has been brought to the form

CH₃ on C[2] and CH₃ on C[4]:

$CH_3{-}CH{-}CH_2{-}CH{-}CH_3$
[1] [2] [3] [4] [5]

is symbolized by the group of type 3 (1, 1, 2), since the substitution takes place on the group $\underset{[4]}{—\text{C—H}}$, i.e. of type 3, and this group is adjacent to two groups of type 1 (viz. side methyl group and group $\underset{[5]}{—CH_3}$) and to one group of type 2 (viz. $\underset{[3]}{—CH_2—}$ group).

Calculation of the contributions		$\phi_a \times 10^5$	ϕ_b
Methane, base value		9509	17·371
Primary substitution of $—CH_3$		7065	6·497
Further substitutions of $—CH_3$;	1 (1)	6388	6·583
(normal chain)	1 (2)	6417	6·686
	1 (2)	6417	6·686
Side methyl groups			
at carbon [2]	2 (1, 2)	5261	5·879
first group at carbon [4]	2 (1, 2)	5261	5·879
second group at carbon [4]	3 (1, 1, 2)	4575	4·455
Double bond	3–1	−801	−1·692
Sum of the contributions		50,092	58·344

$$\Sigma\phi_a = 50{,}092 = a^{0\cdot626}, \quad a = 32\cdot19\times10^6 \left(\frac{\text{cm}^3}{\text{mole}}\right)^2 \text{atm},$$

$$\Sigma\phi_b = 58\cdot344 = b^{0\cdot76}, \quad b = 217\cdot0 \text{ cm}^3/\text{mole}.$$

Calculation of the value of the critical pressure:

From Eq. (1.65) $p_c = a/27\ b^2$

$$p_c = \frac{32\cdot19\times10^6}{27\times217^2} = 25\cdot46 \text{ atm}.$$

Stiel and Thodos[348] have developed a method for estimating the critical constants of aliphatic hydrocarbons, based on the structural formula of the compound. This method, which does not involve summing contributions, is somewhat less accurate than the one described above.

Constants a *and* b (*Method of Forman and Thodos*)

Thodos' method for calculating the contributions to the constants *a* and *b* in van der Waals' equation for aliphatic hydrocarbons, as presented above, was later modified and extended to other classes of compounds by Forman and Thodos.[105]

In contrast with the first method (which yields very good results when applied to aliphatic hydrocarbons), the second method does not employ successive substitution of hydrogen atoms by monocarbon groups, but sums the contributions of such groups. In the first method the exponent 0·626 was assumed for the constant *a*, and 0·76 for the constant *b*. In the second method, the exponent of *a* is 2/3 while that of *b* is 3/4.

Normal Saturated Aliphatic Hydrocarbons. Constants *a* and *b* are found by simple summation of the contributions (Table 1.35)

$$a^{2/3} = \Sigma(\Delta a^{2/3}), \qquad b^{3/4} = \Sigma(\Delta b^{3/4}) \tag{1.67}$$

Branched Aliphatic Hydrocarbons. In this case we employ the correcting factors

$$\begin{aligned} a^{2/3} &= f_a \Sigma(\Delta a^{2/3}) \\ b^{3/4} &= f_b \Sigma(\Delta b^{3/4}) \end{aligned} \tag{1.68}$$

TABLE 1.35. SATURATED ALIPHATIC HYDROCARBONS (n IS THE TOTAL NUMBER OF CARBON ATOMS IN THE MOLECULE) (a IN (cc./g.mole)^2 atm; b IN cc./g.mole)

Group contributions			
Type	Group	Contribution of $\Delta a^{2/3}$	Contribution of $\Delta b^{3/4}$
1	$-CH_3$	$\Delta a^{2/3} = \dfrac{2168}{n} + 14{,}493$	$\Delta b^{3/4} = 11{\cdot}453$
2	$-CH_2-$	$\Delta a^{2/3} = 13{,}678$	$\Delta b^{3/4} = 6{\cdot}262$
3	$-\overset{\mid}{\underset{\mid}{C}}H$	$(n \geqslant 5)\ \Delta a^{2/3} = \dfrac{27{,}560}{n} + 5677$	$\Delta b^{3/4} = \dfrac{23{\cdot}55}{n} - 3{\cdot}824$
4	$-\overset{\mid}{\underset{\mid}{C}}-$	$\Delta a^{2/3} = \dfrac{36{,}013}{n} - 1022$	$\Delta b^{3/4} = \dfrac{52{\cdot}00}{n} - 15{\cdot}337$

To find the correcting factor we must calculate the *Wiener number w*. For each bond between two carbon atoms in the chain we calculate the number of carbon atoms first at one side of the bond and then at the other side and multiply these two numbers by each other. In this way we calculate the products for all C—C bonds and sum them up.

For instance, in the case of 2-methylbutane the number w is

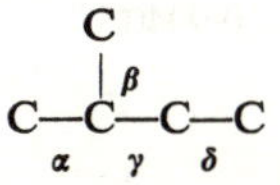

$$\begin{aligned} \alpha\ 1\times4 &= 4 \\ \beta\ 1\times4 &= 4 \\ \gamma\ 3\times2 &= 6 \\ \delta\ 4\times1 &= 4 \\ w_i &= 18 \end{aligned}$$

For a branched chain (iso-) we calculate the number w_i; at the same time, the number w_n is calculated for a normal chain with the same number of carbon atoms.

In the example quoted above

C—C—C—C—C (bonds α, β, γ, δ)

$$\begin{aligned} \alpha\ 1\times4 &= 4 \\ \beta\ 2\times3 &= 6 \\ \gamma\ 3\times2 &= 6 \\ \delta\ 4\times1 &= 4 \\ w_n &= 20 \end{aligned}$$

The correcting factors are obtained from the formulae

$$f_a = \tfrac{1}{3}\left[\frac{w_i}{w_n} + 2 + 0{\cdot}087m + 0{\cdot}0045 \sum_{k=1}^{k=m} k(k-1)\right] \tag{1.69}$$

$$f_b = \tfrac{1}{2}\left[\frac{w_i}{w_n} + 1 + 0{\cdot}101\, m - 0{\cdot}005\, m^2\right] \tag{1.70}$$

In these equations m denotes the number of carbon side-chains in the molecule.

EXAMPLE 1.20.[105] Calculate the van der Waals' constants a and b for 2-methyl-3-ethylpentane

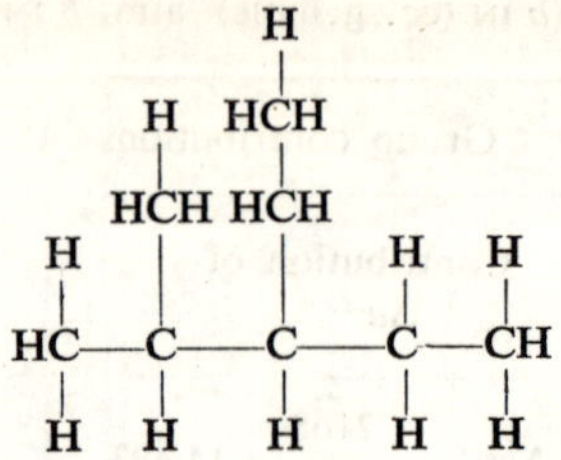

Solution

Number of carbon atoms: Total $n = 8$

Type 1		4
Type 2		2
Type 3		2

Number of side chains $m = 2$

Summation of the contributions to $a^{2/3}$

Type		$\Delta a^{2/3}$
1	$4\times\left(\frac{2168}{8}+14{,}493\right) = 4\times 14{,}794 =$	59,056
2	$2\times 13{,}678 =$	27,356
3	$2\times\left(\frac{27{,}560}{8}+5677\right) = 2\times 9122 =$	18,244
	$\Sigma(\Delta a^{2/3}) =$	104,656

Summation of the contributions to $b^{3/4}$

Type		$\Delta b^{3/4}$
1	$4\times 11{\cdot}453 =$	45·812
2	$2\times 6{\cdot}262 =$	12·524
3	$2\left(\frac{23{\cdot}55}{8}-3{\cdot}824\right) = 2\times(-0{\cdot}880) =$	−1·760
	$\Sigma(\Delta b^{3/4}) =$	56·576

Calculation of the correcting factors.

The Wiener number for 2-methyl-3-ethylpentane $w_i = 67$
n-octane $w_n = 84$ $\quad w_i/w_n = 0{\cdot}798$

$$f_a = \tfrac{1}{3}\left[0{\cdot}798+2+0{\cdot}097\times 2+0{\cdot}0045\sum_{k=1}^{k=2} k(k-1)\right] = 0{\cdot}9940$$

$$f_b = \tfrac{1}{2}[0{\cdot}798+1+0{\cdot}101\times 2-0{\cdot}005\times 2^2] = 0{\cdot}9900$$

$a^{2/3} = 0{\cdot}9940\times 104{,}656 = 104{,}028$ $\qquad b^{3/4} = 0{\cdot}9900\times 56{\cdot}576 = 56{\cdot}010$

$a = 33{\cdot}552\times 10^6\ (\mathrm{cm^3/mole})^2\cdot\mathrm{atm}$ $\qquad b = 214{\cdot}30\ \mathrm{cm^3/mole}$

Unsaturated Aliphatic Hydrocarbons. Van der Waals' constants are found by adding the contributions of double or triple bonds to the sum of the contributions calculated in the same way as for saturated hydrocarbons. These contributions are presented in Table 1.36. In this table the symbol $3u$ denotes the group H—C═ (with a single bond below C).

The numbers characterizing the bond, e.g. 1–1, 2–1, 3–2, etc. denote the types of the groups between which the double (or triple) bond occurs.

EXAMPLE 1.21.[105] Calculate the van der Waals' constants a and b for 2-methyl-1-butene,

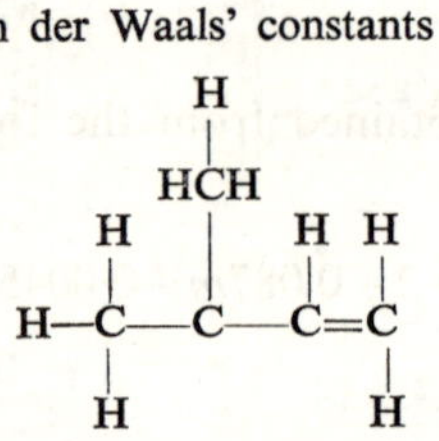

Solution

Values of $a^{2/3}$ and $b^{3/4}$ are calculated for the corresponding saturated hydrocarbon, i.e. for 2-methylbutane, by the method described above, and the contributions to $a^{2/3}$ and $b^{3/4}$ arising from the double bond are then added.

TABLE 1.36. BOND CONTRIBUTIONS FOR CALCULATING THE VALUES OF THE CONSTANTS a AND b IN VAN DER WAALS' EQUATION BY FORMAN AND THODOS' METHOD[105]
(ALIPHATIC UNSATURATED HYDROCARBONS)
(a IN $(cc/g.mole)^2$ atm; b IN cc/g.mole)

Bond	$\Delta a^{2/3}$	$\Delta b^{3/4}$	Bond	$\Delta a^{2/3}$	$\Delta b^{3/4}$
First double bond			Second double bond (*cont.*)		
1–1	−3868	−2·021	3*u*–1	−1332	−1·745
2–1	−3154	−1·895	3*u*–2	−1324	−1·862
2–2	−2551	−2·009	3*u*–3	−1316	−1·979
3–1	−1548	−1·706	3*u* ← 2–1	−1687	−1·399
3–2	−928	−1·820	3*u* ← 2–1	−910	−1·485
3–3	−540	−1·930			
Second double bond			Triple bond		
3–1	−828	−1·259	1–1	−4269	−3·680
3–2	−496	−1·343	1–2	−1934	−3·008
			2–2	−1331	−3·122

It is found that, for 2-methylbutane

$$a^{2/3} = 69{\cdot}370 \quad \text{and} \quad b^{3/4} = 41{\cdot}320$$

The contributions arising from the double bond 2–1 are

$$\Delta a^{2/3} = -3{\cdot}154 \quad \text{and} \quad \Delta b^{3/4} = -1{\cdot}895$$

For 2-methyl-1-butene, therefore,

$$a^{2/3} = 66{\cdot}216 \quad \text{and} \quad b^{3/4} = 39{\cdot}425$$

Hence, $a = 17{\cdot}033 \times 10^6$ $(cm^3/mole)^2$ atm
and $b = 134{\cdot}19$ cm^3/mole.

Hydrocarbons with Naphthenic Rings. The contributions for calculation of the constants a and b for these hydrocarbons are given in Table 1.37 (the numbers referring to the carbon groups in the naphthene ring are denoted by the letter n).

EXAMPLE 1.22. Calculate the van der Waals' constants a and b for 1-ethyl-3-*cis*-n-propylcyclopentane

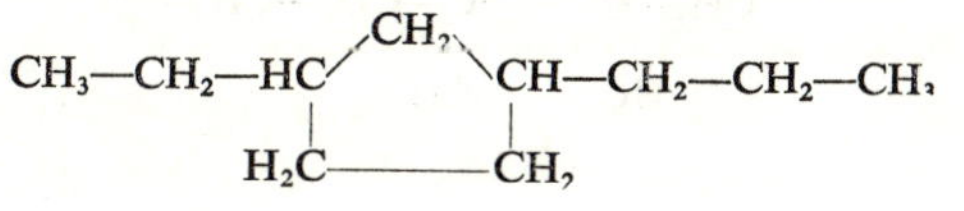

Solution

Summation of the contributions:

		$\Delta a^{2/3}$		$\Delta b^{3/4}$
(a) Naphthene ring				
three groups 2 n:	$3\times(12{,}535) =$	37,605,	$3\times(5{\cdot}338) =$	16·014
two groups 3 n:	$2\times(9910) =$	19,820,	$2\times(0{\cdot}023) =$	0·046
contribution of the naphthene ring		2658		9·073
contribution of position *cis*-1, 3		−2525		−1·493
(b) Side chain, ethyl ($n = 10$)				
one group of type 1		14,710		11·453
one group of type 2		13,678		6·262
(c) Side chain, n-propyl ($n = 10$)				
one group of type 1		14,710		11·453
two groups of type 2	$2\times(13{,}678) =$	27,356	$2\times(6{\cdot}262) =$	12·524
	$\Sigma\Delta a^{2/3} =$	128,012	$\Sigma\Delta b^{3/4} =$	65·332

hence, $a = 45{\cdot}80\times 10^6$ $(cm^3/mole)^2$ atm
$b = 263{\cdot}12$ cm^3/mole.

TABLE 1.37. GROUP CONTRIBUTIONS IN THE NAPHTHENE RING[105]
(a IN $(cc/g.mole)^2$ atm; b IN cc/g.mole)

Type	Group	$\Delta a^{2/3}$	$\Delta b^{3/4}$
2 *n*	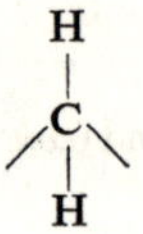	12,535	5·338
3 *n*	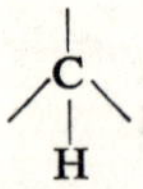	9910	0·023
4 *n*	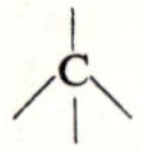	2066	−8·094
Contribution of the naphthene ring		2658	9·073
Position of the groups			
	cis-1,2	−427	−0·866
	-1,3	−2525	−1·493
	trans-1,2	−2525	−1·493
	-1,3	−4195	−2·494

Hydrocarbons with Aromatic Rings. The numbers referring to the types of the carbon groups occurring in aromatic rings are denoted by the letter *a*. The contributions are listed in Table 1.38.

EXAMPLE 1.23.[105] Calculate the van der Waals' constants *a* and *b* for 1-methyl-3-ethylbenzene,

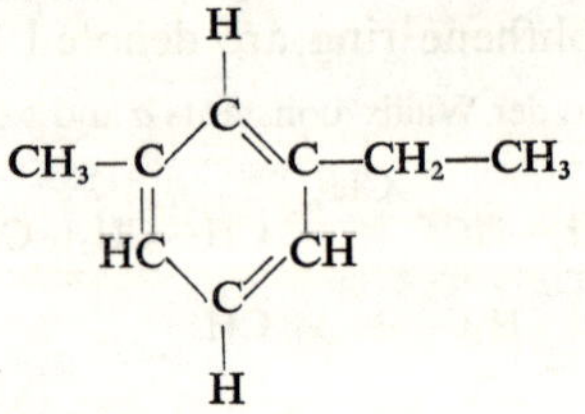

Solution

Summing contributions from

		$\Delta a^{2/3}$		$\Delta b^{3/4}$
(a)	Aromatic ring			
	four groups, type 3 *a* $4\times(11{,}646) =$	46,584	$4\times(5{\cdot}991) =$	23·964
	two groups, type 4 *a* $2\times(11{,}144) =$	22,288	$2\times(1{\cdot}043) =$	2·086
	Contributions for the positions:			
	1–3	−1597		−0·806
	1–5	279		0·254
(b)	Side chain, methyl ($n = 9$)			
	one group of type 1	14,734		11·453
(c)	Side chain, ethyl ($n = 9$)			
	one group of type 1	14,734		11·453
	one group of type 2	13,678		6·262
	$\Sigma\Delta a^{2/3} =$	110,700	$\Sigma\Delta b^{3/4} =$	54·666

hence, $a = 36{\cdot}832\times10^6\ (cm^3/mole)^2$ atm
$b = 207{\cdot}47\ cm^3/mole$.

TABLE 1.38. GROUP CONTRIBUTIONS IN THE AROMATIC RING[105]
(a IN $(\text{cc/g.mole})^2$ atm; b IN cc/g.mole)

Type	Group	$\Delta a^{2/3}$	$\Delta b^{3/4}$
3 a	C—H	11,646	5·991
4 a	C—	11,144	1·043
4 p (junction of two rings)	C	11,561	1·634
Position of the groups			
1–2		−830	−1·253
1–3		−1597	−0·806
1–4		155	0·212
1–5		279	0·254
1–6		488	0·525

Organic Compounds other than Hydrocarbons. The calculation of the contributions $\Delta a^{2/3}$ and $\Delta b^{3/4}$ is based on the following principles:

(1) The size of the molecule is determined by the number n of all atoms, hydrogen atoms excluded. These atoms are known as *functional atoms.* For example, the molecule of methanol consists of 2 such atoms (C, O), hence $n = 2$; for the molecule of acetic acid $n = 4$ (2C, 2O), etc.

(2) The molecule consists of a certain number of carbon atoms forming the hydrocarbon skeleton and of a certain number of atoms of carbon or other elements included in the functional groups. The contributions of carbon atoms of the first kind are calculated in the same way as shown above.

The contributions of the functional groups are calculated from the formulae:

$$\Delta a^{2/3} = \frac{m}{n} + k \tag{1.71}$$

$$\Delta b^{3/4} = \frac{s}{n} + t \tag{1.72}$$

where m, k, s, t are constants characteristic of the functional group. The values of these constants are listed in Table 1.39.

The authors have checked the accuracy of this method on several scores of compounds (non-hydrocarbons), finding that the average error in estimation of the critical temperature is 1·36%, whereas in the case of the critical pressure it is 2·82%. The maximum error for T_c and p_c amounted to 2·91% and 6·78%, respectively.

TABLE 1.39. CONSTANTS FOR FUNCTIONAL GROUPS IN EQS. 1.71 AND 1.72

Type of compound	Functional group	*m*	*k*	*s*	*t*
Alcohols	—OH	30,200	14,000	8·96	7·50
Phenols	—OH	0	8500	0	4·19
Ethers (non-cyclic)	—O—	14,500	6500	0	3·26
Cyclic ethers	—O—	0	9440	0	2·74
Ketones	—C(=O)—	62,800	16,700	27·20	4·55
Carboxylic acids	—C(=O)—OH	142,670	16,730	66·80	5·10
Acid anhydrides	—C(=O)—O—C(=O)—	0	43,880	0	14·78
Formates	H—C(=O)—O—	35,140	26,800	2·29	15·80
Other esters	—C(=O)—O—	37,430	25,500	−3·00	12·20
Primary amines	$—NH_2$	4800	18,900	0	10·15
Secondary amines	—N(H)—	51,800	0	19·60	−1·10
Tertiary amines	—N(—)—	60,200	−4300	29·20	−7·90
Nitriles	—CN	86,000	25,900	39·70	12·10
Aliphatic halogen-derivatives					
Fluorides	—F				
first carbon		2420	12,240	−3·70	10·92
second carbon		−38,500	4510	−48·50	12·86
third carbon		0	3450	0	6·92
Chlorides	—Cl				
first carbon		0	22,580	0	11·54
second carbon		66,000	−5100	19·00	3·90
third carbon		−60,250	29,100	−40·80	19·40
fourth carbon		0	16,500	0	11·46
Bromides	—Br				
first carbon		−2720	23,550	−4·35	11·49
second carbon		0	20,860	0	5·37
Iodides	—I				
first carbon		0	33,590	0	13·91
Aromatic halides					
fluorides	—F	0	4210	0	7·22
chlorides	—Cl	0	17,200	0	10·88
bromides	—Br	0	24,150	0	12·74
iodides	—I	0	34,780	0	15·22

EXAMPLE 1.24. Estimate the constants a and b, and the critical constants T_c and p_c for acetone,

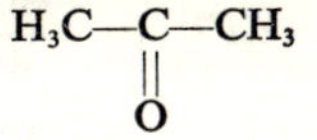

Solution

Acetone contains four functional atoms (3 C, 1 O) $n = 4$

Calculation of the contributions:

two carbons of the aliphatic chain (methyl groups), their contribution is found from Table 1.35, type 1:

$$\Delta a^{2/3} = \frac{2168}{4} + 14{,}493 = 15{,}035, \qquad \Delta b^{3/4} = 11{\cdot}453$$

The contribution of the ketonic (functional) group $-\overset{\text{O}}{\overset{\|}{\text{C}}}-$ is

$$\Delta a^{2/3} = \frac{62{,}800}{4} + 16{,}700 = 32{,}400$$

$$\Delta b^{3/4} = \frac{27{\cdot}20}{4} + 4{\cdot}55 = 11{\cdot}25$$

Summation of the contributions:

two carbons (type 1)	$2 \times 15{,}035 = 30{,}070$	$2 \times 11{\cdot}453 = 22{\cdot}906$
ketonic group	$32{,}400$	$11{\cdot}35$
	$a^{2/3} = 62{,}470$	$b^{3/4} = 34{\cdot}256$
	$a = 15{\cdot}614 \times 10^6$	$b = 111{\cdot}26$

Substitution into Eq. (1.65) gives:

$$T_c = \frac{8 \times 15{\cdot}614 \times 10^6}{27 \times 82{\cdot}055 \times 111{\cdot}26} = 506{\cdot}8°\text{K}$$

$$p_c = \frac{15{\cdot}614 \times 10^6}{27 \times (111{\cdot}26)^2} = 46{\cdot}7 \text{ atm}$$

(calculated values)

The values determined experimentally are:

$T_c = 508{\cdot}7°$K $\quad p_c = 46{\cdot}6$ atm

EXAMPLE 1.25.[105] Estimate the constants a and b, and the critical constants T_c and p_c for fluorotrichloroethane (Freon 113),

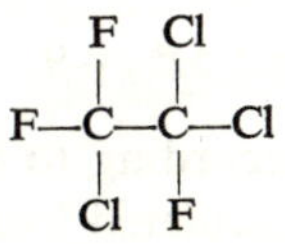

Solution

The compound contains eight functional atoms (2C, 3F, 3Cl) $n = 8$.

Calculation of the contributions:

two carbons of the aliphatic chain $-\overset{|}{\underset{|}{\text{C}}}-$, their contribution is found from Table 1.35, type 4:

$$\Delta a^{2/3} = \frac{36{,}013}{8} - 1022 = 3480, \qquad \Delta b^{3/4} = \frac{52{\cdot}00}{8} - 15{\cdot}337 = -8{\cdot}837$$

two first fluorine atoms:

$$\Delta a^{2/3} = \frac{2420}{8} + 12{,}240 = 12{,}550, \qquad \Delta b^{3/4} = \frac{-3{\cdot}70}{8} + 10{\cdot}92 = 10{\cdot}46$$

one second fluorine atom:

$$\Delta a^{2/3} = \frac{-38{,}500}{8}+4510 = -300, \qquad \Delta b^{3/4} = \frac{-48{\cdot}50}{8}+12{\cdot}86 = 6{\cdot}80$$

one second chlorine atom:

$$\Delta a^{2/3} = \frac{66{,}000}{8}-5100 = 3150, \qquad \Delta b^{3/4} = \frac{19{\cdot}00}{8}+3{\cdot}90 = 6{\cdot}28$$

two third chlorine atoms:

$$\Delta a^{2/3} = \frac{-60{\cdot}250}{8}+29{,}100 = 21{,}570, \qquad \Delta b^{3/4} = \frac{-40{\cdot}80}{8}+19{\cdot}40 = 14{\cdot}30$$

Summation of the contributions:

two carbons	$2\times3480 = 6960$	$2\times(-8{\cdot}837) =$	$-17{\cdot}674$
two first F	$2\times12{,}550 = 25{,}100$	$2\times10{\cdot}46 =$	$20{\cdot}92$
one second F	-300		$6{\cdot}80$
one second Cl	3150		$6{\cdot}28$
two third Cl	$2\times21{,}570 = 43{,}140$		$28{\cdot}60$
	$a^{2/3} = 78{,}050$	$b^{3/4} =$	$44{\cdot}926$
	$a = 21{\cdot}801\times10^6$	$b =$	$159{\cdot}71$

Substitution into Eq. (1.65) gives:

$$T_c = \frac{8\times21{\cdot}801\times10^6}{27\times82{\cdot}055\times159{\cdot}71} = 492{\cdot}9^\circ\text{K}, \qquad p_c = \frac{21{\cdot}801\times10^6}{27\times159{\cdot}71^2} = 31{\cdot}7 \text{ atm}$$

The values determined experimentally are: $T_c = 487{\cdot}3^\circ$K and $p_c = 33{\cdot}7$ atm.

Comparison of the Thodos Method with that of Thodos and Forman

The first method is simpler and the results may be somewhat more accurate, but its application is limited to the estimation of the constants a and b for aliphatic hydrocarbons, while the second method may be applied when other compounds are involved.

c. Calculation of the Critical Parameters from the Molar Refraction and the Parachor

Since the critical parameters, the parachor and the molar refraction all depend on the intermolecular forces, it is possible to estimate the critical parameters from a knowledge of the parachor and the molar refraction. Various equations have been proposed for this, and fairly good results are obtained when Meissner's equations are used.[247] According to Meissner the critical volume can be calculated from the equation:

$$V_c = 0{\cdot}55(1{\cdot}5P_{ch}+9-4{\cdot}34R_D)^{1{\cdot}155} \tag{1.73}$$

Usually the error of the estimation according to this equation does not exceed 5%.

The critical temperature can be calculated from two different equations, the selection being dependent on whether the normal boiling temperature T_b is known or not. When T_b is known we may use the equation:

$$T_c = 20{\cdot}2T_b^{0{\cdot}60}-143-1{\cdot}2P_{ch}+10{\cdot}4R_D+A \tag{1.74}$$

The value of the constant A for hydrocarbons and their halogen-derivatives is zero; for all other compounds A is 10. In the case of hydrocarbons the error of estimation does not exceed 3%, and for all other compounds it is less than 6%. If the boiling temperature T_b is not known, another equation can be employed, viz.

$$T_c = \frac{5850R_D^2-C}{P_{ch}^{1{\cdot}65}} \tag{1.75}$$

The magnitude of the constant C depends upon the type of compound considered; values of C for a number of different groups of compounds are compiled in Table 1.40.

TABLE 1.40. VALUES OF THE CONSTANTS IN THE MEISSNER EQUATIONS

Group of compounds	B, Eq. (1.77)	C, Eq. (1.75)
Monocarboxylic acids	28,000	1,000,000 } large errors
Monohydric alcohols, phenol, cresol, etc.	16,500	450,000 } large errors
Primary amines	6500	200,000
Secondary amines	2000	0
Tertiary amines	−3000	−300,000
Esters of monocarboxylic acids and monohydric alcohols	15,000	700,000
Ethers and mercaptans	4000	290,000
Esters of dicarboxylic acids and monohydric alcohols	30,000	......
Hydrocarbons of acetylene series	−500	−29,000
Aromatic hydrocarbons	−2500	−350,000
Paraffinic and naphthenic hydrocarbons	−2500	−58,000
Olefins	−4500	−140,000
Ketones	15,000	700,000
Monochloroparaffins	4000	760,000
Nitriles	20,000	

According to Meissner, the critical pressure is given by

$$p_c = \frac{60 \cdot 3 T_c}{(1 \cdot 5 P_{ch} + 9 - 4 \cdot 34 R_D)^{1 \cdot 226}} \tag{1.76}$$

If the value of T_c is not available and it must be calculated from Eq. (1.74), the accuracy of the estimation of p_c diminishes and the error may be as high as 15%.

Meissner has also presented an equation for the calculation of the boiling temperature, T_b:

$$T_b = \frac{637 R_D^{1 \cdot 47} + B}{P_{ch}} \tag{1.77}$$

Values of B are listed in Table 1.40 for a number of different groups of compounds.

Meissner's method enables values of the critical parameters to be estimated with an accuracy which is usually sufficient for technological calculations.

d. Additive Estimation of the Critical Pressure

Riedel[309] has noticed that the quantity

$$\varphi = 100 \sqrt{\frac{M}{p_c}} \tag{1.78}$$

where M denotes the molecular weight and p_c the critical pressure) increases by the same value on passing from one compound to the next in a given homologous series. This observation has led to the statement that the quantity φ may be calculated with high accuracy by adding to a basic constant, $A = 33$, the contributions which correspond to the atoms and bonds in the molecule. It is thus possible to find the value of the critical pressure

$$p_c = \frac{10^4 M}{\varphi^2} \tag{1.79}$$

The contributions to the function φ are presented in Table 1.41 (after Reid and Sherwood,[306] and Riedel[309]).

TABLE 1.41. CONTRIBUTIONS FOR CALCULATING φ IN EQ. (1.78) FOR THE ESTIMATION OF THE CRITICAL PRESSURE (p_c, atm)

Atom contributions		Bond contributions	
Basic constant A	33	C=C	−6
H	0	C=O	−12
C	23	C≡N	0
F	23	O—H (alcohols)	−11
Cl	32	O—H (aromatic alcohols)	−21
Br	54	OH (acids)	0
O	18	N—H	−2
N	14	C≡C	−14
S	27	C–S	−5
I	87		
Each ring atom	−4·3		
Condensed rings	−10		

EXAMPLE 1.26. Calculate the critical pressure, p_c, of diphenylmethane, $C_{13}H_{12}$ (molecular weight $M = 168{\cdot}22$).

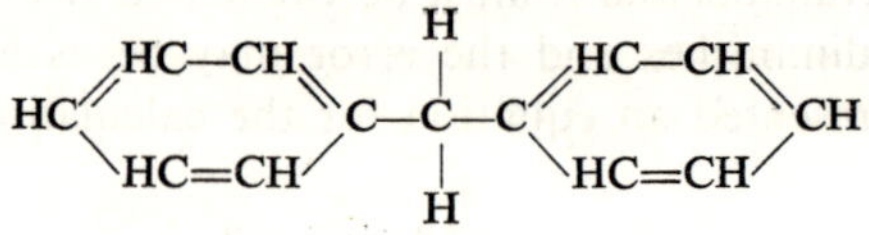

Solution

Summation of the contributions

Basic constant		33
13 C contributions	13×23 =	299
12 H contributions	=	0
6 double bonds	(−6)×6 =	−36
12 ring atoms	(−4·3)×12 =	−51·6
	φ =	244·4

After substituting the value of φ into Eq. (1.79), we find the required value of the critical pressure

$$p_c = \frac{10^4 \times 168{\cdot}22}{(244{\cdot}4)^2} = 28{\cdot}1 \text{ atm}$$

The experimentally determined values are: $\varphi = 244$, $p_c = 28{\cdot}2$ atm. Unfortunately the calculated values of φ do not always agree so closely with the experimental ones.

Lydersen[306] has modified and supplemented the Riedel table of contributions to p_c. According to Lydersen, the value of the critical pressure can be calculated from the formula

$$p_c = \frac{M}{(\phi + 0{\cdot}34)^2} \tag{1.80}$$

where M denotes the molecular weight of the compound, and ϕ the sum of the contributions of the atoms, groups and bonds taken from Table 1.42 (the values of the contributions given in parentheses are not quite certain).

TABLE 1.42. CONTRIBUTIONS FOR CALCULATING THE CRITICAL PRESSURE OF COMPOUNDS (AFTER LYDERSEN[306])

Atoms, groups, bonds	Contributions	Atoms, groups, bonds	Contributions
—CH_3, —CH_2—	0·227	—O—	0·16
—CH_2— (in ring)	0·184	—O— (in ring)	(0·12)
		—OH (in alcohols)	0·06
>CH—	0·210	—OH (in phenols)	(−0·02)
		>CO	0·29
>CH— (in ring)	0·192	>CO (in ring)	(0·2)
=CH—, =CH_2	0·198	—CHO	0·33
		—COO—	0·47
=CH— (in ring)	0·154	—COOH	(0·4)
		—NH_2	0·095
>C<	0·210	>NH	0·135
>C< (in ring)	(0·154)	>NH (in ring)	(0·09)
		>N—	0·17
=C<, =C=	0·198	>N— (in ring)	(0·13)
=C<, =C= (in ring)	0·154	—CN	(0·36)
		—SH, —S—	0·27
=C<, ≡CH	0·153	—S— (in ring)	(0·24)
—F	0·224	=O (for cases other than those given above)	(0·12)
—Cl	0·320		
—Br	(0·50)	—NO_2	(0·42)
—I	(0·83)	=S	(0·24)

It is usually found that critical pressures calculated in this way are rather more accurate than those given by Riedel's original method.

EXAMPLE 1.27. Calculate the critical pressure, p_c, of diphenylmethane using the Lydersen method (cf. Example 1.26).

Solution

Summation of the contributions

10 contributions of H—C(=)—	$10 \times 0{\cdot}154 = 1{\cdot}540$
2 contributions of =C(—)—	$2 \times 0{\cdot}154 = 0{\cdot}308$
1 contribution of —CH_2—	$0{\cdot}227$
	$\phi = 2{\cdot}075$

After substituting ϕ into Eq. (1.80), we find the critical pressure of diphenylmethane

$$p_c = \frac{168{\cdot}22}{(2{\cdot}075+0{\cdot}34)^2} = \frac{168{\cdot}22}{2{\cdot}415^2} = 28{\cdot}8 \text{ atm}$$

The error of estimation is +2·1%.

One of the more accurate methods of calculating the critical pressure has been elaborated by Eduljee[116, 117] who uses the equation

$$p_c = \frac{10^4 M}{(\Sigma \Delta p)^2} \tag{1.81}$$

This equation is similar to Eq. (1.79) presented by Riedel ($\Sigma\Delta p \approx \varphi$), though the contributions Δp, recommended by Eduljee and compiled in Table 1.43, differ from Riedel's contributions to φ.

TABLE 1.43. CONTRIBUTIONS (Δp_c) FOR CALCULATING THE CRITICAL PRESSURE (p_c, atm) OF COMPOUNDS (AFTER EDULJEE[117])

	Δp		Δp
Atom contributions		Bond contributions (*cont.*)	
C	−9·35	$n = 6$	16·4
H	16·20	$n = 7$	11·0
O (in ethers)	17·20	$n = 8$	5·3
N	0·0	$n > 8$	0·0
Cl	48·0	(C)≡(C)	51·1
Br	68·8	(N)—(H) (amines)	−3·15
F	39·9	Rings (also heterocyclic)	
S	27·8	5-membered, saturated	10·5
Si	22·4	6-membered, saturated	7·2
Group contributions		Benzene ring	84·5
—OH (alcohols, phenols)	23·7	Effect of position	
=CO (ketones)	30·2	Single branching or substitution at the second carbon	−1·6
—COOH	57·7		
—C≡N	52·5	As above, at the third carbon or further	−4·75
Bond contributions		The positions are reckoned from the shorter end of the chain	
(C)=(C) (number n of carbons in the chain)			
$n = 2$	28·6	For several branchings or substitutions the contributions are multiplied by the number of branchings	
$n = 3$	27·9		
$n = 4$	25·2		
$n = 5$	21·2		

EXAMPLE 1.28. Estimate the critical pressure, p_c, of diphenylmethane by the Eduljee method (cf. Examples 1.26, 1.27).

Solution

Summation of the contributions

13 C contributions	$13\times(-9{\cdot}35) = -121{\cdot}5$
12 H contributions	$12\times 16{\cdot}20 = 194{\cdot}4$
2 benzene ring contributions	$2\times 84{\cdot}5 = 169$
	$\Sigma\Delta p = 241{\cdot}9$

After substituting the value of $\Sigma\Delta p$ into Eduljee's equation, we obtain

$$p_c = \frac{10^4\times 168{\cdot}22}{241{\cdot}9^2} = 28{\cdot}7 \text{ atm}$$

The error of calculation is $+1{\cdot}7\%$.

Thodos' method, based on the equation

$$p_c = \frac{a}{27b^2} \tag{1.82}$$

has already been discussed in section 5a.

Comparison of several scores of values of the critical pressures calculated by different additive methods, with those determined experimentally, enables an estimate to be made of the accuracy of these methods.[105, 247, 306, 309, 365] The average errors are:

Forman and Thodos's method, Eq. (1.65)	
Hydrocarbons:	
saturated aliphatic	1·06%
unsaturated aliphatic	1·37%
naphthenic	0·20%
aromatic	2·29%
Non-hydrocarbons	2·82%
Riedel's method, Eq. (1.79)	4·3%
Lydersen's method, Eq. (1.80)	3·1%
Eduljee's method, Eq. (1.81)	about 3%
Meissner's method, Eq. (1.76)	about 7%

e. Additive Estimation of the Molal Critical Volume, V_c

It has been seen in section 4a that the molal volume at the normal boiling point, V_b, is an approximately additive quantity and that approximate predictions of this quantity may be made using Kopp's Law and the contributions of Schroeder and Le Bas. It was also seen that the accurate prediction of V_b presented a very complex problem necessitating a consideration of the environment of each group within the molecule. The contribution to be attributed to each atom, bond or group was found to depend not only on the nature of the atom, bond or group but also upon the adjacent bonds and groups. Similar consideration apply to the additive calculation of the molar critical volume, V_c. One method of predicting this quantity is to estimate the parachor and molar refraction using the method of group contributions and then to apply Meissner's equation (1.73).

Lydersen[306] has presented a table of group contributions for the direct additive calculation of V_c (Table 1.44). V_c is obtained by adding these contributions to a base value which is equal to 40.

According to Riedel,[309] the calculation of molal critical volumes by summation of group contributions does not lead to accurate results. He recommends the application of a general method for estimating additive quantities (cf. section 6), consisting in replacing the contributions of the groups or atoms in a standard compound, the structure of which resembles that of the compound under investigation, by other contributions so as to obtain

TABLE 1.44. CONTRIBUTIONS FOR CALCULATING THE MOLAL VOLUME, V_c, AT THE CRITICAL POINT (IN cm^3/mole) (AFTER LYDERSEN[306])

Base value	40	—OH (in alcohols)	(18)
—CH_3 and —CH_2—	55	—OH (in phenols)	(3)
—CH_2— (in ring)	44·5	>C=O	60
>CH—	51	>C=O (in ring)	(50)
>CH— (in ring)	46	—CHO	73
=CH— and =CH_2	45	—COO—	80
=CH— (in ring)	37	—COOH	80
>C<	41	—NH_2	28
>C< (in ring)	(31)*	>NH	(37)
=C< and =C= (also in ring)	36	>NH (in ring)	(27)
≡C— and ≡C—H	(36)	>N—	(42)
—F	18	>N— (in ring)	(32)
—Cl	49	—CN	(80)
—Br	(70)	—SH and —S—	55
—I	(95)	—S— (in ring)	(45)
—O—	20	=O (for cases other than those presented above)	(11)
—O— (in ring)	(8)	—NO_2	(78)
		=S	(47)

* The values given in parenthesis are uncertain.

the critical volume of the desired compound. Values of the contributions to be added to the molal volume of the known compound, when one group in this compound is replaced by a different one, are listed in Table 1.45.

TABLE 1.45. DIFFERENCES OF CONTRIBUTIONS TO THE MOLAL VOLUME AND THE MOLAL VOLUMES OF STANDARD COMPOUNDS AT THE CRITICAL POINT (AFTER RIEDEL[309])

Group introduced	Group replaced	Difference in the contribution	Molal volumes of standard compounds, cm^3/mole	
CH_3	H	+59	Methane	99
C=C	CH—CH	−17		
C≡C	CH_2—CH_2	−34	Ethane	148
F	H	+14	Propane	195
Cl	H	+44		
Br	H	+67	Cyclopentane	260
O	CH_2	−31	Cyclohexane	308
COO	CH_2	+33		
S	CH_2	+11	Naphthalene	400

EXAMPLE 1.29. Estimate, using Riedel's method, the molal volume at the critical point of:

a. 2,2,4-trimethylpentane,

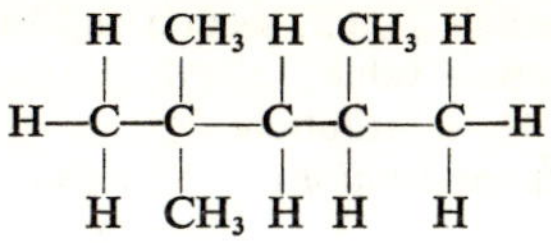

Solution

In the standard compound, i.e. propane, five hydrogen atoms (marked with an asterisk) are substituted by methyl groups

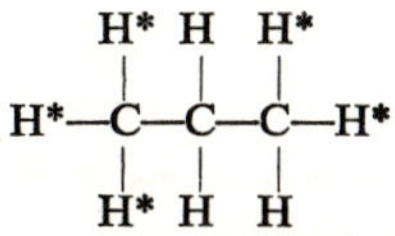

Calculation of the volume V_c

The volume of the standard compound (propane) is	195 cm³/mole
Five substitutions of H by CH_3	$5 \times 59 =$ 295 cm³/mole
The volume of 2,2,4-trimethylpentane	$V_{c(\text{calc})} =$ 490 cm³/mole

V_c determined experimentally is 470 cm³/mole.

The error of estimation is 4·3%.

b. chlorobenzene,

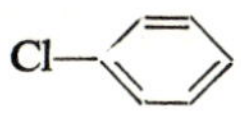

Solution

In the standard compound, cyclohexane, three CH—CH bonds are replaced by C=C bonds and one hydrogen is substituted by Cl.

Calculation of the volume V_c

The volume of the standard compound (cyclohexane) is	308 cm³/mole
Three substitutions of CH—CH by C=C	$3 \times (-17) =$ −51 cm³/mole
One substitution of H by Cl	+44 cm³/mole
The volume of chlorobenzene	$V_{c(\text{calc})} =$ 301 cm³/mole

V_c determined experimentally is 308 cm³/mole.

The error of calculation is −2·3%.

Thodos[365] has suggested a method for estimating the molal critical volume, V_c, of saturated and unsaturated aliphatic hydrocarbons. He uses the equation

$$V_c = 3\beta b \tag{1.83}$$

in which b is van der Waals' constant calculated by summing the contributions according to the scheme presented in section 5a, and β is a coefficient dependent on the number, N_c, of carbon atoms and their configuration in the molecule of the hydrocarbon. According to van der Waals $V_c = 3b$, however, since this relationship is not precise, Thodos introduced the correcting factor β. In the case of normal aliphatic saturated hydrocarbons (with non-branched chains), β is calculated from the formula

$$\beta_{\text{norm}} = 0{\cdot}7849 - 0{\cdot}01337\,N_c \tag{1.84}$$

When the chain of the molecule of a saturated hydrocarbon is branched the following formula should be used:

$$\beta_{\text{iso}} = 0{\cdot}8100 - 0{\cdot}0138\,N_c \tag{1.85}$$

For unsaturated aliphatic hydrocarbons, β can be obtained by multiplying the value for an analogous saturated compound by the factor 0·9931

$$\beta_{\text{unsat}} = 0{\cdot}9931 \times \beta_{\text{sat}} \tag{1.86}$$

EXAMPLE 1.30. Estimate the molal critical volume, V_c, of n-heptane.

Solution

We find the sum of the contributions ϕ_b (cf. section 5a, Table 1.34).

Methane, base value		17·371
First $-CH_3$ substitution		6·497
Five 1 (2) substitutions	$5\times6{\cdot}686 =$	33·430
	$\Sigma\phi_b =$	57·298

After substituting $\Sigma\phi_b$ into Eq. (1.66) we obtain

$57{\cdot}298 = b^{0{\cdot}76}$

hence,

$b = 205{\cdot}8$ cm³/mole

From Eq. (1.84) we get

$\beta_{norm} = 0{\cdot}7849 - 0{\cdot}01337\times7 = 0{\cdot}6913$

where $7 = N_c$, denotes the number of carbon atoms in the molecule

$V_{c(calc)} = 3\times0{\cdot}6913\times205{\cdot}8 = 426{\cdot}8$ cm³/mole

V_c found experimentally is 426 cm³/mole. The error of estimation is about 0·2%.

According to various authors[247, 306, 309] the average error of calculating the critical volume by different methods is:

Meissner's method	Eq. (1.73)	about 3·4%
Lydersen's method	Table 1.44	about 2·1%
Riedel's method	Table 1.45	below 2%
Thodos's method	Eq. (1.83)	below 1·3%

f. Additive Estimation of the Critical Temperature, T_c

Methods for the additive calculation of the critical temperature are usually indirect in that they are based on the additive calculation in the first instance of some other quantities, to which T_c is related in a definite way. For example, in Meissner's equation (1.75), T_c is calculated from a knowledge of the values of two additive quantities: the molar refraction, R_D, and the parachor, P_{ch}.

According to van der Waals, we have the relationship

$$T_c = \frac{8a}{27Rb} \tag{1.87}$$

where T_c is the critical temperature, °K,
R is the gas constant equal to 82·06 cm³.atm/mole.°K,
a is the van der Waals' constant, atm.(cm³/mole)²,
and b is van der Waals' constant, cm³/mole.

Good results are obtained when the critical temperature is calculated from the values of constants a and b in van der Waals' equation, these constants being calculated by the Thodos or Forman and Thodos method (section 5a).

EXAMPLE 1.31. Estimate the critical temperature, T_c, for 3-methylhexane.

Solution

The constants a and b can be estimated, e.g. using the Thodos method (section 5a, Table 1.34), by summing the contributions to $\Sigma\phi_a$ and $\Sigma\phi_b$:

	ϕ_a	ϕ_b
Methane (base value)	9509	17·371
First $-CH_3$ substitution	7065	6·497
Four 1 (2) substitutions	$4\times6417 =$ 25,668	$4\times6{\cdot}686 =$ 26·744
One 2 (2,2) substitution	4865	4·663
	$\Sigma\phi_a =$ 47,107	$\Sigma\phi_b =$ 55·275

$$\Sigma\phi_a = 47{,}107 = a^{0\cdot626}, \qquad a = 29\cdot18\times10^6\left(\frac{\text{cm}^3}{\text{mole}}\right)^2.\text{atm}$$

$$\Sigma\phi_b = 55\cdot275 = b^{0\cdot76}, \qquad b = 196\cdot25\ \text{cm}^3/\text{mole}$$

Hence, from Eq. (1.87), the critical temperature is

$$T_c = \frac{8\times29\cdot18\times10^6}{27\times82\cdot06\times196\cdot25} = 536°\text{K}$$

The critical temperature, T_c, determined experimentally is 535·6°K. The error of estimation is less than 0·1%.

When calculating T_c from Eq. (1.65), where a and b are constants calculated according to the method of Forman and Thodos (section 5a), these authors found the following average error:

Saturated aliphatic hydrocarbons	0·61%
Unsaturated aliphatic hydrocarbons	0·53%
Naphthenic hydrocarbons	0·11%
Aromatic hydrocarbons	0·43%
Organic compounds other than hydrocarbons	1·36%

As can be seen from this comparison, Thodos' method is distinguished for its high accuracy.

Further possibilities of determining the value of the critical temperature by additive calculations making use of the relationships existing between T_c, V_c and p_c exist. Both V_c and p_c can be calculated by additive methods, and it is, therefore, also possible to find the value of T_c.

Special mention should be made of a method for calculating T_c which has been proposed by Riedel.[309] This method involves the use of tabulated contributions to the function $\psi = (T_c/p_c)^{3/4}$ (Table 1.46). p_c is obtained using the methods of Riedel (Eq. (1.79)) or Lydersen (Eq. (1.80)) and T_c is then calculated from the equation

$$T_c = \psi^{4/3} p_c \tag{1.88}$$

TABLE 1.46. CONTRIBUTIONS TO THE FUNCTION $\psi = (T_c/p_c)^{3/4}$

Atoms, bonds	Contributions	Atoms, bonds	Contributions
Basic constant *A*	1·43	Ring atom	−0·13
H	0·20	Single branching	−0·15
C	0·69		
F	0·58	Double branching	−0·40
Cl	1·22		
O (ethers)	0·62	Double bond C=C	+0·10
N (amines)	0·64		
S (sulphides)	1·22	—COO— (esters)	1·96

EXAMPLE 1.32. Estimate the critical temperature, T_c, of 1,3-butadiene, C_4H_6 (molecular weight $M = 54\cdot09$)

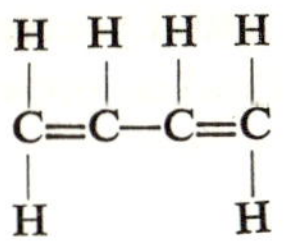

using Riedel's method.

Solution

Summation of the contributions to the functions ψ and φ

	ψ (Table 1.46)	φ (Table 1.41)
Basic constant	1·43	33
Four carbon atoms	$4\times0{\cdot}69 = 2{\cdot}76$	$4\times23 = 92$
Six hydrogen atoms	$6\times0{\cdot}20 = 1{\cdot}20$	$6\times0 = 0$
Two double bonds	$2\times0{\cdot}10 = 0{\cdot}20$	$2\times(-6) = -12$
	$\psi = 5{\cdot}59$	$\varphi = 113$

Calculation of p_c

$$p_c = \frac{10^4\times54{\cdot}09}{113^2} = 42{\cdot}4 \text{ atm}$$

The value of p_c found experimentally is 42·7 atm.

Calculation of T_c

$$T_c = 5{\cdot}59^{4/3}\times42{\cdot}4 = 422°\text{K}$$

The experimental value of T_c is 425·2°K.

The error of estimation is $-0{\cdot}8\%$.

g. Additive Estimation of the Critical Parameters from Experimentally Determined Quantities such as Liquid Density, Boiling Temperature, etc.

The methods discussed above enable us to calculate the values of T_c, p_c and V_c, when only the structural formula of a compound is known. If the values of certain physical constants of a compound (e.g. density, compressibility factor, boiling temperature, vapour pressure, etc.) are known, more accurate predictions of the critical parameters may be made.

Estimation of the Critical Temperature from the Normal Boiling Temperature, T_b

Guldberg (1890) noticed that for the majority of liquids the ratio of the normal boiling temperature of a compound to its critical temperature is approximately constant

$$\theta = \frac{T_b}{T_c} = \frac{2}{3} \tag{1.89}$$

More recent investigations have shown, however, that exceptions exist to Guldberg's rule. Lutskii[232] has shown that the quantity θ is dependent on the form of the molecule as well as on its polarizability and dipole moment. It is found that compounds with molecules of similar shape and with similar dipole moments, obey the approximate relationship

$$\theta = \text{const } R_D^x \tag{1.90}$$

where R_D is the molar refraction

and x is an exponent which is approximately constant for a definite series of compounds.

For normal hydrocarbons $x = 0{\cdot}1$. Since the molar refraction, R_D, is an additive quantity (section 4b), θ can also be calculated by summation of contributions.

Riedel, Vowles, Thomas, Eduljee, and Lydersen have compiled tables of contributions. Riedel as well as Lydersen and Eduljee presented the contributions of atoms and atomic groups.

Riedel's equation[309] is:

$$\theta = 0{\cdot}574+\Sigma\Delta\theta \tag{1.91}$$

Lydersen's equation[306] is:

$$\theta = 0{\cdot}567 + \Sigma\Delta\theta - (\Sigma\Delta\theta)^2 \qquad (1.92)$$

and Eduljee's equation[116] is:

$$\theta = \Sigma\Delta\theta/100 \qquad (1.93)$$

In these equations $\Sigma\Delta\theta$ denotes the sum of the contributions given by Table 1.47. Riedel's method is unsuitable for calculating the values of θ for compounds containing iodine, S= or C≡.

Gambill[116] tabulated the contributions for use in Eduljee's equation (1.93) which are listed in Table 1.48.

The calculation of the critical temperature by various methods is illustrated in the following example.

EXAMPLE 1.33. The boiling temperature of ethyl propyl ether, $C_5H_{12}O$, with the structural formula

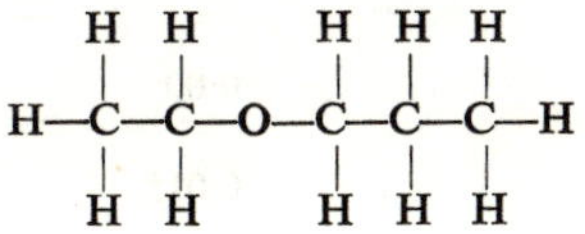

is $T_b = 336{\cdot}8°K$. Estimate the critical temperature.

Solution

a. Riedel's method

Calculation of $\theta = T_b/T_c$

Basic value		$= 0{\cdot}574$
Five $—CH_3$ and $—CH_2$ contributions	$5 \times 0{\cdot}016$	$= 0{\cdot}080$
One —O— contribution		$= 0{\cdot}020$
		$\theta = 0{\cdot}674$

$$T_c = \frac{T_b}{\theta} = \frac{336{\cdot}8}{0{\cdot}674} = 500{\cdot}0°K$$

b. Lydersen's method

Calculation of $\theta = T_b/T_c$

Basic value		$= 0{\cdot}567$
Five $—CH_3$ and $—CH_2$ contributions	$5 \times 0{\cdot}020$	$= 0{\cdot}100$
One —O— contribution		$= 0{\cdot}021$
		$0{\cdot}686$
$(0{\cdot}100 + 0{\cdot}021)^2 = 0{\cdot}121^2 \approx 0{\cdot}014$		$= -0{\cdot}014$
		$\theta = 0{\cdot}672$

$$T_c = \frac{T_b}{\theta} = \frac{336{\cdot}8}{0{\cdot}672} = 501{\cdot}2°K$$

c. Eduljee's method

Calculation of $\theta = T_b/T_c$

Five carbon contributions	$5 \times (-55{\cdot}32)$	$= -276{\cdot}60$
12 hydrogen contributions	$12 \times 28{\cdot}52$	$= 342{\cdot}24$
One oxygen contribution (in ethers)		$= 1{\cdot}59$
		$\theta = 67{\cdot}23$

$$T_c = \frac{T_b}{\theta} = \frac{336{\cdot}8 \times 100}{67{\cdot}23} = 500{\cdot}9°K$$

The critical temperature determined experimentally is $T_c = 500{\cdot}6°K$.

The errors of estimation are:

in Riedel's method	$-0{\cdot}1\%$
in Lydersen's method	$+0{\cdot}1\%$
in Eduljee's method	$+0{\cdot}06\%$

TABLE 1.47. ATOMIC AND STRUCTURAL CONTRIBUTIONS ($\Delta\theta$) FOR CALCULATING $\theta = T_b/T_c$

Group*	Riedel[309]	Lydersen[306]
$—CH_3$, $—CH_2—$	0·016	0·020
$—CH_2—$ (in ring)	—	0·013
—CH< (chain branching)	0·013	0·012
—CH< (in ring)	—	0·012
>C<	0·003	0·00
H	0·00	0·00
=CH—, $=CH_2$	0·015	0·018
=CH— (in ring)	—	0·011
=C<, =C=	0·003	0·00
=C<, =C= (in ring)	—	0·011
≡C—, ≡CH	—	0·005
Ring atom	−0·005	—
—F	0·015	0·018
—Cl	0·013	0·017
—Br	0·010	0·010
—I	—	(0·012)
—O—	0·020	0·021
—O— (in ring)	—	(0·014)
—OH (alcohols)	0·070	0·082
—OH (phenols)	0·029	(0·035)
>CO	0·046	0·040
>CO (in ring)	—	(0·033)
—CHO	0·046	0·048
—COO—	0·039	0·047
—COOH	0·070	0·085
$—NH_2$, >NH	0·027	0·031
>NH (in ring)	—	(0·024)
>N—	0·012	0·014

TABLE 1.47 (*cont.*)

Group*	Riedel[309]	Lydersen[306]
>N— (in ring)	—	0·007
—CN	0·053	(0·060)
—SH, —S—	0·012	0·015
—S— (in ring)	—	0·008
=S	—	(0·003)
$—NO_2$	—	(0·055)

* The dashes stand for bonds with atoms other than hydrogen.

TABLE 1.48. CONTRIBUTIONS ($\Delta\theta$) FOR CALCULATING $\theta = T_b/T_c$ (AFTER EDULJEE[116])

Atoms, groups	Contributions	Remarks
C	−55·32	
H	28·52	
O	1·59	in ethers
N	30·6	in amines
N	−26·29	ring atom
Cl	29·89	
Br	31·15	
F	29·75	
S	1·31	
Si	−54·00	in silanes
—OH	31·63	in phenols
	35·62	in alcohols $n \leqslant 3$
	34·0	$n = 4$
	32·72	$n = 5$
	31·40	$n = 6$
	30·1	$n = 7$
	29	$n = 8$
	28·52	$n > 8$
C=O	31·63	in ketones
—COOH	35·94	
—COO—	4·12	in esters
—C≡N	33·83	
(C)=(C) double bond	56·61	up to three C atoms per double bond
double bond	55·21	four and more C atoms per double bond
(C)≡(C) triple bond	112·9	
(N)—(C) ,,	−19·17	in amines
(N)—(H) ,,	−18·37	in amines
Five-membered ring	54·28	
Six-membered ring	53·52	
Condensation of two rings	0·25	
One bond or substitution at the second carbon atom	−0·34	The branching position or substitution is reckoned according to the conventional numeration of the carbon atoms from the end of the chain
Two branchings or substitutions at the second carbon atom	−1·42	
Branching or substitutions at the third or further carbon atom	−0·96	

In the majority of estimations the average error for Riedel's method is 0·9%, and for Lydersen's method, 0·7%.[306] Eduljee's method yields more accurate results. The average error for various groups of compounds is as follows: normal paraffins 0·75%, branched hydrocarbons 0·27%, normal olefins 0·56%, naphthenes 0·67%, aromatic hydrocarbons 0·80%, alcohols 0·73%, phenols 0·40%, ketones and aldehydes 0·88%, organic acids 1·55%, esters 1·44%, amines 0·69%, chlorides 0·54%, sulphur compounds 1·07%, and heterocyclic compounds 0·98%.

Estimation of the Critical Temperature from the Liquid Densities ρ_1 and ρ_2 at two Temperatures T_1 and T_2

By transformation of Eötvös' and Sugden's equations, Smith, Greenbaum and Rutledge[336] obtained the relationship

$$T_c = \frac{T_2 - T_1}{(\rho^1/\rho^2)^{10/3} - 1} + T_2 + 6 \tag{1.94}$$

A result close to the actual value of T_c is obtained provided that the difference $(T_2 - T_1) \geqslant 20°C$ and that

$$\frac{(T_2 + T_1)}{2} \approx \frac{3}{4} T_c$$

The average error of estimating the value of T_c by this method is about 3%.

EXAMPLE 1.34. Estimate the critical temperature, T_c, for 2-methylbutane.

$\rho_1 = 0{\cdot}63945$ g/cm³ at $T_1 = 273{\cdot}2°K$

$\rho_2 = 0{\cdot}62470$ g/cm³ at $T_2 = 288{\cdot}2°K$

Solution

Substituting the given values into Eq. (1.94), we obtain

$$T_c = \frac{288{\cdot}2 - 273{\cdot}2}{(0{\cdot}63945/0{\cdot}62470)^{10/3} - 1} + 288{\cdot}2 + 6 = 465°K$$

The experimentally determined value for T_c is 461·0°K. The computational error, amounting to +0·9%, is rather small, in spite of the fact that the conditions mentioned above are not fulfilled.

Since it is rather difficult to represent mathematically the relationship between the density and the critical temperature over a wide range of temperatures, Altenburg[2] has developed a graphical method for determining T_c, when the liquid densities ρ_1 at temperature T_1 and ρ_2 at temperature T_2 are available. His method also enables the liquid density to be determined at other temperatures.

This method is based on the application of the equation

$$\rho = \rho_0 \phi(T_r) \tag{1.95}$$

and assumes that the liquid density depends only on the reduced temperature, i.e. that all liquids have the same $\phi(T_r)$ curves. In the case of systems at moderate pressures, this assumption holds for a very large group of compounds.

For the determination of the critical temperature, T_c, the plots presented in Figs. 1.9, 1.10, 1.11 can be used. The calculations should be performed in the following order:

(a) The values of two parameters are calculated

$$P_1 = \rho_1/\rho_2 \quad \text{and} \quad P_2' = T_2/T_1 \tag{1.96}$$

(b) When $P_2' < 1{\cdot}1$, the values of $x = T_2/T_c$ are taken from the plot in Fig. 1.9, and when $P_2' > 1{\cdot}1$ from that in Fig. 1.10.

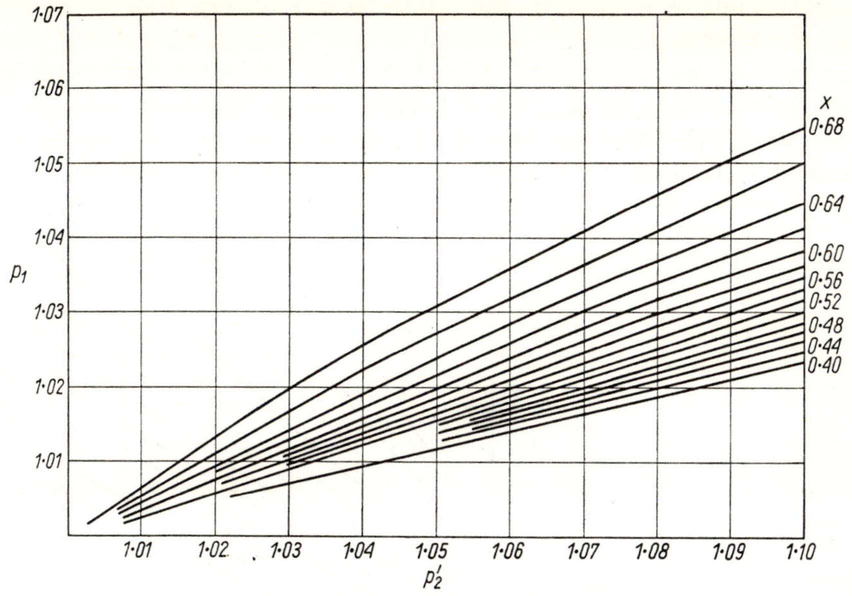

FIG. 1.9. Graph for calculating the critical temperature (T_c) from the liquid density. Plot of x vs. P_1 and P_2' for $P_2' < 1{\cdot}1$ (after Altenburg[2]).

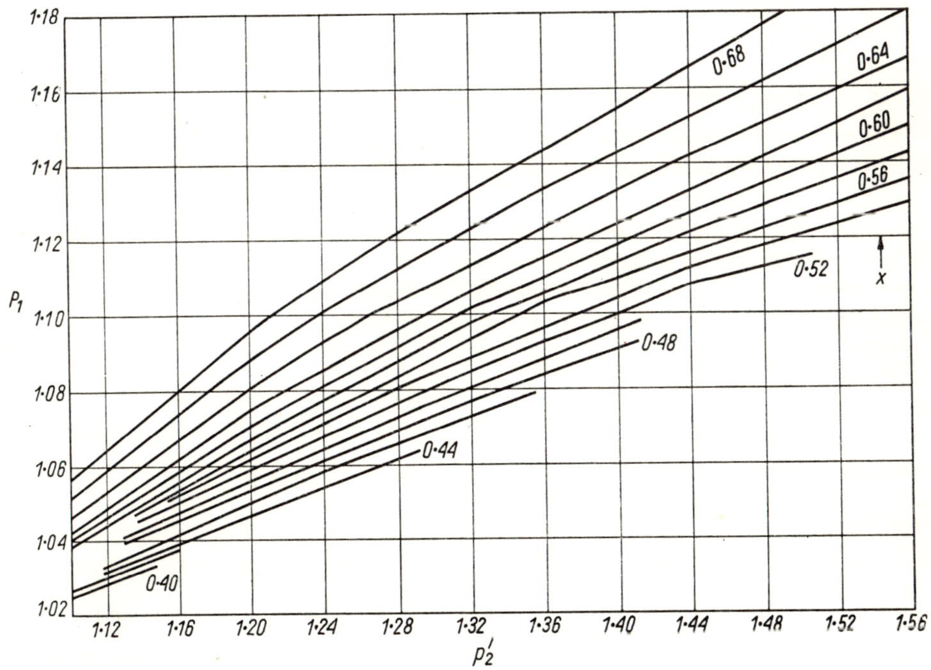

FIG. 1.10. Graph for calculating the critical temperature (T_c) from the liquid density. Plot of x vs. P_1 and P_2' for $P_2' > 1{\cdot}1$ (after Altenburg[2]).

(c) The value of $y = \rho/\rho_0$ is taken from the plot in Fig. 1.11.
(d) The value

$$T_c = T_2/x \tag{1.97}$$

is calculated. Once the value of $\rho_0 = \rho_1/y$ is known we can determine the liquid density at any temperature for which the value of y can be read from the plot. The average error of estimating T_c is about 1·5%, the maximum error does not exceed 5%.

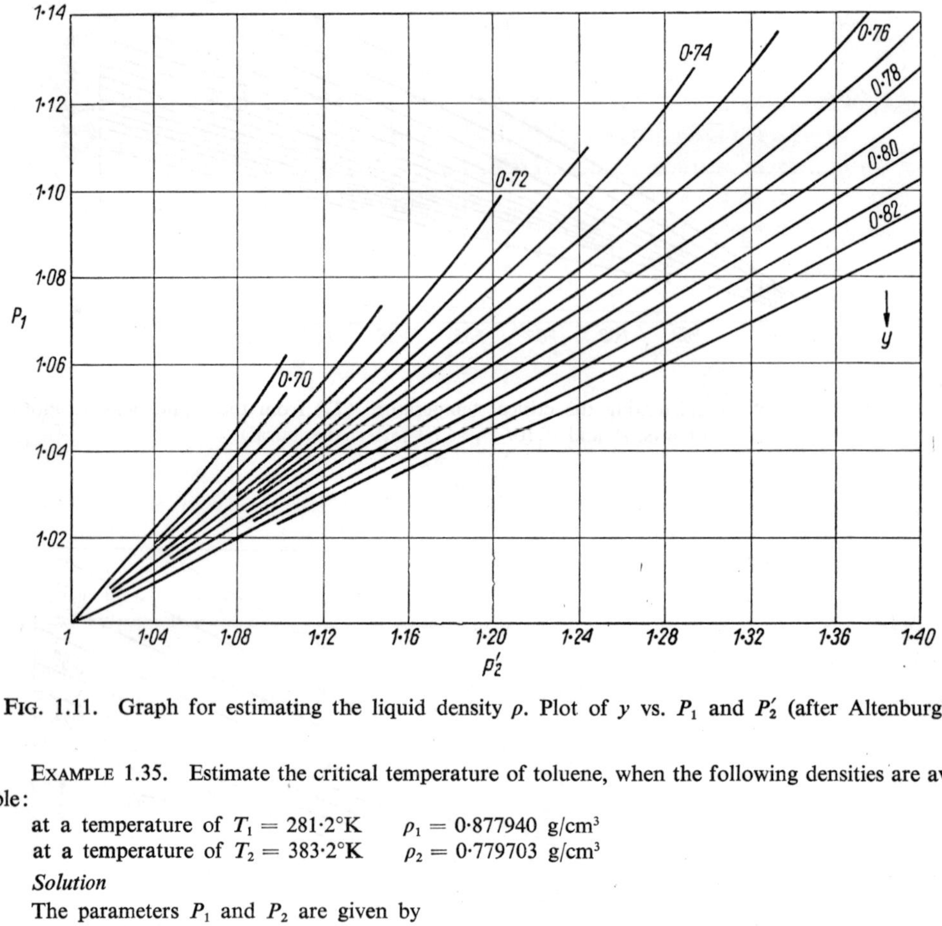

FIG. 1.11. Graph for estimating the liquid density ρ. Plot of y vs. P_1 and P_2' (after Altenburg[2]).

EXAMPLE 1.35. Estimate the critical temperature of toluene, when the following densities are available:

at a temperature of $T_1 = 281{\cdot}2$°K $\quad \rho_1 = 0{\cdot}877940$ g/cm³
at a temperature of $T_2 = 383{\cdot}2$°K $\quad \rho_2 = 0{\cdot}779703$ g/cm³

Solution

The parameters P_1 and P_2 are given by

$$P_1 = \frac{0{\cdot}877940}{0{\cdot}779703} = 1{\cdot}12599 \quad \text{and} \quad P_2' = \frac{383{\cdot}2}{281{\cdot}2} = 1{\cdot}36273$$

It is found from the graphs that
$x = 0{\cdot}642$ and $y = 0{\cdot}766$
whence the calculated values of T_c and ρ_0 are:

$$T_c = \frac{383{\cdot}2}{0{\cdot}642} = 596\text{°K}, \qquad \rho_0 = \frac{0{\cdot}87794}{0{\cdot}766} = 1{\cdot}145$$

The experimental value of T_c is 593·8°K and the error of the estimation is thus +0·37%.

Riedel[312] has proposed a method for predicting T_c, p_c and V_c, and the critical parameter α_c from a knowledge of the vapour pressure at two temperatures T_1 and T_2, and the liquid density ρ at one temperature T. The same author[312] has also given a method for calculating p_c from T_c and the vapour pressures at two temperatures T_1 and T_2. The calculations involved in these methods are, however, rather cumbersome.

Somewhat simpler and more rapid methods have been developed by Filippov:[101] these enable the critical constants to be predicted either from the boiling temperature T_b and the liquid density at two temperatures T_0 and T_1, or from the boiling temperature T_b, the vapour pressure p_1 at T_1 and the liquid density ρ_0 at T_0.

These methods have been discussed in detail and illustrated with examples in a book by the present author.[36]

Filippov[102] has also presented simplified equations for calculating the critical constants, which are based on the assumption, which he has checked in numerous cases, that the reduced density ρ_r of a saturated liquid is dependent only on the reduced temperature T_r and almost independent of the value of any "third parameter" for the substance (p. 106). (The maximum deviation found at $T_r = 0{\cdot}6$ is $\pm 2{\cdot}8\%$.) The assumption is also made that the critical compressibility factor is the same for all substances being equal to $3{\cdot}83^{-1}$. The author used the law of Cailletet and Mathias to extrapolate the density of the supercooled liquid to absolute zero and noticed that the quantity $\varkappa = \rho_0/\rho_c$ is approximately equal to the reciprocal of the critical compressibility factor and amounts to 3·95:

$$\frac{1}{z_c} = \frac{RT_c\rho_c}{Mp_c} \approx 3{\cdot}83, \qquad \frac{\rho_0}{\rho_c} \approx 3{\cdot}95 \tag{1.98}$$

(Filippov claims that the above equations are accurate to within $\pm 3\%$.)

Over the range of temperatures for which the vapour density is very small compared with the density of the liquid, the reduced density of the liquid is approximately a linear function of the reduced temperature being given by

$$\frac{\rho}{\rho_c} = 3{\cdot}95 - 1{\cdot}95\frac{T}{T_c} \tag{1.99}$$

On rearranging equations (1.98) and (1.99), the following equations are obtained for the critical constants:

$$\rho_c = 0{\cdot}253\left(\rho - T\frac{d\rho}{dT}\right) \tag{1.100}$$

$$T_c = -1{\cdot}95\frac{\rho_c}{d\rho/dT} \tag{1.101}$$

$$p_c = \frac{RT_c\rho_c}{3{\cdot}83M} \tag{1.102}$$

where ρ is the liquid density at the temperature T, °K, and $\dfrac{d\rho}{dT} \approx \dfrac{\rho_1 - \rho_2}{t_1 - t_2}$

ρ_1 and ρ_2 are the liquid densities at temperatures t_1 and t_2, respectively.

Filippov has tested his method for numerous substances and finds that the errors in the calculated values of ρ_c do not normally exceed 2%, whereas the error of estimating T_c is lower than 3%.

EXAMPLE 1.36.[102] Estimate the critical constants of n-heptane. Its densities are:

$\rho_{0°C} = 0{\cdot}7005$ g/cm^3, $\rho_{20°C} = 0{\cdot}6836$ g/cm^3

The molecular weight $M = 100{\cdot}2$.

Solution

$$\frac{d\rho}{dT} \approx \frac{0{\cdot}7005 - 0{\cdot}6836}{0-20} = -0{\cdot}000845$$

$$\rho_c = 0{\cdot}253(0{\cdot}7005+273{\cdot}2\times0{\cdot}000845) = 0{\cdot}235 \text{ g/cm}^3$$

$$T_c = -1{\cdot}95\times\frac{0{\cdot}235}{-0{\cdot}000845} = 543°\text{K}$$

$$p_c = \frac{82{\cdot}06\times543\times0{\cdot}235}{100{\cdot}2\times3{\cdot}83} = 27{\cdot}3 \text{ atm}$$

It has been found experimentally that:

$\rho_c = 0{\cdot}234$ g/cm³, $T_c = 540°$K, $p_c = 26{\cdot}8$ atm

In addition to the methods described above, numerous other ways of estimating the critical constants have been proposed. However, they are not described here since they are either less accurate than the above methods or else they are very complicated to use.

Attention should be drawn, however, to the works of Herzog,[149] Jatkar and Laksminarayanan,[282] Vowles,[306] Riedel,[312] and Nokay,[259] in which methods for estimating the critical constants from the structural formula of a compound are discussed.

6. GENERAL METHOD FOR CALCULATING ADDITIVE QUANTITIES

So far only the more important examples of the additive calculation of the constitutive properties of substances have been considered. However, this is one of the most versatile means of determining the values of physico-chemical constants and may be used in many other ways.

In section 4a the factors which complicate the additive calculation of the physico-chemical properties were discussed. It was found that, the contribution, a_i, to be ascribed to one element of the molecule (atom, atomic group or bond), generally depends not only on the kind of the element, but also on its environment in the molecule (e.g. whether it is situated at the end or in the middle of the chain or in a ring, etc.). In general, we can say that a certain additive property, y, of a compound depends on the sum Σa_i of the group contributions; to each contribution, however, a correction p_i should be added (or subtracted), the magnitude of which depends on the position of the element i in the molecule (constitutive effects).

We can write quite generally

$$y = \Sigma(a_i+p_i) \tag{1.103}$$

or

$$y = \Sigma a_i+\Sigma p_i \tag{1.104}$$

Usually the tables of contributions utilized in additive calculations are incomplete, since a prediction of all the possible arrangements of the element in the molecules and the introduction of the corresponding corrections p_i is practically impossible.

A general procedure exists, which enables values of a constitutive and additive physico-chemical quantity, y, to be predicted with fair accuracy even in cases where only a few additive contributions (a_i+p_i) of the individual elements of the molecule are known. This general method requires a knowledge of the values y_A, y_B, y_C ... of the physico-chemical property under consideration for simple compounds A, B, C, ... from which, theoretically, the compound R with the required property, y_R, can be formed.

Let us assume that the compound R could be formally synthesized in a chemical reaction from the chemical compounds A, B, C, according to the equation

$$A+B+C\ldots = R+S+T \tag{1.105}$$

It is not essential that such a reaction can actually occur. The reaction may be chosen so that the majority of the bonds and the configurations of the atoms in the molecule of the product R remain unchanged. Of course, the number of atoms participating in the reaction does not change, either. The reaction is usually chosen so that several comparatively simple bonds for which the values of the contributions are well known, form or are split in the reaction. To summarize:

The reaction constituting the basis for the calculations should be selected in such a way as to ensure that the main configurations of the atoms in the resulting compound R be already contained in the molecules of the reactants A, B, C ... The reactants and the minor products (S, T) should be simple and well-known with known values of the desired physico-chemical property, y, viz.

$$\begin{aligned} y_A &= \Sigma a_A + \Sigma p_A \\ y_B &= \Sigma a_B + \Sigma p_B \\ y_C &= \Sigma a_C + \Sigma p_C \\ y_S &= \Sigma a_S + \Sigma p_S \\ y_T &= \Sigma a_T + \Sigma p_T \end{aligned} \tag{1.106}$$

If the property y is a constitutive additive property, the required value, y_R, can be calculated from the following summation:

$$y_R = y_A + y_B + y_C + \dots - y_S - y_T + \Sigma_b + \Sigma_c \tag{1.107}$$

where Σ_b is the algebraic sum of the contributions of the bonds formed and Σ_c is the sum of the constitutive corrections to the values of Σ_b. Thus Σ_c is a correction term analogous to p which allows for the influence of adjacent groups and bonds on the contributions attributed to each bond formed or broken in the reaction (i.e. each bond which contributes to Σ_b).

In practice the value of Σ_c (which is often unknown) can usually be disregarded as it is very small in comparison with y_R.

The advantage of this procedure is that it eliminates the necessity of taking into account the considerable number of often unknown constitutive contributions, which are equal to the sum

$$p_R = p_A + p_B + p_C + \dots - p_S - p_T \tag{1.108}$$

because these corrections are already contained in the values of $y_A, y_B, y_C \dots y_S, y_T$, which have been used in the calculation.

Examples of the application of the above General Method to the Calculation of Molal Volumes of Organic Compounds. Several methods for calculating additively the molal volume of organic compounds in comparable conditions, e.g. at the normal boiling temperature or at the critical point, are known. Often, however, volumes at other temperatures and pressures are required for technological calculations. Volumes under these more general conditions may be predicted using equation (1.107) in which y now denotes the molal volume. Following Kurtz and Lipkin, the quantity Σ_b (the sum of the contributions arising from the formation and splitting of bonds) is usually calculated from equation (1.44)[97, 214]

EXAMPLE 1.37. Estimate the molal volume of diethyl acetyl succinate at 20°C and atmospheric pressure.

The following reaction may be assumed as the basis of the calculation:

$$3CH_3COOH+2C_2H_5CH = CH_3\text{—}CO\text{—}\underset{\displaystyle |\atop \displaystyle CH_2\text{—}COOC_2H_5}{CH}\text{—}COOC_2H_5+3H_2O+H_2$$

Solution

Available molal volumes (at 20°C)

Acetic acid CH_3COOH	57·3 cm^3/mole
Ethanol C_2H_5OH	58·5 cm^3/mole
Water H_2O	18·0 cm^3/mole

The contribution due to splitting off two terminal hydrogen atoms at the ends of the chains is given by the constant C in equation (1.44) to be 32·6

$$V_{calc} = 3\times 57{\cdot}3+2\times 58{\cdot}5-3\times 18{\cdot}0-32{\cdot}6 = 202{\cdot}3 \text{ cm}^3\text{/mole}$$

The molal volume determined experimentally is V_{exp} = 199·4 cm^3/mole

Error of estimation: 1·45%

EXAMPLE 1.38. Estimate the molal volume of hydroacrylic nitrile in two ways.

a. The following reaction will be assumed as the basis for the first calculation:

$$HO\text{—}CH_2\text{—}CH_2\text{—}OH+HCN = HO\text{—}CH_2\text{—}CH_2\text{—}CN+H_2O$$

Solution

Available molal volumes:

Ethylene glycol $HO\text{—}CH_2\text{—}CH_2\text{—}OH$	55·8 cm^3/mole
Hydrogen cyanide HCN	33·8 cm^3/mole
Water H_2O	18·0 cm^3/mole

$$V_{calc} = 55{\cdot}8+33{\cdot}8-18{\cdot}0 = 71{\cdot}6 \text{ cm}^3\text{/mole}$$

The molal volume determined experimentally is V_{exp} = 67·2 cm^3/mole

Error of estimation: +6·5%.

b. The reaction

$$CH_3OH + C_6H_5\text{—}CH_2\text{—}CN = HO\text{—}CH_2\text{—}CH_2\text{—}CN + C_6H_6$$

will be assumed as the basis for the second calculation.

Solution

Available molal volumes:

Methanol CH_3OH	40·4 cm^3/mole
Benzyl cyanide $C_6H_5\text{—}CH_2\text{—}CN$	115·4 cm^3/mole
Benzene C_6H_6	89·0 cm^3/mole

$$V_{calc} = 40{\cdot}4+115{\cdot}4-89{\cdot}0 = 66{\cdot}8 \text{ cm}^3\text{/mole}$$

Error of estimation: −0·6%.

EXAMPLE 1.39. Estimate the molal volume of diphenyldichloromethane in two ways.

a. The following reaction will be assumed as the basis for the first calculation:

$$2C_6H_5\text{—}Cl + C_6H_{14} = (C_6H_5)_2CCl_2 + C_5H_{12} + H_2$$

Solution

Available molal volumes:

Chlorobenzene C_6H_5Cl	102·0 cm^3/mole
n-Hexane C_6H_{14}	130·7 cm^3/mole
n-Pentane C_5H_{12}	−115·2 cm^3/mole
Splitting off of two hydrogen atoms, by eq. 1.44:	−31·2 cm^3/mole

$$V_{calc} = 2\times 102{\cdot}0+130{\cdot}7-115{\cdot}2-31{\cdot}2 = 188{\cdot}3 \text{ cm}^3\text{/mole.}$$

The molal volume determined experimentally is V_{exp} = 192·0 cm^3/mole.

Error of estimation: −1·9%.

b. The reaction

$$(C_6H_5)_2S + CCl_4 = (C_6H_5)_2CCl_2 + SCl_2$$

will be assumed as the basis of the second calculation.

Solution

Available molal volumes:

Diphenyl sulphide $(C_6H_5)_2S$	166·6 cm^3/mole
Carbon tetrachloride CCl_4	96·4 cm^3/mole
Sulphur dichloride SCl_2	63·5 cm^3/mole

$V_{calc} = 166{\cdot}6 + 96{\cdot}4 - 63{\cdot}5 = 199{\cdot}5\ cm^3/mole$

Error of estimation: $+3{\cdot}9\%$.

EXAMPLE 1.40. Estimate the molal volume of tropane,

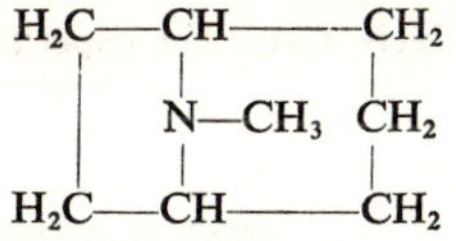

The molal volume determined experimentally is $V_{exp} = 135{\cdot}0\ cm^3/mole$.

The corrections due to the formation and disappearance of bonds are determined from the contributions listed by Kurtz and Lipkin (section 4a, Eq. 1.44). In this equation

N_1 is the number of CH_3, CH_2 and CH groups in open chains; the contribution of each of these groups is $a_1 = 16{\cdot}28$.

N_2 is the number of CH_2 and CH groups in rings (but not at ring junctions); the contribution of these groups is $a_2 = 13{\cdot}15$.

N_3 is the number of CH groups at ring junctions; the contribution of this group is $a_3 = 9{\cdot}7$.

N_4 is the number of double bonds in the molecule; the bond contribution is $a_4 = 6{\cdot}2$.

a. In the first instance the following reaction will be assumed as the basis for the calculation:

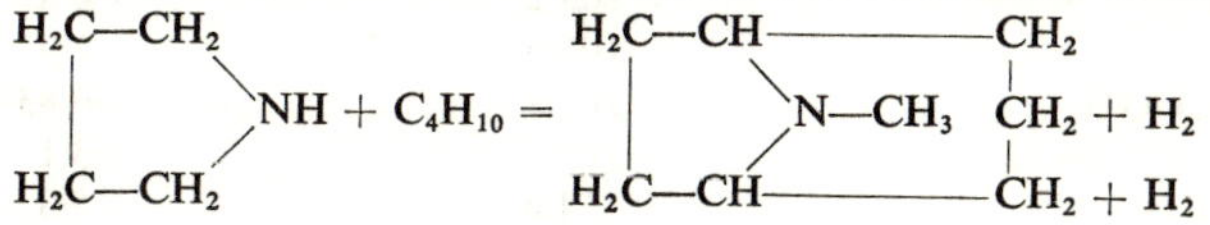

Solution

Splitting off of two terminal hydrogen atoms from the butane: 32·6

Three CH_2 groups pass from the open chain to the ring

$$3(a_1 - a_2) = 3(16{\cdot}28 - 13{\cdot}15) = 3 \times 3{\cdot}13 = 9{\cdot}39 \approx 9{\cdot}4$$

Splitting off of two hydrogen atoms from the two CH_2 groups in the pyrrolidine ring so that they change to CH groups at the ring junctions

$$2(a_2 - a_3) = 2(13{\cdot}15 - 9{\cdot}7) = 2 \times 3{\cdot}45 = 6{\cdot}9$$

Available molal volumes

Pyrrolidine 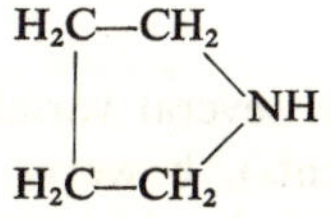 83·4

Butane C_4H_{10} (by extrapolating) 100

$$V_{calc} = 83{\cdot}4 + 100 - 32{\cdot}6 - 9{\cdot}4 - 6{\cdot}9 = 134{\cdot}5\ cm^3/mole$$

Error of estimation: $0{\cdot}37\%$.

b. A second estimate of the molar volume may be obtained on the basis of the following reaction:

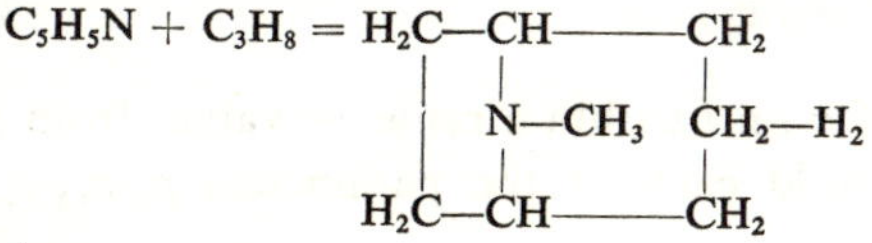

Solution

Changes of bonds:

Three double bonds of the pyridine ring are transformed to single bonds:

$$3a_4 = 3 \times 6{\cdot}2 = 18{\cdot}6$$

Two CH_2 groups pass from the open chain into the ring:

$$2(a_1 - a_2) = 2(16{\cdot}28 - 13{\cdot}15) = 2 \times 3{\cdot}13 = 6{\cdot}26 \approx 6{\cdot}3$$

Two hydrogen atoms are removed from the ends of the chain 32·6

Two CH groups change their position in the pyridine ring to CH groups at the ring junctions:

$$2(a_2 - a_3) = 2(13{\cdot}15 - 9{\cdot}7) = 2 \times 3{\cdot}45 = 6{\cdot}9$$

Available molal volumes

Pyridine 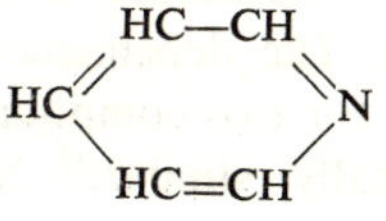 81·0

Propane C_3H_8 (by extrapolating) 82

$V_{calc} = 81{\cdot}0+82+18{\cdot}6-32{\cdot}6-6{\cdot}3-6{\cdot}9 = 135{\cdot}8$ cm³/mole.

Error of estimation: $+0{\cdot}59\%$.

EXAMPLE 1.41. Estimate the molal volume of methylaniline, C_6H_5—NH—CH_3. The experimentally determined molal volume $V_{exp} = 108{\cdot}8$ cm³/mole.

Reactants taken as basis for the calculation	V_{calc}, $\frac{cm^3}{mole}$	Error of estimation %
$C_6H_6 + CH_3NH_2 - H_2$	105·6	−2·9
$C_6H_5CH_3 + CH_3NH_2 - CH_4$	106·1	−2·5
$C_6H_5NH_2 + CH_4 - H_2$	107·4	−1·3
$C_6H_5OH + CH_3NH_2 - H_2O$	120·6	+10·8
$C_6H_5OH + C_2H_5NH_2 - CH_3OH$	114·5	+5·3
$C_6H_5CH_3 + C_6H_5NH_2 - C_6H_6$	107·9	−0·83
$C_6H_5OCH_3 + (CH_3)_2NH - CH_3OCH_3$	101·9	−6·2
$C_6H_5NH_2 + C_2H_5OH - CH_3OH$	109·2	+0·37
$C_6H_6 + C_2H_5NH_2 - CH_4$	105·6	−2·9
$C_6H_5NH_2 + C_2H_5COOH - CH_3COOH$	107·9	−0·83
$C_6H_6 + (CH_3)_2NH - CH_4$	105·7	−3·1
$C_6H_5OH + (CH_3)_2NH - CH_3OH$	114·6	+5·3
$C_6H_5CH_3 + (CH_3)_2NH - C_2H_6$	106·2	−2·4

7. SIMILARITY OF THE PHYSICO-CHEMICAL PROPERTIES OF SUBSTANCES

a. Conditions of Similarity

Usually the quantity y we are interested in depends on several variable parameters (e.g. on pressure, temperature, concentration of components), however, when all the parameters except one are maintained constant, the form of the dependence of the required quantity on the varying parameter can be readily established.

Suppose that

$$y = f(p, t, c_1, c_2, \ldots) \tag{1.109}$$

and that we wish to study the dependence of y on t within a range of values from t_0 to t.

For this purpose, it is convenient to hold each of the parameters $p, c_1, c_2, c_3 \ldots$ constant so that

$$y = f(t) \tag{1.110}$$

y may now be plotted as a function of t on rectangular coordinates for each substance considered, as in Fig. 1.12 where the variation of y with t for two different compounds is compared. The concept of similarity is most simply expressed in terms of these graphs and for this purpose it is convenient to distinguish between the y versus t curves for the various substances considered by using primes. Thus when considering substance 1 we use a single prime and refer to the curve of y' versus t', for substance 2 we use a double prime and refer to the curve of y'' versus t'' and so on. The dependence of a property y on an independent parameter t is said to the "similar" for two compounds if the curves of y versus t for the two compounds are geometrically similar.[35] Such curves are characterized by two scale constants. To every point on curve *1* with co-ordinates (t', y') there corresponds a point on curve *2* with co-ordinates (t'', y'') such that

$$t'/t'' = C_t, \qquad y'/y'' = C_y \tag{1.111}$$

Complete coincidence of the curves $y' = f(t')$ and $y'' = f(t'')$ can be attained by multiplying each value of t'' by the scale constant C_t and each value of y'' by the scale constant C_y. When this is possible, we say that in both cases y varies with t according to the same law. This law is determined by the form of the function $f(t)$.

Certain pairs of corresponding points may readily be distinguished by eye. For example, in Fig. 1.12, the maxima of the curves are a pair of corresponding points: the origins of the co-ordinate systems are also corresponding points.

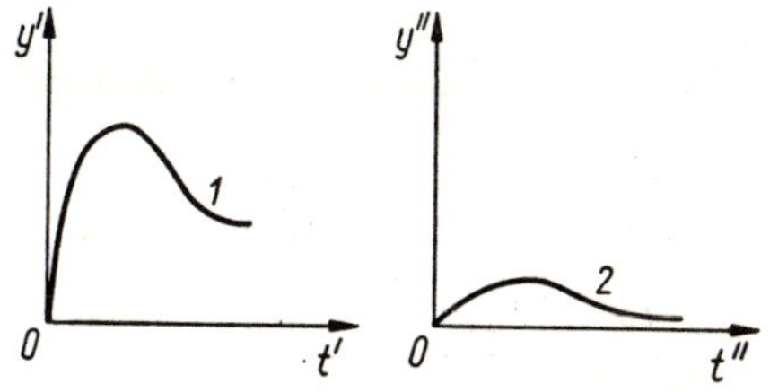

FIG. 1.12. Similarity of the variation of a physico-chemical property (y) with an independent parameter (t) (*1*, *2*—similar curves for two compounds 1 and 2).

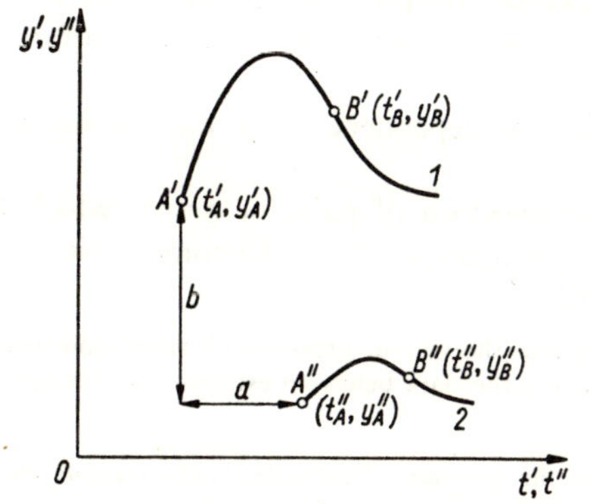

FIG. 1.13. General case of similarity of the variation of a physico-chemical property (y) with parameter (t) for two compounds 1 and 2 (the similar curves *1* and *2* are shifted with respect to each other).

In the general case the curves can be shifted with respect to each other or the origin, as in Fig. 1.13 where the point A'' on curve *2* corresponds to the point A' on curve *1*. The curves in Fig. 1.13 are as in Fig. 1.12 except that the origin of coordinates for curve *2* has been shifted by an amount a in t'' and an amount b in y''. The similarity of the curves *1* and *2* is characterized by the mathematical equations:

$$t'/(t''+a) = C_t, \qquad y'/(y''+b) = C_y \tag{1.112}$$

This is a more general case of similarity.

b. Application of the Principles of Similarity to the Determination of the Physico-chemical Properties of Substances

Experiments show that groups of compounds usually exist such that the compounds within each group show a similar dependence of a given physico-chemical property y on a suitable parameter t. The particular groups of compounds may differ, e.g., in the character of the bonds in the molecule, in molecular structure, in the degree of association of the molecules, etc.

For each pair of compounds (1, 2) belonging to the same group, two scale constants C_y and C_t can be determined. In order to use equations 1.112 it is necessary to evaluate not only these scale constants but also the constants a and b. These quantities may be determined if the coordinates (t, y) of two pairs of corresponding points on the curves *1* and *2* are known. Let us assume, for example, that these are A (coordinates t_A, y_A) and B (coordinates t_B, y_B).

According to Eq. (1.112) four equations can be derived

$$\frac{t'_A}{t''_A+a} = C_t, \qquad \frac{y'_A}{y''_A+b} = C_y$$
$$\frac{t'_B}{t''_B+a} = C_t, \qquad \frac{y'_B}{y''_B+b} = C_y \tag{1.113}$$

Substituting the given values of the coordinates of points (t'_A, y'_A), (t''_A, y''_A), (t'_B, y'_B), (t''_B, y''_B) and solving the system of Eqs. (1.113), it is possible to find the values of the scale constants C_t, C_y as well as the constants a and b.

When the dependence of the property y' of compound 1 on t' is exactly known, the property y'' of compound 2 can be calculated for any arbitrary value of t'' by means of Eqs. (1.112).

Compound 1, for which the dependence of y' on t' is known, is referred to as the reference compound.

If data are limited the selection of pairs of points which truly correspond may be difficult. In this case it may be necessary to make some arbitrary assumption about which pairs of points correspond.

EXAMPLE 1.42. Assuming that the vapour pressures of water and mercury both vary with temperature according to the same law, use the data below to estimate, by extrapolation, the vapour pressure of mercury at 950°F.

Dependence of the pressure of saturated water vapour on temperature

Temperature, °C	10	20	30	40	50	60	70	80	90	100
Vapour pressure, mm Hg	9·21	17·53	31·82	55·32	92·51	149·4	233·7	355·1	526	760

Dependence of the pressure of saturated mercury vapour on temperature

Temperature, °F	678	823
Vapour pressure, lb/sq.in.	15·3	54·8

Solution

We assume arbitrarily as the corresponding points, those points on the vapour pressure curves of water and mercury at which the vapour pressure of water expressed in mm Hg is equal to the vapour pressure of mercury expressed in lb/sq.in. It could just as well be assumed that the values for y' are n or $1/n$ times as large.

According to the tables of saturated water vapour pressure we find

Temperature, °C	17·81	39·82
p, mm Hg	15·3	54·8

Hence, we have

$$t'_A = 17{\cdot}81 \qquad t''_A = 678 \qquad t'_B = 39{\cdot}82 \qquad t''_B = 823$$
$$y'_A = 15{\cdot}3 \qquad y''_A = 15{\cdot}3 \qquad y'_B = 54{\cdot}8 \qquad y''_B = 54{\cdot}8$$

Substituting these values into Eqs. (1.113), we obtain

$$\frac{17{\cdot}81}{678+a} = C_t, \qquad \frac{15{\cdot}3}{15{\cdot}3+b} = C_y, \qquad b = 0, \qquad C_y = 1$$
$$\frac{39{\cdot}82}{823+a} = C_t, \qquad \frac{54{\cdot}8}{54{\cdot}8+b} = C_y, \qquad a = -560{\cdot}5, \qquad C_t = 0{\cdot}152$$

For a temperature of 950°F $= t''_c$, we get $\dfrac{t'_c}{950-560{\cdot}5} = 0{\cdot}152$,

whence

$$t'_c = 59{\cdot}2\text{°C}$$

The vapour pressure of water $y'_c = 143{\cdot}96$ corresponds to this temperature, hence

$$\frac{143{\cdot}96}{y''_c+0} = 1, \qquad y''_c = 143{\cdot}96 \text{ lb/sq.in.}$$

where y''_c is the required value of the vapour pressure of mercury at 950°F. The value of y''_c found experimentally is 146·0 lb/sq.in. Thus the error of calculation is $-1{\cdot}4\%$.

c. Physico-chemical Similarity of Compounds. Graphical Presentation of Relationships in the Form of a Straight Line

In Example 1.42 the conditions have been selected so as to include all the factors complicating the similarity of two curves: the units expressing the properties of the compounds were different, and a considerable shift of the curves with respect to each other occurred. Normally, the problems are simpler; the properties t', t'' of the compounds are expressed in the same units, and likewise for y' and y'' (Fig. 1.13). As corresponding points we choose on the curves, two pairs of points with the same values of y (the selection of two pairs of points with equal values of t leads to an analogous result).

Point A'' with the coordinates (t''_A, y''_A) on the curve of compound 2 corresponds to point A' with the coordinates (t'_A, y'_A) on the curve of compound 1, and

$$y'_A = y''_A$$

In the same way point B'' with the coordinates (t''_B, y''_B) corresponds to point B' with the coordinates (t'_B, y'_B), and $y'_B = y''_B$.

The system of equations (1.113) takes the form

$$\frac{t'_A}{t''_A+a} = C_t, \qquad \frac{t'_B}{t''_B+a} = C_t, \qquad b = 0, \qquad C_y = 1 \tag{1.114}$$

Since the relationship between t' and t'' for two arbitrarily chosen corresponding points A and B is always the same, it will have the same form for any arbitrarily chosen pair of corresponding points and we can write the general relationship

$$t' = C_t t'' + C_t a \tag{1.115}$$

This is an equation of a straight line in the t''/t' coordinate system (Fig. 1.14); the tangent of α (the slope of the straight line) is equal to C_t, and the line intersects the t'

axis at the point at which t' is equal to the product $C_t a$. The result obtained represents a well-known relationship existing between many physico-chemical properties of similar compounds. For example, the boiling temperatures of liquids (Dühring's rule, 1878), the temperatures at which two liquids have the same viscosity (Porter's rule, 1912), and also the temperatures at which the molal concentrations of saturated solutions of two compounds belonging to the same group are equal (Harris' rule, 1932), vary linearly with each other. In all these cases a linear relationship between the temperatures for corresponding points, i.e. for points with equal values of y (vapour pressure, viscosity, solubility of the compared substances), can be observed.

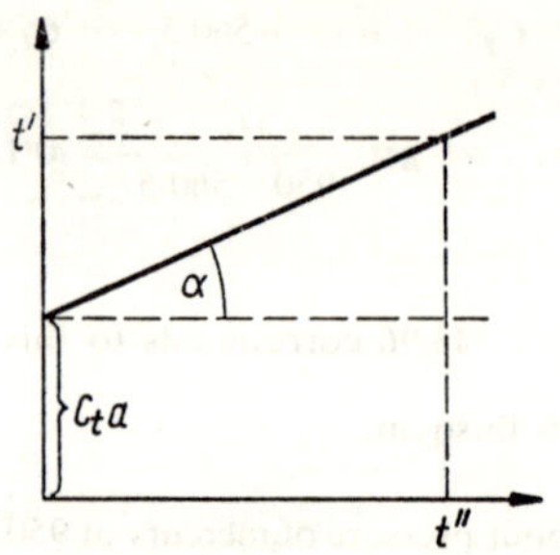

FIG. 1.14. Illustrating a method for comparing a physico-chemical property (y) for two compounds with varying values of the parameter (t). The relationship is expressed in form of a straight line (see Eq. (1.115)).

The first attempt at explaining these linear relationships between the physico-chemical properties of pairs of compounds was presented by Pavlov.[283] Recently a monograph devoted to this problem has been published by Pilař.[286]

Knowledge of the methodology of presenting the dependence of the physico-chemical properties of compounds on various parameters in the form of linear relationships is of great practical importance. The linear plots permit easy interpolation and extrapolation, and enable accurate predictions to be made.

As we have already mentioned, the principles of similarity of physico-chemical properties can only be applied to compounds belonging to the same group. Therefore, in order to obtain a good result, it is necessary to select a suitable reference compound, which in respect to properties and molecular structure is similar to the compound under consideration.

EXAMPLE 1.43. Estimate the value of the viscosity μ'' of chlorobenzene at 60°C, given that the following values of μ'' (cP) for chlorobenzene are available:

$$t_1'' = 10°\text{C}, \quad t_2'' = 100°\text{C}$$
$$\mu_1'' = 0{\cdot}91, \quad \mu_2'' = 0{\cdot}370$$

Solution

We select ethanol as reference compound, as the temperature dependence of viscosity for this compound is known.

Temperature, °C	0	10	20	30	40	50	60	70	80	100
μ', cP	1·78	1·46	1·19	1·00	0·825	0·701	0·591	0·505	0·435	0·326

Interpolating on the basis of these data we determine the temperature at which the viscosities, μ', of ethanol are the same as the values given for chlorobenzene. These temperatures are

$t_1' = 35{\cdot}1$°C, $\mu_1' = 0{\cdot}91$ cP

$t_2' = 86{\cdot}0$°C, $\mu_2' = 0{\cdot}370$ cP

Construction of the graph in Fig. 1.15. On the horizontal axis the temperatures, t', of the reference compound (ethanol) are plotted versus the corresponding temperatures, t'', of chlorobenzene on the vertical axis. The two points are:

point *1* with the coordinates $t_1' = 35{\cdot}1$, $t_2'' = 10$

point *2* with the coordinates $t_2' = 86{\cdot}0$, $t_2'' = 100$

Through these points (*1*, *2*) we construct a straight line. Any point on this line denotes a pair of temperatures, one for each compound, at which the respective viscosities are equal.

On the vertical axis, beginning from point *P*, which corresponds to 60°C (the temperature at which we require the viscosity of chlorobenzene) we construct a horizontal line to the intersection with the straight

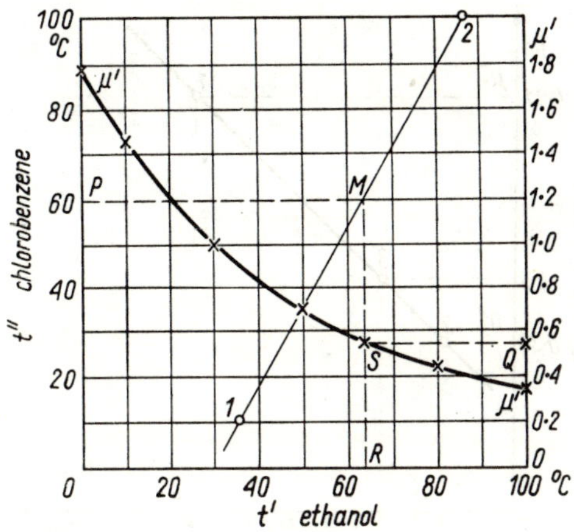

FIG. 1.15. Porter's graph. Comparison of the values of the dynamic viscosities of two compounds at various temperatures. Reference compound ethanol. Compound compared chlorobenzene.

line *1–2* in point *M*, which at the same time corresponds to point *R* on the axis of ethanol temperatures (64°C). This means that the viscosity of chlorobenzene at 60°C is the same as the viscosity of ethanol at 64°C. This value can be interpolated by making use of the viscosity table for ethanol, viz.: $\mu_{64°}' = \mu_{60°}'' = 0{\cdot}52$ cP.

Alternatively, this value can be read directly from the plot in Fig. 1.15, when the curve μ'—μ' representing the dependence of the viscosity of ethanol (reference compound) on temperature t', is plotted in the graph using the viscosity scale μ' (cP) on the right hand side of the diagram. The point *S*, lying at the intersection of the straight line *MR* and the curve μ'—μ', and corresponding to point *Q* on the viscosity scale, determines then the required viscosity of chlorobenzene, $\mu_{60}'' = 0{\cdot}52$ cP. The experimental value of the viscosity of chlorobenzene at 60°C is $\mu_{60}'' = 0{\cdot}52$ cP.

The solution of the set of equations (1.113) leads to the same result:

from the equality $C_t = \dfrac{35{\cdot}1}{10{\cdot}0+a} = \dfrac{86{\cdot}0}{100+a}$ we find $a = 51{\cdot}0$

Substituting this value for a, we obtain

$$C_t = \frac{35{\cdot}1}{10{\cdot}0+51{\cdot}0} = 0{\cdot}573$$

$$t' = 0{\cdot}573\times 60+0{\cdot}573\times 51 = 63{\cdot}8\text{°C}$$

The viscosity of ethanol at 63·8°C is $\mu' = 0{\cdot}52$ cP.

It is equal to the required viscosity of chlorobenzene, hence $\mu_{60°}'' = 0{\cdot}52$ cP.

We have already mentioned that instead of the respectively equal values of the dependent variable $y' = y''$ and $y'_B = y''_B$, the respectively equal values of the independent variable $t'_A = t''_A$ and $t'_B = t''_B$ can be assumed as corresponding points, as in the following example:

EXAMPLE 1.44. Construct a graph for the preceding example (1.43) in the coordinate system μ'—μ''.

The values of μ', the viscosity of the reference compound (i.e. ethanol) are laid off on the abscissa for various temperatures (Fig. 1.16). The temperatures corresponding to the viscosities are denoted by arrows on the same axis.

On the axis of ordinates (μ'' viscosity of chlorobenzene) two points, corresponding to the viscosity of chlorobenzene at 10°C ($\mu_1'' = 0{\cdot}91$) and at 100°C ($\mu_2'' = 0{\cdot}370$), are given.

At the same temperatures the viscosities of the reference compound, ethanol, are $\mu_1' = 1{\cdot}46$ and $\mu_2' = 0{\cdot}326$, respectively. In the graph, point *1* (10°C) corresponds to the coordinates (μ_1', μ_1''), whereas point *2* (100°C) corresponds to the coordinates (μ_2', μ_2'').

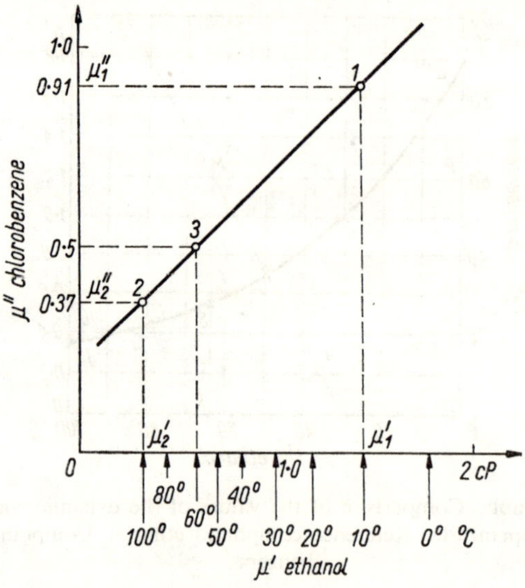

FIG. 1.16. Comparison of the viscosity (μ') of a reference compound (ethanol) and μ'' of another compound at various temperatures. The abscissa bears the temperature scale in addition to the viscosity scale of the reference compound.

A straight line drawn through these two points corresponds to the viscosities of ethanol and chlorobenzene at equal temperatures. Since on the abscissa the temperatures corresponding to the viscosities of ethanol are already plotted, Fig. 1.16 represents the dependence of the viscosity μ'' of chlorobenzene on temperature. We can, therefore, read directly the required value of the chlorobenzene viscosity at 60°C, $\mu_{60}'' = 0{\cdot}5$ cP (point *3*).

For graphs of this type it is often more convenient to employ logarithmic scales on both axes. The values read on the axes of such a graph are sufficiently accurate, but the range is much wider. On the horizontal axis, beside the quantity y' plotted on the logarithmic scale (e.g. viscosity, log μ'), the values of t', corresponding to the values of y', are plotted. Plotting, e.g., the values of the temperatures corresponding to the viscosities (as in Fig. 1.16) we obtain an empirical temperature scale.

d. Comparison of Different Physico-chemical Properties of Different Compounds

In the preceding sections the same physico-chemical property of two different compounds was compared. The rule of similarity can be applied, when some definite physico-chemical property, y, of two different compounds varies with an independent parameter t according to the same law for both compounds. In this case the curves which illustrate the changes are geometrically similar, and the relationship between the properties of the compounds compared may be presented graphically in the form of a straight line (section 7, a to c).

Frequently, however, two different properties of various compounds vary according to the same law, which means that the equations representing the variations of these properties are mathematically homomorphous.[35] If some physico-chemical property y of compound 1 varies with an independent parameter t according to the same law as does another physico-chemical property u of compound 2, also dependent on the changes of the parameter t, then

$$y = F(t), \qquad u = F(t)$$

and the form of the function F is the same in both cases.

The curves in the rectangular coordinate system, representing the relationship between y and t as well as the relationship between u and t, will be geometrically similar. Thus, formally, these curves can be transformed in the same way, as those representing the variations of the same physico-chemical property of two compared compounds. We can construct, for instance, a graph which is analogous to the graphs described above and obtain a rectilinear relationship. Graphs of this type have been applied by Othmer for representing certain relationships between the physico-chemical properties of various compounds.

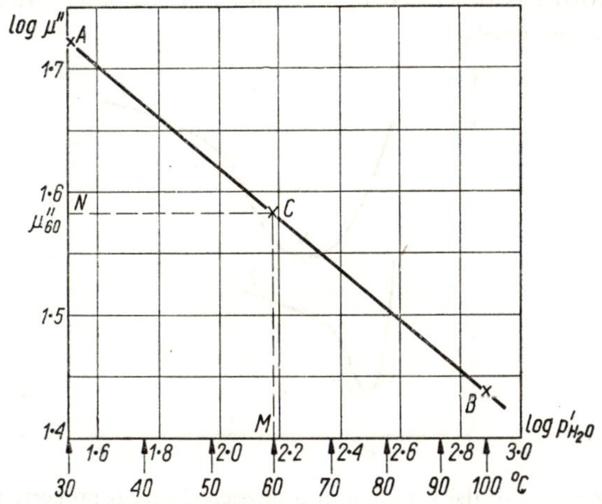

FIG. 1.17. Logarithms of the vapour pressure of a reference compound (water) vs. the dynamic viscosity of another compound (toluene).

EXAMPLE 1.45. Develop the relationship between the logarithm of the vapour pressure of a reference compound (water) and the logarithm of the viscosity of another compound (toluene), using the following data:

Water (reference compound)	Toluene
$t'_A = 30°C$	$t''_A = 30°C$
$\log p'_A = \log 31{\cdot}8 = 1{\cdot}502$	$\log \mu''_A \times 10^2 = \log 52{\cdot}2 = 1{\cdot}719$
$t'_B = 100°C$	$t''_B = 100°C$
$\log p'_B = \log 760 = 2{\cdot}882$	$\log \mu''_B \times 10^2 = \log 27{\cdot}1 = 1{\cdot}432$

Solution

In Fig. 1.17 using values of $\log p'$ and $\log \mu''$ as coordinates we determine point A with the coordinates (1·502, 1·719) and point B with the coordinates (2·882, 1·432). We have to find the value of the viscosity of toluene at 60°C (μ''_{60}).

The vapour pressure of the reference compound (water) at 60°C, $p'_C = 149{\cdot}4$ mm Hg ($\log p'_C = \log 149{\cdot}4 = 2{\cdot}174$), corresponds to point M on the horizontal axis, to point C on the straight line AB and to point N on the axis of the logarithms of the toluene viscosity.

The value of the logarithm of the required toluene viscosity at 60°C, read off at point N, is equal to $\log \mu''_{60} \times 10^2 = 1{\cdot}581$. This corresponds to the value of $\mu''_{60} = 0{\cdot}38$ cP. The viscosity of toluene at 60°C, determined experimentally, is $\mu''_{60} = 0{\cdot}381$ cP. The temperatures corresponding to the logarithms of the vapour pressures of the reference compound at these temperatures have been plotted on the horizontal axis. After changing the scale on the vertical axis to a logarithmic one, the values of the viscosity corresponding to definite temperatures can be taken directly from the diagram.

8. THEORY OF CORRESPONDING STATES

The theory of corresponding states may be regarded as a further development of the theory of similarity discussed in Section 7a. Because of the practical importance of the theory of corresponding states this theory will be discussed separately.

a. Scale Constants and Invariants

It will be assumed, as in section 7a, that the values of all the independent parameters upon which the property of interest (y) depend, remain constant, except one (t) which is varying. We can assume that for the particular group of compounds under consideration the variation of y with t is described by equations of the same type; so that a geometric similarity of the curves exists.

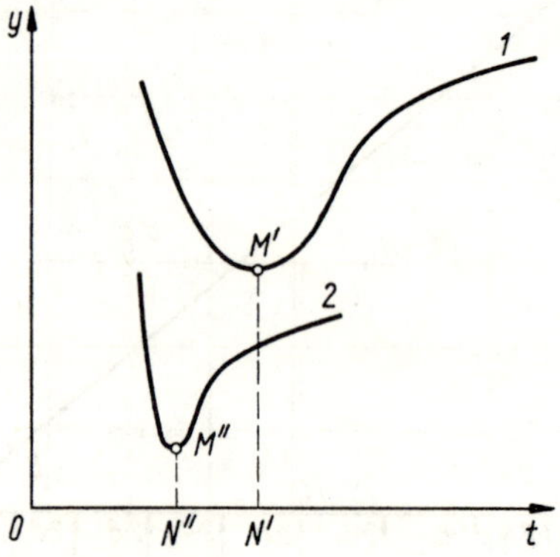

FIG. 1.18. Similar curves of the variation of a physico-chemical property (y) with parameter t for two compounds 1 and 2.

Both curves *1* and *2*, are plotted in the same coordinate system t—y, using the same units and origins. It follows from the theory of similarity of geometric forms that for any arbitrarily chosen pair of corresponding points on the curves *1* and *2* with the co-ordinates (t', y') and (t'', y'') we can write

$$\frac{t'}{t''} = C_t, \qquad \frac{y'}{y''} = C_y \tag{1.116}$$

where C_t and C_y are *scale constants*. (From the condition that the curves are not shifted with respect to each other it follows that the relationship (1.116) is simpler than the general form, viz. $a = 0$ and $b = 0$.)

The dependence of y on t may be illustrated graphically as in Fig. 1.18, where the values of the coordinates of each point on the curve are expressed in physical units. These quantities can also be represented as dimensionless numbers equal to the ratios of the values of the coordinates of the point concerned to the values of the coordinates of some convenient corresponding points. We will refer to these dimensionless quantities as *invariants*; there is one curve representing the relationship between the invariants, which is common for the whole group of compounds.

This is best illustrated by the example in Fig. 1.18. M' and M'' are corresponding points on the similar curves *1* and *2*, respectively; they correspond to a minimum of the function $y = f(t)$. The coordinates of the minima can be chosen as unity, i.e. on the t-axis the unit for curve *1* will be the segment $ON' = n'$, and for curve *2* the segment $ON'' = n''$; on the y-axis the unit for curve *1* will be the segment $M'N' = m'$, and for curve *2* the segment $M''N'' = m''$.

According to Eq. (1.116)

$$\frac{n'}{n''} = C_t, \qquad \frac{m'}{m''} = C_y \tag{1.117}$$

It follows from Eqs. (1.116) and (1.117) that

$$\frac{t'}{n'} = \frac{t''}{n''} = \ldots = \frac{t}{n}, \qquad \frac{y'}{m'} = \frac{y''}{m''} = \ldots = \frac{y}{m} \tag{1.118}$$

The dimensionless fractions t/n and y/m obtained by normalizing t and y in this way are invariants, i.e. dimensionless variables of one curve, which is common to all

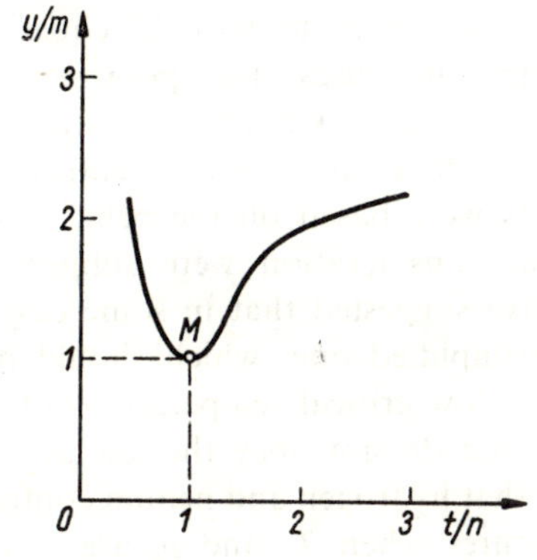

FIG. 1.19. Variation of the normalized value y/m of a physico-chemical property y with the normalised value t/n of the independent parameter t. One common curve is obtained for all compounds belonging to the same group when the variables are normalized by making the coordinates of the minimum on the curve unity.

the compounds of the given group. Such a normalized curve is presented in Fig. 1.19. The presentation of the variation of y with t in this dimensionless system has the great advantage that one graph represents the relationship for a very large number of compounds.

b. Corresponding States

Studying the properties of real gases, van der Waals expressed in 1889 the opinion that the critical points of all compounds are corresponding points. The coordinates of the critical points: critical temperature T_c, critical pressure p_c and critical molal volume

V_c do in fact in many cases fulfil the conditions which are imposed upon the units m and n chosen for calculation of the invariants of similar systems. The calculated values of these invariants, referred to as *reduced parameters,* are equal to

$$T_r = \frac{T}{T_c}, \qquad p_r = \frac{p}{p_c}, \qquad V_r = \frac{V}{V_c} \tag{1.119}$$

For two arbitrarily chosen real gases *1* and *2*, when

$$T_r' = T_r'' \quad \text{and} \quad p_r' = p_r''$$

the equality

$$V_r' = V_r''$$

must be fulfilled.

Thus, for the majority of real gases a common generalized reduced equation can be derived

$$p_r = f(T_r, V_r) \tag{1.120}$$

where T_r, p_r and V_r are reduced parameters of the equation, respectively equal, when the compounds compared are in corresponding states. *A reduced equation of state is universal*—it does not contain any quantities depending on the kind of compound. Starting from the theory of corresponding states, it is possible to obtain many relationships, which are analogous to the generalized dependency (1.120) of the invariants of various physico-chemical quantitics (e.g. the reduced viscosity, reduced thermal conductivity, reduced saturated vapour pressure, etc.).

Curie (1891) has proved that numerous reduced equations of state may exist.

Beside the critical point, in some cases other points may also be considered as corresponding, or approximately so, e.g. the normal boiling temperature† or the melting point. Also related quantities such as, for example entropy, can be involved.[375]

Van der Waals' arguments were based on the belief that the relevant properties of the various compounds under consideration were subject to the simplified similarity rule (1.111). Some workers have suggested that in some cases it is the general similarity rule (1.112), rather than the simplified one, which should be used.

Some substances with very low critical temperatures obey the law of corresponding states at normal temperaturcs but do not obey the law at the critical point itself. Newton[256] has shown, for instance, that hydrogen and helium conform to the law of corresponding states at normal temperatures when T_r and p_r are given by

$$T_r = \frac{T}{T_c+8}, \qquad p_r = \frac{p}{p_c+8} \tag{1.121}$$

where T and p are expressed in °K and atmospheres respectively.

For example, Meyer-Bjerrum (1906) has suggested that better agreement may often be achieved, if we assume

$$V_r = \frac{V-V_0}{V_c-V_0}, \qquad T_r = \frac{T-T_0}{T_c-T_0} \tag{1.122}$$

where the quantity with the subscript zero is a suitably chosen constant. The modified theory of corresponding states, advanced by Bauer, Magat and Surdin, belongs to the

† The normal boiling points are corresponding points for substances with the same critical pressure. It is an empirical fact that the critical pressures of a very large number of compounds do lie in the range 40–60 atmospheres.

same category. In some cases, for example when we want to find the invariants of such quantities as the compressibility factors for liquids or their molal volumes, we should assume as the zero point of the temperature scale, the melting temperature of the compound, and not the temperature of absolute zero. The reduced temperature will then amount (according to Bauer, Magat, Surdin, 1936) to

$$\theta = \frac{T - T_m}{T_c - T_m} \tag{1.123}$$

In this case the two compounds 1 and 2 are in corresponding states, if $\theta_1 = \theta_2$.

However, many other criteria of corresponding states may also be found, e.g. according to Sugden, the molal volumes of two liquids should be compared at their corresponding states, when the values of their surface tensions are equal, and not at the same reduced temperatures; it may be assumed that at equal surface tensions the intermolecular forces are equal, too.

c. Theory of Corresponding States. Theoretical Foundation

In the preceding section the theory of corresponding states was regarded as a consequence of the experimentally proved similarity of certain physico-chemical phenomena.

Pitzer (1939), and later also De Boer (1948) and others, have established theoretical foundations for the theory.[25, 146, 287] When comparing the properties of compounds consisting of non-polar spherical molecules, the interaction of which is expressed by the intermolecular potential equation:

$$E(r)/\varepsilon = F(r)/\sigma \tag{1.124}$$

(cf. section 3, Lennard-Jones equation (1.29)), the properties of the compounds and the state of the system may be represented as a function of the reduced parameters p^*, T^*, v^*, defined by:

reduced pressure	$p^* = p\sigma^3/\varepsilon$
reduced temperature	$T^* = kT/\varepsilon$
reduced volume	$v^* = v/\sigma^3$

where σ and ε are the force constants.

The equation of state may be represented in its reduced form as a function of the reduced parameters

$$p^* = \phi(T^*, v^*) \tag{1.125}$$

The form of the function ϕ is the same for all compounds containing non-polar spherical molecules, for which Eq. (1.124) is valid.

The values of the force constants σ and ε are related to the critical properties of the compounds. These relationships (Eqs. (1.33), (1.34)) have been presented above. Substituting σ and ε as functions of T_c, p_c and V_c into Eq. (1.125), we obtain Eq. (1.120), with

$$p/p_c = f(T/T_c, V/V_c) \tag{1.126}$$

i.e. a generalized equation of state of a real gas in its well-known reduced form.

If the compounds consist of polar molecules, knowledge of the force constants σ and ε is not sufficient for calculation of the potential energy of the system; in such a case an interaction potential equation more complicated than the Lennard-Jones equation is applied. On this basis De Boer derived the following reduced equation of state for polar substances:

$$p/p_0 = \varphi(V/V_0,\ T/T_0\Lambda^*) \tag{1.127}$$

The parameters T_0, V_0 and p_0 depend on the force constants σ and ε. The quantity Λ^* is a third parameter, characteristic of the given kind of polar molecules.

Various other authors have correlated the properties of substances composed of polar molecules, non-polar spherical molecules, polyatomic non-spherical molecules, "flexible" molecules, polar molecules, etc. by introducing similar empirical third parameters into the reduced equation of state. An exact approach to the problem has also been attempted (Prigogine,[301] Bondi and Simkin,[27] Altenburg[4]).

d. The "Third Parameter" in the Modified Theory of Corresponding States

According to the classical assumption adopted by van der Waals, substances are in corresponding states when the reduced parameters: $T_r = T/T_c$, $p_r = p/p_c$, $V_r = V/V_c$ (1.119), are equal to one another. When the classical theory is valid, the quantity K, which is called the *critical coefficient*

$$K = \frac{RT_c}{p_c V_c} \tag{1.128}$$

should have the same value for all compounds.

Experiments show that the condition of equality of the critical coefficients is not always fulfilled and that the simple theory of corresponding states enunciated by van der Waals holds only approximately for certain groups of compounds. Some explanation of this has been provided by the arguments presented above, based both on the general theory of similarity and on a theoretical evaluation of the influence of molecular interactions (cf. section 8c). According to the modified theory of corresponding states, Eq. (1.120) takes on the form

$$p_r = f(T_r,\ V_r, \Lambda) \tag{1.129}$$

The "third parameter" Λ for a given compound may be selected in various ways: usually, the value of this parameter is related to the critical coefficient K or the saturated vapour pressure. When we consider the variation of the reduced vapour pressure with temperature we do not obtain one common curve, which would correspond to a universal reduced equation, but a whole family of similar curves; in this light the differences between the various groups of compounds are obvious.

Riedel[312] studied the dependence of the dimensionless quantity

$$\alpha = \frac{T}{p}\frac{\mathrm{d}p}{\mathrm{d}T} = \frac{\mathrm{d}\ln p}{\mathrm{d}\ln T} = \frac{\mathrm{d}\log p_r}{\mathrm{d}\log T_r} \tag{1.130}$$

on the reduced temperature T_r (in this formula p denotes the pressure of saturated vapour). The plots of α against the reduced temperature T_r for various compounds produce a family of similar curves, and the tangents to these curves at the critical temperature are parallel to the abscissa T_r, i.e.

$$\frac{\mathrm{d}\alpha}{\mathrm{d}T_r} = 0, \quad \text{if} \quad T_r = 1$$

Riedel[312] used as his third parameter the value (α_c) of α at the critical temperature ($T_r = 1$). He called the quantity α_c the *critical parameter*. In this treatment the critical parameter plays the role of the quantity Λ in the reduced equation (1.129). Riedel found that the following relationship between the critical coefficient $K = RT_c/p_cV_c$ and α_c holds for non-associated liquids with an accuracy up to ±2%:

$$K = 3{\cdot}72 + 0{\cdot}26(\alpha_c - 7) \tag{1.131}$$

According to the requirements of the classical theory of corresponding states the value of α_c should be 7. Actually, values of α_c for various compounds lie within the limits 6–8 (Fig. 1.20). The quantity $(\alpha_c - 7)$ is the *measure of deviation* in the behaviour of a compound *from the requirements of the classical theory of corresponding states.*

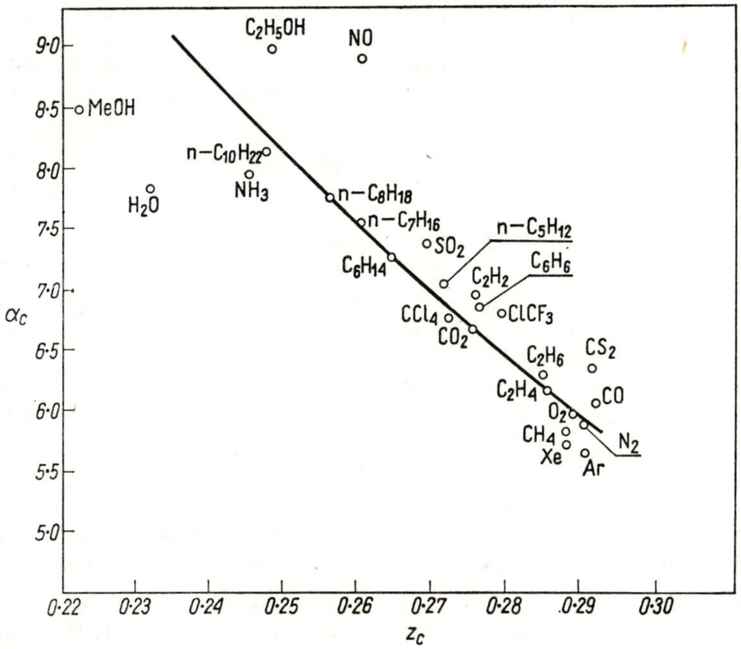

FIG. 1.20. Critical parameter α_c vs. compressibility factor z_c at the critical point (after Hirschfelder *et al.*[156]).

To calculate the value of α_c we need not know the dependence of the vapour pressure on temperature in the vicinity of the critical point. Riedel[312] presented a generalized formula for the dependence of the reduced saturated vapour pressure p_r of a non-associated liquid on the critical parameter α_c and on the reduced temperature T_r

$$\log \frac{1}{p_r} = \phi(T_r) - (\alpha_c - 7)\,\psi(T_r) \tag{1.132}$$

The values of the functions $\phi(T_r)$ and $\psi(T_r)$ for various values of T_r are listed in Tables 1.49 and 1.50. Knowing the vapour pressures of a liquid at one arbitrary temperature (the boiling temperature T_b is usually known, i.e. when the vapour pressure is $p = 1$ atm), it is easy to calculate the value of α_c from Eq. (1.132).

According to Miller,[252] the value of α_c may be calculated from the equation

$$\alpha_c = 0{\cdot}9076\left(1 + \frac{T_{rb} \ln p_c}{1 - T_{rb}}\right) \tag{1.133}$$

TABLE 1.49. DEPENDENCE OF THE FUNCTION $\phi(T_r) = 0{\cdot}1183\ \varphi(T_r) - 7\log T_r$ ON REDUCED TEMPERATURE T_r (AFTER RIEDEL[312])

T_r	0	1	2	3	4	5	6	7	8	9	10
0·35	6·003	5·974	5·944	5·916	5·887	5·859	5·830	5·802	5·774	5·747	5·719
0·36	5·719	5·692	5·664	5·637	5·610	5·583	5·557	5·530	5·504	5·478	5·452
0·37	5·452	5·426	5·400	5·375	5·350	5·324	5·299	5·274	5·250	5·225	5·200
0·38	5·200	5·176	5·152	5·128	5·104	5·080	5·056	5·033	5·010	4·986	4·963
0·39	4·963	4·940	4·917	4·895	4·872	4·849	4·827	4·805	4·783	4·761	4·739
0·40	4·739	4·717	4·696	4·674	4·653	4·632	4·610	4·589	4·568	4·548	4·527
0·41	4·527	4·506	4·486	4·465	4·445	4·425	4·405	4·385	4·365	4·345	4·326
0·42	4·326	4·306	4·287	4·268	4·248	4·229	4·210	4·191	4·173	4·154	4·135
0·43	4·135	4·117	4·098	4·080	4·062	4·044	4·026	4·008	3·990	3·972	3·955
0·44	3·955	3·937	3·919	3·902	3·885	3·868	3·850	3·833	3·816	3·799	3·783
0·45	3·783	3·766	3·749	3·733	3·716	3·700	3·683	3·667	3·651	3·635	3·619
0·46	3·619	3·603	3·587	3·572	3·556	3·540	3·525	3·509	3·494	3·479	3·463
0·47	3·463	3·448	3·433	3·418	3·403	3·388	3·374	3·359	3·344	3·330	3·315
0·48	3·315	3·301	3·286	3·272	3·258	3·243	3·229	3·215	3·201	3·187	3·174
0·49	3·174	3·160	3·146	3·132	3·119	3·105	3·092	3·078	3·065	3·052	3·039
0·50	3·039	3·025	3·012	2·999	2·986	2·973	2·960	2·948	2·935	2·922	2·909
0·51	2·909	2·897	2·884	2·872	2·860	2·847	2·835	2·823	2·810	2·798	2·786
0·52	2·786	2·774	2·762	2·750	2·738	2·726	2·715	2·703	2·691	2·680	2·668
0·53	2·668	2·656	2·645	2·634	2·622	2·611	2·600	2·588	2·577	2·566	2·555
0·54	2·555	2·544	2·533	2·522	2·511	2·500	2·489	2·479	2·468	2·457	2·447
0·55	2·447	2·436	2·425	2·415	2·404	2·394	2·384	2·373	2·363	2·353	2·343
0·56	2·343	2·333	2·322	2·312	2·302	2·292	2·282	2·273	2·263	2·253	2·243
0·57	2·243	2·233	2·224	2·214	2·204	2·195	2·185	2·176	2·166	2·157	2·147
0·58	2·147	2·138	2·129	2·119	2·110	2·101	2·092	2·082	2·073	2·064	2·055
0·59	2·055	2·046	2·037	2·028	2·019	2·011	2·002	1·993	1·984	1·976	1·967
0·60	1·967	1·958	1·950	1·941	1·932	1·924	1·915	1·907	1·899	1·890	1·882
0·61	1·882	1·874	1·865	1·857	1·849	1·841	1·832	1·824	1·816	1·808	1·800
0·62	1·800	1·792	1·784	1·776	1·768	1·760	1·752	1·744	1·737	1·729	1·721
0·63	1·721	1·713	1·706	1·698	1·690	1·683	1·675	1·668	1·660	1·653	1·645
0·64	1·645	1·638	1·630	1·623	1·615	1·608	1·601	1·594	1·586	1·579	1·572
0·65	1·572	1·565	1·558	1·550	1·543	1·536	1·529	1·522	1·515	1·508	1·501
0·66	1·501	1·494	1·487	1·480	1·474	1·467	1·460	1·453	1·446	1·440	1·433
0·67	1·433	1·426	1·420	1·413	1·406	1·400	1·393	1·387	1·380	1·374	1·367
0·68	1·367	1·361	1·354	1·348	1·341	1·335	1·328	1·322	1·316	1·310	1·303
0·69	1·303	1·297	1·291	1·285	1·278	1·272	1·266	1·260	1·254	1·248	1·242
0·70	1·242	1·236	1·230	1·224	1·218	1·212	1·206	1·200	1·194	1·188	1·182
0·71	1·182	1·176	1·170	1·165	1·159	1·153	1·147	1·142	1·136	1·130	1·125
0·72	1·125	1·119	1·113	1·108	1·102	1·096	1·091	1·085	1·080	1·074	1·069
0·73	1·069	1·063	1·058	1·052	1·047	1·041	1·036	1·031	1·025	1·020	1·015
0·74	1·015	1·009	1·004	0·999	0·993	0·988	0·983	0·978	0·973	0·967	0·962
0·75	0·962	0·957	0·952	0·947	0·942	0·936	0·931	0·926	0·921	0·916	0·911
0·76	0·911	0·906	0·901	0·896	0·891	0·886	0·881	0·876	0·872	0·867	0·862
0·77	0·862	0·857	0·852	0·847	0·842	0·838	0·833	0·828	0·823	0·818	0·814
0·78	0·814	0·809	0·804	0·800	0·795	0·790	0·786	0·781	0·776	0·772	0·767
0·79	0·767	0·762	0·758	0·753	0·749	0·744	0·740	0·735	0·731	0·726	0·722

TABLE 1.49 (*cont.*)

T_r	0	1	2	3	4	5	6	7	8	9	10
0·80	0·722	0·717	0·713	0·708	0·704	0·699	0·695	0·691	0·686	0·682	0·677
0·81	0·677	0·673	0·669	0·664	0·660	0·656	0·652	0·647	0·643	0·639	0·635
0·82	0·635	0·630	0·626	0·622	0·618	0·613	0·609	0·605	0·601	0·597	0·593
0·83	0·593	0·588	0·584	0·580	0·576	0·572	0·568	0·564	0·560	0·556	0·552
0·84	0·552	0·548	0·544	0·540	0·536	0·532	0·528	0·524	0·520	0·516	0·512
0·85	0·512	0·508	0·504	0·500	0·496	0·492	0·488	0·484	0·481	0·477	0·473
0·86	0·473	0·469	0·465	0·461	0·457	0·454	0·450	0·446	0·442	0·438	0·435
0·87	0·435	0·431	0·427	0·423	0·420	0·416	0·412	0·409	0·405	0·401	0·398
0·88	0·398	0·394	0·390	0·387	0·383	0·379	0·376	0·372	0·368	0·365	0·361
0·89	0·361	0·358	0·354	0·350	0·347	0·343	0·340	0·336	0·333	0·329	0·325
0·90	0·325	0·322	0·318	0·315	0·311	0·308	0·304	0·301	0·297	0·294	0·290
0·91	0·290	0·287	0·284	0·280	0·277	0·273	0·270	0·266	0·263	0·259	0·256
0·92	0·256	0·253	0·249	0·246	0·242	0·239	0·236	0·232	0·229	0·226	0·222
0·93	0·222	0·219	0·216	0·212	0·209	0·206	0·202	0·199	0·196	0·193	0·189
0·94	0·189	0·186	0·183	0·179	0·176	0·173	0·170	0·166	0·163	0·160	0·157
0·95	0·157	0·153	0·150	0·147	0·144	0·140	0·137	0·134	0·131	0·128	0·124
0·96	0·124	0·121	0·118	0·115	0·112	0·109	0·105	0·102	0·099	0·096	0·093
0·97	0·093	0·090	0·086	0·083	0·080	0·077	0·074	0·071	0·068	0·065	0·061
0·98	0·061	0·058	0·055	0·052	0·049	0·046	0·043	0·040	0·037	0·034	0·031
0·99	0·031	0·027	0·024	0·021	0·018	0·015	0·012	0·009	0·006	0·003	0·000

TABLE 1.50. DEPENDENCE OF THE FUNCTION $\psi(T_r) = 0{\cdot}0364\,\varphi(T_r) - \log T_r$ ON REDUCED TEMPERATURE T_r (AFTER RIEDEL[312])

T_r	0	1	2	3	4	5	6	7	8	9	10
0·35	1·321	1·313	1·306	1·298	1·291	1·284	1·276	1·269	1·262	1·255	1·248
0·36	1·248	1·241	1·234	1·227	1·220	1·213	1·206	1·199	1·193	1·186	1·179
0·37	1·179	1·173	1·166	1·160	1·153	1·147	1·140	1·134	1·128	1·122	1·115
0·38	1·115	1·109	1·103	1·097	1·091	1·085	1·079	1·073	1·067	1·061	1·055
0·39	1·055	1·050	1·044	1·038	1·032	1·027	1·021	1·016	1·010	1·005	0·999
0·40	0·999	0·994	0·988	0·983	0·978	0·972	0·967	0·962	0·956	0·951	0·946
0·41	0·946	0·941	0·936	0·930	0·926	0·921	0·916	0·911	0·906	0·901	0·896
0·42	0·896	0·892	0·887	0·882	0·877	0·873	0·868	0·863	0·859	0·854	0·850
0·43	0·850	0·845	0·840	0·836	0·832	0·827	0·823	0·818	0·814	0·810	0·805
0·44	0·805	0·801	0·797	0·793	0·788	0·784	0·780	0·776	0·772	0·768	0·764
0·45	0·764	0·760	0·756	0·752	0·748	0·744	0·740	0·736	0·732	0·728	0·724
0·46	0·724	0·721	0·717	0·713	0·709	0·706	0·702	0·698	0·695	0·691	0·687
0·47	0·687	0·684	0·680	0·677	0·673	0·670	0·666	0·663	0·659	0·656	0·652
0·48	0·652	0·649	0·645	0·642	0·639	0·635	0·632	0·629	0·626	0·622	0·619
0·49	0·619	0·616	0·613	0·609	0·606	0·603	0·600	0·597	0·594	0·591	0·588
0·50	0·588	0·585	0·582	0·579	0·576	0·573	0·570	0·567	0·564	0·561	0·558
0·51	0·558	0·555	0·552	0·549	0·546	0·544	0·541	0·538	0·535	0·532	0·530
0·52	0·530	0·527	0·524	0·521	0·519	0·516	0·513	0·511	0·508	0·505	0·503
0·53	0·503	0·500	0·498	0·495	0·492	0·490	0·487	0·485	0·482	0·480	0·477
0·54	0·477	0·475	0·472	0·470	0·468	0·465	0·463	0·460	0·458	0·456	0·453

Continued overleaf

TABLE 1.50 (*cont.*)

T_r	0	1	2	3	4	5	6	7	8	9	10
0·55	0·453	0·451	0·449	0·446	0·444	0·442	0·439	0·437	0·435	0·433	0·430
0·56	0·430	0·428	0·426	0·424	0·422	0·419	0·417	0·415	0·413	0·411	0·409
0·57	0·409	0·406	0·404	0·402	0·400	0·398	0·396	0·394	0·392	0·390	0·388
0·58	0·388	0·386	0·384	0·382	0·380	0·378	0·376	0·374	0·372	0·370	0·368
0·59	0·368	0·366	0·364	0·362	0·360	0·359	0·357	0·355	0·353	0·351	0·349
0·60	0·349	0·347	0·346	0·344	0·342	0·340	0·338	0·337	0·335	0·333	0·331
0·61	0·331	0·330	0·328	0·326	0·324	0·323	0·321	0·319	0·318	0·316	0·314
0·62	0·314	0·313	0·311	0·309	0·308	0·306	0·304	0·303	0·301	0·300	0·298
0·63	0·298	0·296	0·295	0·293	0·292	0·290	0·289	0·287	0·286	0·284	0·283
0·64	0·283	0·281	0·280	0·278	0·277	0·275	0·274	0·272	0·271	0·269	0·268
0·65	0·268	0·266	0·265	0·264	0·262	0·261	0·259	0·258	0·257	0·255	0·254
0·66	0·254	0·252	0·251	0·250	0·248	0·247	0·246	0·244	0·243	0·242	0·240
0·67	0·240	0·239	0·238	0·236	0·235	0·234	0·232	0·231	0·230	0·229	0·227
0·68	0·227	0·226	0·225	0·224	0·222	0·221	0·220	0·219	0·218	0·216	0·215
0·69	0·215	0·214	0·213	0·212	0·210	0·209	0·208	0·207	0·206	0·204	0·203
0·70	0·203	0·202	0·201	0·200	0·199	0·198	0·197	0·195	0·194	0·193	0·192
0·71	0·192	0·191	0·190	0·189	0·188	0·187	0·186	0·185	0·184	0·183	0·181
0·72	0·181	0·180	0·179	0·178	0·177	0·176	0·175	0·174	0·173	0·172	0·171
0·73	0·171	0·170	0·169	0·168	0·167	0·166	0·165	0·164	0·163	0·162	0·161
0·74	0·161	0·160	0·159	0·158	0·158	0·157	0·156	0·155	0·154	0·153	0·152
0·75	0·152	0·151	0·150	0·149	0·148	0·147	0·146	0·146	0·145	0·144	0·143
0·76	0·143	0·142	0·141	0·140	0·139	0·138	0·138	0·137	0·136	0·135	0·134
0·77	0·134	0·133	0·133	0·132	0·131	0·130	0·129	0·128	0·128	0·127	0·126
0·78	0·126	0·125	0·124	0·124	0·123	0·122	0·121	0·120	0·120	0·119	0·118
0·79	0·118	0·117	0·116	0·116	0·115	0·114	0·113	0·113	0·112	0·111	0·110
0·80	0·110	0·110	0·109	0·108	0·107	0·107	0·106	0·105	0·104	0·104	0·103
0·81	0·103	0·102	0·101	0·101	0·100	0·099	0·099	0·098	0·097	0·007	0·096
0·82	0·096	0·095	0·094	0·094	0·093	0·092	0·092	0·091	0·090	0·090	0·089
0·83	0·089	0·088	0·088	0·087	0·086	0·086	0·085	0·084	0·084	0·083	0·082
0·84	0·082	0·082	0·081	0·081	0·080	0·079	0·079	0·078	0·077	0·077	0·076
0·85	0·076	0·075	0·075	0·074	0·074	0·073	0·072	0·072	0·071	0·071	0·070
0·86	0·070	0·069	0·069	0·068	0·068	0·067	0·066	0·066	0·065	0·065	0·064
0·87	0·064	0·063	0·063	0·062	0·062	0·061	0·061	0·060	0·059	0·059	0·058
0·88	0·058	0·058	0·057	0·057	0·056	0·056	0·055	0·054	0·054	0·053	0·053
0·89	0·053	0·052	0·052	0·051	0·051	0·050	0·049	0·049	0·048	0·048	0·047
0·90	0·047	0·047	0·046	0·046	0·045	0·045	0·044	0·044	0·043	0·043	0·042
0·91	0·042	0·042	0·041	0·041	0·040	0·040	0·039	0·039	0·038	0·038	0·037
0·92	0·037	0·037	0·036	0·036	0·035	0·035	0·034	0·034	0·033	0·033	0·032
0·93	0·032	0·032	0·031	0·031	0·030	0·030	0·029	0·029	0·028	0·028	0·027
0·94	0·027	0·027	0·026	0·026	0·025	0·025	0·024	0·024	0·023	0·023	0·023
0·95	0·023	0·022	0·022	0·021	0·021	0·020	0·020	0·019	0·019	0·018	0·018
0·96	0·018	0·017	0·017	0·016	0·016	0·016	0·015	0·015	0·014	0·014	0·013
0·97	0·013	0·013	0·012	0·012	0·012	0·011	0·011	0·010	0·010	0·009	0·009
0·98	0·009	0·008	0·008	0·008	0·007	0·007	0·006	0·006	0·005	0·005	0·004
0·99	0·004	0·004	0·004	0·003	0·003	0·002	0·002	0·001	0·001	0·000	0·000

where T_{rb} denotes the reduced normal boiling temperature. This equation has been checked on 85 compounds, and only in 11 cases did the deviation exceed 0·1; it never reached 0·2.

Altenburg[4] found that a relationship exists between the value of the third parameter and the power exponent q, which characterizes the repelling forces in Eq. (1.27) expressing the variation of the potential energy of two molecules with their separation. The value of q depends on the number N_A of the external electrons in the atoms constituting the molecule.

We calculate N_A by summing up the following values:

H	C	N	P	O	S	F	Cl	Br	I
1	4	5	5	6	6	7	7	7	7

Approximately

$$\alpha_c = 3{\cdot}75+5\left(1-\frac{6}{8+0{\cdot}288N_A}\right) \tag{1.134}$$

The deviations (*ca.* $\pm 10\%$) can be due to the effect of the shape of the molecules.

EXAMPLE 1.46.[312] Estimate the value of the critical parameter α_c for propyl acetate (boiling temperature $T_b = 374{\cdot}71$°K, critical temperature $T_c = 549{\cdot}36$°K, critical pressure $p_c = 33{\cdot}20$ atm).

Solution

Reduced temperature $T_r = \dfrac{T_b}{T_c} = \dfrac{374{\cdot}71}{549{\cdot}36} = 0{\cdot}6832$

Reduced pressure $p_r = \dfrac{1}{p_c} = \dfrac{1}{33{\cdot}20}$

In the tables we find: $\phi(0{\cdot}6832) = 1{\cdot}347$, $\psi(0{\cdot}6832) = 0{\cdot}224$.

According to Eq. (1.132)

$$\log\frac{1}{p_r} = \log p_c = \log 33{\cdot}20 = 1{\cdot}521 = 1{\cdot}347+(\alpha_c-7)\times 0{\cdot}224$$

hence,

$$\alpha_c = 7{\cdot}78$$

If the boiling temperature of a liquid is unknown, the approximate value of α_c may be calculated from the Altenburg equation.

The formula of propyl acetate CH_3—COO—C_3H_7 = $C_5H_{10}O_2$

$$N_A = 5\times 4+10\times 1+2\times 6 = 42$$

$$\alpha_c = 3{\cdot}75+5\left(1-\frac{6}{8+0{\cdot}288\times 42}\right)$$

$$\alpha_c = 7{\cdot}25$$

Error of estimation of α_c: $-6{\cdot}5\%$.

When α_c is known for a given substance it is possible to employ the reduced universal equations to calculate many physico-chemical properties of that substance.

Pitzer *et al.*[288] used as their third parameter a quantity referred to as the *acentric factor* ω, defined by:

$$\omega = -1-\log p_r' \tag{1.135}$$

where p_r' denotes the reduced pressure of saturated vapour at the reduced temperature $T_r' = 0{\cdot}7$ (conventionally assumed as the temperature at which the properties of the compounds are compared). If the value of the saturated vapour pressure p_r' is not available, the value of ω may be calculated from the following equation

$$\omega = \frac{3}{7}\left(\frac{\log p_c}{T_c/T_b-1}\right)$$

where T_b denotes the boiling temperature of the liquid (average error $\pm 16{\cdot}5\%$).

The quantities α_c and ω are correlated by the relationships

$$\begin{aligned}\omega &= 0{\cdot}2033\alpha_c - 1{\cdot}181\\ \alpha_c &= 5{\cdot}808 + 4{\cdot}93\omega\end{aligned} \tag{1.136}$$

According to Altenburg[4]

$$\omega \approx 0{\cdot}6\left(1 - \frac{10}{8 + 0{\cdot}288\, N_A}\right) \tag{1.137}$$

where N_A denotes the number of external electrons in the atoms included in the molecule.

Attempts at using analogous third parameters related to the vapour pressure of the compounds, have been made by other authors, e.g. by Filippov.[101]

The *compressibility factor* at the critical point (equal to the reciprocal of the critical coefficient K)

$$z_c = \frac{p_c V_c}{RT_c} = \frac{1}{K}$$

has been used by some authors as the third parameter. According to Lydersen,[160] the value of z_c can be calculated from the following equation (the average error amounting to *ca.* $3{\cdot}8\%$):

$$z_c = \frac{1}{3{\cdot}43 + 0{\cdot}0067\, L^2} \tag{1.138}$$

where L is the heat of vaporization of the liquid at its boiling temperature (kcal/mole).

According to Eq. (1.131), for non-associated liquids a relationship exists between the critical compressibility factor z_c and the critical parameter α_c and ω:

$$\frac{1}{z_c} = 3{\cdot}72 + 0{\cdot}26(\alpha_c - 7) = 1{\cdot}28\omega + 3{\cdot}41 \tag{1.139}$$

The dependence of α_c on z_c for various compounds is presented graphically in Fig. 1.20, where the line corresponds to Eq. (1.139). The values of the parameters z_c, α_c and ω are listed in Table 1.51.

TABLE 1.51. VALUES OF THE THIRD PARAMETERS z_c, α_c, ω (AFTER HOUGEN *et al.*[160])

Defining criterion	Values of the third parameter										
z_c	0·20	0·21	0·22	0·23	0·24	0·25	0·26	0·27	0·28	0·29	0·30
α_c	11·92	11·00	10·17	9·53	8·72	8·08	7·46	6·94	6·43	5·95	5·51
ω	1·242	1·056	0·887	0·756	0·591	0·461	0·341	0·230	0·126	0·030	−0·060

For compounds which are subject to association, e.g. ammonia, ethanol, nitric oxide or water, even the introduction of the third parameter (such as α_c, z_c or ω) does not determine the universal reduced equation explicitly. In such cases Eq. (1.139) cannot be applied, as it is impossible to compute z_c knowing only the value of α_c or ω. Hirschfelder *et al.*[156] have used four parameters, their reduced equation of state taking the form:

$$p_r = F(T_r, V_r, \alpha_c, z_c) \tag{1.140}$$

Equations of this type are distinguished by the wide applicability: they describe exactly the behaviour of a large number of substances, both of gases and liquids.

e. Choice of the Best Method for Estimating the Critical Constants

Critical Constants of Pure Compounds

When applying the theory of corresponding states we must necessarily know the critical constants. The methods now available for calculating these constants are on the whole sufficiently accurate for technological calculations.

When the structural formula of a compound is known, but not its experimentally determined properties, it is usual to use the methods based on the additive summation of contributions. The calculations carried out by the Forman and Thodos method[105] are usually most accurate. The methods recommended by Riedel,[309] Lydersen[306] or Eduljee[116, 117] are less accurate, but they are less troublesome, and the results can be obtained in shorter time.

If some experimentally determined physico-chemical properties are available for the compound of interest, e.g. boiling temperature or liquid density, we may use either the above-mentioned additive methods of Riedel,[309] Lydersen,[306] Eduljee[116] (in which the computation is based on the knowledge of the boiling temperature) or the very rapid graphical method of Altenburg[2] (when the liquid densities at two temperatures are known). As is always the case with semi-empirical methods, it is advisable to perform the calculations by several methods and to compare the predictions. For example, the critical constants calculated according to the method of Forman and Thodos might be compared with those obtained by methods based upon a knowledge of the boiling temperature.

Pseudocritical Parameters

The calculation of the critical parameters T_{cm}, p_{cm} of a mixture of several compounds is a complicated task, and in the case of multi-component mixtures it is often uncertain which method of calculation to choose.[334] Reasonably accurate predictions may, however, be made in the case of multicomponent hydrocarbon mixtures.[135]

Experiments show that the reduced parameters for mixtures calculated by means of experimentally determined or estimated critical parameters of these mixtures: $p_r = p/p_{cm}$, $T_r = T/T_{cm}$, are usually unsuitable for calculations based on the theory of corresponding states. In order to apply the corresponding-states theory to mixtures of compounds it is necessary to find a way of calculating the *pseudocritical values* of the parameters:

pseudocritical temperature	T_{pc}
pseudocritical pressure	p_{pc}
pseudocritical volume	V_{pc}

According to the pseudocritical concept all those single phase properties which may be predicted for pure substances using the law of corresponding states may also be predicted for mixtures provided pseudocritical parameters are substituted for the critical parameters when calculating the reduced quantities. For a mixture therefore the reduced temperature is given by

$$T_{rm} = \frac{T}{T_{pc}}$$

the reduced pressure by $p_{rm} = \frac{p}{p_{pc}}$

and the reduced volume by $V_{rm} = \frac{V}{V_{pc}}$

In other words, the pseudocritical parameters for mixtures should be selected in such a way as to ensure that the investigated physical quantity under consideration (the reduced viscosity, compressibility factor, etc.) has the same value for both mixtures and pure substances, when the reduced parameters of the mixture and the reference compound are equal.

It should be noted that it is exceptional for the pseudocritical parameters to equal the experimentally determined critical parameters of a mixture; usually

$$T_{pc} \neq T_{cm}, \qquad p_{pc} \neq p_{cm}, \qquad V_{pc} \neq V_{cm}$$

where T_{cm}, p_{cm}, V_{cm} are the actual critical parameters of a mixture. Fairly reliable methods exist for calculating T_{cm} for multi-component mixtures of hydrocarbons, when the boiling temperatures of the components are known.[135]

Empirical Equations for Calculating the Pseudocritical Parameters

Kay[193] suggested a simple method for calculating the pseudocritical parameters for hydrocarbon mixtures; this method is based on the assumption that these pseudocritical parameters for a mixture may be calculated by summing up the contributions of the individual components, which are directly proportional both to the critical parameters of these compounds and to their mole fractions in the mixture. If, for example, one mole of a mixture contains x_A moles of the component A (with the critical constants T_{cA}, p_{cA}) and x_B moles of component B (T_{cB}, p_{cB}), then, according to Kay, the pseudocritical parameters T_{pc} and p_{pc} are

$$T_{pc} = x_A T_{cA} + x_B T_{cB} \tag{1.141}$$

and $$p_{pc} = x_A p_{cA} + x_B p_{cB} \tag{1.142}$$

In general, for a mixture of n components

$$T_{pc} = \sum_{i=1}^{n} x_i T_{ci} \tag{1.143}$$

$$p_{pc} = \sum_{i=1}^{n} x_i p_{ci} \tag{1.144}$$

Su, Huang and Chang[351] claim that Kay's method may be applied not only to hydrocarbons, but also to mixtures of other compounds.

Kay's method is widely used because of its unusual simplicity, and most often it yields satisfactory results.

According to Comings, substantial errors may arise if Kay's method is employed for mixtures of chemically unlike molecules or if the critical constants of the constituents of the mixtures differ considerably.

EXAMPLE 1.47. Estimate by Kay's method the pseudocritical constants of a mixture consisting of 60 vol. % of hydrogen, 30% of carbon monoxide and 10% of carbon dioxide.

Solution

Assuming that the mole fractions are proportional to the volume fractions (which is the case in perfect gases), we get

$x_{H_2} = 0{\cdot}6, \quad x_{CO} = 0{\cdot}3, \quad x_{CO_2} = 0{\cdot}1$

In Table 1.31 we find:

$T_{cH_2} = 33{\cdot}2°K, \quad T_{cCO} = 135°K, \quad T_{cCO_2} = 304{\cdot}2°K$

$p_{cH_2} = 12{\cdot}8$ atm, $\quad p_{cCO} = 34{\cdot}5$ atm, $\quad p_{cCO_2} = 73{\cdot}0$ atm

Substituting these values into Eqs. (1.143) and (1.144), we find

$T_{pc} = 0{\cdot}6 \times 33{\cdot}2 + 0{\cdot}3 \times 135 + 0{\cdot}1 \times 304{\cdot}2 = 90{\cdot}9°K$

$p_{pc} = 0{\cdot}6 \times 12{\cdot}8 + 0{\cdot}3 \times 34{\cdot}5 + 0{\cdot}1 \times 73{\cdot}0 = 25{\cdot}33$ atm

Starting from van der Waals' equation, Joffe[176] derived formulae for calculating the pseudocritical constants. A relationship exists between the values of the critical constants of compounds and those of the constants a and b in van der Waals' equation. The latter constants may be calculated additively (cf. section 5a).

If we refer to van Laar's equations as the basis:

$$a_m^{1/2} = x_1 a_1^{1/2} + x_2 a_2^{1/2}$$
$$b_m = x_1 b_1 + x_2 b_2 \tag{1.145}$$

we obtain the relationships

$$\frac{T_{pc}}{p_{pc}^{1/2}} = \frac{x_A T_{cA}}{p_{cA}^{1/2}} + \frac{x_B T_{cB}}{p_{cB}^{1/2}} \tag{1.146}$$

$$\frac{T_{pc}}{p_{pc}} = \frac{x_A T_{cA}}{p_{cA}} + \frac{x_B T_{cB}}{p_{cB}} \tag{1.147}$$

If, instead, after Lorentz another additive relationship is taken for calculating the constant b

$$b_m = x_A^2 b_A - x_B^2 b_B - \tfrac{1}{4}(b_A^{1/3} - b_B^{1/3})^3 x_A x_B \tag{1.148}$$

we obtain for a binary mixture

$$\frac{T_{pc}}{p_{pc}^{1/2}} = \frac{x_A T_{cA}}{p_{cA}^{1/2}} - \frac{x_B T_{cB}}{p_{cB}^{1/2}} \tag{1.149}$$

$$\frac{T_{pc}}{p_{pc}} = \frac{x_A^2 T_{cA}}{p_{cA}} + \frac{x_B^2 T_{cB}}{p_{cB}} + \tfrac{1}{4}\left[\left(\frac{T_{cA}}{p_{cA}}\right)^{1/3} + \left(\frac{T_{cB}}{p_{cB}}\right)^{1/3}\right]^3 x_A x_B \tag{1.150}$$

In the case of a multicomponent mixture we get

$$\frac{T_{pc}}{p_{pc}^{1/2}} = \Sigma_i x_i (T_{ci}/p_{ci}^{1/2}) \tag{1.151}$$

$$\frac{T_{pc}}{p_{pc}} = \tfrac{1}{8}\Sigma_i \Sigma_j x_i x_j \left[\left(\frac{T_{ci}}{p_{ci}}\right)^{1/3} + \left(\frac{T_{cj}}{p_{cj}}\right)^{1/3}\right]^3 \tag{1.152}$$

As may be seen from Table 1.52, compiled by Joffe,[176] these equations permit more accurate predictions than Kay's equations.

The methods for calculating the pseudocritical parameters of mixtures presented by Kay and Joffe, permit calculation of T_{pc}, p_{pc}, and V_{pc}. In order to determine whether two systems are in corresponding states, it is necessary, according to the modified theory of corresponding states (section 8d), to check not only the equality of the reduced

TABLE 1.52. COMPARISON OF METHODS FOR CALCULATING PSEUDOCRITICAL PARAMETERS

System		CH_4–nC_4H_{10}	CH_4–C_3H_8	CH_4–C_2H_6	CH_4–CO_2	CO_2–N_2
First component, mole %		50·44	60	50·71	40·55	50·48
Maximum pressure, atm		191	200	221	272	250
Temperature range, °C		100–300	55–90	21–121	38–238	25–125
Equations	Error, %					
Kay's method	average	2·67	2·63	2·07	3·03	1·83
	maximum	10·2	4·58	6·14	7·54	3·75
(1.146)	average	2·26	2·66	2·04	—	—
(1.147)	maximum	8·80	4·66	5·20	—	—
Joffe's method						
(1.149)	average	1·09	1·62	1·44	2·26	1·07
(1.150)	maximum	3·19	4·17	4·08	4·59	3·17

parameters, but also the equality of the third parameter (e.g. the critical parameter α_c after Riedel, the acentric factor ω after Pitzer or the compressibility factor $z_c = (p_c V_c)/(RT_c)$ at the critical point).

Prausnitz and Gunn[298] have tried to adapt the modified theory of corresponding states to mixtures by introducing a third parameter criterion. They found that the pseudocritical pressure constitutes a rather complicated function, which cannot be calculated by means of simple equations. If on the other hand the ratio of the molal critical volumes of the constituents of a mixture is smaller than 3, the molal pseudocritical volume of a mixture of the components A, B, C may be calculated by means of the simple equation:

$$V_{pc} = x_A V_{cA} + x_B V_{cB} + x_C V_{cC} \tag{1.153}$$

In the same way we can calculate the value of the critical compressibility factor of a mixture

$$z_{pc} = x_A z_{cA} + x_B z_{cB} + x_C z_{cC} \tag{1.154}$$

If the reduced temperature is larger than 1·3, the application of Kay's equation (1.143) to the calculation of the pseudocritical temperature (even if the components differ considerably in their critical temperatures) has given for ten different cases a maximum deviation of about 7% from the values determined experimentally. On the other hand, the pseudocritical pressures calculated by means of Kay's equation (1.144) show considerable deviations of about 20% or even 50% from the experimental values. Therefore, Prausnitz and Gunn recommend Eqs. (1.153) and (1.154) as well as the following simplified equations for computing the pseudocritical constants of a mixture with n components:

Acentric factor

$$\omega_{pc} = \sum_{i=1}^{n} x_i \omega_i$$

Pseudocritical temperature

$$T_{pc} = \sum_{i=1}^{n} x_i T_{ci} \tag{1.155}$$

Pseudocritical pressure

$$p_{pc} = \frac{RT_{pc}}{\sum_{i=1}^{n} x_i V_{ci}} \sum_{i=1}^{n} x_i z_{ci}$$

Generally speaking, these equations give more accurate results than Kay's, if the calculation does not involve a region lying close to the critical point.

Theoretical Equations for the Calculation of Pseudocritical Constants

Calculating the second virial coefficient in the equation of state of real gas, using the Lennard-Jones equation (1.29), Leland and Mueller[218] have derived equations for the pseudocritical temperature and pseudocritical pressure of a mixture

$$T_{pc} = \left[\frac{\sum_{i=1}^{n}\sum_{j=1}^{n} x_i x_j \left(\frac{z_c T_c^{\alpha+1}}{p_c}\right)_i^{1/2} \left(\frac{z_c T_c^{\alpha+1}}{p_c}\right)_j^{1/2}}{\sum_{i=1}^{n}\sum_{j=1}^{n} x_i x_j \left\{\frac{1}{2}\left(\frac{z_c T_c}{p_c}\right)_i^{1/3} + \frac{1}{2}\left(\frac{z_c T_c}{p_c}\right)_j^{1/3}\right\}^3} \right]^{1/\alpha} \tag{1.156}$$

The value of the power exponent α is read as a function of T and p from the table presented by these authors.

$$p_{pc} = \frac{T_{pc} \sum_{i=1}^{n} x_i (z_c)_i}{\sum_{i=1}^{n}\sum_{j=1}^{n} x_i x_j \left[\frac{1}{2}\left(\frac{z_c T_c}{p_c}\right)_i^{1/3} + \frac{1}{2}\left(\frac{z_c T_c}{p_c}\right)_j^{1/3}\right]^3} \tag{1.157}$$

The authors referred to have checked their equations by comparing the calculated values with those determined experimentally. Pure compounds (hydrocarbons from CH_4 to n-C_7H_{16}) were used for which the value of the third parameter z_c is close to the value of this criterion for a mixture. Because of the complicated form of the functions for T_{pc} and p_{pc} an electronic computer was used for these calculations. A high degree of accuracy was achieved, the errors arising from the theoretical equations (1.156) and (1.157) usually being $\frac{1}{2}$ to $\frac{1}{4}$ of those arising from the use of Kay's method.

Arithmetical difficulties restrict the use of the equations of Leland and Mueller. If these equations are applied to a mixture containing polar molecules, the error increases.

Choice of Method for Calculating the Pseudocritical Parameters of Mixtures

Most often Kay's method is used, because of its speed and simplicity. Since under certain conditions this method may give large errors, it is advisable to employ Joffe's equations (1.149)–(1.152), for which the error is several times smaller.

If the modified theory of corresponding states must be applied it is appropriate to use Prausnitz and Gunn's equations (1.155), which are also more accurate than those of Kay. The theoretical equations (1.156) and (1.157) are too complicated and usually unsuitable for technological calculations.

CHAPTER 2

SURFACE TENSION OF LIQUIDS

1. BASIC RELATIONSHIPS

The potential energy of a molecule situated at the surface of a liquid phase is higher than the energy of a molecule in the interior of the liquid.† Therefore, every extension of the liquid area requires an expenditure of work. The energy that is supplied to a liquid in order to increase its surface area by one unit is equal to the *surface tension* of the liquid and is usually expressed in ergs/cm^2.

As a result of the tendency of any system to come to equilibrium in a state for which the external work performed during the formation of the system is a minimum, the liquid shows a tendency to assume a minimum surface area. If the surface involved is flat with a definite perimeter, the effect of this surface tension is to tend to reduce this perimeter. This phenomenon can be attributed to forces acting in the plane of the surface, perpendicularly to the border of the perimeter line.‡ The surface tension may be defined as the magnitude of the *force acting on unit perimeter of the liquid surface area*, and is expressed in dynes/cm. This definition is equivalent to the definition mentioned above. In terms of either of these definitions, the dimensions of the unit of surface tension are: erg/cm^2 = dyne · cm/cm^2 = dyne/cm.

Quite often use is made of the molar surface tension σ_M:

$$\sigma_M = \sigma V^{2/3} \tag{2.1}$$

where σ is the surface tension, in dynes/cm,
and $V = M/\rho_l$ is the volume of one mole of the liquid, cm^3/mole.

Eötvös[89] has found that the relationship

$$\sigma V^{2/3} = K(T_c - T) \tag{2.2}$$

for the molar surface tension is approximately obeyed in many cases. Better agreement with experiment may be obtained, if the critical temperature T_c in Eq. (2.2) is replaced by the quantity $(T_c - 6)$ as in the equation of Ramsay and Shields[304]

$$\sigma V^{2/3} = K(T_c - 6 - T) \tag{2.3}$$

The value of the constant K for non-associated liquids is close to 2·12. For example, for liquefied nitrogen $K_{N_2} = 2$, for n-hexane $K_{C_6H_{14}} = 2{\cdot}11$, benzene $K_{C_6H_6} = 2{\cdot}10$, carbon disulphide $K_{CS_2} = 2{\cdot}04$, carbon tetrachloride $K_{CCl_4} = 2{\cdot}21$, and for propyl acetate $K_{CH_3COOC_3H_7} = 2{\cdot}04$. The values for associated liquids are considerably lower, for example, $K_{H_2O} = 0{\cdot}89$, $K_{CH_3OH} = 0{\cdot}93$, $K_{(CH_3)_2CO} = 1{\cdot}82$, and $K_{HCOOH} = 0{\cdot}90$. If the compound

† In our considerations it has been assumed that above a liquid there is nothing but its saturated vapour.

‡ If the liquid surface is curved, this force acts in a plane tangential to the surface and in a direction which is perpendicular to the perimeter of the surface.

consists of large molecules, the values of the constant K are usually relatively high as for example in the cases of diphenylethane ($K = 2{\cdot}49$), benzophenone ($K = 2{\cdot}63$) and diphenylamine ($K = 2{\cdot}57$).

The magnitude of the surface tension does not depend solely on the structure of the liquid surface, but also on the nature and concentration of the molecules in the gaseous phase above the liquid surface. The fact that the molecules in the liquid surface are subject to the action of the molecules of the liquid phase from one side only is to some extent compensated by the fact that some interactions take place between the surface molecules and molecules in the gaseous phase. The higher the concentration of the latter, the less would the potential energy of the surface molecules be expected to differ from the energy of the molecules in the interior of the liquid and, hence, the lower should be the surface tension. This is confirmed by McLeod's empirical formula:[244]

$$\sigma = C(\rho_l - \rho_n)^4 \tag{2.4}$$

where C is a constant,
ρ_l is the liquid density,
and ρ_n is the density of the saturated vapour over this liquid.

The following data for benzene[97] illustrate this:

t, °C	ρ_l	ρ_n	σ	C
80	0·8127	0·0027	20·28	47·11
100	0·7928	0·0047	18·02	46·70
150	0·7310	0·0144	12·36	46·87
200	0·6610	0·0355	7·17	46·86
250	0·5610	0·0857	2·56	50·20

According to Katayama,[192] more accurate results are obtained by using the corrected equation of Eötvös (2.2), in which the effect of the concentration of a compound present in the vapour on the surface tension is taken into account.

Katayama recommends the formula

$$\sigma = K(T_c - T)\left(\frac{\rho_l - \rho_n}{M}\right)^{2/3} \tag{2.5}$$

By multiplying McLeod's formula (2.4) by M, Sugden obtained a new physical constant, which he called the parachor, and found that it had constitutive and additive properties

$$P_{ch} = \frac{M\sigma^{1/4}}{(\rho_l - \rho_n)} \tag{2.6}$$

The parachor has already been discussed in more detail in Ch. 1, section 4c as an example of a quantity which can be calculated additively by the summation of group contributions. A nomograph for the rapid calculation of P_{ch} has also been presented (Fig. 1.6).

The effect of the concentration and nature of the molecules in the gaseous phase upon surface tension has been clearly demonstrated in some experiments in which surface tension has been measured in the presence of inert gases under pressure.[208] Efremov and Golubev[82] have found for the cyclohexane–hydrogen system (at 20°C):

Pressure, atm	1	100	300
Surface tension, dynes/cm	25·3	22·5	19·2

When investigating the effects of gas pressure on the surface tensions of the n-paraffins C_6–C_9 and of cyclohexane, Gielessen and Schmatz[126] found that the surface tensions of these liquids decreased with increasing pressure of nitrogen or argon, whereas in the case of helium they increased linearly. The results of these investigations as well as those of other authors are presented in Figs. 2.1 and 2.2.

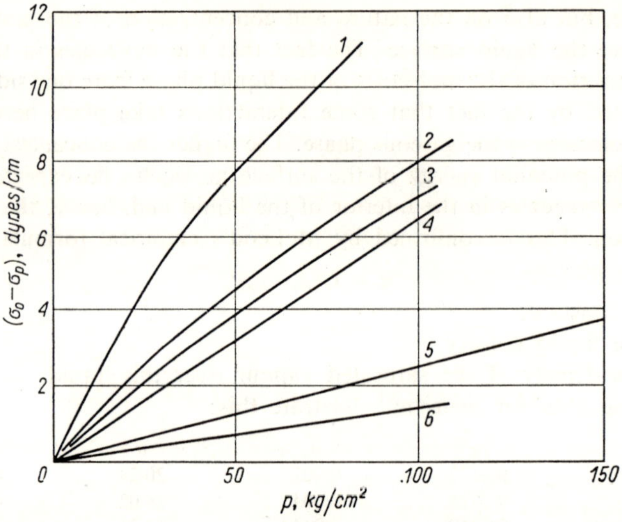

FIG. 2.1. Examples of some cases where surface tension decreases with "inert" gas pressure (after Gielessen and Schmatz[126]). σ_p is the surface tension at pressure p and σ_0 the value to which the surface tension tends as the pressure tends to zero; *1*—CH_4/n-C_6H_{14}; CH_4/H_2O; *2*—N_2/C_9H_{12}; *3*—N_2/n-C_6H_{14}; *4*—N_2/H_2O; *5*—H_2/alcohol, H_2/n-C_6H_{14}; *6*—H_2/H_2O.

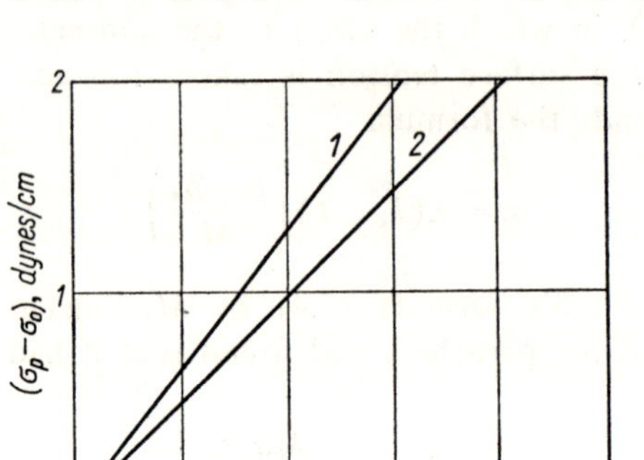

FIG. 2.2. Examples of some cases where surface tension increases with "inert" gas pressure (after Gielessen and Schmatz[126]). Symbols as in Fig. 2.1. *1*—He/n-C_6H_{14}; *2*—He/n-C_9H_{20}.

If surface tension decreases with increase of pressure, as in the case of the nitrogen–n-paraffin system, its variation with pressure may be represented by the equation:[12]

$$\sigma_0-\sigma_p = \frac{abp}{1+bp+cp^2} \tag{2.7}$$

where σ_0 is the surface tension under normal pressure,

σ_p is the surface tension under pressure p,

and a, b and c are constants.

According to Gielessen and Schmatz,[126] any decrease of the surface tension associated with increase in pressure is caused by increase in the concentration of the molecules of the "inert" gas at the surface.

Because of the great importance of surface tension data in technical calculations (e.g. in the calculation of heat transfer coefficients of boiling liquids, or flow rates in two-phase (liquid–vapour) systems as encountered in rectification, numerous attempts have been made to develop methods for the calculation of the value σ. The statistical theory of surface tension makes it possible to calculate this quantity. Eyring *et al.*[323] have calculated surface tensions and Eötvös constants for argon, nitrogen and methane, which are in good agreement with the experimental values. Such calculations, however, cannot be used for the rapid estimation of surface tensions under technological conditions. If a rapid method is required use is usually made of simple empirical or semi-empirical relationships. The relevant methods have been reviewed by Gambill.[111,118]

2. DEPENDENCE OF SURFACE TENSION ON MOLECULAR PROPERTIES

Numerous studies of the influence of molecular shape on surface tension are reported in the literature. Lutskii[229] has attempted to interpret systematically the dependence of surface tension upon such properties of the molecules as mass, volume, shape, dipole moment, and polarizability. His investigations were concerned with normal liquids, composed of simple molecules (i.e. containing no clusters of molecules, complexes, ions, etc.). He divided the liquids into groups so selected that those within any given group differed mainly in one molecular property only (e.g. in mass), whereas the remaining molecular properties were approximately the same. In this way he was able to investigate the effects of all the above factors independently.

Table 2.1 lists the molecular properties of several of the compounds which he considered.

Comparing the surface tension of liquids, the molecules of which differ in mass owing to differences in isotopic composition (e.g. water and heavy water, hydrogen peroxide and deuterium peroxide, etc.) we come to the conclusion that the surface tension is approximately independent of the mass of the molecule, i.e. that

$$(\sigma)_{p,T,\bar{\mu},\alpha,v_M,\phi} = \text{constant} \tag{2.8}$$

where σ is the surface tension. The meaning of the remaining symbols is given in Table 2.1.

Studying the dependence of the surface tension on the volume of the molecule in the isologous series of type I, it has been found that

$$(\sigma v_M^{2/3})_{p,T\bar{\mu},\alpha,\phi} = \text{constant} \tag{2.9}$$

Since the molal volume is proportional to the value of v_M, the result obtained supports Eötvös' rule (the critical temperatures of the compounds compared are alike).

Lutskii showed that the values of $(\sigma v_M^{2/3})$ in the series of isologues of type II increase in direct proportion to the electron polarizability of the molecules, which in turn is proportional to the molar refraction R_D of the compound, raised to the power n, which depends on the temperature. Thus, as shown in Fig. 2.3:

$$(\sigma v_M^{2/3})_{p,T,\bar{\mu},\phi} = \text{constant} \times R_D^n \tag{2.10}$$

TABLE 2.1. CERTAIN MOLECULAR PROPERTIES WHICH INFLUENCE THE SURFACE TENSION OF LIQUIDS (AFTER LUTSKII[229])

Variable under consideration	Compound	Mass $m \times 10^{23}$ g	Volume $v_M \times 10^{24}$ cm^3	Electron polarizability $\alpha \times 10^{24}$ cm^3	Dipole moment μ D	Shape of the molecule ϕ
Molecular weight (isotopes)	H_2O	2·99	18·79	1·47 1·46	1·76 1·86	triangular
	D_2O	3·32	18·79	1·45 1·44	1·78 1·87	”
	NH_3	2·82	21·84	—	1·46	pyramid
	ND_3	3·33	21·84	—	1·49	”
Dipole moment (isoperiodic compounds)	$n\text{-}C_4H_{10}$	9·64	67·88	8·19	0	straight chain
	$C_2H_5\text{—}O\text{—}CH_3$	10·00	62·80	7·01	1·2	”
	$n\text{-}C_3H_7\text{—}NH_2$	9·74	65·85	7·69	1·39	almost straight
	$n\text{-}C_3H_7\text{—}OH$	10·00	62·80	6·98	1·76	chain
	$n\text{-}C_3H_7F$	10·23	57·48	6·26	1·9	”
	$C_2H_5\text{—}CHO$	9·57	57·73	6·35	2·4	”
	$CH_3\text{—}CO\text{—}CH_3$	9·57	57·73	6·35	2·76	”
	$C_2H_5\text{—}CN$	9·08	58·50	6·22	3·66	”
Molecular shape	$n\text{-}C_6H_{14}$	14·29	97·22	11·78	0	chain
	$i\text{-}C_6H_{14}$	14·29	97·22	11·51	0	branched chain
	$C_2H_5\text{—}C(CH_3)_2CH_3$	14·29	97·22	11·83	0	tetrahedral
	cyclo-C_6H_{12}	13·96	88·02	10·96	0	cyclic
Molecular volume (type I isologues)	$n\text{-}C_3H_7I$	28·22	83·97	11·48	1·92	chain
	$n\text{-}C_4H_9Br$	22·74	89·00	11·19	1·95	”
	$n\text{-}C_5H_{11}Cl$	17·68	99·21	11·97	$\approx$ 1·92	”
	$n\text{-}C_6H_{13}F$	17·26	101·49	$\approx$ 11·85	$\approx$ 1·9	”
Electron polarizability (type II isologues)	$n\text{-}C_5H_{11}I$	32·87	113·31	15·13	$\approx$ 1·9	chain
	$n\text{-}C_8H_{17}Br$	32·05	147·68	18·56	$\approx$ 1·9	”
	$n\text{-}C_{19}H_{23}Cl$	31·66	187·23	22·94	$\approx$ 1·9	”

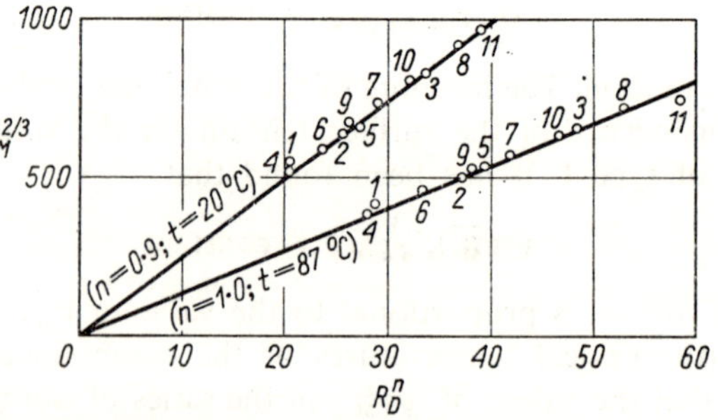

FIG. 2.3. Dependence of the surface tension σ of a liquid compound on the volume v_M and electron polarizability (molar refraction R_D) of the molecules of the compound ($\sigma v_M^{2/3}$ = constant $\times R_D^n$); *1*—$n\text{-}C_3H_7I$; *2*—$n\text{-}C_6H_{13}Br$; *3*—$n\text{-}C_9H_{19}Cl$; *4*—$n\text{-}C_4H_9Br$; *5*—$n\text{-}C_7H_{15}Cl$; *6*—$n\text{-}C_4H_9I$; *7*—$n\text{-}C_7H_{15}Br$; *8*—$n\text{-}C_{10}H_{21}Cl$; *9*—$n\text{-}C_5H_{11}I$; *10*—$n\text{-}C_8H_{17}Br$; *11*—$n\text{-}C_{11}H_{23}Cl$ (after Lutskii[229]).

The dependence of surface tension (at 20°C) upon the shape of the molecules has been represented by Lutskii in the following form:

$$\begin{aligned} \sigma_{\text{(branched chain)}} &= (0{\cdot}93 \pm 0{\cdot}02) \times \sigma_{\text{(non-branched chain)}} \\ \sigma_{\text{(tetrahedron)}} &= (0{\cdot}87 \pm 0{\cdot}03) \times \sigma_{\text{(non-branched chain)}} \\ \sigma_{\text{(cyclic)}} &= (1{\cdot}31 \pm 0{\cdot}03) \times \sigma_{\text{(non-branched chain)}} \end{aligned} \qquad (2.11)$$

The effect of the dipole moment $\bar{\mu}$ on the surface tension of the liquids may be represented by the equation

$$(\sigma)_{p,T,\alpha,v_M,\phi} = (\sigma)_{p,T,\alpha,v_M,\phi,\bar{\mu}=0} + \text{constant} \times \bar{\mu} \qquad (2.12)$$

where $(\sigma)_{p,T,\alpha,v_M,\phi,\bar{\mu}=0}$ denotes a constant equal to the surface tension of a liquid which consists of molecules with properties resembling those of the molecules of the liquid under consideration, but with the dipole moment $\bar{\mu} = 0$.

The dependence of the surface tension σ of liquids on the dipole moment $\bar{\mu}$ of the molecules of the liquid is presented in Fig. 2.4.

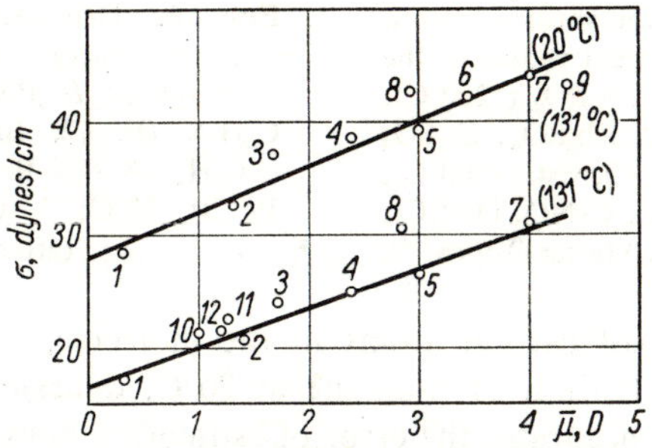

FIG. 2.4. Relation between the surface tension σ of a liquid on the dipole moment $\bar{\mu}$ of its molecules; *1*—n-C_6H_5—C_3H_5; *2*—C_6H_5—O—C_2H_5; *3*—C_6H_5—$N(CH_3)_2$; *4*—$C_6H_4(OH)OCH_3$; *5*—C_6H_5—CO—CH_3; *6*—$C_6H_5CH_2$—CN; *7*—$C_6H_5NO_2$; *8*—*o*-$C_6H_4(OH)CHO$; *9*—*p*-$C_6H_4(OH)CHO$; *10*—*o*-$C_6H_4(OCH_3)CH_3$; *11*—*m*-$C_6H_4(OCH_3)CH_3$; *12*—*p*-$C_6H_4(OCH_3)CH_3$ (after Lutskii[229]).

The simple relationships existing between the surface tension σ, volume v_M, polarizability α (hence also the molar refraction R_D) and dipole moment $\bar{\mu}$ of the molecules of a liquid compound, as found by Lutskii, have enabled him to interpret the dependence of σ on the number of carbon atoms in homologous series, etc.

In the case of associated liquids the relationships are much more complicated and a simple interpretation of these relationships is not possible.

Joliet[178] has shown that for the series: C_6H_5X, *p*-CH_3—C_6H_4—X, PX_3 (where X denotes a halogen atom) as well as for the series of alkyl halides RX an exact linear relationship between the surface tension and the covalent radius of the halogen exists (Chapter 1).

Osipov[266] has found for these series a linear dependence of the surface tension σ on the energy of the bond C—X or P—X. He has also found analogous linear relationships between the surface tension σ and the ionic radius and the bond energy for a series of fused inorganic salts. In Figs. 2.5 and 2.6 examples of the relationships encountered are presented.

For hydrocarbons (n-alkanes), Wiener[111] has presented the empirical formula

$$\sigma = 14{\cdot}6\log(n-3)+11{\cdot}52 \qquad (2.13)$$

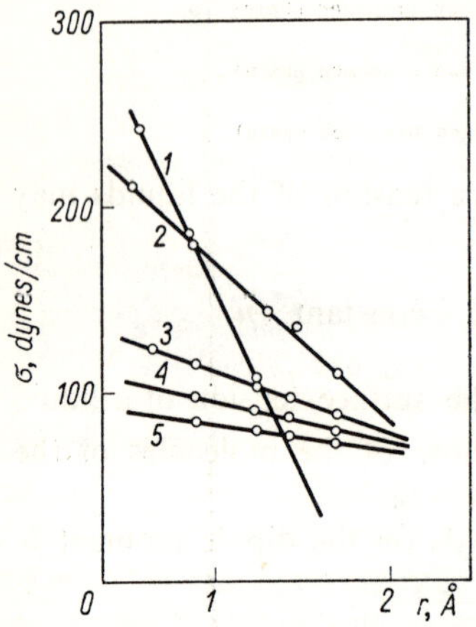

FIG. 2.5. Dependence of the surface tension σ of fused salts on the covalent radius of the ion (Å); *1*—$LiBO_2$, $NaBO_2$, KBO_2 (1200°C); *2*—$LiSO_4$, Na_2SO_4, K_2SO_4, Rb_2SO_4, Cs_2SO_4 (1100°C); *3*—LiCl, KCl, RbCl, CsCl (800°C); *4*—$NaNO_3$, KNO_3, $RbNO_3$, $CsNO_3$ (500°C); *5*—NaI, KI, RbI, CsI (800°C) (after Osipov[266]).

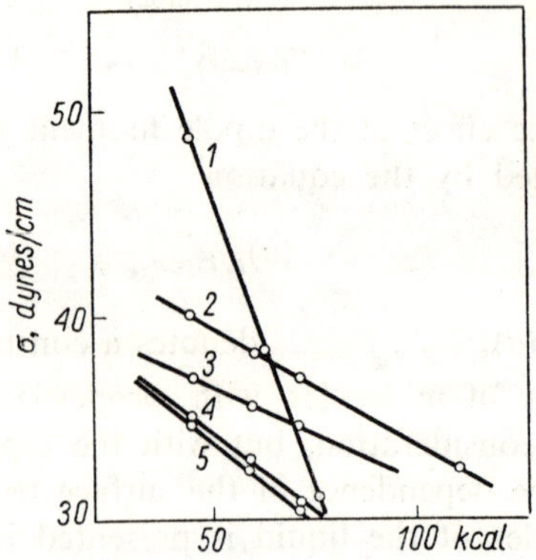

FIG. 2.6. Dependence of the surface tension σ on the energy of the bond R—X (X = = halogen); *1*—PX_3 (Cl, Br, I), 75°C; *2*—C_6H_5X (F, Cl, Br, I), 20°C; *3*—p-CH_3—C_6H_4—X (Cl, Br, I), 60°C; *4*—C_4H_9X (Cl, Br, I), 20°C; *5*—C_3H_7X (Cl, Br, I), 20°C (after Osipov[266]).

where n is the number of carbon atoms in the molecule,
and σ is the surface tension of n-alkane at 20°C, dynes/cm.

This formula is fairly accurate; the error of estimation does not usually exceed ±1%.

3. METHODS OF ESTIMATING THE SURFACE TENSION OF LIQUIDS

a. Estimation of the Surface Tension of a Liquid from the Parachor

Since the parachor is a constitutive and additive quantity and may be estimated with fair accuracy by summing the group contributions, the simplest way of calculating the surface tension of a liquid is to calculate additively the value of the parachor P_{ch} (cf. Ch. 1, section 4c), and to determine the value of the surface tension σ (at 20°C) using the transformed equation (2.6)

$$\sigma = \left[\frac{P_{ch}(\rho_l-\rho_n)}{M}\right]^4 \qquad (2.14)$$

where ρ_l is the liquid density,
ρ_n is the density of the saturated vapour,
and M is the molecular weight.

Since the value of the parachor is, over a wide range of temperatures, almost completely independent of temperature, Eq. (2.14) will determine the value of the surface tension σ at the temperature corresponding to the densities ρ.

At temperatures lying far from the critical temperature, the density of the vapour is very small as compared with the density of the liquid

$$\rho_n \ll \rho_l$$

The quantity ρ_n may then be disregarded, and the simplified equation

$$\sigma = \left(\frac{P_{ch}\rho_l}{M}\right)^4 \tag{2.15}$$

or

$$\sigma = (P_{ch}/V_M)^4 \tag{2.16}$$

may be used, where V_M denotes the molal volume of the liquid.

The latter quantity is constitutive and additive and may be calculated in a similar way to the parachor, i.e. by summing group contributions (cf. Ch. 1, sections 4a and 6). In this way the surface tension σ can be found, when only the structural formula of a compound is known. Such a calculation, however, is usually not very accurate since before σ can be calculated the quantity (P_{ch}/V_M) must be raised to the fourth power. A small error in (P_{ch}/V_M) therefore produces a much larger error in σ. For a rapid calculation of σ according to Eq. (2.14) the nomograph presented in Fig. 1.6 may be used.

More accurate results may be obtained using the methods recommended by Tripathi,[370] Meissner and Michaels,[246] and Kharbanda.[198]

Combining Eqs. (2.6) and (1.47) for calculating the parachor and the molar refraction R_D, we get the equation

$$\sigma = \left[\frac{P_{ch}(n^2-1)}{R_D(n^2+2)}\right]^4 \tag{2.17}$$

where n is the refractive index (sodium D line). Usually the errors in estimating σ from the latter equation do not exceed $\pm 2{\cdot}5\%$. The accuracy of calculating the surface tension by means of Eq. (2.17) is illustrated by the following comparison:[282]

Compound	Acetone	Ethyl iodide	Ethanol	Octane	Nitromethane
Surface tension, calculated	23·35	28·05	21·92	21·30	35·38
Surface tension, determined experimentally	23·09	28·10	22·03	21·32	35·60

b. Surface Tensions of Liquid Mixtures

As Michaels, Alexander and Becker[248] have shown, Eq. (2.17) may be used to estimate the surface tensions of liquid mixtures.

The parachor and molar refraction of a mixture are calculated additively

$$\begin{aligned} P_{ch,m} &= P_{ch,1}x_1+P_{ch,2}x_2+P_{ch,3}x_3 \\ R_{D,m} &= R_{D,1}x_1+R_{D,2}x_2+R_{D,3}x_3 \\ n_m &= n_1x_1+n_2x_2+n_3x_3 \end{aligned} \tag{2.18}$$

σ_m the surface tension of the mixture is then calculated using the equation:

$$\sigma_m = \left[\frac{P_{ch,m}(n_m^2-1)}{R_{D,m}(n_m^2+2)}\right]^4 \tag{2.19}$$

The subscripts 1, 2 and 3 refer to the first, second and third components of the mixture and the subscript m refers to the mixture.

The error of estimating surface tensions of binary mixtures by Eq. (2.19), is about 3·4–7%, whereas for ternary mixtures it may be as high as 19%. This method is not suitable for estimating the surface tensions of liquid mixtures differing considerably in degree of polarity.

A drawback to Eqs. (2.17) and (2.19) is the necessity of knowing the refractive index n of the substances involved. Kharbanda[198] has constructed a nomograph (Fig. 2.7) for the rapid calculation of the surface tension σ based on Eqs. (2.17) and (2.19). When

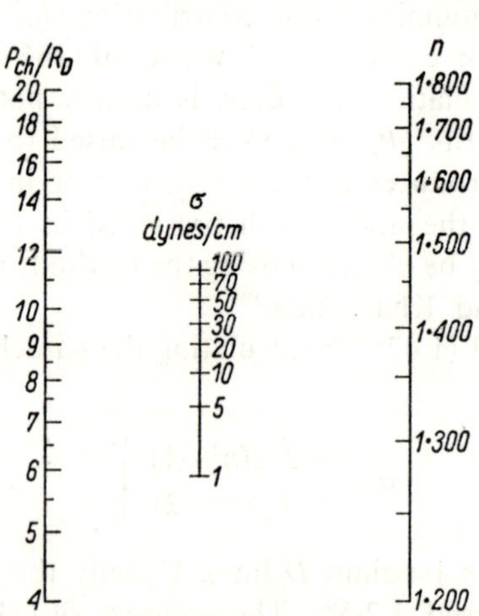

FIG. 2.7. Nomograph for calculating the surface tensions of liquids (after Ketelaar[196]).

using this nomograph, one should calculate the values of the parachor P_{ch} and of the molar refraction R_D by summing their contributions, find the value of the P_{ch}/R_D ratio (by division), and connect the calculated value of P_{ch}/R_D (left scale) by means of a straight line with the experimental value of the refractive index n. The required value (dynes/cm) is read from the middle scale.

The approximate formula (2.15) is also suitable for calculating the surface tensions of mixtures of non-polar and non-associated liquids at low pressures. The values of P_{ch}, ρ_l and M are calculated additively for the pure components and the values for the mixture are then computed from these as averages weighted according to the molar composition. According to Hammick and Andrew[141] for non-associated liquids the error of estimation amounts to 0·5%, maximally to 1%, for associated liquids to about 3·5%.

These methods for estimating surface tensions of liquid mixtures are only suitable for application to systems subjected to low pressures. If a system is under high pressure, up to 100 atm, as with a liquid in equilibrium with its vapour at a temperature well in excess of its normal bubble point, good results may be obtained by calculating the surface tension using an equation, which has been derived by Weinaug and Katz[396] and by Reno and Katz

$$\sigma_m^{1/4} = P_{ch,1}\left(\frac{\rho_l x_1}{M_l} - \frac{\rho_n y_1}{M_n}\right) + P_{ch,2}\left(\frac{\rho_l x_2}{M_l} - \frac{\rho_n y_2}{M_n}\right) + \ldots \qquad (2.20)$$

In this equation σ_m is the surface tension of the mixture,
$P_{ch,1}$ and $P_{ch,2}$ are the parachors of components 1 and 2,
ρ_l and ρ_n are the densities of the liquid and of the saturated vapour,
M_l and M_n are the mean molecular weights of the liquid and of the saturated vapour,
x_1 and x_2 are the mole fractions of components 1 and 2 in the liquid,
and y_1 and y_2 are the mole fractions of components 1 and 2 in the vapour.

d. Variation of Surface Tension with Temperature and Pressure

The effect of temperature T (°K) upon the surface tension of liquids is expressed by the equations presented in section 1 of this chapter, which show that the surface tension decreases with increase in temperature.

The linear relationship between the molar surface tension $\sigma V^{2/3}$ and the difference of temperature $(T_c - T)$, as found by Eötvös, tends to break down in the region close to the critical temperature T_c. Equation (2.3) modified by Ramsay and Shields, as well as Katayama's equation (2.5), give fairly good results.

In order to use the above equations it is necessary to know either the liquid density ρ_l (for estimating the volume V of 1 mole of the liquid) or the densities ρ_l and ρ_n of the liquid and the saturated vapour. Values of the constant K, and the critical temperature are also required. If these quantities are not available, an approximate estimation of σ may be made from the value $\sigma_{T'}$ of the surface tension at any temperature T'. Using this it is possible to calculate the value of the constant K. The approximate values of V and T_c may be calculated as constitutive and additive quantities (cf. Chapter 1).

Another approximate method for estimating the value of the surface tension σ_T at temperature T, when one value of the liquid density ρ_l at this temperature is known, is based on Eq. (2.15). If the liquid density at another temperature is available, the approximate value of ρ_l at temperature T can be calculated by means of the methods described in Chapter 1.

Van der Waals made use of the theory of corresponding states to correlate the variation of the surface tension σ with temperature T (°K). His equation[377] has the following form:

$$\sigma = \sigma_0(1 - T/T_c)^n \qquad (2.21)$$

where T_c denotes the critical temperature, and σ_0 and n are constants, characteristic of the liquid.

In the case of non-associated liquids, the average value of the power exponent $n = 1{\cdot}23$. For normal alkanes, Markov[236] has obtained excellent agreement with experiment, assuming that $\sigma_0 = 54{\cdot}292$ and $n = 1{\cdot}259$. However, the rounded-off value $n = 1{\cdot}2$ is often used.

It follows from Eq. (2.21) that for two temperatures T_1 and T_2

$$\frac{\sigma_2}{\sigma_1} = \left(\frac{T_c \;\; T_2}{T_c - T_1}\right)^{1{\cdot}2} \qquad (2.22)$$

For a rapid solution of this equation a nomograph constructed by Kharbanda[198] may be used (Fig. 2.8).

If in Eq. (2.22) we assume that $T_1 = T_b$ (the boiling temperature), we obtain

$$\sigma = \sigma_b \left(\frac{T_c - T}{T_c - T_b} \right)^{1.2} \tag{2.23}$$

By substituting the approximate value $(3P_{ch}T_b)/(4V_b)$ for T_c in this equation Bowden[32] showed that:

$$\sigma = \left(\frac{P_{ch}}{V_b} \right)^4 \left(\frac{3P_{ch}T_b - 4V_b T}{3P_{ch}T_b - 4V_b T_b} \right)^{1.2} \tag{2.24}$$

where V_b denotes the molal volume at the boiling temperature T_b. The quantities P_{ch} and V_b are calculated additively by summing the group contributions. Bowden's formula

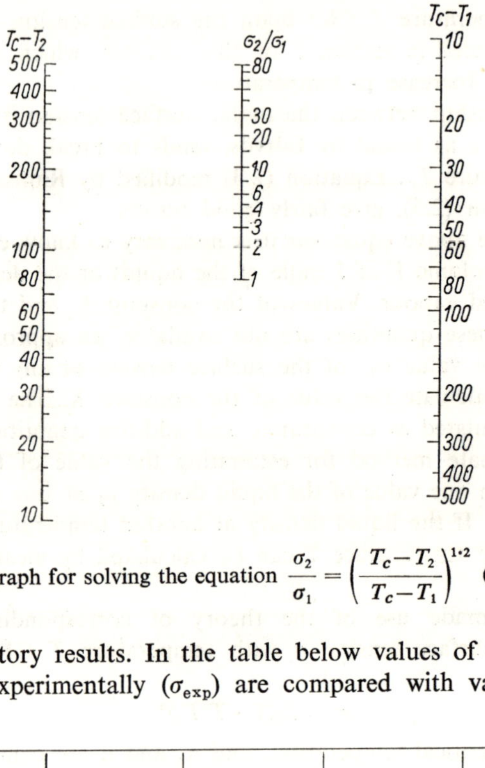

FIG. 2.8. Nomograph for solving the equation $\dfrac{\sigma_2}{\sigma_1} = \left(\dfrac{T_c - T_2}{T_c - T_1} \right)^{1.2}$ (after Ketelaar[196]).

generally gives satisfactory results. In the table below values of the surface tension of benzene, determined experimentally (σ_{exp}) are compared with values (σ_{calc}) calculated from Eq. (2.24).

Temperature, °C	20	40	60	80	100	160
σ_{exp} (dynes/cm)	29	26·3	23·6	21·2	18·2	9·4
σ_{calc} ,,	28·5	26·1	23·7	21·2	18·9	10·0

The exponent in Eq. (2.22) is higher than 1·2 in the case of liquids with associated molecules. This equation may also be employed in the case of liquid mixtures, though in this case the pseudocritical temperature T_{pc} of the mixture, calculated, e.g. by means of Kay's equation (1.144),[111] is used instead of the critical temperature T_c.

The dependence of the surface tension on specific volume has been investigated. Since the specific volume is itself a function of temperature, the equation obtained may be regarded as providing an alternative method for predicting surface tension as a function of temperature. Such a relationship is represented, for example, by the equation put forward by Volyak:[388]

$$\sigma = A(v'^2 e^{-xv'} - v''^2 e^{-xv''}) \tag{2.25}$$

where v' is the specific volume of the liquid,
v'' is the specific volume of the vapour,
and A and x are constants characterizing a given compound.

Usually the second term is extremely small, and may be neglected. The values of the constants A and x are calculated from the values of σ at two different temperatures. A few values of A and x are given below:

	H_2O	CO	C_6H_6	C_2H_5OH	n-C_6H_{14}	$(CH_3)_2CO$
A	16440	27220	7914	3567	4710	4594
x	2·38792	3·54382	2·38899	1·89552	1·7856	2·4242

The average deviations of the calculated values of σ from those determined experimentally are: at 330°C < 2·7%, at 340°C < 5%.

By using the modified theory of corresponding states (Prigogine[301]) it is possible to represent the reduced surface tension of many weakly polar liquids, as a unique function of the modified reduced density.[27] This relationship, though interesting from a theoretical point of view, does not provide a simple method for calculating the surface tension.

c. Prediction of Surface Tensions Using the Theory of Similarity

In order to represent the effect of temperature upon the surface tension of a liquid, use may be made of the similarity between the variation with temperature of the surface tension of the investigated liquid and of a suitably chosen reference liquid (it is desirable that the liquids be similar).

Carr and Wolczynski[56] have developed this method and present the relationship in the form of a graph similar to the Dühring plot (Ch. 1, section 7c). Denoting by T_1 and T_1' the temperatures at which the surface tensions of the investigated liquid and of the reference liquid have the same value σ_1, and by T_2 and T_2' the temperatures at which the values of the surface tension are σ_2, we may write

$$\frac{(T_2 - T_1)}{(T_2' - T_1')} = K \tag{2.26}$$

where K is constant.

Hence, plotting the temperatures T_1', T_2', T_3' ... of the reference liquid on the abscissa, and the temperatures T_1, T_2, T_3 ... at which the surface tension of the investigated liquid attains the same value as in the case of the reference liquid on the ordinate a straight line should be obtained. Experiments show that this graph may only be used for extrapolation over a rather restricted range of temperatures, since over wide temperature ranges the lines become somewhat curved. The deviations are particularly high in the vicinity of the critical temperature. To construct the graph, it is sufficient to know two values of the surface tension of the investigated liquid: σ_1 at temperature T_1 and σ_2 at temperature T_2. If the critical temperatures of both liquids are known, it is sufficient to know only one value of σ, say σ_1, at temperature T_1, since at temperature T_c the surface tension $\sigma_c = 0$.

More accurate results can be obtained by the graphical method recommended by Othmer.[270] This method is based on the same principle of similarity between the variations of surface tension of the investigated and the reference liquids.

Denoting by:

- σ the surface tension of the investigated liquid at temperature T (°K), dynes/cm,
- σ' the surface tension of the reference liquid at temperature T' (°K), dynes/cm,
- T_c and T_c' the critical temperatures of the investigated liquid and of the reference liquid,

and writing:

$$\left.\begin{aligned} T_D &= T_c - T \\ \text{and} \quad T_D' &= T_c' - T' \end{aligned}\right\} \tag{2.27}$$

Othmer has found that when $T_D = T_D'$

$$\log \sigma = m \log \sigma' + c \tag{2.28}$$

where m and c are constants for the pair of liquids.

If two values of the surface tension of the investigated liquid σ_1 and σ_2 at temperatures T_1 and T_2, respectively, are known and the differences of temperature

$$T_c - T_2 = T_c' - T_2'$$

$$T_c - T_1 = T_c' - T_1'$$

are kept identical, it is possible to determine the values of the surface tensions σ_2' and σ_1' of the reference liquid, and, hence, to calculate the values of the constants m and c, or to draw the rectilinear relationship between $\log \sigma$ and $\log \sigma'$.

The variation of the surface tension with temperature at temperatures close to the critical temperature T_c is represented by Othmer's graph much better than by the graph described previously. On the basis of Eq. (2.28) a point nomograph (Fig. 2.9) for about 100 liquids has been constructed. Instead of placing the points, corresponding to the individual liquids, on the nomograph, the coordinates of these points are given in Table 2.2, and the nomograph has been provided with a suitable net facilitating location of the required point. When using the nomograph given in Fig. 2.9, it is necessary to determine the location of a point with the coordinates (x, y), corresponding to the investigated liquid, and then to construct a straight line through this point and through the point on the right hand axis, which corresponds to the temperature t (°C) of the liquid. This straight line gives, on the left axis, the required value of the surface tension σ (dynes/cm) at temperature t. If the coordinates x and y for the liquid of interest are not available, they may be deduced provided the surface tension of the liquid at two temperatures is available. Denote the surface tensions at t_1 and t_2 by σ_1 and σ_2 and rule two straight lines on Figure 2.9, the first joining σ_1 to t_1 and the second joining σ_2 to t_2. The point of intersection of these two lines gives the coordinate x and y.

EXAMPLE 2.1. Estimate the surface tension of methyl ethyl ketone at 75°C.

Solution

According to Table 2.2, for this compound $x = 23{\cdot}6$ and $y = 97$. The value read from the nomograph (Fig. 2.9) is $\sigma = 18$ dynes/cm. The experimentally determined value is 17·8 dynes/cm.

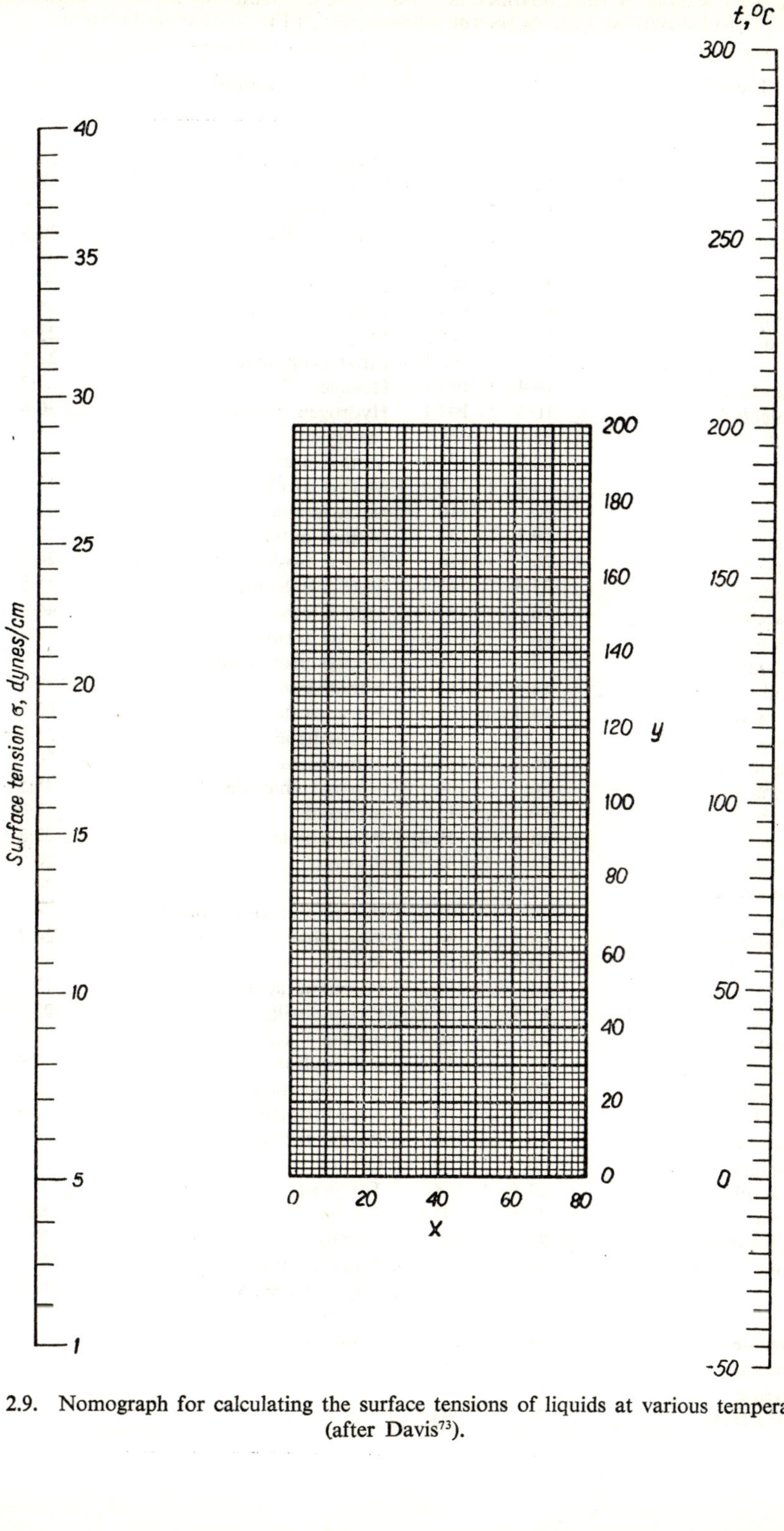

FIG. 2.9. Nomograph for calculating the surface tensions of liquids at various temperatures (after Davis[73]).

TABLE 2.2. VALUES OF THE COORDINATES x AND y FOR LOCATING THE POINTS CORRESPONDING TO INDIVIDUAL LIQUIDS ON THE NOMOGRAPH IN FIG. 2.9 (AFTER DAVIS[73])

Liquid	x	y
Acetal	19	88
Acetaldehyde	33	78
Acetamide	17	192·5
Acetic anhydride	25	129
Acetone	28	91
Acetonitrile	33·5	111
Acetophenone	18	163
Allyl alcohol	12	111·5
Ammonia	56·2	63·5
iso-Amyl acetate	16·4	103·1
iso-Amyl alcohol	16·4	103·1
Anethole	13	158·1
Aniline	22·9	171·8
Anisole	24·4	138·9
Benzene	30	110
Benzonitrile	19·5	159
Benzylamine	25	156
Bromobenzene	23·5	145·5
iso-Butyl acetate	16	97·2
n-Butyl alcohol	26	107·5
iso-Butyl alcohol	5	103
Butyric acid	14·5	115
iso-Butyric acid	14·8	107·4
Butyronitrile	20·3	113
Carbon disulphide	35·8	117·2
Carbon tetrachloride	26	104·5
Chloral	30	113
Chlorine	45·5	59·2
Chlorobenzene	23·5	132·5
p-Chlorobromobenzene	14	162
Chloroform	32	101·3
p-Chlorotoluene	18·7	134
o-Cresol	20	161
m-Cresol	13	161·2
p-Cresol	11·5	160·5
Cyclohexane	42	86·7
Diethylaniline	17	142·6
Diethyl axalate	20·5	130·8
Diethyl sulphate	19·5	139·5
Dimethylamine	16	166
Dimethylaniline	20	149
Dimethyl sulphate	23·5	158
Ethyl acetate	27·5	92·4
Ethyl acetoacetate	21	132
Ethyl alcohol	10	97
Ethylamine	11·2	83
Ethylbenzene	22	118
Ethyl benzoate	14·8	151
Ethyl bromide	31·6	90·2
Ethyl butyrate	17·5	102
Ethyl iso-butyrate	20·9	93·7
Ethylene chloride	32	120
Ethylene oxide	42	83
Ethyl ether	27·5	64
Ethyl formate	30·5	88·8
Ethyl iodide	28	113·2
Ethyl ketone	20	101
Ethyl mercaptan	35	81
Ethyl methyl ketone	23·6	97
Ethyl propionate	22·6	97
Hexane	22·7	72·2
Hydrogen cyanide	30·6	66
Mesitylene	17	119·8
Methyl acetate	34	90
Methyl alcohol	17	193
Methylamine	42	58
Methyl butyrate	25	88
Methyl iso-butyrate	24	93·8
Methyl chloride	45·8	53·2
Methyl ether	44	37
Methyl formate	38·5	88
Methyl propionate	29	95
Naphthalene	22·5	165
Nitric oxide	62·5	0·5
Nitroethane	25·4	126·1
Nitromethane	30	139
Nitrosyl chloride	38·5	93
n-Octane	17·7	90
Paraldehyde	22·3	103·8
Phenetole	20	134·2
Phenol	20	168
Phosphorus oxychloride	26	125·2
Piperidine	24·7	120
Pyridine	34	138·2
Propionic acid	17	112
Propionitrile	23	108·6
n-Propyl acetate	23	97
n-Propyl alcohol	8·2	105·2
n-Propylamine	25·5	87·2
p-iso-Propyltoluene	12·8	121·2
Propyl formate	24	97
Quinoline	19·5	183
Tiophene	35	121
Toluene	24	113
Triethylamine	20·1	83·9
Trimethylamine	21	57·6
Triphenylmethane	12·5	182·7
m-Xylene	20·5	118
p-Xylene	19	117

EXAMPLE 2.2. The measured values of the surface tension of myristic acid $CH_3—(CH_2)_{12}—COOH$ are:

at temperature $t_1 = 56{\cdot}8°C$ $\sigma_1 = 28{\cdot}6$ dynes/cm
at temperature $t_2 = 149{\cdot}3°C$ $\sigma_2 = 21{\cdot}3$ dynes/cm

Estimate the values of the characteristic coordinates x, y as well as the value of the surface tension σ_3 at temperature $t_3 = 100°C$.

Solution

Connecting $t_1 = 56{\cdot}8°C$ (right axis of the nomograph 2·9) with $\sigma_1 = 28{\cdot}6$ (left axis) and $t_2 = 149{\cdot}3$ with $\sigma_2 = 21{\cdot}3$ by means of straight lines, we get at the intersection of these lines the point with the coordinates $x = 4$, $y = 147$.

A straight line constructed through this point and through the point $t_3 = 100°C$ (right axis), determines on the left axis the required value of $\sigma_3 = 25{\cdot}5$ dynes/cm.

4. RELATIONSHIP BETWEEN THE SURFACE TENSION AND VARIOUS OTHER PHYSICAL CONSTANTS

Many authors have established the existence of relationships between the values of various physical constants and the surface tension of a liquid. A detailed review of these works may be found in Partington's work (vol. 1).[282]

Particular attention has been paid to the relation between the surface tension and heat of vaporization of liquids. Theoretical considerations indicate that such a relationship exists. On passing from the bulk of the liquid into the surface the molecule covers, according to Ostwald, about half the path of the passage to the gas phase; therefore, the surface energy (which is equal to the surface tension expressed, e.g. in ergs/cm^2) per molecule should be approximately equal to one-half of the vaporization energy (also per molecule). In fact, such a calculation performed, e.g. for mercury at 360°C, affords the expected value of the energy: (1) calculated on the basis of the surface tension $E_\sigma = 3{\cdot}4 \times 10^{-13}$ ergs, (2) on the basis of the heat of vaporization $E_L = 4{\cdot}8 \times 10^{-13}$ ergs; we see that the order of magnitude is in both cases the same.[326]

According to the investigations carried out by Harkins and Roberts,[142] the ratio of the molar surface energy to the molal heat of vaporization is not constant; it is larger for unsymmetrical than for symmetrical molecules, and increases with rise of temperature.

Ianson[166] has established a relationship between the constant K in the Eötvös equation (2.2) and Trouton's constant L_b/T_b (L_b is the heat of vaporization at the normal boiling temperature T_b). Approximately

$$L_b/T_b = 9{\cdot}76\ K \tag{2.29}$$

This relationship is reasonably accurate for some liquids, and in general deviations do not exceed 10%.

Walden[389] has formulated some approximate empirical relationships between the molal heat of vaporization L_b, the molal heat of fusion L_m, the molal volume V, the densities ρ_b and ρ_m at the boiling and melting temperatures T_b and T_m and the surface tensions σ_b and σ_m at T_b and T_m, respectively. In these equations M denotes the molecular weight of the compound.

$$\rho_b \frac{L_b}{M} = 35{\cdot}8\ \sigma_b \tag{2.30}$$

$$L_b = 3{\cdot}64\ V\sigma_b \tag{2.31}$$

$$\rho_m \frac{L_m}{M} = 7{\cdot}3\ \sigma_m \tag{2.32}$$

$$M\sigma_b = 0{\cdot}581\ \rho_b T_b \tag{2.33}$$

$$\frac{V\sigma_b}{T_b} = 5{\cdot}67 \tag{2.34}$$

$$\frac{L_b}{\sigma_b V^{2/3}} = 18{\cdot}0 \tag{2.35}$$

$$\frac{\rho_b L_b}{\sigma_b M} = \frac{92{\cdot}0}{\log T_b} \tag{2.36}$$

According to Steinhardt,[343] a relationship exists between the surface tension of a liquid, its heat of vaporization and the shape of the molecules.

Approximately:

$$\frac{\Delta H_v - RT}{\sigma^{0\cdot5}} = k \tag{2.37}$$

where ΔH_v is the total molal latent heat of vaporization.

In the case of liquids consisting of tetrahedral or cyclic molecules, k is approximately constant (at 25°C $k \approx 1{\cdot}4$), e.g.

	CCl_4	C_6H_6	cyclohexane	cyclohexene	cyclopentane
k	1·44	1·43	1·48	1·49	1·37

The more the molecule deviates from the spherical shape, the higher is the value of k, e.g.

$n\text{-}C_5H_{12}$	$n\text{-}C_6H_{14}$	$n\text{-}C_7H_{16}$	$n\text{-}C_8H_{18}$	$n\text{-}C_9H_{20}$
1·51	1·66	1·84	2·02	2·22

In addition to equations (2.30)–(2.36) Walden has proposed some equations relating surface tensions at the normal boiling point (σ_b) to the critical constants p_c (critical pressure), T_c (critical temperature) and V_c (critical volume). For the critical pressure he obtains:

$$p_c = \frac{5{\cdot}72\sigma_b}{\log T_b} \tag{2.38}$$

or, more accurately

$$p_c = \frac{6{\cdot}5\sigma_b}{\log T_b} - 0{\cdot}012 T_b \tag{2.39}$$

He also proposes the relationship:

$$\sigma_b / p_c = 0{\cdot}45 \tag{2.40}$$

and

$$\frac{V_c \sigma_b}{T_c} = 9{\cdot}8 \tag{2.41}$$

Richards and Mathews[308] have found that a relationship exists between the surface tension σ and the isothermal compressibility β

$$\beta\sigma^{4/3} = 0{\cdot}00253 \tag{2.42}$$

This relation has been checked on more than thirty organic liquids over a pressure range from 100 to 500 atm.

Rao and Rao[305] have presented a somewhat different form of the relationship existing between σ and β. According to them

$$\sigma = \left(\frac{A}{B}\right)^{1/\tau} \tag{2.43}$$

where on the average the constants $A \approx 0{\cdot}21$ ergs $\tau/(\text{atm.cm}^2)\tau$, and $1/\tau = 0{\cdot}627$. Considerable deviations from this relationship have been found for such liquids as n-propanol and aniline.

Auerbach[16] has found a relationship between the surface tension σ (dynes/cm), liquid density ρ (g/cm^3) and velocity of sound in liquids u (cm/sec),

$$\sigma = 6{\cdot}3 \times 10^{-4}\ \rho_u^{3/2} \tag{2.44}$$

This relationship is supported by the fact that both the parachor, correlated with the surface tension, and the molar velocity of sound (Eq. (1.55)) are constitutive and additive quantities.

According to Buehler,[48] the surface tension σ is related to the dynamic viscosity μ of a liquid by the following equation:

$$\sigma^{1/4} = \frac{\log(\log\mu)+2{\cdot}9}{(I/P_{ch})} \tag{2.45}$$

where I is the structural constant of viscosity, introduced by Souders,
and P_{ch} is the parachor.

The values of the structural constant I, which is an additive constant, may be calculated by summing the relevant contributions (cf. Ch. 4, section 5a). Alternatively, it may be estimated from the molar refraction R_D by means of Lagemann's nomograph (Fig. 4.13). In general, the value of the fraction I/P_{ch} ranges from 1·16 to 1·30. The average value, most frequently encountered, is 1·22. For solving equation (2.45), the nomograph[73] constructed by Davis[72] may be used. Equation (2.45) must not be employed for liquids for which the value of the parachor increases only slightly with rise of temperature. For example, it should not be used for such liquids as phenols, carboxylic acids, alcohols and amines.

Several equations, illustrating the relationships between the surface tension and other properties of liquids, have been compiled by Gambill.[111]

The above equations ((2.30)–(2.45) provide only approximate values for the surface tension, but they may be helpful in obtaining rough estimates of this property or indeed of other physico-chemical properties connected with the surface tension.

CHAPTER 3

VISCOSITIES OF GASES

1. SOME BASIC CONCEPTS OF THE KINETIC THEORY OF GASES

The final chapters of this book are concerned with the *transport properties of fluids*, namely the internal friction (or viscosity) and the coefficients of heat conduction and diffusion. The respective processes are referred to as *transport processes.* This name is derived from the fact that the molecule of a gas, being in random motion and colliding with other molecules, transfers its momentum (the phenomenon of viscosity) and its kinetic energy (the phenomenon of heat conduction); an exchange of the positions of the molecules (the phenomenon of diffusion) can also take place.

In order to comprehend the essence of the transport phenomena, it may be apposite to recall certain basic concepts and definitions of the kinetic theory of gases in its most simple and elementary form†.

One of the basic equations of the kinetic theory of gases is the Clausius equation, according to ..hich the pressure of a gas is given by

$$p = \frac{2}{3}\left(\frac{\overline{mw^2}}{2}\right)n \tag{3.1}$$

where p is the pressure in dynes/cm^3,

$\left(\frac{\overline{mw^2}}{2}\right)n$ is the total kinetic energy of the molecules contained in 1 cm^3, ergs,

m is the mass of one molecule, grams,

w is the molecular velocity, cm/sec,

and n is the number of molecules in 1 cm^3.

The equation of state of 1 mole of a perfect gas

$$pV = RT$$

may be represented as

$$pV = NkT \tag{3.2}$$

where V is the volume occupied by one mole of the gas, cm^3,

N is the number of molecules in 1 mole of the gas, i.e. Avogadro's number,

$$N = 6{\cdot}023\times10^{23}\left(\frac{1}{\text{mole}}\right),$$

k is Boltzmann's constant, $k = \frac{R}{N} = \frac{8{\cdot}31\times10^7}{6{\cdot}02\times10^{23}} = 1{\cdot}38\times10^{-16}$ ergs/°K,

and T is the absolute temperature in °K.

† For derivations of the equations given here the reader is referred to textbooks of physics or physical chemistry, e.g. by Partington,[282] Shtrauf,[326] Jellinek,[175] or Eucken.[92]

From Eqs. (3.1) and (3.2) we obtain the Boltzmann equation

$$\left(\overline{\frac{mw^2}{2}}\right) = \frac{3}{2} kT \tag{3.3}$$

It follows from this equation that the mean translational kinetic energy of gas molecules in random thermal motion is proportional to the absolute temperature of the gas. It also follows that the *root-mean-square velocity* $\sqrt{\overline{w^2}}$ of the molecules is

$$\sqrt{\overline{w^2}} = \sqrt{\frac{3kT}{m}} = \sqrt{\frac{3kNT}{M}} = \sqrt{\frac{3RT}{M}} \tag{3.4}$$

where M is the molecular weight, g/mole.

For example, for nitrogen at a temperature of 17°C the root-mean-square velocity is

$$\sqrt{\overline{w^2_{N_2}}} = \sqrt{\frac{3\times 8{\cdot}31\times 10^7\times 290}{2}} \approx 5\times 10^4 \text{ cm/sec} = 500 \text{ m/sec}$$

The mean molecular velocity for hydrogen, calculated in a similar way, amounts to $\sqrt{\overline{w^2_{H_2}}} \approx 1800$ m/sec. These are of course only average values. The individual molecules in a gaseous system are, at any instant, all moving at different speeds, some much faster than others. According to Maxwell (1860), in a system containing n molecules the number dn with velocities lying in the range w to $(w+\mathrm{d}w)$, is given by

$$\mathrm{d}n = \frac{4}{\sqrt{\pi}} n \frac{w^2}{w_m^2} \mathrm{e}^{-w^2/w_m^2} \frac{\mathrm{d}w}{w_m} \tag{3.5}$$

w_m in the above equation is the *most probable value* of the velocity of the molecules in the given system, corresponding to the maximum of the Maxwell curve representing the relationship between $(\mathrm{d}n/\mathrm{d}w)$ and w, i.e. between the variation of the number of molecules with velocity w and the velocity itself (Fig. 3.1).

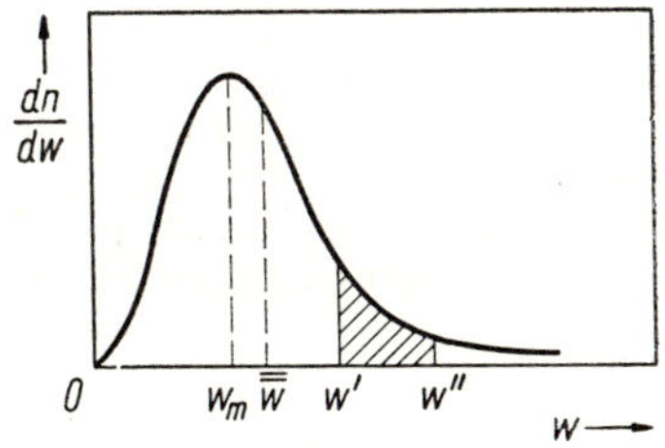

FIG. 3.1. Maxwell's function.

The shaded area in the graph corresponds to the number of molecules with velocities between w' and w''.

The most probable molecular velocity w_m is not equal to the root-mean-square velocity $\sqrt{\overline{w^2}}$, but is somewhat lower

$$w_m = \frac{2kT}{m} = \frac{2RT}{M} \tag{3.6}$$

In addition to the velocities w_m and $\sqrt{\overline{w^2}}$ it is sometimes necessary to know the value of the *arithmetic mean velocity* $\overline{w}$ of the molecules in the system

$$\overline{w} = \sqrt{\frac{8kT}{\pi m}} = \sqrt{\frac{8RT}{\pi M}} \tag{3.7}$$

The very large root-mean-square velocities, $\sqrt{\overline{w^2}}$, obtained in the above examples may seem at first sight to be incompatible with the known fact that the velocities of the transfer processes (e.g. diffusion) are comparatively small. The reason for this discrepancy arises from the fact that, at normal pressures, the rates at which transport phenomena can take place are limited by intermolecular collisions. Although each molecule may move at a considerable speed between two successive collisions, at each collision it changes its direction with the result that it moves in an irregular and random fashion, sometimes in one direction, sometimes in another, and the resultant distance travelled, measured along a straight line, may, even after a long period of time, be very small. The quantity which characterizes the random motion of the gas molecules is the *mean free path,* i.e. the average distance a molecule traverses between two successive collisions.

The kinetic theory of gases yields the following expression for the mean free path $\overline{l}$ of a molecule:

$$\overline{l} = \frac{1}{\sqrt{2}\pi\sigma^2 n} = \frac{0{\cdot}707}{n\pi\sigma^2} \tag{3.8}$$

where σ is the active diameter of the molecule (cf. Ch. 1, section 2c),
and n is the number of molecules in unit volume.

Since the gas density ρ is given by $\rho = mn$ (where m is the mass of one molecule), and $p = nkT$, the mean free path $\overline{l}$ may also be expressed in the form

$$\overline{l} = \frac{m}{\sqrt{2}\pi\sigma^2\rho} \tag{3.9}$$

or

$$p\overline{l} = \frac{kT}{\sqrt{2}\pi\sigma^2}$$

where k is Boltzmann's constant.

The mean free path of a molecule is very small. For example for nitrogen,[326] at 0°C and atmospheric pressure, $\sigma = 3{\cdot}18\times10^{-8}$ cm and $n = 2{\cdot}7\times10^{19}$ 1/cm³,

so

$$\overline{l} = \frac{1}{\sqrt{2}\times3{\cdot}14\times10^{-15}\times2{\cdot}7\times10^{19}} \approx 0{\cdot}9\times10^{-5}\ \text{cm}$$

On transforming Eq.(3.8) it is found that z, the number of molecular collisions occurring in unit volume per unit time, is given by

$$z = \frac{\sqrt{2}}{2}\pi\sigma^2 n^2 \overline{w} \tag{3.10}$$

It follows from this equation that the number of molecular collisions, taking place in unit time in a given volume, is proportional to the square of the number of molecules enclosed within this volume.

In the simple picture presented above, molecules are regarded as hard spheres. In practice, as shown in Chapter 1, molecules repel each other when the distance between them becomes very small and attract each other at larger separations. Owing to the existence of the attractive forces, two molecules, which would not have collided had they followed rectilinear paths, may do so since the result of their mutual attraction is to produce a curvature of their paths. The result of this phenomenon is that intermolecular collisions occur more frequently than would be expected on the basis of geometric considerations, and that the free path of the molecules is shorter. The net effect is thus equivalent to an increase in the molecular diameter. The effect is most pronounced at low temperatures since the average velocity of the molecules is then comparatively small and the path curvature produced by the attractive forces is appreciable.

It may be shown that

$$\sigma_T = \sigma_0 \sqrt{1+\frac{C}{T}} \tag{3.11}$$

where σ_T is the "active diameter" of the molecule at temperature T,
σ_0 is the "active diameter" of the molecule at a reference temperature as discussed later,
C is a constant for a given gas,
and T is the absolute temperature.

The constant C is given by $C = 2E_p/3k$ where E_p is the potential energy arising from the mutual attraction of the molecules.

The mean free path of the molecules is therefore

$$\bar{l} = \left[\frac{1}{\sqrt{2}\,\pi\sigma_0^2 n}\right]\left[\frac{1}{1+\left(\frac{C}{T}\right)}\right] \tag{3.12}$$

The equations presented above have been developed from the elementary kinetic theory of gases; they form a sufficient basis for a simple treatment of the mechanisms of transfer phenomena.

2. PHENOMENON OF INTERNAL FRICTION (VISCOSITY) IN FLUIDS

In order to understand the phenomenon of internal friction in a fluid, it is convenient to imagine a series (Fig. 3.2) of very thin parallel layers of fluid moving in the direction of the X-axis, each successive layer moving somewhat faster than the layer immediately below and somewhat slower than the layer above.

Let us assume that the thickness of each of these layers is equal to dz and that the velocity of each layer exceeds that of the layer below it by an amount dw. In terms of this picture the velocity gradient is dw/dz, and the speed of the fluid varies linearly along the Z-axis. (This axis is perpendicular to the direction of the motion of the fluid.)

Application of a force R acting along the X-axis will increase the velocity of motion of the upper layer, and will accelerate the motion of the layers lying below, since these layers tend to be dragged along by those lying above them; the lower layers, on the other hand, will tend to retard the motion of the upper layers. The magnitude of the interaction forces between two adjacent layers is proportional to the magnitude of the area of contact (A) between them.

This tendency for one layer of fluid to drag adjacent layers along with it gives rise to the phenomenon of internal friction (viscosity). According to Newton

$$R = -\mu A(\mathrm{d}w/\mathrm{d}z) \tag{3.13}$$

This equation states that the force R of internal friction is proportional to the area A of contact of two layers moving over each other, to the velocity gradient $\mathrm{d}w/\mathrm{d}z$ and to

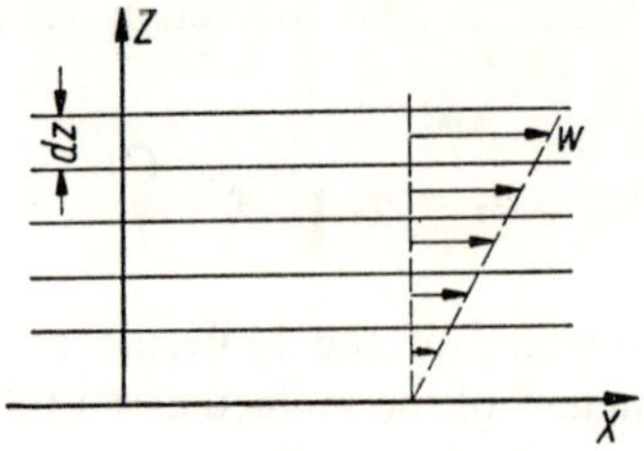

FIG. 3.2. Illustration of the phenomenon of internal friction in fluids.

the coefficient μ; the latter is variously referred to as the *coefficient of internal friction*, the *dynamic coefficient of viscosity* or, sometimes, the *dynamic viscosity* or simply, the *viscosity*.

The negative sign in Eq. (3.13) indicates that the viscous force acts in an opposite direction to the motion of the fluid.

Maxwell[240] explained the phenomenon of internal friction in terms of the kinetic theory of gases. According to Maxwell, the motion of the gas molecules in a given layer is obtained by superimposing on the flow velocity w, which is directed along the X-axis in a random motion in all directions. Molecules will therefore be moving back and forth along the Z-axis (perpendicular to the line of flow) though there is no net flow in this direction. Owing to these random motions molecules continually pass from any given layer to the less rapidly moving layer which lies beneath it. On average, the momentum of these molecules, measured along the direction of flow, is greater than that of the molecules in the layer below. When these molecules collide with those in the slower layer, they transfer their excess momentum to the molecules in this layer, thus tending to accelerate this layer. Conversely molecules which move upwards from the lower layer to the layer above tend to retard the motion of this layer along the X-axis. The net result of the random motion of the molecules is thus to retard the faster flowing upper layer and to accelerate the slower moving lower layer, as if a frictional force existed between the two fluid layers moving over each other. A further consequence of the above picture should be mentioned: in order to maintain the flow of fluid an external force must be applied to the faster moving upper layer, because this layer is continually the x component of the velocity of the molecules from lower layers. Molecules which leave the upper layer in a downward direction have a comparatively large x component of velocity which is converted, in the lower layer, into a random motion. As a result part of the energy of flow is converted into thermal energy, i.e. into the random motion of the molecules. A mathematical approach to this phenomenon has been presented by Stefan.[341]

It follows from Eq. (3.13) that the viscosity is numerically equal to the force acting upon unit area of contact between two layers moving over each other when the velocity gradient is unity.

Since the ratio of the force acting tangentially to the surface to the magnitude of the surface area has the character of a shearing stress, the viscosity may be defined as being equal to the shearing stress for unit velocity gradient.

In the c.g.s. system the units of viscosity are

$$\left[\frac{\dfrac{\text{g.cm}}{\text{sec}^2}\,.\text{cm}}{\text{cm}^2.\dfrac{\text{cm}}{\text{sec}}}\right]=\left[\frac{\text{g}}{\text{cm.sec}}\right]$$

This unit, called the *poise* (symbol P), is too large for convenience and in practice, viscosity is usually expressed in smaller units such as the *centipoise* (symbol cP), the millipoise (mP) or the micropoise (μP); $1\text{P} = 10^2\ \text{cP} = 10^3\ \text{mP} = 10^6\ \mu\text{P}$.

Technical units are also sometimes employed:

$$1\ \text{poise} = 360\,\frac{\text{kg(mass)}}{\text{m.hr}}$$

$$1\,\frac{\text{kg(force).sec}}{\text{m}^2} = 98{\cdot}1\text{P}$$

Instead of using the coefficient of viscosity μ, it is sometimes convenient to use the *kinematic coefficient of viscosity* ν which is defined by the equation

$$\nu = \mu/\rho \tag{3.14}$$

where ρ is the density.

The unit of kinematic viscosity in the c.g.s. system is 1 *stoke* (S) equal to 1 cm^2/sec though Engineers sometimes use a unit of $1(\text{m}^2/\text{hr}) = 2{\cdot}777$ S.

Simplified considerations based on the kinetic theory of gases lead to the following equation for the viscosity of a gas:

$$\mu = \frac{\rho\bar{w}\bar{l}}{3} \tag{3.15}$$

where ρ is the density of the gas,
$\bar{w}$ is the mean molecular velocity,
and $\bar{l}$ is the mean free path of the molecules.

A more rigorous treatment developed by Chapman,[57] leads to the equation

$$\mu = 0{\cdot}499\rho\bar{w}\bar{l} \tag{3.16}$$

The gas density ρ is directly proportional to the number (n) of molecules in unit volume, i.e.

$$\rho = nm \tag{3.17}$$

where m is the mass of a molecule. Now according to Eq. (3.8) the mean free path $\bar{l}$ of the molecules is inversely proportional to n, hence after substituting the values of $\bar{l}$ and ρ given by Eqs. (3.8) and (3.17) into Eq. (3.15) or Eq. (3.16) the quantity n will disappear.

Since the mean velocity $\overline{w}$ of the molecules is independent of n (Eq. (3.7)), it follows from (3.15) or (3.16) that the dynamic viscosity μ of a gas at constant temperature does not depend on the number of molecules in unit volume, and hence it should not depend on the pressure, either.

Within a certain pressure range, experiments support this unexpected conclusion.

Transformation of Eq. (3.16) leads to an equation representing the effect of temperature on the viscosity μ of gases. After substituting in this equation the value of ρ (given by Eq. (3.17)), the mean free path $\overline{l}$ of the molecules (Eq. 3.12) and the mean molecular velocity (Eq. 3.7), we obtain the expression

$$\mu_T = \frac{K\sqrt{T}}{\left(1+\dfrac{C}{T}\right)} = K\frac{T^{3/2}}{(T+C)} \tag{3.18}$$

where K and C are constants, characteristic of a given gas. At a temperature of 0°C, i.e. 273°K,

$$\mu_0 = K\frac{273^{3/2}}{(273+C)} \tag{3.19}$$

Dividing Eq. (3.18) by Eq. (3.19), we obtain

$$\mu_T = \mu_0\frac{(273+C)}{(T+C)}\left(\frac{T}{273}\right)^{3/2} \tag{3.20}$$

This equation, which was derived by Sutherland,[353] represents the dependence of gas viscosity on temperature. The so-called *Sutherland constant* C is found to be approximately constant only at temperatures substantially exceeding the critical temperature, e.g. for air at 0°C $C = 112$, at 1600°C $C = 124$. In the low-temperature region the values of C depend strongly on the temperature, as may be seen from Fig. 3.3.

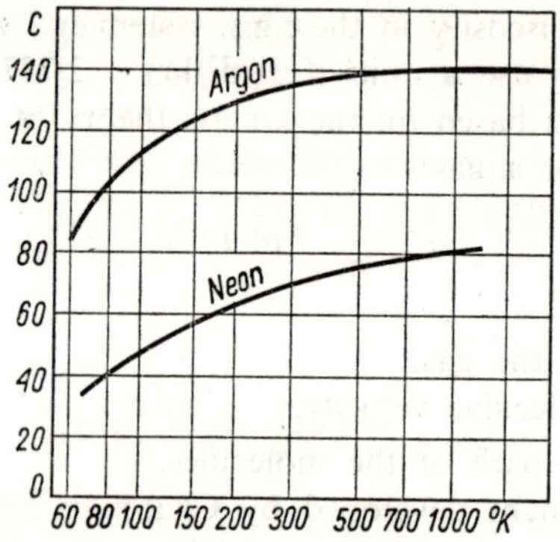

FIG. 3.3. Sutherland's constants C for argon and helium as functions of temperature (after Andrussov[7]).

For hydrogen and helium at temperature ups ot 20°K Chapman[58] has derived the corrected equation

$$\frac{\mu_0}{\mu_T} = \left(\frac{T_0}{T}\right)^{0.5}\frac{1+(C/T)-(C'/T)^2}{1+(C/T_0)-(C'/T_0)^2} \tag{3.21}$$

where C and C' are constants characteristic of the gas.

3. METHODS OF ESTIMATING THE VISCOSITY OF GASES WHICH ALLOW FOR INTERMOLECULAR FORCES

The equations for the viscosity of gases developed in section 2 have been known for a long time and were derived on the basis of the kinetic theory of gases. In more recent years our knowledge of the nature and magnitude of the forces of molecular interaction has increased and it has now become possible to calculate the viscosity of gases allowing for these forces by means of statistical mechanics. Such calculations are most readily performed for substances with non-polar spherical molecules, the intermolecular force fields for which are adequately represented by the Lennard-Jones equation.

It follows from the considerations presented in section 3c (Ch. 1) that it should be possible to calculate the viscosity μ, if the values of the force constants σ and ε in the Lennard-Jones equation (1.29) are known. Values of these constants for several score of compounds are compiled in Table 1.16. In section 3c approximate formulae have also been given, which enable values of σ and ε for certain types of compound or their mixtures to be estimated, if they are not included in the table.

According to Hirschfelder, Bird and Spotz,[153,154,25] the value of the viscosity for a non-polar pure gas at a moderate pressure is given approximately by the following equation:

$$\mu^{\circ}\times 10^{7} = 226{\cdot}93\,\frac{\sqrt{MT}}{\sigma^{2}\Omega^{(2,2)*}} \tag{3.22}$$

where μ° is the viscosity within the pressure range in which the viscosity does not depend on pressure, g/cm·sec,
M is the molecular weight, g/mole,
T is the temperature, °K,
σ is a force constant defined in terms of the Lennard-Jones equation, Å,
and $\Omega^{(2,2)*}$ is a function of the variable $T^{*} = kT/\varepsilon$. Numerical values of this function are listed in Table 3.1.

A more precise value of the viscosity μ' may be obtained by multiplying the value of μ, as found from Eq. (3.22), by a correction factor f_{μ} which is dependent on T^{*} (values of f_{μ} are listed in Table 3.3)

$$\mu' = \mu f_{\mu} \tag{3.23}$$

Bromley and Wilke[46] have expressed Eq. (3.22) in the following form

$$\mu^{\circ} = \text{constant}\times f_{1}\left(\frac{kT}{\varepsilon}\right) \tag{3.24}$$

At high temperatures (say 10,000°C) the viscosities calculated from the Hirschfelder equation are subject to appreciable errors (Amdur[10]).

For a given compound the constant in Eq. (3.24) is given by

$$\text{constant} = \frac{5}{16\sigma^{2}}\sqrt{\frac{M\varepsilon}{N\pi}} \tag{3.25}$$

The function f_{1} (Table 3.2) is related with the function $\Omega^{(2,2)*}$ by the relationship

$$f_{1}\left(\frac{kT}{\varepsilon}\right) = \frac{1}{\Omega^{(2,2)*}}\sqrt{\frac{kT}{\varepsilon}} \tag{3.26}$$

TABLE 3.1. FUNCTIONS FOR ESTIMATING THE PROPERTIES OF NON-POLAR GASES (AFTER HIRSCHFELDER[153])

T^*	$\Omega^{(1,1)*}$	$\Omega^{(2,2)*}$
0·30	2·662	2·785
0·35	2·476	2·628
0·40	2·318	2·492
0·45	2·184	2·368
0·50	2·066	2·257
0·55	1·966	2·156
0·60	1·877	2·065
0·65	1·798	1·982
0·70	1·729	1·908
0·75	1·667	1·841
0·80	1·612	1·780
0·85	1·562	1·725
0·90	1·517	1·675
0·95	1·476	1·629
1·00	1·439	1·587
1·05	1·406	1·549
1·10	1·375	1·514
1·15	1·346	1·482
1·20	1·320	1·452
1·25	1·296	1·424
1·30	1·273	1·399
1·35	1·253	1·375
1·40	1·233	1·353
1·45	1·215	1·333
1·50	1·198	1·314
1·55	1·182	1·296
1·60	1·167	1·279
1·65	1·153	1·264
1·70	1·140	1·248
1·75	1·128	1·234
1·80	1·116	1·221
1·85	1·105	1·209
1·90	1·094	1·197
1·95	1·084	1·186
2·00	1·075	1·175
2·10	1·057	1·156
2·20	1·041	1·138
2·30	1·026	1·122
2·40	1·012	1·107
2·50	0·9996	1·093
2·60	0·9878	1·081
2·70	0·9770	1·069
2·80	0·9672	1·058
2·90	0·9576	1·048
3·00	0·9490	1·039
3·10	0·9406	1·030
3·20	0·9328	1·022
3·30	0·9256	1·014
3·40	0·9186	1·007
3·50	0·9120	0·9999
3·60	0·9058	0·9932
3·70	0·8998	0·9870
3·80	0·8942	0·9811
3·90	0·8888	0·9755
4·00	0·8836	0·9700
4·10	0·8788	0·9649
4·20	0·8740	0·9600
4·30	0·8694	0·9553
4·40	0·8652	0·9507
4·50	0·8610	0·9464
4·60	0·8568	0·9422
4·70	0·8530	0·9382
4·80	0·8492	0·9343
4·90	0·8456	0·9305
5·0	0·8422	0·9269
6·0	0·8124	0·8963
7·0	0·7896	0·8727
8·0	0·7712	0·8538
9·0	0·7556	0·8379
10·0	0·7424	0·8242
20·0	0·6640	0·7432
30·0	0·6232	0·7005
40·0	0·5960	0·6718
50·0	0·5756	0·6504
60·0	0·5596	0·6335
70·0	0·5464	0·6194
80·0	0·5352	0·6076
90·0	0·5256	0·5973
100·0	0·5170	0·5882
200·0	0·4644	0·5320
300·0	0·4360	0·5016
400·0	0·4170	0·4811

TABLE 3.2. VALUES OF THE AUXILIARY FUNCTION f_1 FOR ESTIMATING THE VISCOSITY FROM EQS. (3.24) AND (3.30) (AFTER BROMLEY[46])

T^*	f_1	T^*	f_1	T^*	f_1
0·30	0·1969	1·65	1·0174	4·0	2·0719
0·35	0·2252	1·70	1·0453	4·1	2·1090
0·40	0·2540	1·75	1·0729	4·2	2·1457
0·45	0·2834	1·80	1·0999	4·3	2·1820
0·50	0·3134	1·85	1·1264	4·4	2·2180
0·55	0·3440	1·90	1·1529	4·5	2·2536
0·60	0·3751	1·95	1·1790	4·6	2·2888
0·65	0·4066	2·00	1·2048	4·7	2·3237
0·70	0·4384	2·1	1·2558	4·8	2·3583
0·75	0·4704	2·2	1·3057	4·9	2·3926
0·80	0·5025	2·3	1·3547	5·0	2·4264
0·85	0·5346	2·4	1·4028	6·0	2·751
0·90	0·5666	2·5	1·4501	7·0	3·053
0·95	0·5985	2·6	1·4962	8·0	3·337
1·00	0·6302	2·7	1·5417	9·0	3·607
1·05	0·6616	2·8	1·5861	10	3·866
1·10	0·6928	2·9	1·6298	20	6·063
1·15	0·7237	3·0	1·6728	30	7·880
1·20	0·7544	3·1	1·7154	40	9·488
1·25	0·7849	3·2	1·7573	50	10·958
1·30	0·8151	3·3	1·7983	60	12·324
1·35	0·8449	3·4	1·8388	70	13·615
1·40	0·8744	3·5	1·8789	80	14·839
1·45	0·9036	3·6	1·9186	90	16·010
1·50	0·9325	3·7	1·9576	100	17·137
1·55	0·9611	3·8	1·9962	200	26·80
1·60	0·9894	3·9	2·0343	400	41·90

TABLE 3.3. FUNCTIONS FOR ESTIMATING THE VISCOSITY μ, THERMAL CONDUCTIVITY λ AND DIFFUSIVITY D OF PURE NON-POLAR GASES AND THEIR MIXTURES IN THE PRESSURE-INDEPENDENT REGION (AFTER BIRD[25] AND HIRSCHFELDER[153])

T^*	$\overline{A}^*$	$\overline{B}^*$	f_μ	f_λ	f_D
0·30	1·046	1·289	1·0014	1·0022	1·0001
0·50	1·093	1·284	1·0002	1·0003	1·0000
0·75	1·105	1·233	1·0000	1·0000	1·0000
1·00	1·103	1·192	1·0000	1·0001	1·0000
1·25	1·099	1·165	1·0001	1·0002	1·0002
1·50	1·097	1·143	1·0004	1·0006	1·0006
2·00	1·094	1·119	1·0014	1·0021	1·0016
2·50	1·094	1·106	1·0025	1·0038	1·0026
3·00	1·095	1·101	1·0034	1·0052	1·0037
4·00	1·098	1·095	1·0049	1·0076	1·0050
5·00	1·101	1·092	1·0058	1·0090	1·0059
10·0	1·110	1·094	1·0075	1·0116	1·0076
50·0	1·130	1·095	1·0079	1·0124	1·0080
100·0	1·138	1·095	1·0080	1·0125	1·0080
400·0	1·154	1·095	1·0080	1·0125	1·0080

The notation used in the equations presented above is the same as in Eq. (3.22); k is the Boltzmann constant, N is Avogadro's number and the force constant ε is defined by the Lennard-Jones equation.

In the section below we shall use the following notation:

$$T^* = \frac{kT}{\varepsilon}$$
$$\theta = \frac{1}{\sigma^2}\sqrt{\frac{M\varepsilon}{k}} \tag{3.27}$$

Bromley and Wilke[46] have constructed nomographs which enable values of the viscosity of a gas to be predicted from the values of T^* and θ. The latter quantities may be calculated, if the force constants σ and ε for the gas are known. These may be obtained from Table 1.16; alternatively they may be estimated from the relationships

$$\varepsilon/k \approx 0{\cdot}75\ T_c$$
$$\varepsilon/k \approx 1{\cdot}39\ T_b$$
$$\sigma \approx 0{\cdot}833\ V_c \tag{3.28}$$
$$\frac{kT}{\varepsilon} \approx 1{\cdot}33 T_r$$

where T_c is the critical temperature, °K,
T_b is the boiling temperature, °K,
V_c is the critical volume,
and T_r is the reduced temperature†.

These authors provide a table with values of T^* and θ computed for 46 compounds. The nomographs of Bromley and Wilke are presented in Figs. 3.4 and 3.5.

These nomographs are very simple to use. After calculating, the values of T^* and θ for the desired temperature and compound, a straight line is drawn through the appropriate points on the T^* and θ axes of the nomograph. The point where this line intersects the $\mu°$ axis gives the required value of the viscosity $\mu°$ (in cP). Figure 3.4 is for low- and Fig. 3.5 for high-molecular-weight gases.

If θ for a given gas is unknown, but the viscosity at one temperature is available then, using the nomograph, it is possible to determine θ and to calculate the viscosities of this gas at other temperatures.

The above methods are strictly applicable only to gases consisting of non-polar molecules. In the case of gases with polar molecules, it is possible to use methods which are based on the assumption that the energy of the system of two molecules is described by the Stockmayer equations (1.31) and (1.32).[25]

Joshi and Saxena[180] have calculated the viscosities of a number of polar gases using the Stockmayer equation as modified by Krieger[206] and in many cases their calculated values are in good agreement with the experimental ones. The method which they used is, however, too complicated to be useful for rapid technological calculations.

Although strictly applicable only to non-polar gases, the nomographs in Figs. 3.4 and 3.5 may sometimes be used to predict the temperature dependence of viscosities of polar gases with little error, as in the example below.

† Bird, Hirschfelder and Curtiss[25] recommend somewhat different relationships for calculating ε/k (Ch. 1): $\varepsilon/k \approx 0{\cdot}77 T_c$, $\varepsilon/k \approx 1{\cdot}15 T_b$.

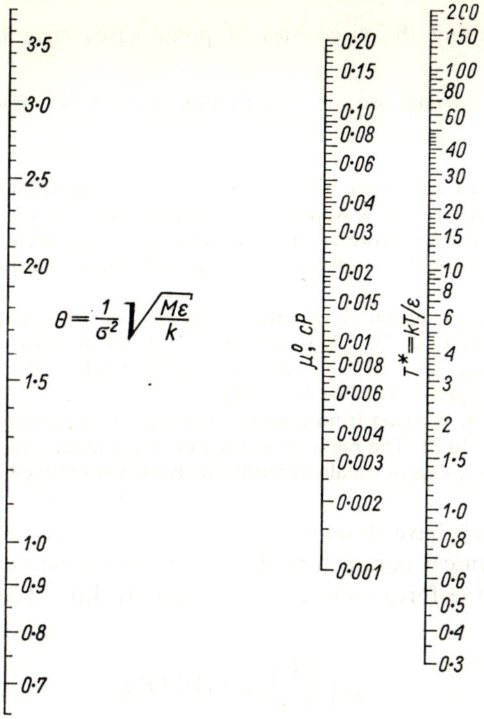

FIG. 3.4. Nomograph for calculating the viscosities of low-molecular-weight gases (after Bromley[46]).

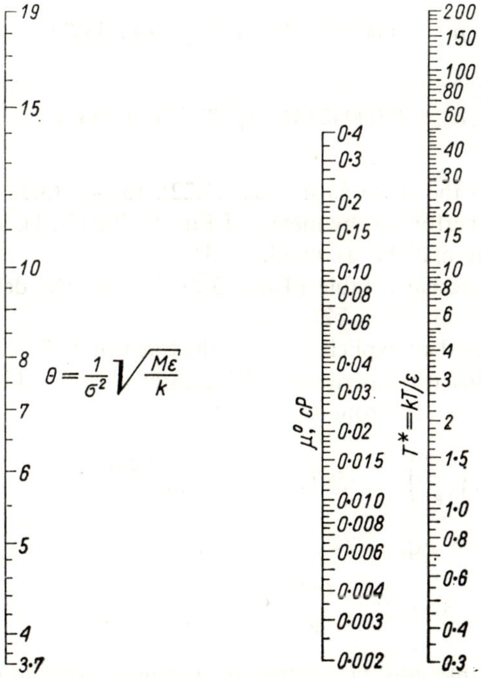

FIG. 3.5. Nomograph for calculating the viscosities of high-molecular-weight gases (after Bromley[46]).

Methods of predicting the viscosities of polar gases have been discussed by J. R. Sutton.[5]

EXAMPLE 3.1.[46] The dynamic viscosity of water vapour at 0°C is 0·0094 cP. Estimate the viscosity at 534·5°K.

Solution

For the purpose of this calculation ε/k is arbitrarily calculated from the equation $\varepsilon/k = 1{\cdot}39T_b$. (In view of the strong polarity of water vapour ε/k thus calculated can only be regarded as having an approximate physical significance. However, errors introduced on this account are largely compensated by calculating θ, not from values of σ but from an experimental viscosity at 0°C combined with the above estimate for ε/k—ED).

Since the normal boiling temperature of water, T_b, is 373·2°K, ε/k is estimated to be 519°K and T^* at 0°C is 273·2/519, i.e. 0·516. T^* at 534·5°K is 543·5/519, i.e. 1·028. A straight line is constructed through the points $\mu_0^\circ = 0{\cdot}0094$ cP and $T_0^* = kT_0/\varepsilon = 0{\cdot}516$ on the nomograph in Fig. 3.5: the point of intersection of this line with the θ axis gives the value $\theta = 10{\cdot}22$.

The viscosity at 543·5°K is found from the same nomograph by drawing a straight line through the points $T^* = 1{\cdot}028$ and $\theta = 10{\cdot}22$. The point at which this line intersects the μ° axis gives the viscosity at 534·5°K to be 0·018 cP. The experimentally determined viscosity is 0·019 cP so the error in the predicted value is 5·3%.

Bromley and Wilke[46] have developed equations giving μ° in terms of the reduced temperature and the critical parameters. They obtained these equations by substituting approximate values of the force constants, as given by Eq. (3.28), into Eqs. (3.22) and (3.24) and writing

$$f_1\left(\frac{kT}{\varepsilon}\right) \approx f_1(1{\cdot}33T_r) \tag{3.29}$$

where T_r is the reduced temperature.

As a result of these substitutions the equations

$$\mu^\circ = 0{\cdot}00333\ (MT_c)^{1/2}V_c^{-2/3}f_1(1{\cdot}33T_r) \tag{3.30}$$

and

$$\mu^\circ = 0{\cdot}000423M^{1/2}p_c^{2/3}T_c^{-1/6}f_1(1{\cdot}33T_r) \tag{3.31}$$

are obtained.

The notation used is the same as in Eqs. (3.22), (3.24), (3.28) and (3.29). The average error of estimations performed by means of Eqs. (3.30), (3.31), is of the order of 4·5% though for polar gases it may be as much as 15%.

The above authors present a table (Table 3.2) showing the dependence of the function $f_1(kT/\varepsilon) \approx f_1(1{\cdot}33T_r)$ on $T^* = kT/\varepsilon$.

Scheibel[320] has proposed an explicit form for the function $f_1(T^*)$ occurring in Eq. (3.24). He has shown that at reduced temperatures, T_r, greater than 0·3 this function may be calculated with sufficient accuracy from the equation

$$f_1\left(\frac{kT}{\varepsilon}\right) = 1{\cdot}58T_r^{0{\cdot}645} - \frac{0{\cdot}261}{(1{\cdot}9T_r)^{0{\cdot}9\log(1{\cdot}9T_r)}} \tag{3.32}$$

After substitution into Eq. (3.24), we obtain

$$\mu^\circ = 32{\cdot}5T_r^{0{\cdot}645}\frac{M^{0{\cdot}5}T_c^{0{\cdot}5}}{V_c^{0{\cdot}667}} - \frac{8{\cdot}7}{(1{\cdot}9T_r)^{0{\cdot}9\log(1{\cdot}9T_r)}} \tag{3.33}$$

The accuracy of estimations performed by means of the latter equation is of the same order as that of Eq. (3.24), the average error being 4·4%. A similar formula has also been derived by Gambill.[112] In practice, however, the methods for estimating the viscosity which are described above, e.g. Eqs. (3.30) and (3.31), are more convenient and no less accurate than those put forward by Scheibel or Gambill.

The function f_1 may be used directly to estimate the viscosity μ_2° of a gas at temperature T_2 from a knowledge of the viscosity μ_1° of this gas at some other temperature T_1. For this purpose either of the following relationships may be used

$$\frac{\mu_1^\circ}{\mu_2^\circ} = \frac{f_1\left(\dfrac{kT_1}{\varepsilon}\right)}{f_1\left(\dfrac{kT_2}{\varepsilon}\right)} \tag{3.34}$$

or

$$\frac{\mu_1^\circ}{\mu_2^\circ} \approx \frac{f_1(1{\cdot}33T_{r1})}{f_1(1{\cdot}33T_{r2})} \tag{3.35}$$

where $T_{r1} = T_1/T_c$, $T_{r2} = T_2/T_c$.

Alternatively, if the viscosity of a gas is known at two temperatures, then it is possible to calculate ε/k from (3.34) using Table 3.2.

It should be remembered that although the estimate of ε/k obtained from the approximate equation (3.28) may be subject to considerable error, the error in μ becomes only an insignificant one when formula (3.35) is used; e.g., if the error in the determination of ε/k is 10%, and $\mu_2/\mu_1 = 2$, then the error of estimation will amount only to 1·5%. For this reason Eqs. (3.34) and (3.35) may even be used to extend viscosity data for polar substances with surprisingly little error, as in the example below.

EXAMPLE 3.2. Solve the preceding example (3.1) using Eq. (3.35) instead of the nomograph given in Fig. 3.4. The critical temperature of water, T_c, is 647·3°K.

Solution

$$T_{r1} = \frac{273{\cdot}2}{647{\cdot}3} = 0{\cdot}422, \qquad T_{r2} = \frac{534{\cdot}5}{647{\cdot}3} = 0{\cdot}826$$

By interpolating, we get from Table 3.2

$$f_1(1{\cdot}33T_{r1}) = f_1(0{\cdot}561) = 0{\cdot}3509$$
$$f_1(1{\cdot}33T_{r2}) = f_1(1{\cdot}100) = 0{\cdot}6928$$

It follows from Eq. (3.35) that

$$\mu_2^\circ = \frac{\mu_1^\circ f_1(1{\cdot}33T_{r2})}{f_1(1{\cdot}33T_{r1})} = \frac{0{\cdot}0094\times 0{\cdot}6928}{0{\cdot}3509} = 0{\cdot}0185 \text{ cP}$$

The experimental value of μ_2° is 0·019 cP so the error in the predicted value is −2·6%.

EXAMPLE 3.3. Estimate the dynamic viscosity μ° of chloroform vapour $CHCl_3$ at a temperature of 250°C:

a. Using the force constants σ and ε in conjunction with Eq. (3.22). The molecular weight of chloroform, M, is 119·4.

Solution

It is seen from Table 1.16 (Ch. 1) that, for chloroform,

$\sigma = 5{\cdot}430$ Å, $\varepsilon/k = 327$°K

hence, the value of $T^* = kT/\varepsilon$ calculated for $T = 250+273 = 523$°K is

$$T^* = \frac{523}{327} = 1{\cdot}600$$

The corresponding value of the function $\Omega^{(2,2)*}$ is found from Table 3.1 to be 1·279. Substituting the values of M, T, σ and Ω in Eq. (3.22), we obtain

$$\mu^\circ \times 10^7 = 266{\cdot}93\,\frac{\sqrt{119{\cdot}4\times 523}}{5{\cdot}430^2\times 1{\cdot}279} = 1768, \text{ i.e. } 0{\cdot}01768 \text{ cP}$$

The viscosity of chloroform vapour determined experimentally is 0·1788 cP so the predicted value is in error by 0·56%.

b. Estimate the viscosity of chloroform vapour at 250°C using the critical parameters in conjunction with Eq. (3.31).

It is found from Table 1.31 that the critical constants of chloroform are:

$$t_c = 260{\cdot}9°\text{C}, \quad T_c = 260{\cdot}9 + 273{\cdot}2 = 534{\cdot}1°\text{K}, \quad P_c = 54{\cdot}9 \text{ atm.}$$

The reduced temperature is

$$T_r = \frac{523}{534} = 0{\cdot}980, \quad \text{so} \quad 1{\cdot}33T_r = 1{\cdot}302$$

and $f_1(1{\cdot}33T_r) = 0{\cdot}8163$ (interpolated from Table 3.2).

By substituting in Eq. (3.31), we obtain

$$\mu° = 0{\cdot}000423 \times 119{\cdot}4^{1/2} \times 54{\cdot}9^{2/3} \times 534^{-1/6} \times 0{\cdot}8163 = 0{\cdot}0191 \text{ cP}$$

The error of estimation is +7·3%.

Strictly, the methods used above are only applicable to a gas consisting of non-polar molecules since only in this case is the intermolecular potential energy adequately described by the Lennard-Jones equation. Chloroform, considered in example 3.3, is a polar compound with a dipole moment of about 1 Debeye unit. It would appear from this example that the prediction methods developed for non-polar compounds can be applied with reasonable success to chloroform vapour though in example 3.3b the error in the predicted viscosity exceeds seven per cent. (Great theoretical significance should not perhaps be attached to the good agreement obtained in example 3.3a since the force constants used for this calculation were themselves derived from viscosity data. The calculation in this case therefore amounts to the calculation of viscosity at one temperature from its value at a different temperature: it has already been seen that this type of prediction can be carried out with greater accuracy than the prediction of the absolute magnitude of the viscosity).

Krieger[206] has developed a method for predicting the viscosities of polar gases. His method is based on an equation for the intermolecular potential energy, which differs from the Lennard-Jones equation by a term depending on the dipole moment of the molecules. His method should be more accurate than those listed above, when applied to the prediction of the viscosities of polar gases.

4. EXPERIMENTAL STUDIES OF THE VISCOSITIES OF GASES

The viscosities of a considerable number of gases have been determined experimentally over wide ranges of temperature and pressure: these data have been collected by Golubev.[131]

Table 3.4 shows values of the dynamic viscosity $\mu°$ for about fifty organic and inorganic gases and vapours together with values of the constant C in the Sutherland equation (3.20). Values of $\mu°$ are given for a series of temperatures† and the viscosity $\mu_c°$ at the critical temperature is also shown in each case. If a given measurement refers to a temperature differing from that given in the column heading, the temperature at which the measurement was carried out is shown in parentheses.

For technological calculations, it is often necessary to know the viscosity of gases at high temperatures: values of the dynamic viscosity $\mu°$ of gases at temperatures ranging from 0 to 1000°C are given in Table 3.5.

† All values shown refer to pressures in the range for which the viscosity $\mu°$ is constant. The superscript ° is used to indicate this.

TABLE 3.4. VALUES OF THE VISCOSITIES OF GASES ($cP \times 10^5$) (AFTER PARTINGTON[282])

Compound	Temperature, °C							μ_c°	C
	0	20	50	100	150	200	250		
Air	1708	1807	—	2171	—	2569	2751	—	112
Ammonia NH_3	916	982	1092	1278	1463	1645	1813	1380	503
Argon	2114	2217	—	2695	—	3223	3444	1209	142
Bromine Br_2	1464	1526(19°)	1528(25°)	1921(94°)	—	2369(190°)	—	2958	533
Carbon monoxide CO	1659	1763	1864	2083	2276	2463	2632	878	101·2
Carbon dioxide CO_2	1375	1493(26·9°)	1725(76·9°)	—	2160(126·9°)	2354(226·9°)	2556(276·9°)	1510	254
Carbon disulphide CS_2	886	—	—	1303(114·3°)	1434(152·5°)	1561(190·2°)	1830(267·3°)	—	499·5
Chlorine Cl_2	1227	1312	1470(55·1°)	1651(99·4°)	1845(145·4°)	2081(199·5°)	2244	1839	330
Cyanogen C_2N_2	928	987	—	1271	—	—	—	—	320
Helium He	1885	1941	—	2281	—	2672	2853	97·1	80
Hydrogen H_2	835	876	937	1032	1126	1212	1297	153	84·4
Hydrogen bromide HBr	1697	1819	—	2344	—	—	—	2305	357
Hydrogen chloride HCl	1313	1434(21°)	1600(54°)	1832(99°)	2093(154°)	2303	2530	1581	362
Hydrogen iodide HI	1726	1857(20·5°)	—	2310(96·6°)	2631(149·4°)	2922(198·6°)	3228(251°)	2640	331
Hydrogen sulphide H_2S	1166	1241	—	1587	—	—	—	—	331
Iodine I_2	1228	—	—	1785(102·2°)	—	2319(232·2°)	2561(319·2°)	3615	568
Krypton Kr	2327	2460(15°)	—	—	—	—	—	1780	188
Neon Ne	2962	3111	—	3646	4177(184·5°)	4248	4532	705	56
Nitrogen N_2	1652	1728(16·1°)	1880(51·6°)	2084	2287(151·9°)	2461	2929	835	104
Nitric oxide NO	1777	1876	2036	2272	2474	2682	2870	1208	128
Nitrous oxide N_2O	1381	1488(26·9°)	1723(76·9°)	1943(126·9°)	2158(176·9°)	2355(226·9°)	2555(276·9°)	1536	260
Oxygen O_2	1911	1985(14·8°)	—	2568(127·7°)	—	3017	3220	1154	125
Sulphur dioxide SO_2	1158	1266	1352(40°)	1612	1716(120°)	2038	—	1884	416

Continued overleaf

TABLE 3.4 (*cont.*)

Compound	Temperature, °C							μ_c°	C
	0	20	50	100	150	200	250		
Water vapour H_2O	861	937	—	1255	1445	1635	1827	—	650
Xenon Xe	2107	2250(16·5°)	—	2871(99·4°)	—	—	—	2249	252
Acetone $(CH_3)_2CO$	660	—	—	991(119°)	1101(159·5°)	1186(190·4°)	1334(247·7°)	—	541
Acetylene C_2H_2	995	1020	1113	1254	1318(120°)	—	—	1073	215
Benzene C_6H_6	677	738	918	1031(131·3°)	1110(161·3°)	1198(194·6°)	1343(252·5°)	1423	447·5
n-Butane n-C_4H_{10}	686	739	787(40°)	947	998(120°)	—	—	—	358
iso-Butane i-C_4H_{10}	692	744	792(40°)	947	995(120°)	—	—	—	330
Carbon tetrachloride CCl_4	898	—	—	1334(127·9°)	1463(169·3°)	1562	1808(282·2°)	1750	365
Chloroform $CHCl_3$	929	1001	—	1267	1491(161·3°)	1579(189·1°)	1776	1780	373
Ethane C_2H_6	851	909	998	1142	1278	1408	1525	935	252
Ethanol C_2H_5	752	—	—	1173(130·2°)	1293(180·7°)	1355(191·8°)	1519(251·7°)	—	407
Ethyl chloride C_2H_5Cl	937	—	—	—	—	—	—	—	411
Ethylene C_2H_4	944	1008	1103	1256	1403	1541	1666	962	241
Ethyl ether $(C_2H_5)_2O$	679	—	—	955	1079(159·7°)	1222(217·7°)	1300(251·0°)	—	404
n-Hexane n-C_6H_{14}	586	—	—	866(120·7°)	958(160·8°)	1021(189·9°)	1144(248·0°)	—	436·1
Methane CH_4	1022	1087	1182	1331	1471	1603	1725	726	164
Methanol CH_3OH	865	—	—	1259(111·3°)	1408(153·9°)	1527(188·8°)	1725	—	487
Methyl bromide CH_3Br	1229	1327	1457	1797(120°)	—	—	—	2115	402
Methyl chloride CH_3Cl	981	1061	1175	1357	—	1769(219·4°)	—	1512	454
Methylene chloride CH_2Cl_2	910	991(22·1°)	—	1267	—	1667(219·1°)	—	1752	395
Methyl ether $(CH_3)_2O$	846	909(19·5°)	1044(60°)	1167	1228(120°)	—	—	1255	426
n-Pentane n-C_5H_{12}	621	—	—	911(121·6°)	995(158·9°)	1126(219·1°)	1191(249·5°)	—	383
Propane C_3H_8	746	801	880	1008	1132	1253	1362	990	278
n-Propanol n-C_3H_7OH	682	—	—	1025(121·7°)	1102(149°)	1267(209·7°)	1350(243·2°)	—	515·6
iso-Propanol i-C_3H_7OH	698	—	—	1028(119·2°)	1112(149·2°)	1248(198·4°)	1382(251·0°)	—	459·9
Propylene C_3H_6	775	835	959(60°)	1071	1122(120°)	—	—	1045	362

TABLE 3.5. VARIATION OF THE VISCOSITIES OF GASES WITH TEMPERATURE (VISCOSITY IN kg/m.hr) (AFTER HOBLER[159])

t, °C	H_2	N_2	O_2	CO_2	CO	CH_4	C_2H_6	C_2H_4	NH_3	NO
0	0·030	0·0605	0·0706	0·051	0·583	0·0367	0·0313	0·0346	0·0327	0·0648
100	0·039	0·0767	0·0907	0·069	0·0738	0·0479	0·0410	0·0457	0·0460	0·0817
200	0·045	0·0907	0·1080	0·083	0·0868	0·0587	0·0504	0·0562	0·0557	0·0965
300	0·051	0·1033	0·1235	0·0961	0·0986	0·0673	0·0590	0·0655	0·0716	0·1100
400	0·0558	0·1148	0·1375	0·1087	0·1094	0·0763	0·0677	0·0742	0·0841	0·1224
500	0·0605	0·1253	0·1505	0·1206	0·1192	0·0839	0·0756	0·0821	0·0950	0·1332
600	0·0648	0·1350	0·1627	0·1318	0·1285	0·0922	0·0828	0·0896	0·1053	0·1444
700	0·0691	0·1444	0·1742	0·1422	0·1372	0·0990	0·0896	0·0965	0·1156	0·1539
800	0·0731	0·1530	0·1850	0·1523	0·1454	0·1062	0·0961	0·1033	0·1253	0·1633
900	0·0761	0·1613	0·1951	0·1616	0·1530	0·1123	0·1019	0·1094	0·1346	0·1721
1000	0·0803	0·1692	0·2048	0·1706	0·1606	0·1188	0·1076	0·1156	0·1436	0·1827

It is seen from these tables that the dynamic viscosities of gases increase with rise in temperature. An increase of temperature from 0 to 1000°C may result in a 3- or even 4-fold growth in $\mu°$.

Viscosities at temperatures other than those listed may be predicted using the Sutherland formula (3.20); it should be remembered, however, that the Sutherland constants C may vary with temperature; most of those listed in Table 3.4 refer to the temperature range 0–300°C.

Figure 3.6 shows dynamic viscosities of some common gases as functions of temperature over the range −220 to +100°C.

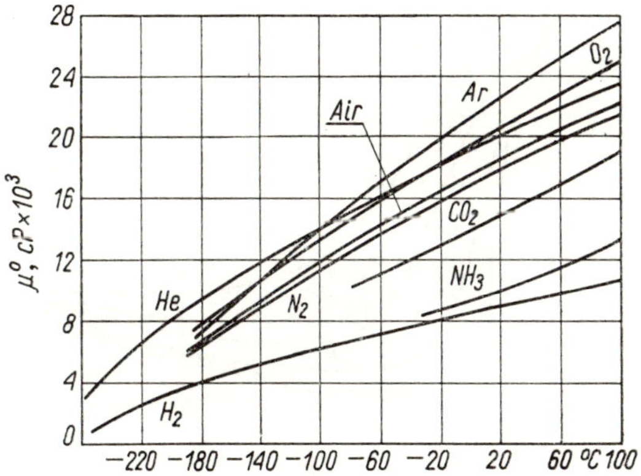

FIG. 3.6. The viscosities of some common gases as functions of temperature in the low temperature region (after Erk[175]).

According to simple kinetic theory, the viscosities of gases should be independent of pressure (cf. section 2). The following data have been obtained for air[282] at sub-atmospheric pressures.

p, mm Hg	750	380	20·5	2·4	1·53	0·63
$\mu/\mu°$	1	1·01	1·004	0·978	0·956	0·908

In this table μ is the viscosity of air under pressure p at 15°C, and $\mu°$ is the viscosity of air under a pressure of 1 atm at 15°C.

It is seen that substantial variations in viscosity occur at very low pressures, i.e. at pressures of a few mm Hg or less. If the pressure is raised substantially above atmospheric (say to 20 atm), variations in the viscosity again occur. These latter variations arise because, at these pressures, gases do not obey the perfect gas laws.

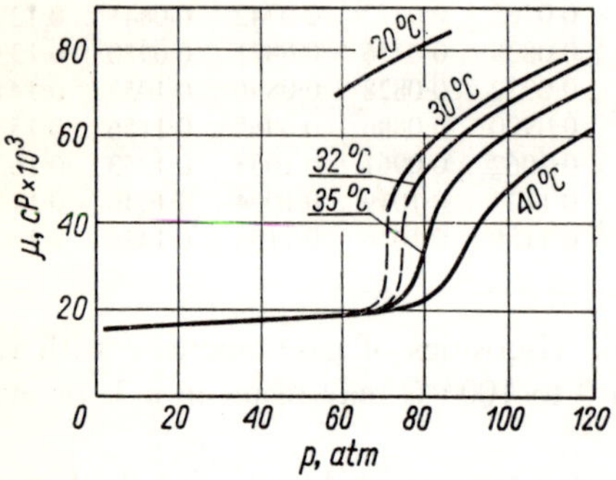

FIG. 3.7. The viscosity of gaseous carbon dioxide as a function of pressure and temperature (after Eucken[91]).

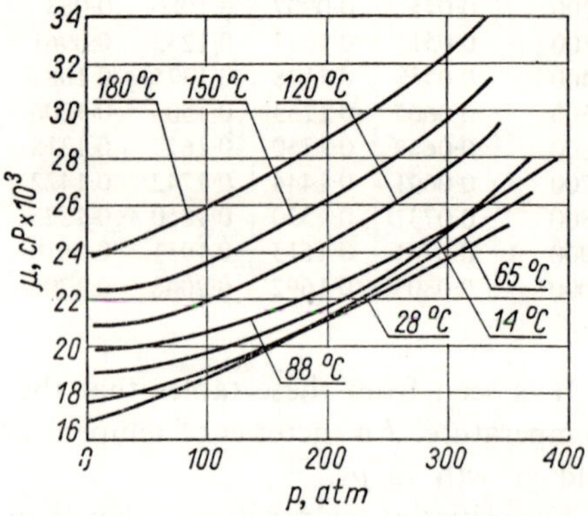

FIG. 3.8. The viscosity of gaseous carbon monoxide as a function of pressure and temperature (after Glaser[129]).

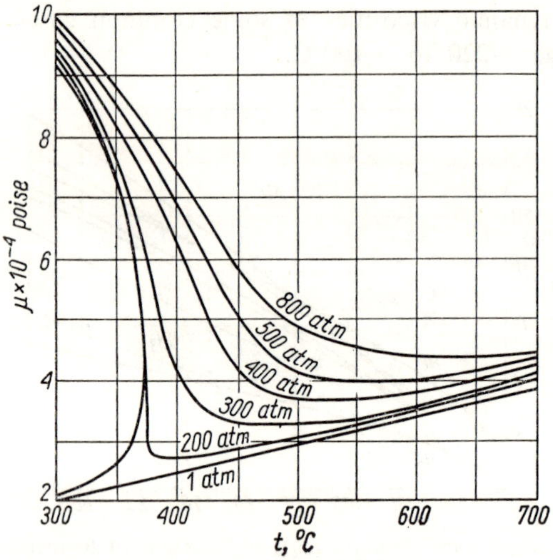

FIG. 3.9. The viscosity of water vapour as a function of temperature and pressure (after Mayinger[242]).

Figure 3.7 shows how the viscosity of carbon dioxide varies with pressure and temperature for temperatures in the range 20 to 40°C and pressures up to 120 atm.

The effects of pressures up to 400 atm upon the viscosity of carbon monoxide are shown in Fig. 3.8 for temperatures between 14 and 180°C.

It is seen from Fig. 3.7 that, over quite a wide range of pressures (up to about 60–70 atm) the viscosity of carbon dioxide is virtually independent of pressure. The higher the temperature, the higher is the pressure at which variations in viscosity become noticable.

The variation of the viscosity of water vapour with temperature and pressure is shown in Fig. 3.9.

If the pressure is raised sufficiently the viscosity of any gas begins to increase sharply with further rise of pressure. Within the high-pressure range and at high temperatures the viscosity of a gas increases with increase in temperature, whereas at low temperatures a rise of temperature is accompanied by a decrease of viscosity (the intersection of the isotherms in Fig. 3.8 arises from this fact).

The fact that the viscosities of gases depend upon the pressure may be explained in terms of the kinetic theory of gases if it is remembered that Eqs. (3.15) and (3.16) were derived on the assumption that only bimolecular collisions occur in the system. At sufficiently low pressures a significant proportion of the collisions undergone by the molecules are with the walls of the containing vessel and the theory upon which Eqs. (3.15) and (3.16) are based becomes inapplicable.

As the pressure rises the average distance between the molecules decreases and the forces of mutual attraction play an increasingly significant part in determining the viscosity. Under these conditions a more realistic model for the intermolecular forces is required than the "hard sphere" model upon which Eqs. (3.15) and (3.16) are based. Under these conditions also the role of tri-molecular collisions becomes more important.

5. SEMI-EMPIRICAL EQUATIONS FOR ESTIMATING GAS VISCOSITIES AND THEIR TEMPERATURE DEPENDENCE

All the equations presented in this section refer to pressures in the range over which μ° is virtually constant. Several equations for the viscosity have been discussed in section 3. In these equations some allowance is made for the forces of molecular interaction. Equations (3.22), (3.23) and (3.24) can only be applied when values of the force constants are known for the gas under consideration. If these constants are not available (e.g. from Table 1.16), it is necessary to use empirical relationships (e.g. Eqs. (3.28)) which enable them to be estimated from critical or boiling point data. By combining expressions, such as Eq. (3.28), for the force constants with Eqs. (3.22) and (3.24) semi-empirical equations may be obtained for the viscosities μ°, of gases. The equations of Bromley and Wilke, (3.30) and (3.31), have already been described. Other equations will be discussed later in this chapter.

The temperature dependence of the dynamic viscosities of gases is given with fair accuracy by the Sutherland equation (3.20). According to this equation

$$\mathring{\mu}_T = \mathring{\mu}_0 \frac{(273+C)}{(T+C)} \left(\frac{T}{273}\right)^{3/2} \tag{3.36}$$

where $\mathring{\mu}_T$ is the viscosity of the gas at temperature T, °K,

$\mathring{\mu}_0$ is the viscosity of the gas at temperature 0°C = 273°K,

and C is the Sutherland constant which is characteristic of the gas under consideration.

When applied over an appropriate temperature range† the Sutherland equation (3.36)

† See section 2.

generally gives fairly accurate results, as, for example, in the following comparison of the estimated and observed values of the viscosity of ethylene.[282] (For ethylene $\mu_0^\circ = 96{\cdot}13\ \mu p$ and $C = 225{\cdot}9$.)

t, °C	$-21{\cdot}2$	$+15{\cdot}0$	$+99{\cdot}5$	$+182{\cdot}4$	$+302{\cdot}0$
μ_{exp}°	89·1	100·6	127·8	153·0	182·6
μ_{calc}°	89·0	101·2	127·8	151·9	183·3

Vogel[386] has found that an approximate estimate of the constant C may be obtained from the relationship

$$C = 1{\cdot}47T_b \tag{3.37}$$

where T_b is the normal boiling temperature of the compound, °K.

This relationship holds quite well, as may be seen from the following typical examples

Compound	O_2	Cl_2	CO	CH_3OH	C_2H_4
T_b, °K	90·2	239·1	81·7	337·9	169·5
$C_{\text{calc}} = 1{\cdot}47T_b$	132·5	351	120	496	249
C_{exp}	125	350	118	487	226

In order to use Sutherland's equation, we have to know the viscosity of the gas under consideration at one temperature. The value of the constant C for a given gas hardly depends on temperature provided this is sufficiently high.† For example, for air at 0°C $C = 113$, and at 1600°C $C = 124$.

Sutherland's equation (3.36) is usually found to give more accurate results than do other equations. (Other equations which are sometimes used include those recommended by Reinganum and Trautz and also the exponential equation.) Reviews of various other formulae are to be found in Reid and Sherwood's book[306] and in Gambill's report.[113]

Andrussov[5,9] has put forward a method for estimating the viscosities of gases with high accuracy at temperatures from -50°C to $+1000$°C (and in some cases even up to 1600°C). According to this author, if no association or decomposition of the gas takes place, the viscosity is given by the equation

$$\mu_T = \frac{191{\cdot}8\left(1+\sqrt{2M}\right)}{V^{2/3}}\,\xi\left(\frac{T}{273{\cdot}16}\right)^{n_\mu} \tag{3.38}$$

where μ_T is the viscosity of the gas at temperature T, μP,
M is the molecular weight,
ξ is a compensating parameter for the viscosity,
n_μ is an exponent given by Eq. (3.39),
and V is the molal volume.

† See section 2.

The value of the ratio $\xi/V^{2/3}$ may be estimated from one experimental value of the viscosity. The exponent n_μ depends on the temperature t (°C), being given by

$$n_\mu = A - B \times 10^{-4} t + C \times 10^{-7} t^2 \tag{3.39}$$

Table 3.6 lists values of $V^{2/3}$, ξ, A, B, and C for a number of gases.

TABLE 3.6. VALUES OF $V^{2/3}$, ξ, A, B AND C (IN EQS. (3.38) AND (3.39)) (AFTER ANDRUSSOV)

Gas	$V^{2/3}$	ξ	A	B	C
H_2	4·00	0·583	0·674	0·034	—
He	2·97	0·750	0·66	0·03	—
D_2	4·285	0·692	0·678	0·04	—
CH_4	9·49	0·757	0·88	2·10	—
NH_3	8·94	0·636	1·06	1·04	—
H_2O	6·05	0·390	1·082	0·25	—
C_2H_2	10·69	0·644	0·96	1·8	—
CO	7·90	0·806	0·762	1·35	0·5
N_2	8·41	0·859	0·753	1·3	0·4
C_2H_4	11·97	0·691	0·95	2·0	—
NO	8·09	0·863	0·79	1·9	1·1
C_2H_6	13·91	0·712	0·94	1·3	—
O_2	6·30	0·700	0·76	1·4	0·64
CH_3OH	11·09	0·559	1·18	0·6	—
H_2S	10·29	0·678	1·02	2·4	—
HCl	9·30	0·680	1·05	2·1	—
Ar	6·83	0·750	0·84	2·5	1·04
CO_2	9·55	0·662	0·942	2·1	0·56
N_2O	11·42	0·786	0·96	2·6	0·8
C_3H_8	17·62	0·663	0·99	3·2	—
CH_3Cl	13·66	0·637	1·06	1·7	—
SO_2	11·76	0·579	1·06	2·6	0·9
Cl_2	13·47	0·656	1·07	2·8	—
CS_2	16·32	0·570	1·33	2·1	—
CCl_4	23·40	0·588			

Licht and Stechert[223] express Sutherland's equation (3.36) in the form

$$\frac{\mu^\circ}{\mu_c^\circ} = \frac{1+C_r}{T_r+C_r} T_r^{3/2} \tag{3.40}$$

where $C_r = C/T_c$. According to these authors, the viscosity μ_c° of a gas at its critical temperature and under a moderate pressure may be estimated approximately from the formula

$$\mu_c^\circ = 3{\cdot}5 \times 10^{-4} M^{1/2} p_c^{2/3} T_c^{-1/6} \tag{3.41}$$

The reduced Sutherland constant C_r for various gases ranges from 0·6 to 1·0, the average value being about 0·8. Combining this value for C_r with Eqs. (3.40) and (3.41), Licht and Stechert obtain the formula

$$\mu^\circ = 0{\cdot}000630 \frac{M^{1/2} p_c^{2/3}}{T_c^{1/6}} \frac{T_r^{3/2}}{T_r+0{\cdot}8} \tag{3.42}$$

It may be noted that this equation enables the dynamic viscosity of a gas to be estimated from a knowledge only of its critical constants. In this respect it resembles Bromley and Wilke's equation.

Chapman's equation (3.16), which was derived from the simple kinetic theory of gases, has also been made the basis for a prediction method. By assuming that the density of a gas is proportional to its molecular weight M, that the molecular velocity is inversely proportional to $\sqrt{M}$, that the mean free path is related to the diameter of the molecule by (3.8), and that the molecular diameter may be deduced from the critical parameters, using van der Waals' equation, Falkovskii[96] established that

$$\mu^\circ = 1{\cdot}286\times 10^{-4} M^{1/2} p_c^{2/3} T_r \tag{3.43}$$

This formula is simpler than the equations discussed above, and it is almost as accurate.

It sometimes happens that the boiling temperature and the critical temperature of a compound are known, whereas the critical pressure and critical volume are not. In this case Arnold's equation[12] may be used. This equation resembles that of Licht and Stechert (3.42), which has already been described.† It differs in that Vogel's equation (3.37) is used for C_r and that p_c has been eliminated, effectively by assuming a proportionality between V_b and V_c and that $p_c V_c / R T_c$ = constant.

According to Arnold's equation,

$$\mu^\circ = 0{\cdot}0027 \frac{M^{1/2} T_c^{1/2}}{V_b^{2/3}} \frac{T_r^{3/2}}{T_r + 1{\cdot}47 T_{rb}} \tag{3.44}$$

where $T_{rb} = T_b/T_c$ and μ is expressed in cP.

Arnold's equation is somewhat less accurate than the equation presented earlier.

Flynn and Thodos[104] and Stiel and Thodos[5, 346] have estimated the viscosities of a number of gases and, using dimensional methods have developed a series of equations which enable the viscosities of gases to be calculated with relatively high accuracy. The above authors find that for all non-polar gases except hydrogen and helium, the relationship between the viscosity and the reduced temperature can be represented by two equations:

When $T_r < 1{\cdot}50$

$$\mu^\circ \zeta = 34\times 10^{-5} T_r^{0{\cdot}94} \tag{3.45}$$

where

$$\zeta = T_c^{1/6} M^{-1/2} p_c^{2/3} \tag{3.46}$$

When $T_r > 1{\cdot}50$

$$\mu^\circ \zeta = 17{\cdot}78\times 10^{-5} (4{\cdot}58 T_r - 1{\cdot}67)^{5/8} \tag{3.47}$$

In these equations

μ° is the viscosity at atmospheric pressure, cP,

and $T_r = T/T_c$ is the reduced temperature.

Equations (3.45) to (3.47) have been checked in 785 cases, including noble gases, diatomic gases, CO_2, CCl_4, and paraffinic, olefinic, naphthenic, acetylenic and aromatic hydrocarbons; the mean deviation from the experimentally determined values is 1·8%.

† Licht and Stechert's equation was actually derived many years after Arnold's equation.

In the case of polar gases where hydrogen bonding can occur between the molecules, Flynn, Stiel and Thodos find that there is a relationship between the product $\mu^\circ\zeta$ and the value of z_c (the compressibility factor at the critical point). When $T_r < 2$

$$(\mu^\circ\zeta)z_c^{5/4} = (7{\cdot}55T_r - 0{\cdot}55)\times 10^{-5} \tag{3.48}$$

If the gas is polar but hydrogen bonding does not occur, the following formula is found to be obeyed:

$$(\mu^\circ\zeta)z_c^{2/3} = (1{\cdot}90T_r - 0{\cdot}29)^{4/5}\times 10^{-4} \tag{3.49}$$

On inserting the value of ζ given by Eq. (3.46), Eq. (3.49) may be rewritten in the form

$$\mu^\circ = 188{\cdot}8\times 10^{-5}\,\frac{\sqrt{MT_c}}{V_c^{2/3}}\,(1{\cdot}90T_r - 0{\cdot}29)^{4/5} \tag{3.50}$$

The mean deviations from the values determined experimentally are 1·5 and 2·6% for Eqs. (3.48) and (3.49), respectively.

a. Viscosities of the Saturated Vapour and Liquid Phases

Novikov[261] finds that the viscosities μ_v and μ_l of the saturated vapour and liquid phases in the vicinity of the critical point are connected by the relationship

$$\frac{1}{\mu_v} + \frac{1}{\mu_l} = 6330 + 13(t_c - t) \tag{3.51}$$

where t is the temperature in degrees centigrade, t_c is the critical temperature also in °C.

A number of equations for calculating μ have been proposed by Gambill.[112]

b. Extension of Viscosity Data Using Graphical Methods Based on Reference Compounds

Methods based on reference compounds depend on discovering straight line relationships between the viscosity of one compound and some convenient property of a second reference compound taken at the same temperature. Exhaustive discussions of such methods are given by Pilar,[286] and by Othmer and Josefowitz.[269]

In practice two relationships have been found particularly useful as a basis for graphical constructions. These are

$$\log\mu = A\log p_s + B \tag{3.52}$$

and

$$\log\mu = A'\log\mu' + B' \tag{3.53}$$

where μ is the viscosity of the gas under consideration at temperature T,
μ' is the viscosity of a reference gas at the same temperature,
and p_s is the vapour pressure of a reference compound at temperature T;
A, B, A' and B' are constants characterizing the chosen pair of compounds (the compound under consideration and the reference compound).

The first equation is the more convenient of the two and it forms the basis of a graphical method for extending viscosity data which has been recommended by Othmer. This method involves constructing a graph on log–log paper. The abscissa for each point is the vapour pressure of the reference liquid (usually water) at some convenient temperature and the ordinate is the viscosity of the gas under consideration at the same temperature. Points plotted in this way for a given gas and reference compound fall on a virtually straight line. The temperatures to which the vapour pressures refer are usually also indicated along the abscissa, so that the viscosity corresponding to any particular

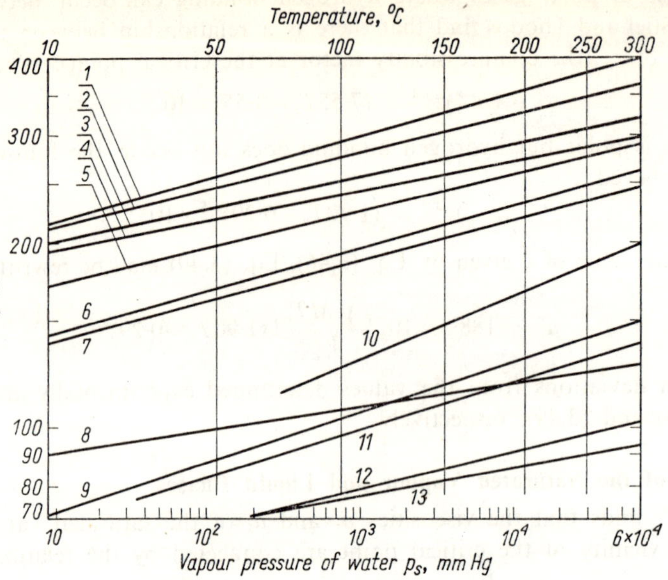

FIG. 3.10. Othmer graph showing the viscosities of some common gases as functions of temperature. The vapour pressure of the reference substance (water) is shown along the lower abscissa and is plotted against the viscosity (which is shown along the ordinate) using a log–log scale. The temperature is shown along the upper abscissa (after Othmer[269] and Pilar[286]). *1*—xenon, *2*—argon, *3*—oxygen, *4*—helium, *5*—air, *6*—carbon dioxide, *7*—nitrous oxide, *8*—hydrogen, *9*—propane, *10*—ethylene, *11*—diethyl ether, *12*—n-butane, *13*—isobutane.

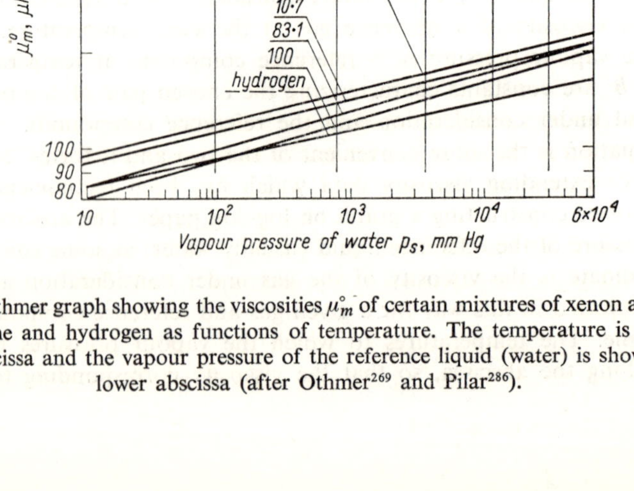

FIG. 3.11. Othmer graph showing the viscosities μ_m° of certain mixtures of xenon and hydrogen, and of propane and hydrogen as functions of temperature. The temperature is shown along the upper abscissa and the vapour pressure of the reference liquid (water) is shown along the lower abscissa (after Othmer[269] and Pilar[286]).

temperature may be read directly from the graph. The viscosities of 13 gases are shown as functions of temperature (with water as reference compound) in Fig. 3.10. In order to draw the line for any given gas it suffices to know the viscosity of the gas at two temperatures. This method enables viscosity data to be interpolated and extrapolated very rapidly as a function of temperature. The same construction may be used for gas mixtures of fixed composition (Fig. 3.11).

The viscosities shown in Fig. 3.10 are for pressures of about one atmosphere. If more elevated pressures are considered the viscosity becomes noticeably pressure de-

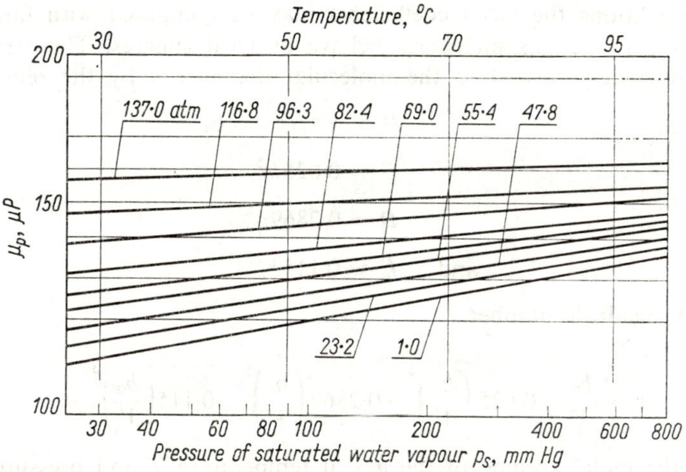

FIG. 3.12. Othmer graph showing the viscosity μ_p of methane as a function of pressure and temperature. The temperature is shown along the upper abscissa and the vapour pressure of the reference compound (water) along the lower abscissa (after Othmer[269] and Pilar[286]).

pendent. A similar plot to that employed in Fig. 3.10 may still be employed, but isobaric values of the viscosity are now plotted. Figure 3.12 shows such a graph for methane in the temperature range 30–100°C and in the pressure range 1–137 atm.

6. VARIATION OF THE VISCOSITY OF A GAS WITH PRESSURE

According to simple kinetic theory (Eqs. (3.15) and (3.16)) the viscosity of a gas should not depend on pressure. This result is found to be approximately true only at moderate pressures. The upper limit to the pressure range for which μ is constant for a given temperature depends upon the temperature and the gas, varying from several to a few dozen atmospheres. When this limit is exceeded, the viscosity of the gas begins to increase with rise in pressure (cf. section 4).

a. Calculations Based on the Equation of State of a Real Gas

According to Kammerlingh-Onnes[189] the equation of state of a real gas can be represented in the form

$$\frac{pV}{RT} = 1 + \frac{B}{V} + \frac{C}{V^2} + \frac{D}{V^3} + \frac{E}{V^4} + \dots \tag{3.54}$$

or

$$\frac{pV}{RT} = 1+y \tag{3.55}$$

where B, C, D, E ... are the second, third, fourth, fifth... virial coefficients respectively. The magnitudes of these coefficients depend upon the gas considered and the temperature.

At sufficiently high temperatures the attractive part of the intermolecular potential becomes negligible by comparison with the translational kinetic energy of the molecules. Under these conditions the virial coefficients may be computed with fair accuracy on the assumption that the gas molecules behave as hard spheres. The virial coefficients are then approximately related to the molecular diameter σ by the relationships

$$\begin{aligned} B &= \tfrac{2}{3}\pi N\sigma^3 = b_0 \\ C &= 0{\cdot}625\, b_0^2 \\ D &= 0{\cdot}2869\, b_0^3 \\ \text{and} \quad E &\approx 0{\cdot}115\, b_0^4 \end{aligned} \tag{3.56}$$

where N is Avogadro's number.

Hence,

$$y = \frac{b_0}{V} + 0{\cdot}625\left(\frac{b_0}{V}\right)^2 + 0{\cdot}2869\left(\frac{b_0}{V}\right)^3 + 0{\cdot}115\left(\frac{b_0}{V}\right)^4 + \ldots \tag{3.57}$$

where V is the molal volume of the gas at temperature T and pressure p. Values of the function y are listed as a function of b_0/V in the literature.[25]

As the starting point for his studies of the viscosities of gases, Enskog[88, 140, 306] considered the pressure dependence of the viscosity of a hypothetical gas consisting of perfectly rigid spherical molecules. He showed that, for such a gas

$$\mu_p = \mu^\circ\left(\frac{2\pi N\sigma^3}{3V}\right)\left(\frac{1}{y} + 0{\cdot}80 + 0{\cdot}761y\right) \tag{3.58}$$

where μ_p is the viscosity of the gas under pressure p,
and μ° is the viscosity of the gas under a moderate pressure.
y in this equation is given by Eq. (3.55)

$$y = \frac{PV}{RT} - 1 = z - 1 \tag{3.59}$$

where z is the compressibility factor of the "hard spheres" gas.

Enskog suggested that (3.58) could be used to represent the effects of pressure on the viscosities of gases composed of real molecules if the pressure P occurring in the definition of y (Eq. (3.59)) were replaced by the "thermal pressure" $T(\partial P/\partial T)_V$. The value of y to be used in Eq. (3.58) is therefore given by

$$y = \frac{V}{RT}\left[T\left(\frac{\partial P}{\partial T}\right)_V\right] - 1 \tag{3.60}$$

Hiroshi, Shimotake and Thodos have used Enskog's equation to calculate the viscosities of ammonia and sulphur dioxide at high densities with reasonable accuracy.[152]

However, the practical applications of Enskog's equation are rather limited, since in order to use it, it is necessary to know the molecular diameter and also the *PVT* relationships for the gas considered.

Damasius and Thodos[69] have suggested a convenient method for applying Enskog's equation; they find that the value of y at the critical point is given by

$$y_c = z_c \alpha_c - 1 \tag{3.61}$$

where $z_c (= p_c V_c / RT_c)$ is the critical compressibility factor,
and $\alpha_c (= \mathrm{d} \ln p / \mathrm{d} \ln T_c)$ is the Riedel parameter (cf. Ch. 1, section 8d).

The above authors have constructed a graph (Fig. 3.13), showing the variation of the reduced quantity $y_r = y/y_c$ with reduced temperature (T_r) and reduced pressure (P_r). (This graph is based on the *PVT* properties of argon.) After calculating the value of α_c, e.g. from Eq. (1.132) or (1.133), and the value of y_c from Eq. (3.61), the value of y_r at any desired values of T_r and p_r can be read from Fig. 3.13; Enskog's modulus, y, may then be found from the equation

$$y = y_r y_c$$

McLaughlin[5] has calculated the viscosity of nitrogen at 50°C at various pressures and has compared his calculated values with observed viscosities, as in the following table.

Pressure p, atm	15·37	57·60	104·5	212·4	320·4	541·7	742·1	965·8
$\mu \times 10^6$, calc., poises	181	190	208·8	224	266	348	418	492
$\mu \times 10^6$, exp., poises	191·3	198·1	205	237·3	273·7	350·9	416·3	491·3

Mention should also be made of a method developed by Gardner[124] who transformed and simplified Enskog's equation to give

$$\frac{\mu_p}{\mu^\circ} = 1 + 0{\cdot}175(b\rho) + 0{\cdot}8651(b\rho)^2 \tag{3.62}$$

where ρ is the gas density, g/cm³,
and b is a volume correction, cm³/g, given by

$$b = 1{\cdot}783 \times 10^{-7} T^{0{\cdot}75} \mu^{\circ -1{\cdot}5} M^{-0{\cdot}25} \tag{3.63}$$

(μ° is expressed in g/cm.sec).

Although Eqs. (3.62) and (3.63) describe the pressure dependence of the viscosities of nitrogen, water vapour and other gases with fair accuracy, they are inaccurate when applied to compounds with more complex molecules.

A review of work on the variation of the viscosity of gases with pressure has been presented by Liley.[5, 224]

EXAMPLE 3.4. Estimate by the methods presented above the value of the dynamic viscosity μ_p of carbon monoxide at 50°C ($T = 323{\cdot}2$°K) and pressure $p = 700$ atm.

Solution

a. *Using Eqs.* (3.57) *and* (3.58) (*i.e. without taking account of the intermolecular attractive forces*).
The value of the compressibility factor may be found from tables.[160] From Table 1.31, the critical constants of carbon monoxide are found to be

$t_c = -138{\cdot}7$°C, ($T_c = 134{\cdot}5$°K), $p_c = 34{\cdot}6$ atm

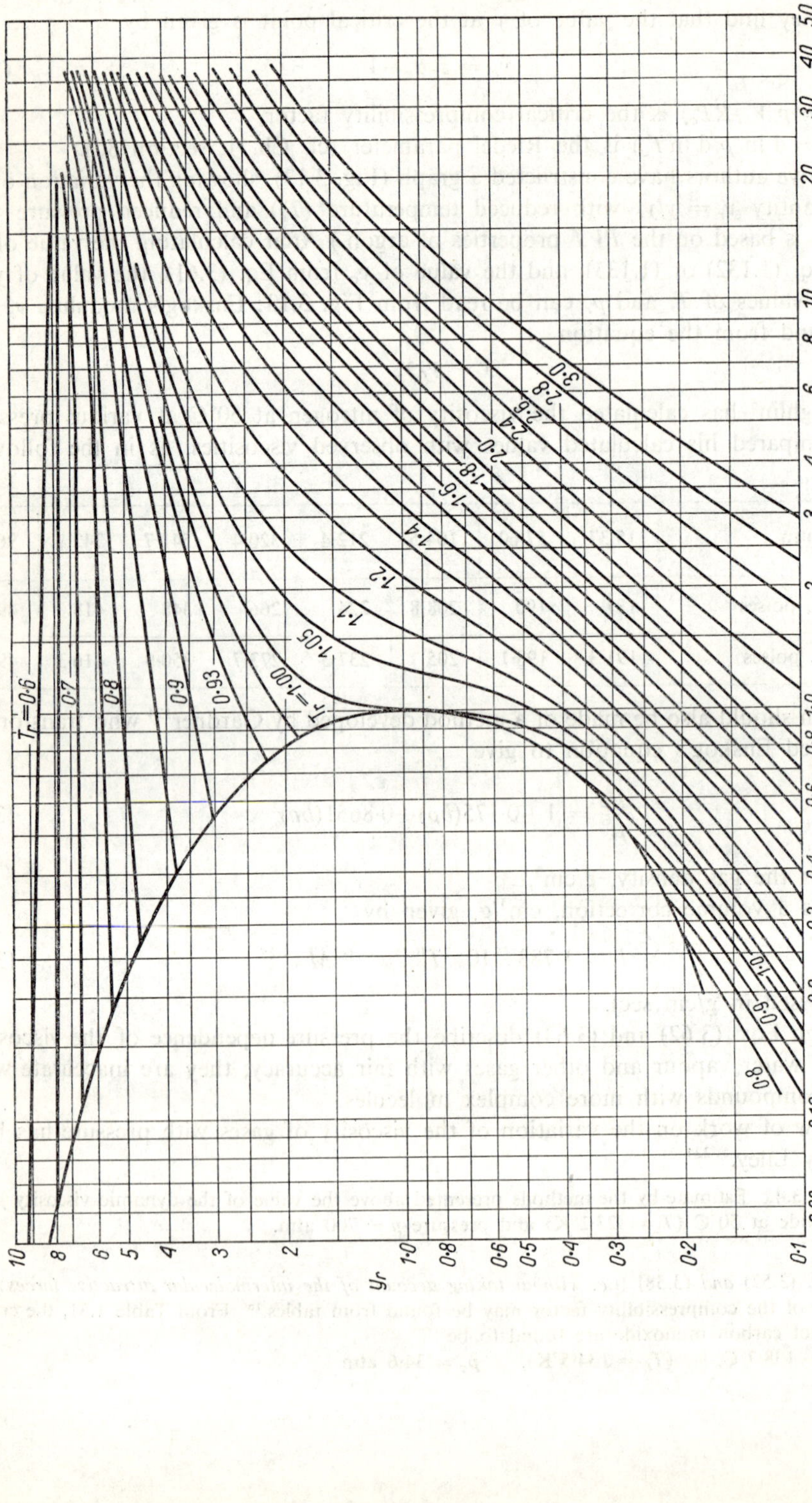

FIG. 3.13. The reduced Enskog modulus y_r as a function of reduced pressure and reduced temperature.

The reduced temperature and pressure are therefore

$$T_r = \frac{323 \cdot 2}{134 \cdot 5} = 2 \cdot 40 \quad \text{and} \quad p_r = \frac{700}{34 \cdot 6} = 20 \cdot 2$$

The value of z corresponding to $T_r = 2\cdot40$ and $p_r = 20\cdot2$ is found from tables[160] to be 1·66. To apply Eq. (3.58), the volume V of the gas is required.

From the definition of z

$$V = \frac{zRT}{p} = \frac{1 \cdot 66 \times 82 \cdot 06 \times 323 \cdot 2}{700} = 63 \cdot 0 \text{ cm}^3/\text{mole}$$

We must also know the diameter σ of the molecule. From Table 1.16 it is seen that, for carbon monoxide $\sigma = 3\cdot763$ Å

Hence

$$b_0 = \frac{2}{3}\pi N\sigma^3 = 67 \cdot 22 \text{ cm}^3/\text{mole}$$

The ratio (b_0/V) to be used in Eq. (3.57) is therefore

$$\frac{b_0}{V} = 1 \cdot 065$$

so

$$y = 1 \cdot 065 + 0 \cdot 625 \times 1 \cdot 065^2 + 0 \cdot 2869 \times 1 \cdot 065^3 + 0 \cdot 115 \times 1 \cdot 065^4 = 2 \cdot 272$$

The value of $\mu°$, i.e. the viscosity of carbon monoxide at 50°C and a moderate pressure, is taken from Table 3.4, $\mu° = 186\cdot4$ μP.

Hence, from Eq. (3.58)

$$\mu_p = 186 \cdot 4 \times 1 \cdot 065 \left(\frac{1}{2 \cdot 272} + 0 \cdot 80 + 0 \cdot 761 \times 2 \cdot 272 \right) = 595 \ \mu\text{P}$$

b. *Using the graph presented in Fig.* 3.13

The value of y_r corresponding to $T_r = 2\cdot40$ and $p_r = 20\cdot2$ is found from Fig. 3.13 to be 1·95. From Fig. 1.20 it is found that, for carbon monoxide

$\alpha_c = 6\cdot35$ and $z_c = 0\cdot292$

Hence,

$y_c = 6\cdot35 \times 0\cdot292 - 1 = 1\cdot022$ and $y = 1\cdot95 \times 1\cdot022 = 2\cdot0$

Substituting this value in Eq. (3.58) we obtain

$$\mu_p = 186 \cdot 4(1 \cdot 065) \left(\frac{1}{2 \cdot 0} + 0 \cdot 80 + 1 \cdot 522 \right) = 557 \ \mu\text{P}$$

c. *Using Eqs.* (3.62) *and* (3.63)

Substituting $\mu° = 186\cdot4 \times 10^{-4}$ P, $T = 323\cdot2$°K and $M = 28\cdot01$ in Eq. (3.63) gives

$$b = \frac{1 \cdot 783 \times 10^{-7} \times 323 \cdot 2^{3/4}}{186 \cdot 4^{3/2} \times (10^{-6})^{3/2} \times 28 \cdot 01^{1/4}} = 2 \cdot 32$$

The molal volume $V = 63\cdot0$ cm³/mole (from (a))

The density $\rho = \frac{M}{V} = \frac{28\cdot01}{63\cdot0} = 0\cdot445$ g/cm³

so $b\rho = 2\cdot32 \times 0\cdot445 = 1\cdot032$

On substituting the above values for $\mu°$ and $b\rho$ in Eq. (3.62) we obtain

$$\mu_p = 186 \cdot 4(1 + 0 \cdot 175 \times 1 \cdot 032 + 0 \cdot 8651 \times 1 \cdot 032^2) = 392 \ \mu\text{P}$$

d. *Using the "corresponding states"† graph (Fig. 3.14) constructed by Uyehara and Watson (see section 6b below)*

For $T_r = 2\cdot40$ and $p_r = 20\cdot2$, we find $\mu_r = 2\cdot50$. Hougen and Watson[160] report that the value of the viscosity μ_c of carbon monoxide at the critical point (μ_c) is 190 μP.

Hence,

$$\mu_p = 2 \cdot 50 \times 190 = 475 \ \mu\text{P}$$

It is seen from the above examples that when the various prediction methods described in this section are applied to a particular case, they may lead to very different results. Clearly these methods can only be used for approximate calculations.

† The theory of reference compounds (Ch. 1) is closely connected with the theory of corresponding states. The application of this theory to the presentation of the variation of gas viscosity with pressure has been described in section 5a (Fig. 3.11).

b. Use of the Theory of Corresponding States to Predict Gas Viscosities at Elevated Pressures

In 1894 Onnes,[264] on the basis of the van der Waals equation of state and the kinetic theory of gases, came to the conclusion that the function†

$$\mu\left(\frac{T_c}{M^3 p_c^4}\right)^{1/6}$$

should be the same for all compounds when their states corresponded, i.e. when their reduced temperatures (T/T_c) were equal and their pressures (p/p_c) were equal. From this result there follow certain relationships, e.g. (3.31), (3.41), (3.42) and (3.44), from which the viscosities μ° of gases in the pressure-independent region may be predicted. In addition Onnes' theory offers a simple and rapid means of estimating the effects of pressure on the viscosities of gases.

Two types of diagram may be used to represent the effects of reduced pressure and temperature upon gas viscosity: the first type of diagram shows the reduced viscosity μ_r as a function of T_r and p_r (μ_r is given by

$$\mu_r = \mu/\mu_c \tag{3.64}$$

where μ_c is the viscosity of a compound at the critical point).

The curves representing the relationship between μ_r, T_r, and p_r for different gases are found approximately to coincide, so a generalized graph of μ_r as a function of T_r and p_r may be drawn.

The second and simpler type of diagram shows as ordinate the ratio of the gas viscosity μ_p under a high pressure to the viscosity μ° in the region in which gas viscosity does not depend on pressure (the viscosities μ_p and μ° are measured at the same temperature T). This ratio is plotted as a function of p_r for integral values of T_r. It is found that the curves representing (μ_p/μ°) as a function of T_r and p_r for different gases coincide, thus producing a generalized viscosity diagram.

The first method has been elaborated by Uyehara and Watson.[376] These authors present a set of curves (Fig. 3.14) showing reduced viscosity μ_r as a function of reduced temperature for integral values of the reduced pressure. The lowest curve corresponds to μ_r°, i.e. to the value of the reduced dynamic viscosity in the region in which viscosity does not vary with pressure.

Knowing the critical parameters T_c and p_c of a gas, we can calculate the reduced parameters and read from the graph the value of the reduced viscosity μ_r. In order to calculate μ, the viscosity μ_c of the gas at the critical point is required. This is a practical drawback to the method since μ_c has been measured for only a few gases. The values of μ_c for some compounds are listed in Table 3.7.

If an experimental value for μ_c is not available, Uyehara and Watson recommend that it be calculated from the molecular weight and critical constants using one of the empirical equations presented below (μ_c is given in micropoises, μP)

$$\mu_c = 61{\cdot}6\,\frac{M^{1/2}T_c^{1/2}}{V_c^{2/3}} \tag{3.65}$$

† In this function μ is the viscosity of the gas, T_c, p_c are critical constants, and M is the molecular weight.

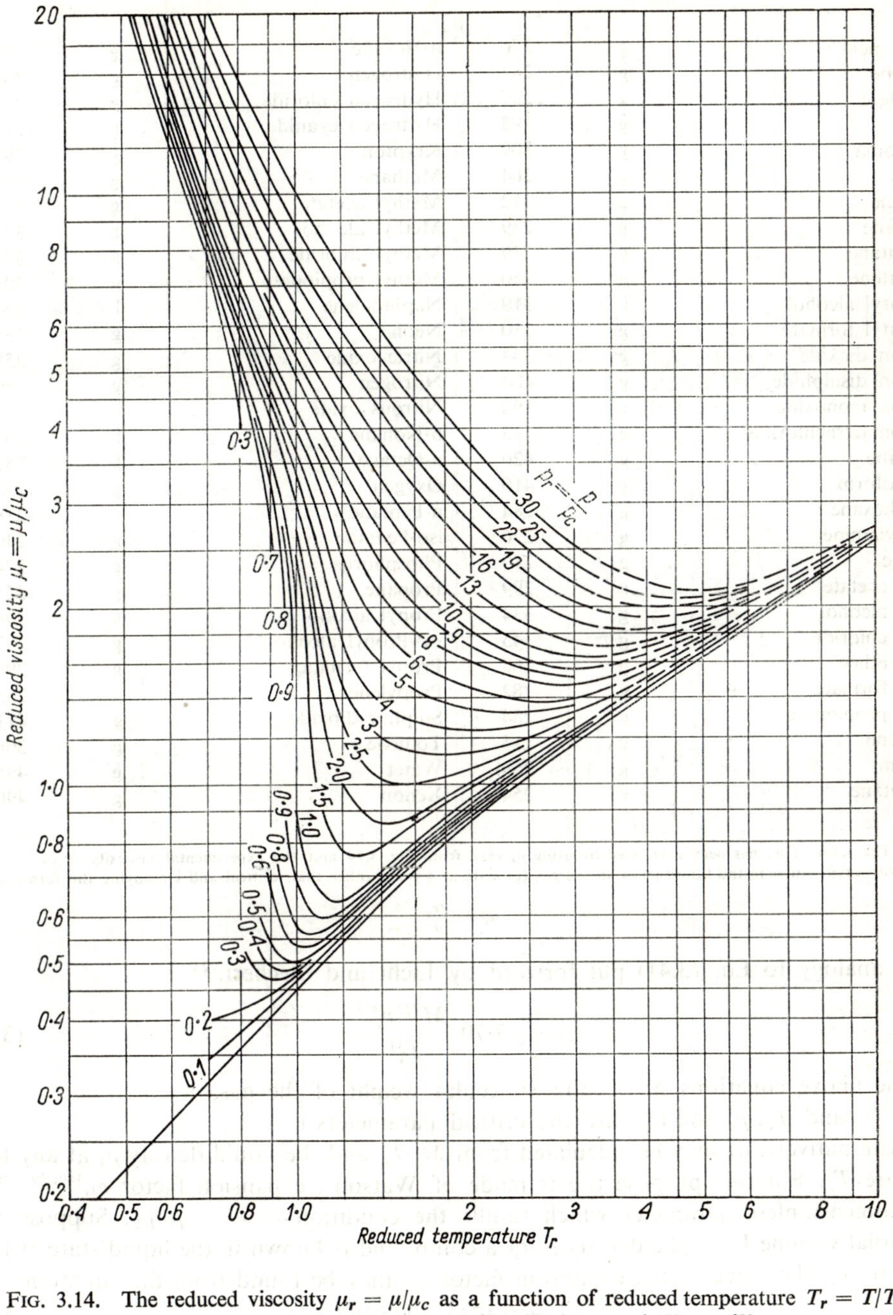

FIG. 3.14. The reduced viscosity $\mu_r = \mu/\mu_c$ as a function of reduced temperature $T_r = T/T_c$ and reduced pressure $p_r = p/p_c$ (after Uyehara and Watson[376]).

TABLE 3.7. VALUES OF VISCOSITIES μ_c AT THE CRITICAL POINT*

Compound	Phase	μ_c, μP	Compound	Phase	μ_c, μP
Acetic acid	g	376	n-Hexane	g	248
Acetone	g	285	Hydrogen	g	34·7
Acetylene	g	237	Hydrogen chloride	g	353
Air	g	193	Hydrogen cyanide	g	316
Ammonia	g	309	Krypton	g	396
Argon	g	264	Methane	g	159
Benzene	g	312	Methyl acetate	g	307
n-Butane	g	239	Methyl alcohol	g	375
iso-Butane	g	239	Methyl chloride	g	338
iso-Butene	g	250	Methyl propionate	g	301
iso-Butyl alcohol	l	319	Naphthalene	l	340
iso-Butyl formate	g	310	Neon	g	156
Carbon dioxide	g	343	Nitric oxide	g	258
Carbon disulphide	g	404	Nitrogen	g	180
Carbon monoxide	g	190	Nitrous oxide	g	332
Carbon tetrachloride	g	413	n-Nonane	l	265
Chlorine	g	420	n-Octane	l	259
Chloroform	g	410	Oxygen	g	250
Cyclohexane	g	284	n-Pentane	g	238
Diethylamine	g	276	iso-Pentane	g	240
Ethane	g	210	Phosphorus	g	274
Ethyl acetate	g	299	Propane	g	228
Ethyl alcohol	g	334	Propyl alcohol	g	298
Ethyl chloride	g	345	iso-Propyl alcohol	g	297
Ethyl ether	g	268	Propyl formate	g	296
Ethyl formate	g	282	Propylene	g	233
Ethyl propionate	g	284	Sulphur dioxide	g	411
Ethylene	g	215	Toluene	g	306
Helium	g	25·4	Water	g	495
n-Heptane	l	254	Xenon	g	490

* The value of μ_c has been calculated by using μ_r read from Fig. 3.14 and the experimental viscosity μ of the compound measured, either in the liquid (l) or in the gas (g) state at a point other than critical and employing the formula

$$\mu_c = \frac{\mu_r}{\mu}$$

or in analogy to Eq. (3.41) put forward by Licht and Stechert,[223]

$$\mu_c = 7{\cdot}70\,\frac{M^{1/2}p_c^{2/3}}{T_c^{1/6}} \tag{3.66}$$

(In the above equations M is the molecular weight of the gas,
and T_c, p_c and V_c are the critical parameters.)

Alternatively, μ_c may be calculated from M, T_c and the liquid density ρ_1 at any temperature T_1. For this purpose use is made of Watson's expansion factor ω.[160, 393] This is a dimensionless parameter which fulfills the condition $\omega_1/\omega_2 = \rho_1/\rho_2$. Suppose that the molal volume V_1 or the density ρ_1 of a compound is known in the liquid state at temperature T_1, the value of the expansion factor ω_1 may be found from the equation†

$$\omega_1 = 0{\cdot}1745 - 0{\cdot}0838\,T_r \tag{3.67}$$

† This equation is only applicable at reduced temperatures less than 0·65 and pressures not exceeding 10 atm (Ed.).

Since $\omega_c = 0{\cdot}044$, Eq. (3.66) may be rewritten

$$\mu_c = 7{\cdot}70\frac{M^{1/2}T_c^{1/2}}{(V_1\omega_1)^{2/3}} = 7{\cdot}70\frac{T_c^{1/2}(\rho_1/\omega_1)^{2/3}}{M^{1/6}} \tag{3.68}$$

If viscosity data are available for the substance considered but these refer to temperatures or pressures which differ from those at which these data are required Fig. 3.14 may be used to calculate the viscosity at the correct temperature and pressure. In this case, the best procedure is to deduce μ_c from the viscosities which are available, Fig. 3.14 being used for this purpose. Suppose, for example, that the viscosity μ_1 is available at some temperature and pressure, T_1 and p_1, i.e. at a reduced temperature T_{r1} and a reduced pressure p_{r1}. The value μ_{r1} of the reduced viscosity under these conditions may be read from Fig. 3.14 and μ_c may then be calculated from the expression

$$\mu_c = \mu_1/\mu_{r1} \tag{3.69}$$

The value of μ_c calculated in any of the above ways may be used in conjunction with Fig. 3.14 to estimate the viscosity μ at any temperature T and pressure p.

The use of Uyehara and Watson's method is demonstrated in the following examples.

EXAMPLE 3.5. Estimate the viscosity μ_p of carbon monoxide at a temperature of 120°C and a pressure of 300 atm.

Solution

The critical constants for CO are taken from Table 1.31:

$$T_c = -138{\cdot}7+273{\cdot}2 = 134{\cdot}5\text{°K}, \qquad p_c = 34{\cdot}6 \text{ atm},$$

$$\rho_c = 0{\cdot}31 \text{ g/cm}^3, \qquad V_c = \frac{28{\cdot}01}{0{\cdot}31} = 90{\cdot}5 \text{ cm}^3\text{/mole}, \qquad M = 28{\cdot}01 \text{ g/mole}$$

It follows from Eq. (3.65) that

$$\mu_c = 61{\cdot}6\times\frac{28{\cdot}01^{1/2}\times 134{\cdot}5^{1/2}}{90{\cdot}5^{2/3}} = 186\ \mu\text{P}, \qquad T = 120+273{\cdot}2 = 393{\cdot}2\text{°K}$$

$$T_r = \frac{393{\cdot}2}{134{\cdot}5} = 2{\cdot}92, \qquad p_r = \frac{300}{34{\cdot}6} = 8{\cdot}68$$

Using these values for T_r and p_r it is found from Fig. 3.14 that

$$\mu_r = 1{\cdot}46$$

μ_p is given by Eq. (3.64) to be

$$\mu_p = \mu_r\mu_c = 1{\cdot}46\times 186 = 272\ \mu\text{P}$$

The value of μ_p determined experimentally is 274 μP (see Fig. 3.8) and the difference between the predicted and observed values is 0·75%.

EXAMPLE 3.6. Estimate the viscosity μ_c of carbon dioxide at the critical point, when: (a) the critical constants T_c and p_c are known, (b) the value of the viscosity μ_0° at 0°C and a moderate pressure is known.

Solution

a. From Table 1.31 it is found that the critical parameters for carbon dioxide are

$$T_c = 304{\cdot}1\text{°K} \quad \text{and} \quad p_c = 73{\cdot}0 \text{ atm}$$

The molecular weight, M, is 44·01 g/mole

According to Eq. (3.66), therefore,

$$\mu_c = 7{\cdot}7\times\frac{44{\cdot}01^{1/2}\times 73{\cdot}0^{2/3}}{304{\cdot}1^{1/6}} = 342\ \mu\text{P}$$

b. The viscosity μ_0 of carbon dioxide at 0°C or 273·2°K, is known to be 137·5 μP, the reduced temperature is then

$$T_r = \frac{273{\cdot}2}{304{\cdot}1} = 0{\cdot}898 \text{ (Table 3.4).}$$

The reduced viscosity μ_r° corresponding to this value of T_r is found from the lower curve in Fig. 3.14 to be 0·401. μ_c is now calculated from Eq. (3.69)

$$\mu_c = \mu_0/\mu_r = 137{\cdot}5/0{\cdot}401 = 343\ \mu\text{P}$$

(according to Hougen and Watson[160] $\mu_c = 343$ μP).

EXAMPLE 3.7. Estimate the viscosity of cyclohexane vapour at the critical point, knowing the critical temperature of cyclohexane, $t_c = 281{\cdot}02°C$, and the density at 20°C, $\rho = 0{\cdot}7791$ g/cm³. The molecular weight (M) of cyclohexane is 84·15.

Solution

At 20°C the reduced temperature of cyclohexane is given by

$$T_r = \frac{273{\cdot}2+20}{281{\cdot}02+273{\cdot}2} = \frac{293{\cdot}2}{554{\cdot}22} = 0{\cdot}530$$

It follows from Watson's equation (3.67) that at 20°C

$$\omega = 0{\cdot}1745 - 0{\cdot}0838\times 0{\cdot}530 = 0{\cdot}1301$$

By Eq. (3.68), therefore,

$$\mu_c = 7{\cdot}70\,\frac{554{\cdot}22^{1/2}\left(\dfrac{0{\cdot}7791}{0{\cdot}1301}\right)^{2/3}}{84{\cdot}15^{1/6}} = \frac{7{\cdot}70\times 23{\cdot}5\times 3{\cdot}29}{2{\cdot}09} = 285\ \mu\text{P}$$

This figure is in very good agreement with the value (284 μP) listed in Table 3.7.

The disadvantage of the method recommended by Uyehara and Watson is that it involves a knowledge of the value of the dynamic viscosity μ_c at the critical point. Experimentally determined values of this quantity are rarely available and empirically based estimations of it may be subject to considerable error.

A simpler method of applying the theory of corresponding states consists in estimating the value of the ratio of $\mu_p/\mu°$, i.e. the ratio of the viscosity under a high pressure p to the viscosity $\mu°$ at the same temperature but within the pressure range over which the viscosity does not depend on pressure. Values of $\mu°$ have been determined experimentally for many gases (Table 10.3) and, unlike μ_c, $\mu°$ may usually be found without difficulty.

The variation of $\mu_p/\mu°$ with the reduced parameters T_r and p_r has been studied by several authors (Comings, Egly and Mayland,[64] Bromley and Wilke,[46] Carr, Parent and Peck[55]). All these workers find that when μ_p/μ_0 is plotted against T_r and p_r the curves for the various gases studied coincide. Reid and Sherwood[306] have constructed a composite graph (Fig. 3.15), based on the results of all the works quoted above.

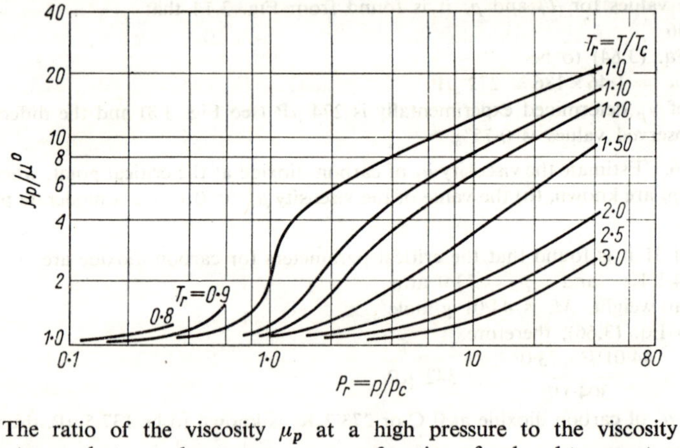

FIG. 3.15. The ratio of the viscosity μ_p at a high pressure to the viscosity $\mu°$ at the same temperature and at a moderate pressure, as a function of reduced temperature $T_r = T/T_c$ and reduced pressure $p_r = p/p_c$ (after Reid and Sherwood[306]).

This graph and the graph presented by Uyehara and Watson (Fig. 3.14) are of course interrelated. Thus, as seen from Fig. 3.15, $\mu_c/\mu_c° = 2{\cdot}2$ where $\mu_c°$ is the viscosity in the pressure-independent region at temperature T_c (Table 3.4).

An approximate simple relationship is thus obtained†

$$\mu_c = 2{\cdot}2\mu_c^\circ \qquad (3.70)$$

In the table below, values of the viscosity μ_c at the critical point, calculated from Eq. (3.70) are compared with some values given by Hougen and Watson[160] (μ_c° for use in equation 3.70 was taken from Table 3.4 for each gas). The values given by Hougen and Watson were inferred from gas phase data using Eq. (3.69) and a reduced viscosity diagram.

Gas	NH_3	CO	Cl_2	N_2	SO_2	C_6H_6
$\mu_c = 2{\cdot}2\ \mu_c^\circ$†	304	193	404	184	415	313
μ_c‡	309	190	420	180	411	312

† Calculated from Eq. (3.70) using values of μ_c° listed in Table 3.4.
‡ Values calculated from Eq. (3.69) (see Hougen and Watson[160]).

Predictions based on Figs. 3.14 and 3.15 are not very accurate, the average error amounting to about 10%. These predictions are subject to particularly large errors when applied to gases in the critical region. For this reason Grunberg and Nissan[137] have constructed a nomograph for this case (Fig. 3.16). This nomograph shows μ_p/μ° as a function of reduced temperature $T_r = T/T_c$ and reduced density $\rho_r = \rho/\rho_c$. The nomograph is most suitable for the critical region and only applicable if $p_r < 4$.

(The use of the nomograph follows standard practice. After calculating the reduced parameters T_r and ρ_r, a straight line is drawn through the respective points laid off on the appropriate nomograph axes to intersect with the μ_p/μ° axis: the desired value of μ_p/μ° is given by this intersection point.) As in the case of the preceding methods, the viscosity μ° of the gas in the pressure-independent region at temperature T should be known.

A drawback to the use of this method lies in the fact that it is necessary to know the density ρ of the gas at the temperature T considered and also the density ρ_c at the critical point.

Jossi, Stiel and Thodos[181] have shown how equations based on (3.45)–(3.50) may be used to estimate the viscosity of gases in the high-pressure region. As the starting point for their treatment they point out that the relationship (6.2) presented by Abas-Zade[1] for the thermal conductivity of liquids, may be extended to viscosities. They find that an analogous relationship exists between the *residual viscosity*, $\mu - \mu^\circ$, and the density and that, in the cases which they have investigated, this relationship is closely obeyed both for compressed and for liquefied gases. These authors plotted the quantity $(\mu-\mu^\circ)\zeta$ as a function of the reduced density ($\rho_r = \rho/\rho_c$) for each of the substances argon, nitrogen, oxygen, carbon dioxide, sulphur dioxide, methane, ethane, propane, n-butane, iso-butane and n-pentane and found that the data for all these substances lay on a common curve (μ denotes the viscosity under a high pressure, μ° the viscosity in the region of moderate pressures, cP and ζ is given by Eq. (3.46) to be $\zeta = T_c^{1/6}M^{-1/2}p_c^{-2/3}$). Thodos and his colleagues represent their curve by the empirical equation

$$[(\mu-\mu^\circ)\zeta+10^{-4}]^{1/4} = 0{\cdot}10230+0{\cdot}023364\rho_r+0{\cdot}58533\rho_r^2-$$
$$-0{\cdot}040758\rho_r^3+0{\cdot}0093324\rho_r^4 \qquad (3.71)$$

† In the case of diatomic gases Brebach and Thodos[34] recommend the use of coefficient 2·11 instead of 2·2, while Filippova[103] has found the value of this coefficient to be equal to 2·14.

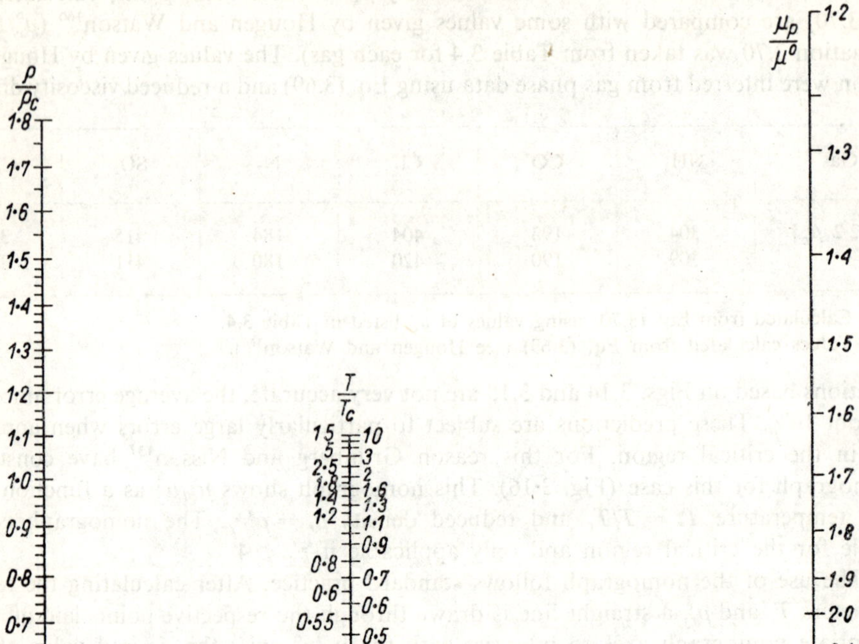

FIG. 3.16. Nomograph for determinating the viscosities of vapours and gases in the pressure range close to the critical region when $p_r < 4$ (after Grunberg and Nissan[137]).

Over limited ranges of reduced density this equation may be replaced by simpler relationships. Thus over the range $0{\cdot}3 \leqslant \rho_r \leqslant 2{\cdot}0$, the equation

$$(\mu-\mu^\circ)\zeta = (23{\cdot}12\, e^{1{\cdot}079\rho_r} - 25)\times 10^{-5} \tag{3.72}$$

is sufficiently accurate.

If $\rho_r < 0{\cdot}3$, the following equation can be applied:

$$(\mu-\mu^\circ)\zeta = 11{\cdot}10\times 10^{-5}\,(e^{1{\cdot}58\rho_r}-1) \tag{3.73}$$

Equations (3.71)–(3.73) are satisfactory for all the compounds investigated, except for water, ammonia and hydrogen. In the remaining cases, Jossi, Stiel and Thodos report good agreement between the calculated values and those determined experimentally.

Some examples of the application of the graphs presented in Figs. 3.15 and 3.16, as well as of Eqs. (3.71) and (3.73) for estimating the viscosities of gases under high pressures, are given at the end of this section.

A simple and accurate method of calculating the viscosity of gases under high pressures has been presented by Filippova and Ishkin.[103] This method is based on the application of the principle of similarity to the residual viscosity. According to the results presented by Filippov[99] and Predvoditelev,[300] the residual viscosity

$$\Delta\mu = \mu_{p,T} - \mathring{\mu}_T$$

(where $\mu_{p,T}$ is the gas viscosity at pressure p and temperature T,
and $\mathring{\mu}_T$ is the gas viscosity in the region of moderate pressures at temperature T), complies well with the equation

$$\Delta\mu = \frac{C\rho^{4/3}}{(1-b\rho)} \tag{3.74}$$

where ρ is the gas density and b and C are constants. When applying these equations, Filippova and Ishkin find it convenient to introduce two parameters σ and ω defined by

$$\sigma = \frac{\rho^{4/3} T_c^{7/6}}{\Delta\mu\, p_c^{2/3} M^{5/6}} \tag{3.75}$$

and

$$\omega = \frac{\rho T_c}{p_c M} \tag{3.76}$$

(In these equations T_c and p_c are critical temperature and pressure, and M is the molecular weight).

In terms of σ and ω Eq. (3.74) becomes

$$\sigma = \alpha - \beta\omega \tag{3.77}$$

where α and β are dimensionless parameters which are found to be equal for certain groups of compounds. Thus for N_2, O_2, CO, CO_2, CH_4, C_2H_4, C_2H_6 and C_3H_8

$$\sigma \triangleq 5{\cdot}33 - 0{\cdot}0387\omega \tag{3.78}$$

This linear relationship between σ and ω predicts viscosities with a mean deviation of the order of 3–5%, the maximum deviation being about 12%. The equation cannot be used for argon. In the table below a comparison is made of measured viscosities of carbon dioxide at 373°K with values predicted by means of Eq. (3.78).

Pressure, atm	100	250	400	600	800
$\mu\times10^7$, exp., g/cm.sec	2240	4700	6780	8800	10400
$\mu\times10^7$, calc., g/cm.sec	2249	4645	6695	8825	10475
Deviation, %	+0·5	−1·2	−1·3	+0·3	+0·8

The agreement in this case is very good, but the authors do not state clearly to which gases their method may be applied.

It is also interesting to note that good results may be obtained when the viscosities of compressed gases are estimated by means of Panchenkov's equation (4.19) for the viscosity of liquids. Gonikberg[132] has calculated the viscosity of carbon dioxide at 100°C in the range of 500–800 atm and found the average deviation to be about 3%.

EXAMPLE 3.8. Estimate the viscosity μ_p of carbon monoxide at 50°C ($T = 323{\cdot}2$°K) under a pressure $p = 700$ atm (cf. Example 3.4). The critical constants, taken from Table 1.31, are: $t_c = -138{\cdot}7$°C ($T_c = 134{\cdot}5$°K) and $p_c = 34{\cdot}6$ atm.

Solution

At a temperature of 50°C and a moderate pressure, the viscosity of carbon dioxide is found from Table 3.4 to be $\mu° = 186{\cdot}4$ μP.

The reduced temperature and pressure are

$$T_r = \frac{323{\cdot}2}{134{\cdot}5} = 2{\cdot}40 \quad \text{and} \quad p_r = \frac{700}{34{\cdot}6} = 20{\cdot}2$$

For these values of T_r and p_r, $\mu_p/\mu° = 2{\cdot}35$ (Fig. 3.15). Hence, $\mu_p = 2{\cdot}35\times186{\cdot}4 = 438$ μP.

The value of μ_p, calculated using Uyehara and Watson's graph (Fig. 3.14), is 476 μP. The difference between these estimates amounts to 8·7%.

EXAMPLE 3.9. Estimate, using the nomograph in Fig. 3.16, the viscosity μ_p of ethylene under a pressure p of 55 atm and at a temperature t of 40°C ($T = 313{\cdot}2$°K); under these conditions the density ρ is 0·091 g/cm³.

Solution

From Table 1.31 the critical constants of ethylene are seen to be:

$$t_c = 9{\cdot}7\text{°C } (T_c = 282{\cdot}9\text{°K}), \quad p_c = 50{\cdot}9 \text{ atm} \quad \text{and} \quad \rho_c = 0{\cdot}22 \text{ g/cm}^3.$$

The viscosity $\mu°$ of ethylene in the pressure-independent region at 40°C, interpolated from Table 3.4, is $\mu° = 109{\cdot}3$ μP. The reduced temperature and pressure are

$$T_r = \frac{313{\cdot}2}{282{\cdot}9} = 1{\cdot}106 \quad \text{and} \quad \rho_r = \frac{0{\cdot}091}{0{\cdot}22} = 0{\cdot}414$$

The nomograph (Fig. 3.16) is now used:

A straight line passing through the point 0·414 on the ρ_r scale, and 1·106 on the T scale of the nomograph, intersects the $\mu_p/\mu°$ axis at $\mu_p/\mu° = 1{\cdot}26$.

Hence

$$\mu_p = 1{\cdot}26\times109{\cdot}3 = 138 \text{ μP}$$

The experimentally determined value of μ_p is 141 μP so the error in this estimate is −2·1%.

EXAMPLE 3.10. Estimate by means of the method put forward by Jossi, Stiel and Thodos,[181] the viscosity of ethylene at 50°C and a pressure of 89·3 atm. The density ρ at this pressure is 0·184 g/cm³, the critical constants are $T_c = 282{\cdot}4$°K, $p_c = 50{\cdot}0$ atm, $\rho_c = 0{\cdot}227$ g/cm³ and the molecular weight, M, 28·05.

Solution

$$\zeta = \frac{(282{\cdot}4)^{1/6}}{(28{\cdot}05)^{1/2}\cdot(50{\cdot}0)^{2/3}} = 0{\cdot}356$$

The reduced temperature and density are

$$T_r = \frac{273{\cdot}2+5}{282{\cdot}4} = 1{\cdot}144 \quad \text{and} \quad \rho_r = \frac{0{\cdot}184}{0{\cdot}227} = 0{\cdot}811$$

The value of $\mu^\circ\zeta$ at moderate pressures is calculated from Eq. (3.45)

$$\mu^\circ\zeta = 34\times10^{-5}\times1{\cdot}144^{0{\cdot}94} = 38{\cdot}6\times10^{-5}$$

From Eq. (3.72)

$$(\mu-\mu^\circ)\zeta = (23{\cdot}12e^{1{\cdot}079\times0{\cdot}811})\times10^{-5} = 30{\cdot}5\times10^{-5}$$

so

$$\mu\zeta = (30{\cdot}5+38{\cdot}6)\times10^{-5} = 69{\cdot}1\times10^{-5}$$

and

$$\mu = \frac{69{\cdot}1\times10^{-5}}{0{\cdot}0356} = 1{\cdot}941 \text{ cP}$$

(the value determined experimentally is 1·975 cP).

7. COMPARISON OF THE ACCURACY OF THE VARIOUS METHODS FOR ESTIMATING THE VISCOSITY OF GASES AND THE CHOICE OF THE MOST SUITABLE METHOD

The methods for estimating the viscosity of gases, which have been discussed above, may be classified in four groups:

(a) Estimation of the viscosity μ° of a gas in the pressure-independent region (i.e. at moderate pressures) when no viscosity data for the gas under consideration are available.

(b) Estimation of the viscosity μ°_{T1} of a gas in the pressure-independent region at temperature T_1 from the viscosity μ°_{T2} of this gas at temperature T_2.

(c) Estimation of the viscosity μ_p of a gas under pressure p (which is high enough for the viscosity to be dependent on pressure), if one value of the viscosity of this gas is known at some other conditions of temperature and pressure.

(d) Interpolation and extrapolation of gas viscosities.

Reid and Sherwood[306] have calculated viscosities by various methods for 270 compounds and have compared the results obtained with the values determined experimentally, thus assessing the accuracy of each method. The methods considered were those of Bromley and Wilke, Licht and Stechert, and Arnold. The findings of Reid and Sherwood[306] are summarized in Table 3.8 which also includes the findings of some other authors. The notation used in Table 3.8 is as follows:

σ, ε force constants (cf. Ch. 1),
k Boltzmann constant,
T_b boiling temperature,
ρ_c, V_c, T_c, p_c density, molal volume, temperature and pressure at the critical point,
z_c compressibility factor at the critical point,
μ° dynamic viscosity of a gas in the region of moderate pressures where viscosity does not depend on pressure,
μ°_0 viscosity in pressure-independent region at 0°C,
μ°_T viscosity in pressure-independent region at temperature T (°K),
ρ gas density,
ρ_l liquid density,
σ molecular diameter (cf. Eq. (3.58)).

Stiel and Thodos[346] have checked various equations for the viscosity in more than 200 cases, and have found the following average deviations:

Bromley–Wilke Eq. (3.30)	2·62%
Licht–Stechert Eq. (3.42)	2·17%
Stiel–Thodos Eqs. (3.45), (3.47)	1·64%

TABLE 3.8. COMPARISON OF METHODS FOR ESTIMATING THE VISCOSITIES OF GASES

Authors	Equation	Data necessary for the estimation	Average error	Remarks
a. Prediction of the gas viscosity in the pressure-independent region when no experimental values of the viscosity are available.				
Hirschfelder, Bird and Spotz	(3.22)	σ, ε/k		accurate method for non-polar gases if ε and σ are known
Bromley and Wilke	(3.24)	σ, ε/k		approximate calculations for non-polar gases
Bromley and Wilke approximate equations	(3.24) (3.28)	T_b, T_c, T_r or V_c		
Bromley and Wilke approximate equations	(3.30)	T_c, V_c	3·2%	for non-polar gases
	(3.31)	T_c, p_c	4·9%	for polar gases, error up to 15%
Scheibel	(3.33)	T_c, V_c	4·4%	
Licht and Stechert	(3.42)	T_c, p_c	4·7%	
Falkovskii	(3.43)	T_c, p_c	5%	
Arnold	(3.44)	T_c, p_c	5·5%	
Stiel and Thodos	(3·45)–(3·47)	T_c, p_c	1·8%	for non-polar gases
Stiel and Thodos	(3·48)–(3·50)	T_c, p_c, z_c	2·6%	for polar gases
b. Variation of gas viscosity μ° with temperature				
Sutherland	(3.20) (3.36) (3.37)	μ_0°, C or μ_0°, T_b	< 2%	
Othmer *et al.*		μ_{T_1}, μ_{T_2}		graphical method of interpolation and extrapolation
Bromley and Wilke	nomographs (Figs. 3.4, 3.5)	$\mu_{T_1}^\circ$, T_b or T_c		
Bromley and Wilke	(3.34), (3.35)	$\mu_{T_1}^\circ$, T_c	< 2%	accurate method when non-polar gases are involved; for polar gases the error may be appreciable (10%)
Andrussov	(3.38)	constants in Table 3.6		accurate method
c. Variation of gas viscosity with pressure				
Theoretical estimations based on Enskog's formula	(3.58), (3.62)	σ, z		approximate value
	(3.62)	μ°, ρ		
Uyehara and Watson	Fig. 3.14	μ°, T_c, p_c or ρ_1, T_c, p_c or T_c, p_c	*ca.* 10%	
Comings, Egly *et al.*	Fig. 3.15	μ°, T_c, p_c	*ca.* 10%	
Grunberg and Nissan	Fig. 3.16	ρ_1, ρ_c, μ°, T_c	7%	
Othmer *et al.*		μ_{p1}, μ_{p2}		graphical method of interpolation and extrapolation
Jossi, Stiel and Thodos	(3.71)–(3.73)	T_c, p_c, ρ_c, μ°		
Filippova and Ishkin	(3.75)–(3.78)	μ°, ρ, T_c, p_c	3–5%	checked on few gases only

For $T_r > 10$, the equation (3.42) of Licht and Stechert is unreliable, giving errors of up to 10%; the equation (3.47) of Stiel and Thodos yields fairly accurate results.

As a result of these comparisons, Reid and Sherwood,[306] Licht and Stechert[223] and other authors recommend the use of the following methods.

a. Prediction of gas viscosity, μ°, at moderate pressures when no experimentally determined viscosity data are available. If the gas is non polar and the force constants are known, Eq. (3.22) may be used. If the critical constants of a gas are known, it is possible to use Bromley and Wilke's equation (3.30), provided the values of T_c and V_c (or ρ_c) are known. If all the critical constants T_c, p_c, V_c (or ρ_c), and, hence, also z_c, are known, Eqs. (3.45)–(3.50) of Stiel and Thodos, which are accurate and easily operable, are recommended.

Note. In the majority of cases the values of p_c and T_c are known so Licht and Stechert's equation (3.42) should be applied.

b. Calculation of gas viscosity, μ°_{T2}, for a moderate pressure at temperature T_2 if the gas viscosity at temperature T_1 is available.

Sutherland's equation (3.36) may be employed when the viscosity μ°_0 at 0°C and Sutherland's constant C for the given gas are known. If the viscosity μ°_{T1} is known at temperature T_1, and not at 0°C, the value of μ°_0 may be estimated. If the constant C is unknown, it may be estimated from Vogel's equation (3.37).

For a rapid (but somewhat less accurate) calculation of the effect of temperature upon gas viscosity, the use of the nomographs presented in Figs. 3.4 and 3.5 is recommended. The method of Bromley and Wilke (Eqs. (3.34) and (3.35)) may be applied for non-polar gases over wide ranges of temperature. For polar gases the error in the calculation will increase.

In the region of temperatures from −50°C to +1000°C very good results may be obtained by using Andrussov's equation (3.38). For this purpose, however, a knowledge of the constants in the equation is required.

c. Estimation of gas viscosity μ_p under a high pressure if the viscosity μ° of the same gas is known under a moderate pressure.

Reid and Sherwood[306] conclude that all predictions of gas viscosities at high pressures are rather uncertain because of the lack of a developed theory and of accurate experimental data. They recommend the following procedure:

(1) Calculation of the value of μ° (at the same temperature T for which the value of μ_p is to be estimated) by means of the methods given under (a) and (b).

(2) Determination of μ_p/μ° from Fig. 3.15 if $p_r > 4$ or from the nomograph in Fig. 3.16 if $p_r < 4$. (It should be mentioned here that the use of Fig. 3.14 (instead of Fig. 3.15) combined with a value of μ_c obtained from (3.69) or from one of the other equations recommended by Uyehara and Watson leads to results of similar accuracy.)

On the basis of more recent studies it appears that the methods of Jossi, Stiel and Thodos (Eqs. (3.71)–(3.73)), and of Filippova and Ishkin (Eqs. (3.75)–(3.78)) may be more accurate and that, if the critical constants of a gas and its viscosity in the region of moderate pressures are known, these methods should be employed. Unfortunately, they have not yet been checked in a sufficiently large number of cases.

d. If the gas viscosity is known at two different temperatures, it is possible to perform an interpolation or extrapolation with fairly high accuracy (both at moderate and high pressures) by using Othmer's plots (cf. section 5). Alternatively use may be made of the equations discussed above, which represent the dependence of gas viscosity upon temperature and pressure. Usually, however, this procedure is more tedious: frequently it is less accurate also.

8. VISCOSITY OF GAS MIXTURES

a. Experimental Data

It is found that the dynamic viscosities μ_m of gaseous systems do not vary linearly with composition, i.e. that the viscosity is not an additive property. The viscosity of the gaseous sulphur dioxide–hydrogen system is shown as a function of composition in Fig. 3.17. It is seen that at each temperature the viscosity/vol % curve shows a maximum,

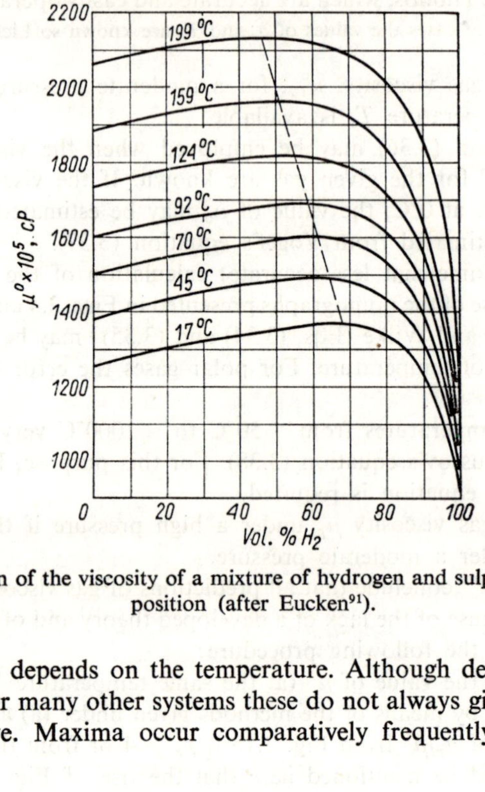

FIG. 3.17. Variation of the viscosity of a mixture of hydrogen and sulphur dioxide with composition (after Eucken[91]).

the location of which depends on the temperature. Although deviations from additivity have been observed for many other systems these do not always give rise to any extremum on the viscosity curve. Maxima occur comparatively frequently in systems containing hydrogen.

b. Estimation of the Viscosity of Gas Mixtures

Simplified considerations based on the kinetic theory of gases lead to the expression

$$\mu_m^\circ = \tfrac{1}{3}(\rho_1 \bar{w}_1 \bar{l}_1 + \rho_2 \bar{w}_2 \bar{l}_2 + \rho_3 \bar{w}_3 \bar{l}_3 + \ldots) \tag{3.79}$$

for the viscosity of a gaseous mixture in the pressure-independent region. In this equation which is analogous to Eq. (3.15), the subscripts 1, 2, 3, ... refer to the components present. Thus ρ_1 is the mass of component 1 present in 1 cm^3 of mixture, etc.

Sutherland,[354] Thiesen,[364] and others,[282] having introduced appropriate values for the mean free paths of the molecules in a binary mixture (components 1 and 2), obtained the following relationship:

$$\mu_{1,2}^\circ = \frac{\mu_1^\circ}{1+k_{1,2}n_2/n_1} + \frac{\mu_2^\circ}{1+k_{2,1}n_1/n_2} \tag{3.80}$$

where n_1 and n_2 are the concentrations of components 1 and 2 expressed in mole/cm^3,
and $k_{1,2}$ and $k_{2,1}$ are constants related to the molecular weights of the constituents and the diameters of the molecules.

If the constants are suitably chosen, these equations are found to represent viscosity data for binary mixtures with an error† of the order of only 0·6%.[50]

By taking into consideration the forces of molecular interaction it is possible to calculate the dynamic viscosities μ_m for gas mixtures. The procedure recommended by Bird, Hirschfelder and Curtiss[25] is, however, quite complex.

To estimate the viscosity of a mixture of two gases, 1 and 2, with non-polar molecules, it is necessary first to find the viscosity $\mu_{1,2}$ of a hypothetical gas with an average molecular weight equal to $2M_1M_2/(M_1+M_2)$, and force constants $\sigma_{1,2}$ and $\varepsilon_{1,2}$ by means of the equation

$$\mu_{1,2}\times 10^7 = \frac{266{\cdot}93}{\sigma_{1,2}^2\Omega^{(2,2)*}}\sqrt{\frac{2M_1M_2T}{M_1+M_2}} \tag{3.81}$$

where M_1 is the molecular weight of gas 1,
M_2 is the molecular weight of gas 2,
and $\Omega^{(2,2)*}$ is a function of T* ($= kT/\varepsilon_{1,2}$). This function is tabulated in Table 3.1,
$\varepsilon_{1,2}$ and $\sigma_{1,2}$ are force constants describing the interaction between molecules of types 1 and 2 (they may be estimated from Eq. (1.39)).

The following relationship exists between the value of the viscosity $\mu_{1,2}$ of the hypothetical gas and the diffusivity $D_{1,2}$:

$$\mu_{1,2} = \frac{5}{3}\frac{M_1M_2}{M_1+M_2}\frac{pD_{1,2}}{\bar{A}_{1,2}^*RT} \tag{3.82}$$

where $\bar{A}_{1,2}^*$ denotes the function, the values of which are listed in Table 3.3.

Knowing the viscosity $\mu_{1,2}$ and the viscosities μ_1 and μ_2 of the pure components of a mixture, we can calculate the required dynamic viscosity μ_m of the mixture using the equation

$$\frac{1}{\mu_m} = \frac{X_\mu+Y_\mu}{1+Z_\mu} \tag{3.83}$$

In this equation, X_μ, Y_μ and Z_μ are related to the mole fractions of the components of the mixture. They are calculated from the equations

$$X_\mu = \frac{x_1^2}{\mu_1}+\frac{2x_1x_2}{\mu_{1,2}}+\frac{x_2^2}{\mu_2} \tag{3.84}$$

$$Y_\mu = \frac{3}{5}\bar{A}_{1,2}^*\left[\frac{x_1^2}{\mu_1}\frac{M_1}{M_2}+\frac{2x_1x_2}{\mu_{1,2}}\frac{(M_1+M_2)^2}{4M_1M_2}\frac{\mu_{1,2}^2}{\mu_1\mu_2}+\frac{x_2^2}{\mu_2}\frac{M_2}{M_1}\right] \tag{3.85}$$

$$Z_\mu = \frac{3}{5}\bar{A}_{1,2}^*\left\{x_1^2\frac{M_1}{M_2}+2x_1x_2\left[\frac{(M_1+M_2)^2}{4M_1M_2}\left(\frac{\mu_{1,2}}{\mu_1}+\frac{\mu_{1,2}}{\mu_2}\right)-1\right]+x_2^2\frac{M_2}{M_1}\right\} \tag{3.86}$$

The value of the function $\bar{A}_{1,2}^*$ is taken from Table 3.3 for $T^* = kT/\varepsilon_{1,2}$.

† See Eq. (3.88).

In the case of a mixture of ν gases with non-polar molecules, the simple kinetic theory of gases leads to the relationship:[25]

$$\mu_m = \sum_{i=1}^{\nu} \frac{x_i^2}{\dfrac{x_1^2}{\mu_i} + \sum\limits_{\substack{k=1\\k\neq 1}}^{\nu} 2x_i x_k \dfrac{RT}{pM_i D_{ik}}} \tag{3.87}$$

Values of μ_m estimated from this equation do not always conform exactly with those found experimentally.

Hirschfelder, Taylor, Kihara and Rutherford[157] have employed the exact Chapmann–Enskog theory for the calculation of the viscosities of binary gasmixtures and for the determination of the composition at which the viscosity attains its maximum value at a given temperature. Such maxima occur quite frequently (i) when one of the gaseous constituents of the mixture is polar, and the other non-polar, (ii) when the viscosities of the pure gases are approximately equal, and (iii) when the molecular weights of the constituents differ considerably from each other.

The calculated viscosities were found to be in very good agreement with those determined experimentally.

c. Semi-Empirical Equations for Estimating the Viscosities of Gas Mixtures at Moderate Pressures

The equations given under (b) do not find wide application for technical calculations, because either they are not accurate enough or they entail the use of a quantitative characteristic of the compounds which are usually unknown.[43]

These theoretically based equations are, however, valuable, since they have served as the basis for simpler semi-empirical relationships which have subsequently been derived and which make it possible to estimate the viscosity of gaseous mixtures rapidly and with sufficient accuracy.

Thus Buddenberg and Wilke[49] have pointed out that Eq. (3.87) may be used to calculate multicomponent viscosities with acceptable accuracy if the coefficient 2 in the denominator is replaced by a coefficient 1·385, which they found empirically. Their formula reduces to the following form for a binary mixture of components 1 and 2:

$$\mu_m^\circ = \frac{\mu_1^\circ}{1 + 1{\cdot}385 \dfrac{x_2}{x_1} \dfrac{\mu_1}{\rho_1 D_{1,2}}} + \frac{\mu_2^\circ}{1 + 1{\cdot}385 \dfrac{x_1}{x_2} \dfrac{\mu_2}{\rho_2 D_{1,2}}} \tag{3.88}$$

where μ_m° is the viscosity of a mixture of gases 1 and 2,
x_1 and x_2 are the mole fractions of components 1 and 2,
μ_1° and μ_2° are the viscosities of these components,
ρ_1, ρ_2 are the densities,
and $D_{1,2}$ is the diffusivity.

This equation has the same form as the earlier equations of Sutherland, Thiesen and others (3.80).

All the viscosities in (3.88) refer to the region of moderate pressures, where viscosity does not depend on pressure.

In order to use this equation it is necessary to know the diffusivity $D_{1,2}$. If this quantity is known, however, the equation is found to be quite accurate, the average error of estimation amounting to about $\pm 2{\cdot}2\%$; at higher temperatures (100–225°C) the error may be higher, viz. of the order of $\pm 7\%$.

Buddenberg and Wilke have extended their treatment to multi-component mixtures with good results.[50] (In this case the diffusivities $D_{1,2}$, $D_{1,3}$, $D_{2,1}$, etc. appear in the denominators of the particular terms of the equation.) For example the same authors calculated the viscosity at 20°C of a mixture containing (in mole fractions): CO_2 0·104, CO 0·285, H_2 0·016, and N_2 0·595. Their calculated viscosity was 171·8 μP. The value determined experimentally was 173·8 μP so the error in estimation amounted to only 1·1%.

A rapid, and comparatively accurate method of estimating the viscosities of gaseous mixtures has been proposed by Wilke.[400] The method is based on the equation

$$\mu_m^\circ = \frac{\mu_1^\circ}{1+\dfrac{x_2}{x_1}\phi_{1,2}} + \frac{\mu_2^\circ}{1+\dfrac{x_1}{x_2}\phi_{2,1}} \tag{3.89}$$

The notation in this equation is the same as in Eq. (3.88). $\phi_{1,2}$ and $\phi_{2,1}$ are functions of the viscosities and molecular weights of pure components, viz.:

$$\phi_{1,2} = \frac{\left[1+\left(\dfrac{\mu_1^\circ}{\mu_2^\circ}\right)^{0\cdot5}\left(\dfrac{M_2}{M_1}\right)^{0\cdot25}\right]^2}{2\sqrt{2}\left[1+\dfrac{M_1}{M_2}\right]^{0\cdot5}} \tag{3.90}$$

$$\phi_{2,1} = \frac{\left[1+\left(\dfrac{\mu_2^\circ}{\mu_1^\circ}\right)^{0\cdot5}\left(\dfrac{M_1}{M_2}\right)^{0\cdot25}\right]^2}{2\sqrt{2}\left[1+\dfrac{M_2}{M_1}\right]^{0\cdot5}}$$

These equations involve neither the diffusivity $D_{1,2}$ nor the densities ρ_1, ρ_2 of the pure gases (for gases which deviate considerably from perfect gas behaviour Wilke introduces correction terms into Eqs. (3.90). These terms involve the density ratios ρ_1/ρ_2 and ρ_2/ρ_1, but usually this correction may be omitted).

In order to eliminate the somewhat cumbersome calculation of the values of the functions $\phi_{1,2}$ and $\phi_{2,1}$, Bromley and Wilke have constructed plots of $\phi_{1,2}$ vs. M_1/M_2 and vs. μ_1°/μ_2° (Figs. 3.18 and 3.19).

The estimation of the viscosity of a mixture by this method is simple and rapid: having evaluated the ratios μ_1/μ_2, μ_2/μ_1, M_1/M_2 and M_2/M_1, corresponding values of $\phi_{1,2}$ and $\phi_{2,1}$ are read from Figs. 3.18 and 3.19. These values for the function ϕ are then substituted together with the viscosities μ_1 and μ_2 of the pure gases, and the mole fractions x_1 and x_2 into Eq. (3.89). For 17 different mixtures, Wilke has found the average error of estimation to be equal to 0·97% (or 0·47%, if the mixture H_2/A, for which more significant deviations have been found, is not taken into account).

Jackson[169] has compared the measured values of the viscosity of a mixture of methane and carbon dioxide with values calculated from Eq. (3.89). He finds:

% CH_4	0	10·3	18·3	29·7	42·1	53·7	65·1
$\mu_{exp.}$	151·0	149·4	147·7	145·3	141·4	137·3	132·8
$\mu_{calc.}$	—	149·1	147·4	144·6	141·0	137·0	132·3
% CH_4	73·0	78·9	85·0	90·5	93·3	100	
$\mu_{exp.}$	129·1	124·8	122·6	119·2	117·4	111·4	
$\mu_{calc.}$	128·5	125·4	121·8	118·3	116·6		

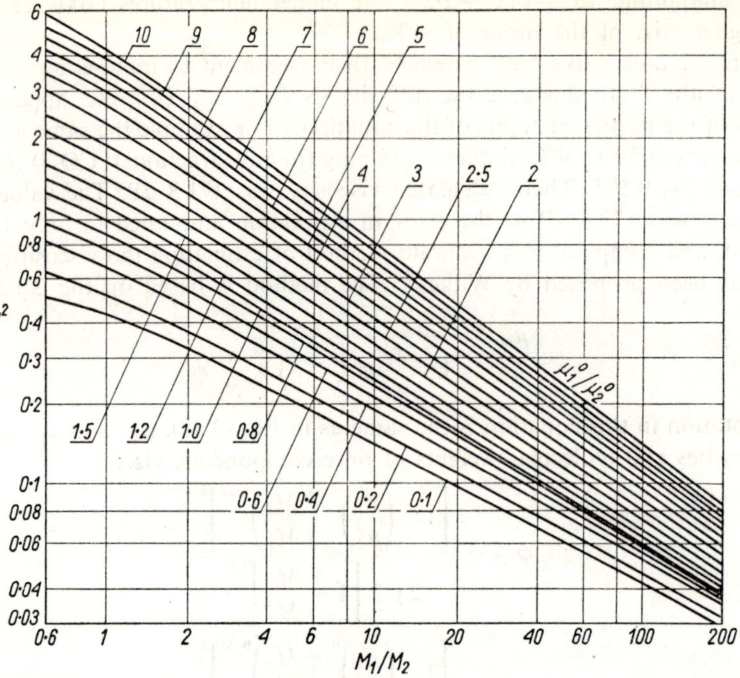

FIG. 3.18. Function $\phi_{1,2}$ for calculating the viscosities of gas mixtures using Wilke's equation (3.89) (after Bromley[46]).

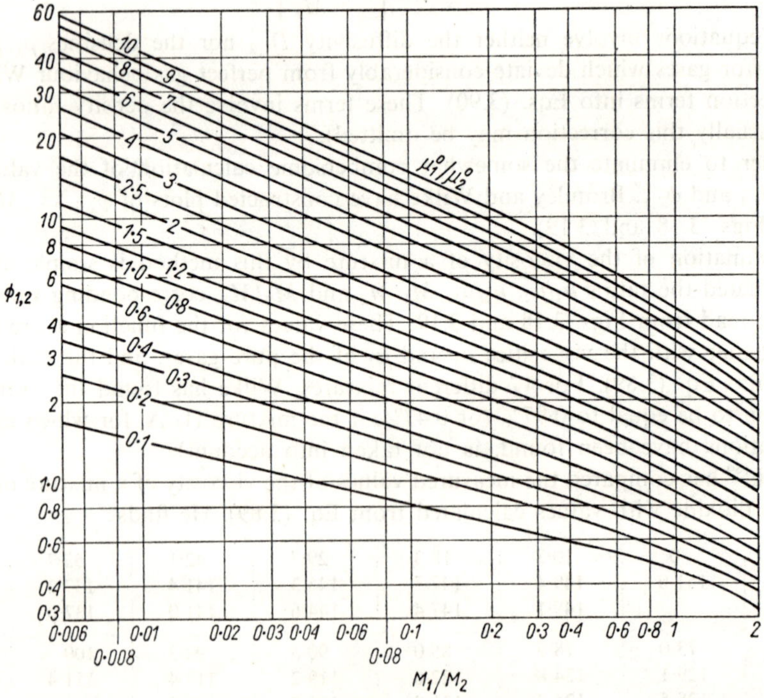

FIG. 3.19. Function $\phi_{1,2}$ for calculating the viscosities of gas mixtures using Wilke's equation (3.89) (after Bromley[46]).

Wilke's formula for multi-component mixtures is as follows:

$$\mu_m^\circ = \frac{\mu_1^\circ}{1+(x_2/x_1)\phi_{1,2}+(x_3/x_1)\phi_{1,3}+\ldots} + \frac{\mu_2^\circ}{1+(x_1/x_2)\phi_{2,1}+(x_3/x_2)\phi_{2,3}+\ldots} + \ldots \quad (3.91)$$

The application of this equation is illustrated in Example 3.11 (taken from the work of Bromley and Wilke).

In the case of gaseous mixtures containing more than two components use may be made of a method recommended by Saxena and Narayanan,[315] provided the viscosities $\mu_1, \mu_2, \mu_3 \ldots$ of the pure compounds and two values of the viscosity (for different compositions) of each binary mixture: $\mu'_{1,2}$, $\mu''_{1,2}$; $\mu'_{1,3}$, $\mu''_{1,3}$; $\mu'_{2,3}$, $\mu''_{2,3}$ ... are known.

On substituting the values of $\mu'_{1,2}$, μ_1, μ_2 and $\mu''_{1,2}$, μ_1, μ_2 in Wilke's equation (3.89), two equations are obtained containing two quantities $\phi_{1,2}$ and $\phi_{2,1}$ which are unknown. After solving these equations the values of these unknown quantities are found.

In the same way, by substituting $\mu'_{1,3}$, μ_1, μ_3 and $\mu''_{1,3}$, μ_1, μ_3, we can find the values of $\phi_{1,3}$ and $\phi_{3,1}$, etc.

The values of ϕ thus determined may now be used in conjunction with Eq. (3.91) to calculate the viscosity μ_m of a multi-component mixture. This method is somewhat more accurate than Wilke's method which involved determining the value of ϕ from Eq. (3.90) or from the plots given in Figs. 3.18 and 3.19. Saxena and Narayanan present the following values for the Ar–Ne–He mixture (373°K):

Mole fractions			$\mu \times 10^7$, g/cm.sec		
Ar	Ne	He	exp.	Wilke (3.89)	Method of the authors
0·2670	0·5576	0·1754	3237	3223	3215
0·3213	0·3193	0·3594	3044	2966	3024
0·5851	0·2166	0·1983	2886	2829	2895
0·2382	0·2189	0·5429	2957	2867	2938

EXAMPLE 3.11.[46] Estimate the viscosity of a mixture of the composition given below at a temperature $T = 293$°K.

Component	Mole or volume per cent $x \times 10^2$	Viscosity of the pure component μP	Molecular weight M
CO_2	4·8	147·2	44·01
O_2	0·3	201·9	32·00
CO	26·4	174·9	28·01
H_2	17·2	87·55	2·02
CH_4	2·6	108·7	16·04
N_2	48·2	178·1	28·04
Higher hydrocarbons (C_2H_6)	0·5	90·9	30·07

Solution

We may consider Eq. (3.91) to be the sum of the contributions to the viscosity arising from each particular component:

$$\mu_m = \sum_{i=1}^{i=n} [\mu_i] \quad (3.92)$$

where
$$[\mu_i] = \frac{\mu_i}{1+\dfrac{1}{x_i}\sum_{\substack{j=1\\j\neq i}}^{n} x_i\phi_{ij}} \tag{3.93}$$

The calculation is presented in tabular form below:

Component i	j	$\frac{\mu_i}{\mu_j}$	$\frac{M_i}{M_j}$	ϕ_{ij} (Figs. 3.18 and 3.19)	x_i	x_j	$x_j\phi_{ij}$
CO_2	O_2	1·731	1·375	0·72	0·048	0·003	0·0022
	CO	0·884	1·571	0·73	0·048	0·264	0·1930
	H_2	1·682	21·83	0·18	0·048	0·172	0·0310
	CH_4	1·358	2·744	0·67	0·048	0·026	0·0174
	N_2	0·828	1·571	0·72	0·048	0·482	0·3470
	C_2H_6	1·620	1·464	1·03	0·048	0·005	0·0052

$\Sigma x_j\phi_{ij} = 0·5958$

Similarly, taking CO as component i, $\Sigma = 0·6235$
and likewise, for O_2 $= 0·8963$
for H_2 $= 1·6099$
for CH_4 $= 0·8954$
for N_2 $= 0·4188$
and for C_2H_6 $= 0·6379$

Calculation and summation of the viscosity contributions

Component	$\frac{\Sigma x_j\phi_{ij}}{x_i}$	$[\mu_i] = \frac{\mu_i}{1+\frac{\Sigma x_j\phi_{ij}}{x_i}}$
CO_2	12·42	11·00
O_2	29·88	0·65
CO	2·36	52·00
H_2	9·36	8·45
CH_4	34·50	3·06
N_2	0·87	95·30
C_2H_6	127·6	0·71

The calculated value of the viscosity of the mixture is $\mu_m = \Sigma[\mu_i] = 171·1$ μP, and the experimental value is $\mu_m = 171·4$ μP. The agreement between the two values is excellent.

The method of Wilke discussed above gives fairly good results, though in the case of mixtures composed of many compounds the calculation is tedious.

A less accurate, though considerably more rapid method, is provided by the empirical formula of Henning and Zipperer[148]

$$\mu_m^\circ = \frac{x_1\mu_1^\circ\sqrt{M_1T_{c1}}+x_2\mu_2^\circ\sqrt{M_2T_{c2}}+\ldots}{x_1\sqrt{M_1T_{c1}}+x_2\sqrt{M_2T_{c2}}+\ldots} \tag{3.94}$$

where μ_m° is the viscosity of the gaseous mixture at t °C,
μ_1° and μ_2° are the viscosities of pure gaseous compounds at t °C,
x_1 and x_2 are the volume fractions of the components in the mixture,
M_1 and M_2 are the molecular weights of the constituents of the mixture,
and T_{c1} and T_{c2} are the critical temperatures of the constituents of the mixture.

Schmidt[322] has checked this equation for several industrial gas mixtures (two producer gases and three flue gases of different compositions). He finds that the average error of estimation for temperatures in the range 0–1000°C was 1·9%; the error never exceeded ±4%. On the other hand, when the above equation was used to predict the viscosities of coke-oven gases with five different compositions (up to 600°C) it was found that the calculated viscosities of gases rich in hydrogen were too large at high temperatures.

TABLE 3.9. VALUES OF $\sqrt{MT_c}$ FOR USE WITH EQ. (3.94)

Gas	$\sqrt{MT_c}$	Gas	$\sqrt{MT_c}$	Gas	$\sqrt{MT_c}$
H_2O	108	H_2	8·13	C_3H_8	128
Air	61·9	CO	61·4	n-C_4H_{10}	157
CO_2	115·5	CH_4	55·1	n-C_5H_{12}	184
N_2	59·5	C_2H_4	89·0	n-C_6H_{14}	209
O_2	70·2	C_2H_6	95·6		

When estimating viscosities of industrial gas mixtures Table 3.9,[284] which gives $\sqrt{MT_c}$ for a number of gases, may be found to be useful.

For predicting the viscosity of a mixture of gases $j, k, l \ldots$ as a function of temperature Andrussov[5] suggests the following modified version of Eq. (3.38)

$$\mu_{j,k,l\ldots} = \frac{191{\cdot}8\left(1+\sqrt{2\bar{M}}\right)}{\overline{V^{2/3}}}\,\bar{\xi}\left(\frac{T}{273{\cdot}16}\right)^{\bar{n}_\mu} \tag{3.95}$$

where $\mu_{j,k,l\ldots}$ is the viscosity of the mixture at temperature T and a moderate pressure, μP,

$\overline{V^{2/3}}$ is the mean value of the molal volume raised to the power 2/3,

$\bar{M}$ is the molecular weight,

$\bar{\xi}$ is a compensating parameter,

and $\bar{n}_\mu$ is the mean power exponent (Eq. (3.39)).

The mean value, $\overline{V^{2/3}}$, is calculated from the pure component molal volumes (Table 3.6) using the equation

$$\overline{V^{2/3}} = c_j V_j^{2/3} + c_k V_k^{2/3} + c_l V_l^{2/3} + \ldots$$

where c_j, c_k, c_l are the mole fractions of the components j, k and l.

For a mixture of two gases j and k at a temperature of 273·16°K Eq. (3.95) takes on the form

$$\mu_{j,k} = \frac{191{\cdot}8\left[1+\sqrt{2(c_j M_j + c_k M_k)}\right]}{c_j V_j^{2/3} + c_k V_k^{2/3}}\,(c_j \xi_j + c_k \xi_k) \tag{3.96}$$

EXAMPLE 3.12. Estimate the viscosity of the mixture considered in Example 3.11 by Andrussov's method.

Data:

Component	Concentration (volume or mole fraction of the component) c	Molecular weight M	$V^{2/3}$	Parameter ξ	A	B	C
CO_2	0·048	44·01	9·55	0·662	0·942	2·1	0·56
O_2	0·003	32·00	6·30	0·700	0·76	1·4	0·64
CO	0·264	28·01	7·90	0·806	0·762	1·35	0·5
H_2	0·172	2·02	4·00	0·583	0·674	0·034	—
CH_4	0·026	16·04	9·49	0·757	0·88	2·10	—
N_2	0·482	28·04	8·41	0·859	0·753	1·3	0·4
C_2H_6	0·005	30·07	13·91	0·712	0·94	1·3	—

Solution

	c_jM_j	$c_jV_j^{2/3}$	$c_j\xi_j$	c_jA_j	c_jB_j	c_jC_j
CO_2	2·120	0·460	0·0318	0·0453	0·101	0·0269
O_2	0·096	0·019	0·0021	0·0023	0·0042	0·0019
CO	7·400	2·082	0·2126	0·2010	0·356	0·1320
H_2	0·347	0·688	0·1005	0·1159	0·0051	—
CH_4	0·416	0·246	0·0197	0·0229	0·0546	—
N_2	13·695	4·060	0·4140	0·3631	0·627	0·1926
C_2H_6	0·167	0·069	0·0036	0·0047	0·0065	—
Σ	24·24	7·62	0·784	0·755	1·154	0·353

$$\bar{n}_\mu = 0{\cdot}755-1{\cdot}154\times10^{-4}\times20+0{\cdot}353\times10^{-7}\times20^2$$

$$\bar{n}_\mu = 0{\cdot}755-0{\cdot}002308+0{\cdot}00001412 = 0{\cdot}753$$

$$\mu_{\text{calc}} = \frac{191{\cdot}8\,(1+\sqrt{2\times24{\cdot}24})}{7{\cdot}62}\times0{\cdot}784\left(\frac{293{\cdot}16}{273{\cdot}16}\right)^{0{\cdot}753}$$

$\mu_{\text{calc}} = 165{\cdot}8$ μP, $\mu_{\text{exp}} = 171{\cdot}4$ μP
(error 3·27%)

EXAMPLE 3.13. Estimate the viscosity of the mixture considered in Example 3.12 using Henning and Zipperer's equation (3.94)

Data

Component	Volume (or mole) fraction of the component x_i	Viscosity of the component μ_i, μP	$\sqrt{M_iT_{ci}}$ (from Table 3.9)
CO_2	0·048	147·2	115·5
O_2	0·003	201·9	70·2
CO	0·264	174·9	61·4
H_2	0·172	87·55	8·13
CH_4	0·026	108·7	55·1
N_2	0·482	178·1	59·5
C_2H_6	0·005	90·9	95·6

Solution

The values compiled above are inserted into Eq. (3.94).

The calculation of the values of the numerator and denominator in Eq. (3.94) is tabulated below:

Gas	$x_i \mu_i \sqrt{M_i T_{ci}}$	$x_i \sqrt{M_i T_{ci}}$
CO_2	817	5·54
O_2	42	0·21
CO	2833	16·21
H_2	123	1·40
CH_4	156	1·43
N_2	5108	28·68
C_2H_6	43	0·48
Σ	9122	53·95

$$\mu_{calc} = \frac{\Sigma x_i \mu_i \sqrt{M_i T_{ci}}}{x_i \sqrt{M_i T_{ci}}} = \frac{9122}{53 \cdot 95} = 169 \ \mu P \qquad \mu_{exp} = 171 \cdot 4 \ \mu P$$

Error: $-1 \cdot 4\%$

The viscosity of a mixture at temperature T_2 can be predicted from its viscosity at temperature T_1 using the Andrussov equation (3.95). For this purpose the equation presented by Ulibin[374] may be used,

$$\mu_2 = \mu_1 \left(\frac{T_2}{T_1} \right)^{\sum_{i=1}^{n} x_i m_i} \tag{3.97}$$

where μ_1 and μ_2 are the viscosities of the mixture at temperatures T_1 and T_2,
x_i mole fraction of component i,
and m_i power exponent in Eq. (3.97) applying to pure component i.

It follows from this equation that it is necessary to know the value of the exponent m_i, which may be calculated from Eq. (3.39) and Table 3.6.

The average error of estimation by Ulibin's equation is about 1%.

d. Calculation of the Viscosities of Gas Mixtures at High Pressures

Carr *et al.*[54, 55] have found that the viscosities of multi-component mixtures of hydrocarbons may be estimated from the generalized graphs (Fig. 3.15) used for pure substances. In order to find the values of the reduced parameters T_r and p_r for use with these graphs they used a pseudocritical temperature T_{pc} and a pseudocritical pressure p_{pc} calculated from Kay's rules[193] (Eqs. (1.141) and (1.142)).

At pressures up to 680 atm and temperatures up to 110°C, the predicted viscosities agreed with the experimental ones to within 8%. To carry out the calculation, it is necessary to predict, using one of the methods described above, the value of μ_m°. The pseudocritical parameters T_{pc} and p_{pc} are then calculated and the reduced parameters $T_r = T/T_{pc}$, $p_r = p/p_{pc}$ are obtained. The value of μ_p/μ_m° is read from Fig. 3.15 and hence the value of μ_p, i.e. the viscosity of the given mixture under pressure p and at temperature T, is obtained.

According to Gambill,[114] who has reviewed various methods for estimating the viscosities of gas mixtures, the graph represented in Fig. 3.15 is more suitable for calculating the viscosities of mixtures under high pressures than that given in Fig. 3.14.†

† When using Uyehara and Watson's graph (Fig. 3.14) the pseudocritical viscosity μ_{pc} is taken to be the molar average of the critical viscosities of the pure components ($\mu_{pc} = x_1 \mu_{pc1} + x_2 \mu_{pc2} + \ldots$).

If the pressures are not too high, the error of estimation in the latter case is of the order of 6%; for high pressures it reaches 10%. Gambill also discusses various other equations, e.g. that of Johnson for calculating viscosities, which are not discussed in this chapter.

e. Recommended Methods for Estimating the Viscosities of Gas Mixtures

Of the more accurate methods for calculating the viscosity of a gas mixture in the pressure-independent region, the most convenient is that recommended by Wilke (Eq. (3.89) and Figs. 3.18 and 3.19). The method of Saxena and Narayanan is somewhat more accurate, but it requires a knowledge of the viscosities of the binary mixtures which may be made up from the components of the multi-component mixture under consideration so the calculation of a viscosity by this method is very cumbersome.

Andrussov's formula (3.94) also gives very good results, provided the constants of this equation for the components forming the given mixture are known (Table 3.6).

A rapid, approximate determination of the viscosity of a gas mixture under a moderate pressure may be performed using Eq. (3.94) and Table 3.9.

When estimating the viscosity of a gas mixture at a high pressure it is best to use the graph presented in Fig. 3.15 in conjunction with the pseudocritical parameters, calculated from Kay's equations.

For interpolation and extrapolation purposes the graphs of Othmer (Fig. (3.12)) are suitable.

CHAPTER 4

VISCOSITIES OF LIQUIDS

1. EXPERIMENTAL DATA

The viscosity is one of several macroscopic properties which determine the behaviour of liquids and which are consequently of great importance to the engineer. A knowledge of liquid viscosity is required in calculations connected with fluid flow, heat transfer, filtration, sedimentation, and many other operations. Numerous measurements of liquid viscosity have been made under various conditions and a wealth of experimental data now exists. The viscosities of liquids are found to decrease with increase in temperature; this is in contrast with the behaviour of gases, the viscosities of which increase as the temperature is raised.

The viscosity increases (with very few exceptions) as the pressure exerted on the liquid is raised.

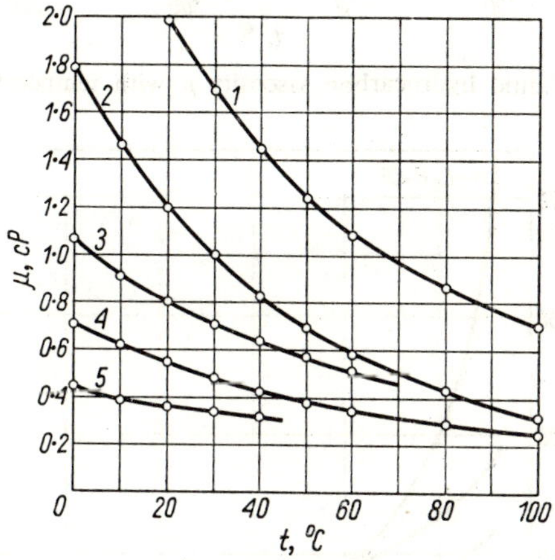

FIG. 4.1. Liquid viscosity μ (cP) vs. temperature t (°C); *1*—nitrobenzene, *2*—ethanol, *3*—chlorobenzene, *4*—n-octane, *5*—carbon disulphide.

A few examples of the variation of liquid viscosity with temperature are shown in Figs. 4.1, 4.2 and 4.3.

It is seen from these figures that the decrease in liquid viscosity with increase in temperature can be quite rapid.

Thorpe and Rodger[367] have shown that the variation of liquid viscosity with temperature can be represented by the following empirical equation

$$\mu_t = \frac{C}{1+\alpha t+\beta t^2} \tag{4.1}$$

where μ_t is the viscosity at temperature t (°C), in poises,
and C, α and β are constants characteristic of the given liquid.

The results of viscosity measurements on a number of liquids at different temperatures are summarized in Table 4.1, values of the constants in Eq. (4.1) being given where these are available.

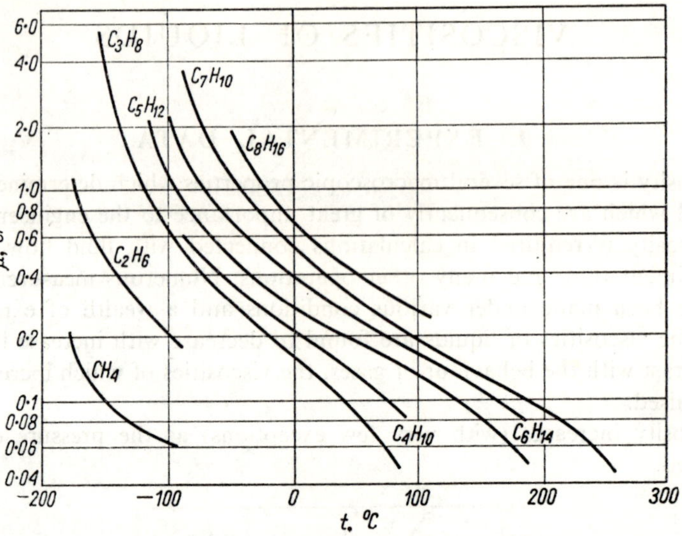

FIG. 4.2. Variation of liquid hydrocarbon viscosity μ with temperature t (after Parisot[281]).

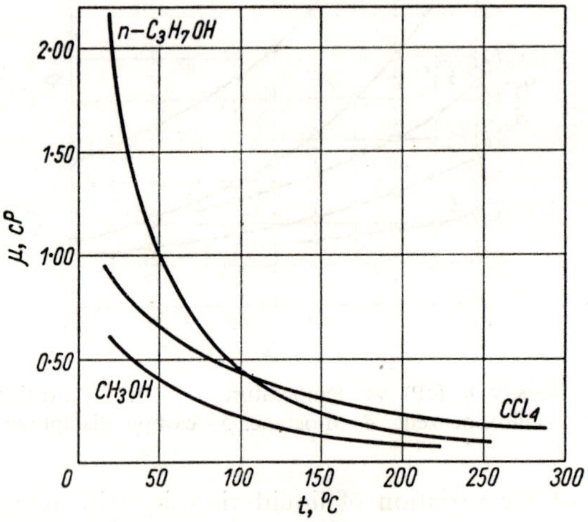

FIG. 4.3. Variation of viscosity μ of methanol, n-propanol and carbon tetrachloride with temperature t (after Kopylov[203]).

The data presented in Table 4.1 are taken partly from the book by Hatschek,[145] and partly from other references.[306,282,11,184]

In Table 4.2, values of the viscosity of water in the temperature range 0–100°C are presented. Values of the viscosity of water at higher temperatures have been listed by Mayinger.[242]

TABLE 4.1. VISCOSITIES OF LIQUIDS

Compound	t, °C	μ, cP	Constants in Eq. (4.1)		
			α	β ($\times 10^4$)	C
Acetaldehyde	20	0·2215	0·00963	0·2953	0·002671
Acetic acid	20	1·219	0·01826	0·8537	0·016867
Acetic anhydride	20	0·902	0·01735	0·6122	0·012416
Acetone	20	0·3225	0·01064	0·3115	0·003949
Acetophenone	50	1·246			
″	80	0·734			
Acetylacetone	0	1·09			
Allyl alcohol	20	1·361			
Allyl bromide	20	0·4955	0·01177	0·2871	0·006190
Allyl chloride	20	0·3295	0·01111	0·2629	0·004059
Allyl iodide	20	0·7265	0·01316	0·3361	0·009296
Ammonia, liquid	−69	0·475			
″ ″	−40	0·276			
″ ″	−33·5	0·255			
Amyl alcohol	15	4·65			
″ ″	30	2·99			
iso-Amyl alcohol	0	8·6			
″ ″	20	4·36			
″ ″	100	0·63			
Amyl butyrate	0	1·77			
″ ″	20	1·21			
Aniline	0	10·2			
″	20	4·40			
″	100	0·80			
Anisole	0	1·78			
″	20	1·32			
″	60	0·97			
Benzaldehyde	25	1·39			
Benzene	20	0·649	0·01861	0·6181	0·009055
Bromobenzene	0	1·52			
″	20	1·13			
″	100	0·52			
o-Bromotoluene	0	2·21			
″	20	1·51			
m-Bromotoluene	0	1·73			
″	20	1·25			
n-Butane, liquid	−90	0·63			
″ ″	−60	0·403			
″ ″	−30	0·282			
″ ″	0	0·210			
iso-Butane, liquid	−80	0·628			
″ ″	−10	0·239			
Butene, liquid	−80	0·45			
″ ″	−40	0·26			
Butyl acetate	20	0·732			
″ ″	80	0·366			
Butyl alcohol	−40	22·4			
″ ″	0	5·19			
″ ″	20	2·95			
″ ″	80	0·54			
iso-Butyl alcohol	−40	51·3			
″ ″	0	8·3			

Continued overleaf

TABLE 4.1 (*cont.*)

Compound	t, °C	μ, cP	Constants in Eq. (4.1)		
			α	β ($\times 10^{-4}$)	C
iso-Butyl alcohol	20	3·95			
,, ,,	80	0·78			
Butyric acid	20	1·538	0·02109	1·1073	0·022747
iso-Butyric acid	20	1·315	0·01917	0·9215	0·018872
Carbon dioxide, liquid	0	0·099			
,, ,, ,,	10	0·085			
,, ,, ,,	20	0·071			
,, ,, ,,	30	0·053			
Carbon disulphide	20	0·367	0·00820	0·1302	0·004294
Carbon tetrachloride	20	0·969	0·01801	0·6747	0·013466
Chlorine, liquid	−76·5	0·729			
,, ,,	−52·4	0·566			
,, ,,	0	0·385			
Chlorobenzene	0	1·06			
,,	20	0·80			
,,	80	0·435			
Chloroform	20	0·564	0·01149	0·2588	0·007006
Copper, liquid	1085	3·36			
,, ,,	1200	3·12			
o-Cresol	20	9·8			
,,	60	2·3			
m-Cresol	20	21			
,,	60	3·2			
p-Cresol	20	20·2			
,,	60	3·5			
Cyclohexane	20	0·97			
,,	60	0·54			
Cyclohexanol	20	68·0			
,,	60	7·8			
Decalin	20	2·40			
,,	50	1·58			
Diallyl	20	0·274	0·01118	0·3017	0·003388
Diethyl ether	−40	0·47			
,, ,,	0	0·296			
,, ,,	20	0·243			
Diethyl ketone	20	0·4655			
Diethyl malonate	15	2·38			
,, ,,	30	1·75			
Dimethylaniline	20	1·41			
,,	80	0·64			
2,2-Dimethylpropane	−15	0·431			
Dioxane	20	1·26			
,,	60	0·685			
Diphenyl	80	1·24			
,,	100	0·97			
Ethane, liquid	−175	0·985			
,, ,,	−120	0·257			
,, ,,	−85	0·162			
Ethyl acetate	0	0·578			
,, ,,	20	0·449			
,, ,,	80	0·248			
Ethyl alcohol	−40	4·79			

Continued overleaf

TABLE 4.1 (*cont.*)

Compound	t, °C	μ, cP	Constants in Eq. (4.1) α	β (×10⁻⁴)	C
Ethyl alcohol	−20	2·38			
,, ,,	0	1·78			
,, ,,	20	1·19			
,, ,,	50	0·71			
,, ,,	80	0·435			
Ethylaniline	20	2·25			
,,	60	1·01			
Ethylbenzene	20	0·665	0·01448	0·4530	0·008745
Ethyl bromide	20	0·392	0·01064	0·1822	0·004776
Ethyl butyrate	15	0·711			
,, ,,	30	0·595			
Ethyl chloride	−20	0·392			
,, ,,	0	0·320			
,, ,,	40	0·224			
Ethylene, liquid	−170	0·70			
,, ,,	−140	0·31			
,, ,,	−100	0·15			
Ethylene bromide	20	1·716	0·02007	0·7018	0·024579
Ethylene chloride	20	0·833	0·01653	0·5451	0·011269
Ethyl formate	20	0·402			
,, ,,	30	0·358			
Ethyl iodide	20	0·583	0·01113	0·2658	0·007190
Ethyl propionate	15	0·564			
,, ,,	30	0·473			
Formamide	0	7·3			
,,	20	3·75			
,,	100	0·83			
Formic acid	20	1·782	0·02870	1·695	0·029280
Glycerol	−42	$6·71 \times 10^6$			
,,	0	12110			
,,	20	1490			
Heptane	20	0·4105	0·01214	0·4004	0·005180
iso-Heptane	20	0·379	0·01199	0·3863	0·004767
Hexane	20	0·320	0·01122	0·3337	0·003965
iso-Hexane	20	0·300	0·01110	0·3599	0·003713
Hydrogen, liquid		0·011			
Hydrogen chloride, liquid	−85	0·458			
Hydrogen sulphide, liquid	−65	0·443			
Iodobenzene	20	1·49			
,,	100	0·69			
Iron (2·5% C), liquid	1400	2·25			
Isoprene	20	0·2155	0·01002	0·1542	0·002600
Lead, liquid	350	2·58			
,, ,,	600	1·38			
Mercury	−20	1·855			
,,	0	1·685			
,,	20	1·554			
,,	100	1·240			
,,	200	1·052			
,,	300	0·950			
Methane, liquid	−185	0·226			
,, ,,	−160	0·115			

Continued overleaf

TABLE 4.1 (*cont.*)

Compound	t, °C	μ, cP	Constants in Eq. (4.1)		
			α	β ($\times 10^{-4}$)	C
Methyl acetate	20	0·381			
,, ,,	80	0·217			
Methyl alcohol	−40	1·75			
,, ,,	0	0·817			
,, ,,	20	0·584			
,, ,,	60	0·351			
Methyl chloride	10	0·202			
,, ,,	30	0·166			
,, ,,	60	0·129			
Methylene chloride	−20	0·68			
,, ,,	0	0·537			
,, ,,	20	0·435			
Methyl ethyl ketone	20	0·423	0·01284	0·3639	0·005383
Methyl formate	0	0·43			
,, ,,	20	0·345			
Methyl iodide	20	0·487	0·01067	0·1719	0·005940
Methyl propyl ketone	20	0·501	0·01325	0·3965	0·006464
Naphthalene	80	0·967			
,,	100	0·776			
Nitric acid	0	2·275			
,, ,,	10	1·770			
Nitrobenzene	0	3·09			
,,	20	2·01			
,,	80	0·87			
Nitromethane	0	0·844			
,,	20	0·657			
,,	80	0·357			
o-Nitrotoluene	0	3·83			
,,	20	2·37			
,,	80	0·94			
m-Nitrotoluene	20	2·33			
,,	80	0·92			
p-Nitrotoluene	60	1·20			
,,	100	0·76			
Nonane	0	0·97			
,,	60	0·44			
Octane	20	0·538	0·01394	0·4926	0·007025
Pentane	20	0·232	0·01044	0·2301	0·002827
iso-Pentane	0	0·272			
,,	20	0·223			
Phenol	20	11·6			
,,	60	2·56			
,,	140	0·69			
Phenyl cyanide	20	1·33			
,, ,,	100	0·515			
Phenyl propyl ketone	20	2·36			
Phosphorus, liquid	21·5	2·34			
,, ,,	79·9	1·21			
Propane, liquid	−190	13·8			
,, ,,	−140	0·984			
,, ,,	−80	0·327			
,, ,,	−40	0·205			

Coutinued overleaf

TABLE 4.1 (*cont.*)

Compound	t, °C	μ, cP	Constants in Eq. (4.1)		
			α	β ($\times 10^{-4}$)	C
Propionic acid	20	1·090	0·01720	0·6941	0·015199
Propionic anhydride	20	1·116	0·02005	0·8315	0·016071
Propyl acetate	20	0·58			
,, ,,	80	0·304			
Propyl alcohol	−40	13·5			
,, ,,	0	3·85			
,, ,,	20	2·20			
,, ,,	80	0·63			
iso-Propyl alcohol	−40	23·2			
,, ,,	0	4·60			
,, ,,	20	2·39			
,, ,,	80	0·52			
Pyridine	0	1·33			
,,	20	0·95			
,,	60	0·58			
Sodium chloride, liquid	841	1·30			
,, ,, ,,	924	0·97			
Sulphur, liquid	123	10·94			
,, ,,	149	7·09			
,, ,,	200	21,500			
,, ,,	220	18,600			
Sulphur dioxide, liquid	−33·5	0·5508			
,, ,, ,,	0·1	0·3936			
Sulphuric acid	0	48·4			
,, ,,	20	25·4			
,, ,,	50	8·82			
,, ,,	80	5·19			
Tetrachloroethane	0	2·66			
,,	20	1·75			
,,	60	0·97			
Tetrachloroethylene	0	1·14			
,,	20	0·88			
,,	60	0·60			
Tetralin	20	2·02			
,,	50	1·3			
Tin, liquid	240	2·12			
,, ,,	400	1·43			
Thiophene	20	0·659	0·01518	0·4358	0·008708
Toluene	20	0·586	0·01462	0·4220	0·007684
o-Toluidine	0	10·2			
,,	20	4·35			
,,	80	1·11			
m-Toluidine	0	8·7			
,,	20	3·81			
,,	80	1·00			
p-Toluidine	50	1·75			
,,	80	1·00			
,,	120	0·50			
Trichloroethylene	0	0·71			
,,	20	0·58			
,,	60	0·41			
Valeric acid	20	2·236			

Continued overleaf

TABLE 4.1 (*cont.*)

Compound	t, °C	μ, cP	Constants in Eq. (4.1)		
			α	β ($\times 10^{-4}$)	C
Valeric acid	50	1·25			
o-Xylene	20	0·807	0·01701	0·5636	0·008457
m-Xylene	20	0·615	0·01418	0·3923	0·008019
p-Xylene	20	0·6435	0·01472	0·4578	0·008457
Zinc, liquid	280	1·68			
" "	389	1·31			

TABLE 4.2. THE VISCOSITY OF WATER (WHERE μ IS THE VISCOSITY AS ORDINARILY DEFINED, AND ν IS THE KINEMATIC VISCOSITY) (AFTER D'ANS[11])

t, °C	μ, cP	ν, cSt	t, °C	μ, cP	ν, cSt	t, °C	μ, cP	ν, cSt
0	1·7887	1·7890	35	0·7205	0·7248	70	0·4062	0·4154
5	0·5155	1·5156	40	0·6533	0·6584	75	0·3794	0·3892
10	1·3061	1·3065	45	0·5958	0·6017	80	0·3556	0·3659
15	1·1406	1·1416	50	0·5497	0·5564	85	0·3341	0·3451
20	1·0046	1·0064	55	0·5072	0·5146	90	0·3146	0·3259
25	0·8941	0·8968	60	0·4701	0·4781	95	0·2981	0·3099
30	0·8019	0·8054	65	0·4359	0·4445	100	0·2821	0·2944

Figure 4.2 shows viscosity versus temperature curves for liquid hydrocarbons (at temperatures higher than the normal boiling points of these liquids the data were obtained at elevated pressures, while Fig. 4.3 illustrates the temperature dependence of the viscosities of methanol, n-propanol and carbon tetrachloride.

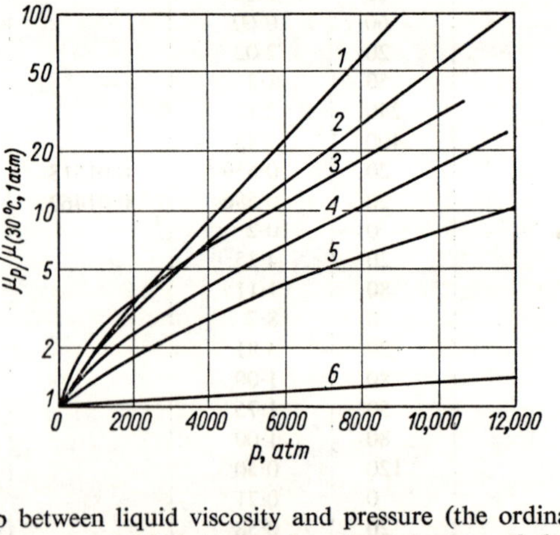

FIG. 4.4. Relationship between liquid viscosity and pressure (the ordinate denotes the ratio of the viscosity at pressure p and 30°C, to the viscosity at 1 atm and the same temperature; *1*—glycerol, *2*—ethylacetate, *3*—ethyl ether, *4*—ethanol, *5*—methanol, *6*—mercury (after Bridgmann[91]).

Values of the viscosity μ for several scores of liquids, not mentioned in Table 4.1, may be found from the alignment chart in Fig. 4.5 and the coordinates listed in Table 4.3. The use of this chart is described in the following section (section 2).

An increase in pressure generally produces an increase in liquid viscosity (Fig. 4.4), but the magnitude of the effect is strongly dependent on the liquids considered. As a result of Bridgman's investigations[38] it has been found that two pressure ranges may be distinguished: the range up to about 1000–4000 atm, within which the liquid viscosity rises proportionally to the pressure, and the range of higher pressures at which the viscosity increment becomes larger (exponential).

The more complex the structure of the liquid molecule the greater is the effect of pressure, e.g. at a pressure of 11,700 atm, the ratio $\mu_p/\mu_{1\,\text{atm}}$ amounts to 1·32 for mercury, 10 for methanol, and 10^7 for eugenol (4-allyl-2-methoxyphenol).

As a rough rule it may be assumed, when estimating the pressure effect, that a pressure increase of *ca.* 330 atm causes a rise of liquid viscosity equivalent to that caused by a temperature decrease of 1°C (Bondi). At pressures commonly used in industry (below 40 atm), the influence of pressure on liquid viscosity is unnoticeable.[119]

2. GRAPHICAL INTERPOLATION AND EXTRAPOLATION OF LIQUID VISCOSITY DATA

One graphical method for interpolating and extrapolating liquid viscosity data is to use the alignment chart presented in Fig. 4.5. This is a point nomograph. The coordinates of the points corresponding to different liquids are given in Table 4.3. In order to find the viscosity of a given liquid at temperature t_1 a straight line is drawn through the point corresponding to the given liquid and through the point t_1 on the temperature axis (which is on the left of the chart): this line is then extended to intersect the viscosity axis (which is on the right of the chart). The point of intersection gives the required viscosity. If the coordinates for the liquid under consideration are not listed in Table 4.3, they may be determined if the viscosity at two different temperatures is known. For this purpose two lines are drawn connecting the points denoting the viscosities (on the right-hand axis) with the points for the corresponding temperatures (left-hand axis). The desired point is given by the intersection of these lines in the coordinate grid in the centre of the nomograph. The coordinates are read off from this grid. The nomograph may then be used to predict the viscosity of the liquid at other temperatures.

EXAMPLE 4.1. Estimate the viscosity μ of ethyl acetate at 140°C using the nomograph in Fig. 4.5.

Solution

From Table 4.1, the viscosities of ethyl acetate at 0°C and 80°C are seen to be:

at 0°C $\mu_0 = 0{\cdot}578$ cP

at 80°C $\mu_{80} = 0{\cdot}248$ cP

The straight lines connecting the points $\mu = 0{\cdot}578$ with $t = 0$, and $\mu = 0{\cdot}248$ with $t = 80$ intersect at a point with coordinates $X = 13{\cdot}7$ and $Y = 9{\cdot}1$ in the grid of the nomograph (Fig. 4.5). A straight line drawn through this point and through the point $t = 140$°C on the temperature axis intersects the viscosity axis at the point $\mu = 0{\cdot}15$ cP. The experimental value of μ is 0·152 cP.

The graphical methods most frequently used for interpolating and extrapolating liquid viscosity data are based on the principle of similarity (cf. Ch. 1). Several methods based on this principle may be used, e.g. the Dühring line method (Ch. 1, section 7c) or Othmer's method (Ch. 1, section 7d). In both cases straight-line relationships are obtained, which enable interpolations and extrapolations to be carried out with convenience and safety.

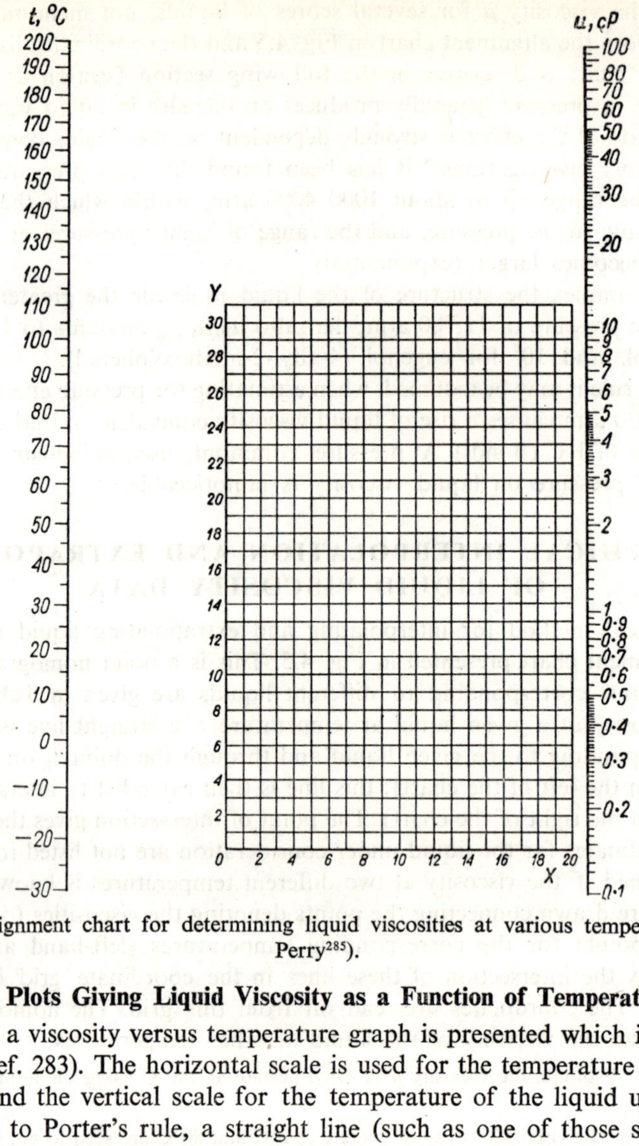

FIG. 4.5. Alignment chart for determining liquid viscosities at various temperatures (after Perry[285]).

a. Dühring-type Plots Giving Liquid Viscosity as a Function of Temperature or Pressure

In Fig. 4.6, a viscosity versus temperature graph is presented which is based on Porter's rule (cf. Ref. 283). The horizontal scale is used for the temperature of the reference liquid (water) and the vertical scale for the temperature of the liquid under consideration. According to Porter's rule, a straight line (such as one of those shown) results if the temperature at which the reference liquid has a given viscosity is plotted against the temperature at which the liquid under consideration has the same viscosity. The falling curve on the graph gives the viscosity of the reference liquid, i.e. water, as a function of temperature. Viscosities are given on the right-hand vertical, and water temperatures on the horizontal axis.

To find the viscosity of octane, for instance, at a temperature of 30°C, we draw a straight line, parallel to the abscissa, to intersect with the straight line corresponding to octane. From the point of intersection, a vertical line is dropped to the curve for the reference compound. The viscosity corresponding to this second point of intersection is that required. Thus it is predicted that the viscosity of octane at 30°C is the same as that of the reference liquid (water) at 58°C, i.e. 0·48 cP. The experimental value is 0·479 cP.

TABLE 4.3. PREDICTION OF THE VISCOSITIES OF LIQUIDS (COORDINATES[285, 324] FOR USE WITH FIG. 4.5)

Compound	Coordinates		Compound	Coordinates	
	X	Y		X	Y
Acetic acid, 70%	9·5	17·0	Freon-11	14·4	9·0
Acetic anhydride	12·7	12·0	Freon-12	16·8	5·6
Acetone, 35%	7·9	15·0	Freon-21	15·7	7·5
Acetonitrile	14·4	7·4	Freon-22	17·2	4·7
Acrylic acid	12·3	13·9	Freon-113	12·5	11·4
Allyl alcohol	10·2	14·3	Glycerol, 50%	6·9	19·6
Ammonia, 26%	10·1	13·9	Hydrochloric acid, 31·5%	13·0	16·6
Amyl acetate	11·8	12·5	Kerosene	10·2	16·9
Arsenic trichloride	13·9	14·5	Linseed oil, raw	7·5	27·2
Brine, $CaCl_2$, 25%	6·6	15·9	Methanol, 90%	12·3	11·8
Brine, NaCl, 25%	10·2	16·6	Methanol, 40%	7·8	15·5
Bromine	14·2	13·2	Methyl acrylate	13·0	9·5
n-Butyl acrylate	11·5	12·6	Methyl propionate	13·5	9·0
Butyl n-butyrate	13·2	10·3	Nitric acid, 95%	12·8	13·8
Butyl iso-butyrate	12·3	9·7	Nitric acid, 60%	10·8	17·0
iso-Butyl acetate	13·1	11·3	Nitrogen dioxide, liquid	12·9	8·6
iso-Butyl bromide	13·1	10·9	Octyl alcohol	6·6	21·1
iso-Butyl iodide	13·3	12·4	Pentachloroethane	10·9	17·3
Chlorosulphonic acid	11·2	18·1	Phosphorus tribromide	13·8	16·7
o-Chlorotoluene	13·0	13·3	Phosphorus trichloride	16·2	10·9
m-Chlorotoluene	13·3	12·5	Propyl bromide	14·5	9·6
p-Chlorotoluene	13·3	12·5	Propyl chloride	14·4	7·5
Cyclohexane	9·8	12·9	Propyl formate	13·1	9·7
Dibromoethane	12·7	15·8	Propyl iodide	14·1	11·6
Dichloroethane	13·2	12·2	iso-Propyl bromide	14·1	9·2
Dichloromethane	14·6	8·9	iso-Propyl iodide	13·7	11·2
Diethyl ketone	13·5	9·2	Propylene glycol	3·1	23·7
Diethyl oxalate	11·0	16·4	Sodium	16·4	13·9
Diethylene glycol	5·0	24·7	Sodium hydroxide, 50%	3·2	25·8
Dimethyl oxalate	12·3	15·8	Stannic chloride	13·5	12·8
Dipropyl ether	13·2	8·6	Sulphuric acid, 110%	7·2	27·4
Dipropyl oxalate	10·3	17·7	Sulphuric acid, 100%	8·0	25·1
Ethyl acrylate	12·7	10·4	Sulphuric acid, 98%	7·0	24·8
Ethyl alcohol, 95%	9·8	14·3	Sulphuric acid, 60%	10·2	21·3
Ethyl alcohol, 40%	6·5	16·6	Sulphuryl chloride	15·2	12·4
Ethyl butyl acrylate	11·2	14·0	Titanium tetrachloride	14·4	12·3
Ethyl hexyl acrylate	9·0	15·0	Triethylene glycol	4·7	24·8
Ethyl propyl ether	14·0	7·0	Turpentine spirits	11·5	14·9
Ethylene glycol	6·0	23·6	Vinyl acetate	14·0	8·8
Fluorobenzene	13·7	10·4	Water	10·2	13·0

An analogous chart for predicting the effect of pressure on the viscosity of a liquid (in this case diethyl ether), is shown in Fig. 4.7[283]. The reference substance is carbon disulphide, the viscosity–pressure curve for which appears as curve *1* in the figure. This graph enables the viscosity of ethyl ether to be read off for a given pressure; for example, as the

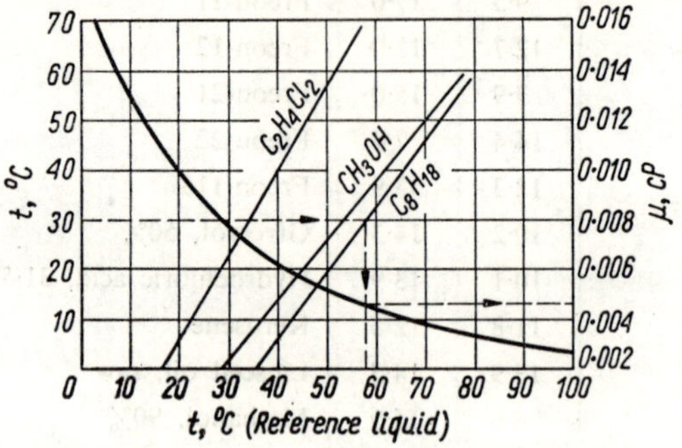

FIG. 4.6. Graph illustrating Porter's rule.

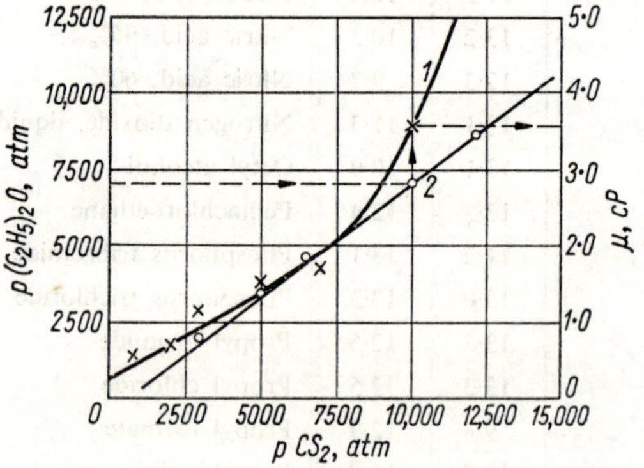

FIG. 4.7. Effect of pressure on liquid viscosity.

broken line on the figure shows, at 7000 atm, the viscosity of ethyl ether is equal to that of carbon disulphide at 10,000 atm, viz., 3·5 cP. This predicted viscosity agrees closely with the experimentally determined value. The viscosity of any other liquid for which data are available at two pressures may be obtained in a similar manner after drawing the appropriate straight line on the figure.

b. Othmer-type Plot Giving the Variation of Liquid Viscosity with Temperature[268,286]

The variation of liquid viscosity μ with temperature T is expressed approximately by an equation analogous to that relating the vapour pressure P_n to temperature:

$$\mathrm{d}\log\mu = -\frac{\mathrm{d}T}{RT^2}E_\mu \tag{4.2}$$

This may be compared with the familiar expression for the vapour pressure:

$$\mathrm{d}\log p_n = \frac{\mathrm{d}T}{RT^2}L \tag{4.3}$$

where R is the gas constant,
E_μ is the "activation energy",
and L is the molal heat of vaporization.
When the temperature range is not too large, the quantities E_μ and L may be considered to be independent of temperature.

Dividing the first of these equations by the second, we obtain

$$\frac{d \log \mu}{d \log p_n} = \frac{E_\mu}{L}$$

hence, upon integrating,

$$\log \mu = -\frac{E_\mu}{L} \log p_n + C \tag{4.4}$$

Thus a log–log plot of the viscosity of a given liquid versus the vapour pressure of the reference liquid (at the same temperature) should be linear. A knowledge of the liquid viscosity at two different temperatures is sufficient for the construction of this straight line, and hence for the estimating of the viscosity of the liquid at other temperatures.

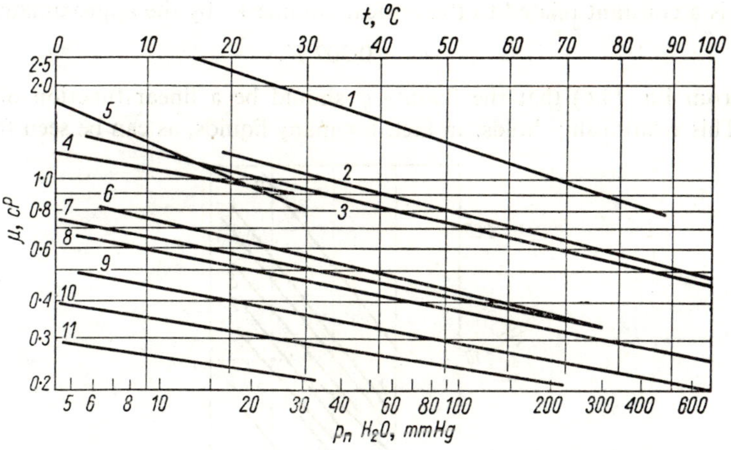

FIG. 4.8. Othmer's plot showing the effect of temperature on liquid viscosity; *1*—valeric acid, *2*—acetic acid, *3*—propionic acid, *4*—water, *5*—bromine, *6*—benzene, *7*—toluene, *8*—n-octane, *9*—n-heptane, *10*—n-hexane, *11*—n-pentane (after Pilař[286]).

Viscosity data for liquids may be interpolated and extrapolated in this manner. Othmer plots for the viscosities of a number of liquids are shown in Fig. 4.8. The vapour pressure of the reference liquid (water), mm Hg, is plotted from the lower logarithmic scale, and temperatures, °C, corresponding to these vapour pressures, are shown along the upper abscissa. The liquid viscosity μ, cP, is read from the logarithmic ordinate. As an illustration of the accuracy with which interpolations may be carried out, the viscosity of acetic acid (line 2) at 60°C is given by Fig. 4.8 to be $\mu = 0{\cdot}70$ cP. This is the same as the value determined experimentally.

Slavyanskii[332] has also developed a graphical method for presenting the liquid viscosity–temperature relationship. He obtains a straight line plot by using a uniform temperature scale as the abscissa, and a non-uniform viscosity scale (selected empirically on the basis of comparison with the viscosity of some reference substance) as the ordinate.

However, straight lines are only obtained on this type of diagram when the compounds considered and the reference substance all belong to the same group of related compounds.

c. Bachinskii's Equation

In 1912, Bachinskii[17] drew attention to a simple relationship between the viscosity of a non-associated liquid and its density

$$\frac{1}{\mu} = \frac{A}{\rho} - B \tag{4.5}$$

where A and B are constants,
$1/\mu$ is the reciprocal of liquid viscosity referred to as liquid fluidity,
and $1/\rho$ is the specific volume.

This equation is often presented in another form, viz.,

$$v = \omega + C\varphi \tag{4.6}$$

where C is a constant,
v is the specific volume ($v = 1/\rho$),
and φ is the fluidity ($\varphi = 1/\mu$).
ω is a constant related to the critical volume V_c by the approximate relationship

$$\omega = 0{\cdot}307\, V_c \tag{4.7}$$

It follows from Eq. (4.6) that the fluidity φ should be a linear function of the specific volume v. This relationship holds, in fact, for many liquids, as can be seen from Fig. 4.9.

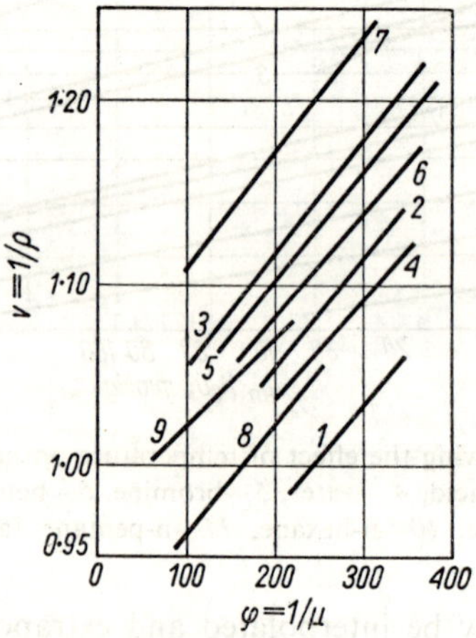

FIG. 4.9. Specific volume v vs. fluidity φ of liquid; *1*—methyl formate, *2*—ethyl formate, *3*—propyl formate, *4*—methyl acetate, *5*—ethyl acetate, *6*—methyl propionate, *7*—benzene, *8*—acetic acid, *9*—carbon tetrachloride (after Panchenkov[278]).

Bachinskii's formula is not obeyed by alcohols, organic acids, some halogen compounds (e.g. ethylene dichloride), water, or mercury.[278] With the exception of water these compounds give convex curves, instead of straight lines; the curve for water is concave.

Equations (4.5) and (4.6) cannot be used for calculating liquid viscosities at high pressures (say upwards of a thousand atmospheres), since the value of the constant C varies with pressure.

It should be mentioned that theoretical attempts to justify Bachinskii's rule have contributed much to our knowledge of the phenomenon of liquid viscosity.[278]

3. DEPENDENCE OF LIQUID VISCOSITY ON MOLECULAR PROPERTIES

The viscosity of a liquid depends on the properties of the molecules of which it is composed. *A priori* one would expect that molecular properties, such as molecular volume v_m, shape factor ϕ, weight m, polarizability α and dipole moment μ, should all influence the viscosity. This problem has been studied by Lutskii,[230] who compared the viscosities of groups of normal, non-associated liquids selected so that their molecules differed mainly in one property (e.g. polarizability, or dipole moment).

Comparison of the viscosities of isotopic compounds, the molecules of which differ only in mass m (e.g. liquefied methane and deuteriomethane), shows that the viscosities are proportional to $\sqrt{m}$†

$$(\mu)_{p,T,\phi,\mu,\alpha} = \text{constant}\sqrt{m} \tag{4.8}$$

Comparison of the ratio $\mu/\sqrt{m}$ for the isologues whose molecules differ in volume v_m shows that the viscosity does not depend on the size of the molecule. Theoretical considerations suggest that the liquid viscosity should depend on the free volume in the liquid, i.e. intermolecular free space, and not on the co-volume. Lutskii does not exclude the possibility that the apparent independence of liquid viscosity on the size of the molecule may be the result of the compensating effects of various factors related to that size.

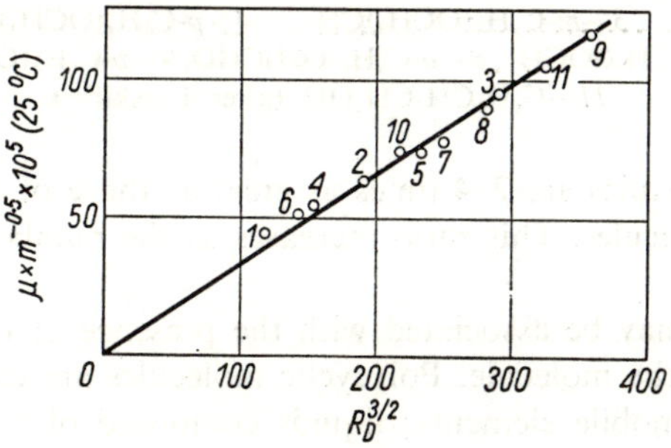

FIG. 4.10. Dependence of liquid viscosity on molecular mass and polarizability‡; *1*—C_2H_5I, *2*—n-$C_5H_{11}Br$, *3*—n-$C_8H_{17}Cl$, *4*—n-C_3H_7I, *5*—n-$C_6H_{13}Br$, *6*—n-C_4H_9Br, *7*—n-$C_7H_{15}Cl$, *8*—n-$C_6H_{13}I$, *9*—n-$C_9H_{19}Br$, *10*—n-$C_5H_{11}I$, *11*—n-$C_8H_{17}Br$ (after Lutskii[230]).

In order to investigate the effect of polarizability upon liquid viscosity, Lutskii considered the relationship between the quantity $(\mu/\sqrt{m})_{p,T,\phi,\mu}$ (thus eliminating the effect of the differences between the masses m of the molecules) and the molar refraction R_D which is proportional to the electron polarizability α_{el} of the molecule (Ch. 1, section 2e) and showed that

$$(\mu/\sqrt{m})_{p,T,\phi,\mu} = \text{constant} \times R_D^{3/2} \tag{4.9}$$

The validity of this relationship is evident from Fig. 4.10. When plotted in the $R_D^{3/2}$ vs. $\mu m^{-0.5}$ coordinate system, the points corresponding to compounds with similar dipole moments and similar shape factors fall on a straight line.‡

† Smith and Brown[335] had previously came to the same conclusion.

‡ The substances considered in Fig. 4.10 have similar dipole moments (they all lie in the range 1.85 to 1.96 Debeye). That they have similar shape factors may not be so clear: Lutskii considers that the shape factors of the higher members of a homologous series may be considered to be similar, at least in the context of viscosity calculations (see section 10b).

Lutskii also examined, in an analogous way, the relationship between liquid viscosity and dipole moment $\bar{\mu}$. He found that the viscosity varied linearly with $\bar{\mu}$ for compounds with similar molecular weights, similar molecular shapes and similar values of R_D (Fig. 4.11),† i.e. that

$$\mu_{(p,\,T,\,\phi,\,m,\,\alpha)} = \text{constant} \times \bar{\mu} + A \tag{4.10}$$

An investigation of the dependence of the liquid viscosity μ on the shape factor ϕ of the molecules showed that, at 25°C, the viscosities of tertiary compounds are 10–20% greater than those of compounds with straight-chain molecules. The viscosities of com-

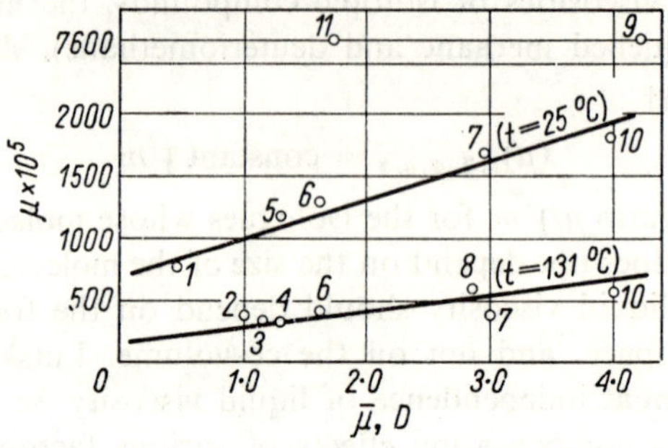

FIG. 4.11. Dependence of liquid viscosity on dipole moment†; *1*—n-$C_6H_5C_3H_7$, *2*—*o*-$C_6H_4(OCH_3)CH_3$, *3*—*m*-$C_6H_4(OCH_3)CH_3$, *4*—*p*-$C_6H_4(OCH_3)CH_3$, *5*—$C_6H_5OC_2H_5$, *6*—$C_6H_5N(CH_3)_2$, *7*—$C_6H_5COCH_3$, *8*—*o*-$C_6H_4(OH)CHO$, *9*—*p*-$C_6H_4(OH)CHO$, *10*—$C_6H_5NO_2$, *11*—$C_6H_5CH_2CH_2OH$ (after Lutskii[230]).

pounds with cyclic molecules are 2–4 times as great as those of the corresponding liquids composed of chain molecules. This ratio increases as the number of carbon atoms in the molecule is raised.

These regularities may be associated with the presence or otherwise of bonds which permit free rotation in the molecule. Polycyclic molecules are comparatively rigid having a fixed shape with no mobile elements; liquids composed of this type of molecule have high viscosities. On the other hand, the viscosities of liquids made up of molecules with a straight-chain structure (in particular normal aliphatic hydrocarbons) are low. In the latter case, the presence of a considerable number of bonds, permitting the rotation of one part of the molecule with respect to the other, may enable the molecules to arrange and fit themselves so as to facilitate the process of flow.

It follows from Lutskii's analysis that, for molecules with similar shape factors ϕ, the following relationship should exist:

$$\left(\frac{\mu}{\sqrt{m}\,R_D^{3/2}}\right)_{p,\,T,\,\phi} = A + \text{constant} \times \bar{\mu} \tag{4.11}$$

This relationship is exemplified for the series C_6H_5X in Fig. 4.12. It is evident from this plot that the points corresponding to different compounds of the series lie approximately in a straight line. Large deviations are observed for the derivatives containing —OH, —COOH and —NH_2 as the X-group.

† The substances considered in Fig. 4.11 would be expected to have similar shape factors since they all have one dominant group (the benzene ring) in common. They also have very similar molecular weights and similar values of R_D (and hence of α).

In the work referred to,[230] Lutskii has explained the deviations from his rules, which arise in the case of associated liquids, as caused by the formation of intermolecular or intramolecular hydrogen bonds. For such liquids, the observed values of viscosity may

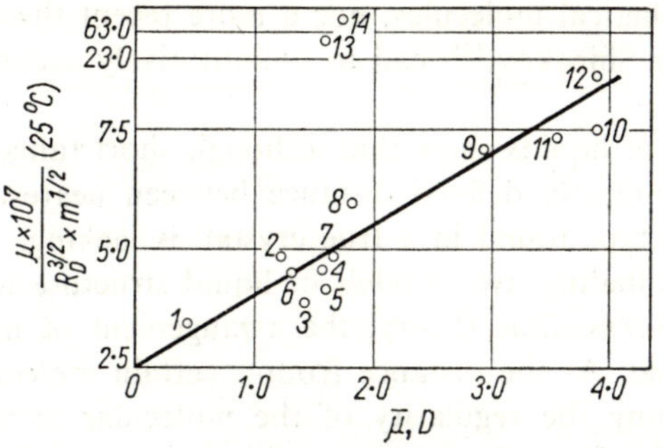

FIG. 4.12. $\left[\frac{\mu \times 10^7}{R_D^{3/2} \times m^{1/2}}\right]_{p,T,\phi}$ vs. dipole moment for compounds of the type C_6H_5X, where X denotes: *1*—CH_3, *2*—OCH_3, *3*—F, *4*—Br, *5*—Cl, *6*—I, *7*—$N(CH_3)_2$, *8*—$COOC_2H_5$, *9*—$COCH_3$, *10*—CN, *11*—NC, *12*—NO_2, *13*—NH_2, *14*—OH (after Lutskii[230]).

be as much as 5–10 times as great as those calculated on the basis of the regularities encountered in isologous series.

Discrepancies caused by differences in the shape factor ϕ are lower.

4. LIQUID VISCOSITY: A TENTATIVE THEORETICAL APPROACH

While theories of the gaseous and solid states, based on considerations of statistical mechanics and thermodynamics, are well established, our understanding of the liquid state is comparatively poor. The lack of a general theory of the liquid state and insufficient knowledge of the structure of liquids make it impossible to develop any exact, theoretically justified, computational method for estimating liquid viscosity, the mechanism of which is not yet sufficiently known.†

In contradistinction to the mechanism of internal friction in gases (which depends on the transfer of momentum from one molecule to another through molecular collisions), the main contribution to the viscous forces which occur in liquids arises from the existence of attractive forces between the molecules. In terms of this simplified picture it is possible to explain the difference between the effect of temperature on the viscosity of a gas and that of a liquid. In the case of a gas, the internal friction rises as the temperature increases because of the increase in the momentum of the molecules. In the case of a liquid, on the other hand, a decrease in viscosity takes place as the temperature rises because the role of the attractive forces becomes less important at high temperatures.‡

This behaviour is in agreement with Bachinskii's rule described above, according to which as the temperature of a liquid is raised, the volume and the intermolecular

† This problem is discussed in more detail by Panchenkov.[278]

‡ At high temperatures the attractive part of the intermolecular potential becomes negligible compared with the kinetic energy of the molecules.

distances increase and, therefore, the forces of molecular interaction decrease, while the fluidity of the liquid increases (section 2b).

Efforts to refine this qualitative interpretation have led to various theories of the structure of liquids. Neither the older theory of Jäger,[170] dealing with an *ideal liquid* made up of perfectly elastic spherical molecules, nor a more recent theory given by Brillouin[39] and further developed by Shirokov,[325] lead to quantitatively accurate expressions for the liquid viscosity.

X-ray investigations of liquids show that although short range order exists in liquids (in that there is a fairly clearly defined distance between nearest neighbour molecules) long range order, of the type found in a true crystal, is lacking.

Partly based on this finding, two models of liquid structure have been put forward. According to the *pseudo-crystalline* theory, the arrangement of molecules in a liquid is related to that in a crystal. As the distance from a certain molecule, assumed to be the starting point in estimating the regularity of the molecular arrangement, is increased, the deviations of molecules from the normal positions in a regular arrangement become increasingly large. Because of the molecular motions free spaces, not occupied by molecules, are constantly appearing and disappearing in the lattice. Numerous authors[302] have used this model to provide an interpretation of the results of X-ray investigations as well as an explanation of the phenomenon of liquid viscosity.

According to the *microcrystalline* theory, the liquid consists of very tiny crystals, constituting aggregates of several tens or hundreds of molecules. These assemblies are unstable, change their positions, decompose and form anew, so that a given molecule incessantly changes its membership in the assembly, and even enters into new ones. This concept was developed by Stewart.[345]

At temperatures not too far removed from the melting point of a substance, its structure in the liquid state approaches, as far as the molecular arrangement is concerned, that of a solid crystal. At high temperatures (when the boiling point is approached), the condition of the liquid more closely resembles that of a gas.

This picture of the structure of liquids constitutes the basis for Andrade's considerations[6] as well as for Eyring's theory[94] of the variation of liquid viscosity with temperature.

Andrade pictured the flowing liquid as composed of a series of layers each moving longitudinally with the direction of flow (cf. Fig. 3.2). He considered that the molecules in adjacent liquid layers could be regarded, for the purposes of calculating the viscosity, as vibrating about their average positions with a frequency ν. At every extreme libration the molecules of one layer come into contact with those of an adjacent layer and share some of their momentum, measured along the direction of flow, with this layer. He further assumed that attractive forces between the molecules tended to place them in a position favourable for communication of momentum when at extreme libration. By allowing the average intermolecular potential energy at extreme libration to be $-E$ he derived the relationship

$$\mu = C_1\left(\frac{\nu}{x}\right) e^{E/kT}$$

If ν and (EV) are taken to be independent of temperature the following equation for the viscosity results:

$$\mu = \frac{C_2}{v^{1/3}} e^{D/vT} \tag{4.12}$$

where C_1, C_2 and D are constants (for many compounds $C_2 \approx 0{\cdot}4\times 10^{-3}$),
x is the mean distance between two adjacent molecules in the liquid,
v is the specific volume,
k is Boltzmann's constant,
and T is the absolute temperature.

The above equation is often referred to as Andrade's equation. An empirical equation of similar form has long been known (Reynolds, 1886):

$$\mu = A\,e^{B/T} \tag{4.13}$$

If the coefficients A and B are taken to be independent of temperature, then, as has been found, formula (4.13) often fails to express correctly the variation of liquid viscosity with temperature. Andrade's equation differs from (4.13) in that he regards A and B as functions of the specific volume, v. As a result of this modification Andrade was able to obtain good agreement between the calculated and observed viscosities for many liquids.

Eyring considered the activation energy required to enable a molecule to pass from a given site in the liquid to an adjacent vacant site (cf. Ref. 130). He arrived at the following expression for the viscosity:

$$\mu = A\,e^{-\frac{\Delta H - T\Delta S}{RT}} = A\,e^{-\frac{\Delta H}{RT}}\,e^{\frac{\Delta S}{R}} \tag{4.14}$$

where A and R are constants,
ΔH is the heat of activation,
and ΔS is the entropy of activation.

This equation is based on the assumption that the process of flow in a viscous liquid involves the energy of activation required to make the molecule "jump" from the position in which it is placed to that corresponding to the next point of the pseudocrystal lattice, passing over the energy barrier.

By combining these theoretical considerations with an empirical relationship between the activation energy for viscous flow, E_v, and the energy of liquid vaporization ΔE_v, Eyring found, after introducing some simplifying assumptions, that:

$$\mu = 1{\cdot}090\times 10^{-3}\,\frac{M^{1/2}T^{3/2}}{V^{2/3}\Delta E_v}\,e^{\frac{\Delta E_v}{nRT}} \tag{4.15}$$

where μ is the liquid viscosity,
M is the molecular weight,
T is the absolute temperature,
R is the gas constant,
V is the molal volume,
ΔE_v is the internal energy of vaporization†,
and n $(= \Delta E_v/E_v)$ is a coefficient which is similar in magnitude for most substances (for molecules with spherical symmetry, $n = 3$; for oblong or polar molecules, $n \approx 4$).

† The internal energy of vaporization, ΔE_v, is calculated on the basis of the relationship:
$\Delta E_v = L - p\Delta V$, which simplifies to
$\Delta E_v = L - RT$ if the vapour is assumed to behave as an ideal gas,
where L is the molal heat of vaporization of the liquid,
p is the pressure,
and ΔV is the change of volume.

It may be noted that the fraction preceding the exponential term in Eq. (4.15) corresponds to the coefficient A in Eq. (4.13). Since this fraction contains the quantity $T^{3/2}$, Eyring's theory would appear to suggest that the frequently made assumption that the coefficient A does not depend on temperature, may be wrong (see section 8).

Although Eq. (4.15) may sometimes be used to obtain a rough estimate of the way in which viscosity varies with temperature, it often leads to erroneous results if used to calculate absolute values of liquid viscosity. Only in the simplest cases are the results consistent with experiment (Eyring *et al.* (cf. Ref. 5)), and the possibility of applying this method for practical purposes is very limited. Qualitatively, however, agreement is sufficiently good to suggest that the theoretical basis for Eq. (4.15) may be largely correct. This is also suggested by the work of Pratap and Narsimhan.[297] These authors have discussed certain relationships between the viscosity, surface tension, molal volume and molal energy of vaporization of liquids and have confirmed the validity of some conclusions resulting from the theory. When deriving Eq. (4.15) from Eq. (4.14) a proportionality was assumed between the energy of activation E_v and the internal energy of vaporization, ΔE_v. If, in place of this assumption, it is assumed that ΔE_v is proportional to the *free energy of activation*, $\Delta F^{\ddagger}$, the equation

$$\mu = \frac{hN}{V} e^{\left(\frac{\Delta E_v}{mRT}\right)} \tag{4.16a}$$

is obtained, where $m = \Delta E_v/\Delta F^{\ddagger}$. The above equation may be used to calculate m from viscosity and latent heat data. A study of such data for nearly 100 substances (including associated liquids but excluding metals) shows that

$$\Delta E_v/\Delta F^{\ddagger} = 2{\cdot}45 \tag{4.16b}$$

Combining Eq. (4.16b) with the above expression for μ, (4.16a),

$$\mu = \frac{hN}{V} e^{\frac{\Delta E_v}{2{\cdot}45RT}} \tag{4.16c}$$

where h is Planck's constant,
N is Avogadro's number,
and the other symbols are as in Eq. (4.15).

Equation (4.16b) enables the free energy of activation for viscous flow to be calculated to within $\pm 6\%$: values of the viscosity μ may be predicted from Eq. (4.16c) to within $\pm 30\%$.[130]

Higgins and Raw[150] have calculated the activation energy for viscous flow using the Lennard-Jones equation for the intermolecular potential and also the Sutherland equation. They find that neither of these equations predict correct values for E_v. On the other hand, good results are obtained from the equations of Slater and Kirkwood[329] (ignoring, however, the repulsive forces of the molecules). According to these authors the potential energy $\phi(r)$ of two molecules whose centres are a distance r apart is given by

$$\phi(r) = -0{\cdot}75\ (a_0 p)^{1/2} e^2 \alpha^{3/2} r^{-6} \tag{4.17}$$

where $a_0 = 0{\cdot}529$ Å (the radius of a hydrogen atom after Bohr),
e is the charge of the electron,
p is the number of electrons in the outer shell,
and α is the polarizability of the molecule.

Using this equation, the activation energy for the flow of a viscous liquid is found to be:

$$E_v = 0{\cdot}75\ N\ (a_0\ p)^{1/2} e^2 \alpha^{3/2} r_0^{-6} \tag{4.18}$$

According to some authors, the exponent of the quantity α should be higher than 1·5 (1·69 and 1·86 have been suggested). The order of magnitude of the values calculated by means of Eq. (4.18) agrees with those obtained from experimental data, though discrepancies are considerable.

Born and Green[30] have derived a rigorous expression for the liquid viscosity in terms of the intermolecular force fields. Before their expression, which is based upon statistical considerations, can be used it is necessary to solve an extremely complicated integro-differential equation. Born and Green make no serious attempt to solve this equation though they do derive simple approximations. Thus according to Green[134] the viscosity μ is proportional to $e^{-\phi(r_0)/kT}$, where the function $\phi(r_0)$ is the energy of interaction between two molecules the centres of which are a distance r_0 apart. (r_0 is slightly greater than the molecular diameter.)

Although the works referred to above are of major theoretical importance, most of them do not lead to equations which are readily applicable in practice. In contrast, the investigations of Panchenkov[277, 278, 279] have led to results which may find application also in technical calculations. By considering a comparatively simple model for the liquid state, Panchenkov finds that

$$\mu = 3\sqrt{6R}\sqrt[3]{\frac{\omega^2}{N}}\, M^{-5/6} \rho^{4/3}\, T^{1/2} f_s e^{\varepsilon/RT}(1-e^{-\varepsilon/RT})^2 \tag{4.19}$$

where μ is the liquid viscosity,
R is the gas constant,
ω is the effective volume of molecules in one mole,
N is Avogadro's number,
M is the molecular weight,
ρ is the liquid density,
T is the absolute temperature,
ε is the energy of the "bonds" between liquid molecules, calculated on the basis of the heat of vaporization in vacuum,
and f_s is a steric factor.

Calculations based on the assumption that $f_s = 1$ give a correct picture of the dependence of liquid viscosity on temperature; on the other hand, the absolute values of the viscosity μ are not given correctly.

The steric factor is given by $f_s = e^{\Delta S'/R}$, $\Delta S'$ being the entropy change in formation of a single "bond" (with reference to one mole). Panchenkov has suggested that $\Delta S'$ may be calculated from the equation

$$\Delta S' = -\frac{L_T N}{q} - \frac{N(S_{id} - S_n)}{q} \tag{4.20}$$

where L_T is the latent heat of vaporization at temperature T,
q is the total number of "bonds" in one mole of the liquid,
S_{id} is the entropy of one mole of the compound under consideration in the ideal gas state,
and S_n is the entropy of one mole of the compound in the state of the saturated vapour.

After introducing the factor f_s into Eq. (4.19), good agreement is obtained between the calculated and observed viscosities, deviations ranging from tenths of per cent to a few per cent only. The validity of the assumptions made in the above equations was discussed in subsequent papers.[280,369]

Although it is quite possible to calculate viscosities using Panchenkov's equation, its practical value is limited by the fact that the determination of some of the quantities occurring in the equation is tedious and may be rather more difficult than the measurement of the liquid viscosity.

5. METHODS OF ESTIMATING THE LIQUID VISCOSITY WHICH ARE BASED ON THE PRINCIPLE OF ADDITIVITY

It is possible to estimate the viscosity of a liquid with reasonable accuracy only when at least one value, determined experimentally under known conditions of temperature and pressure, is known. If, however, such a value is not available, it is necessary to resort either to the troublesome and inaccurate methods described in the previous section, or to equally inaccurate, though less awkward, methods based on the application of the principle of additivity.

According to Gambill,[118] all these methods fail for liquids with high viscosities, say above 30 cP. Liquids composed of large asphetical molecules in which intermolecular hydrogen bonding takes place fall within this category. Reasonably accurate results may be obtained, however, when the liquid viscosity does not exceed 5 cP.

a. Souders' Method

Souders[337] has found that the coefficient I (called the viscosity constant) in the equation

$$\log(\log 10\mu) = \frac{I}{M}\rho_1 - 2{\cdot}9 \tag{4.21}$$

can be calculated additively by summing up contributions corresponding to the respective groups and bonds which are present in the molecules of the liquid considered. A list of these contributions is given in Table 4.4 which, however, does not include sulphur compounds. In Eq. (4.21), μ denotes the liquid viscosity expressed in cP, ρ_1 the liquid density in g/cm^3 and M the molecular weight. Using Souders' method it is possible to predict approximate values of liquid viscosity even at temperatures up to the normal boiling point. The accuracy of the calculation depends to a great extent on a precise knowledge of the liquid density.

A relation exists between the constant I in Souders' equation and the molar refraction R_D. Since the refraction, which is an additive quantity, can be calculated with good accuracy by the method of group contributions (cf. Ch. 1, section 4b) the constant I can also be determined in this way with fair accuracy. A point nomograph (Fig. 4.13) constructed by Lagemann (cf. Ref. 118) can be used for this purpose for compounds of seven homologous series. In order to find the value of the constant I for a substance belonging to one of the series included in the nomograph, a straight line is drawn joining the point corresponding to the value of the molar refraction R_D of this substance (left-hand scale) to the point in the centre of the nomograph corresponding to the required homologous series. This

TABLE 4.4. CONTRIBUTIONS FOR CALCULATING THE VISCOSITY CONSTANT I IN SOUDERS' EQUATION

Atom	H	O	C	N	Cl	Br	I
Contribution	+2·7	+29·7	+50·2	+37·0	+60	+79	+110

Contributions of groups and bonds

Double bond	−15·5	H—C(=O)—R	+10
5-member ring	−24	—CH=CH—CH_2—X†	+4
6-member ring	−21	R_2CH—X†	+6
Side groups on a 6-member ring:		—OH	+57·1
Molecular weight < 17	−9	COO	+90
Molecular weight > 16	−17	COOH	+104·4
Ortho or *para* position	+3	NO_2	+80
Meta position	−1		
R_2CH—CHR_2	+8		
R—C(R)(R)—R	+10		
—CH_2—	+55·6		

† X is a negative group.

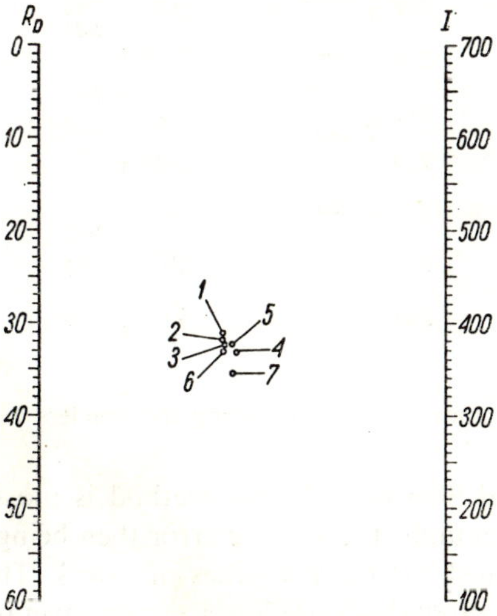

FIG. 4.13. Lagemann's alignment chart for determining the constant I in the Souders equation (4.21); *1*—monohydric alcohols, *2*—monocarboxylic acids, *3*—esters of acetic acid, *4*—aliphatic ketones, *5*—ethyl esters, *6*—paraffinic hydrocarbons, *7*—alkyl iodides.

line is then extended to intersect the right-hand axis, from which the value of the constant I is read. To avoid finding the antilogarithm twice, use can be made of the nomograph in Fig. 4.14, where z is given by

$$z = \log(\log 10\mu)$$

If the value of z, obtained by substituting a value for I calculated as above into Eq. (4.21), is negative it is laid off on the left-hand scale of the left axis and the viscosity μ is read from the right-hand scale of the left axis. If z is positive it is laid off on the left-hand scale of the right axis, and the viscosity μ is read on the right-hand scale of the right axis.[71]

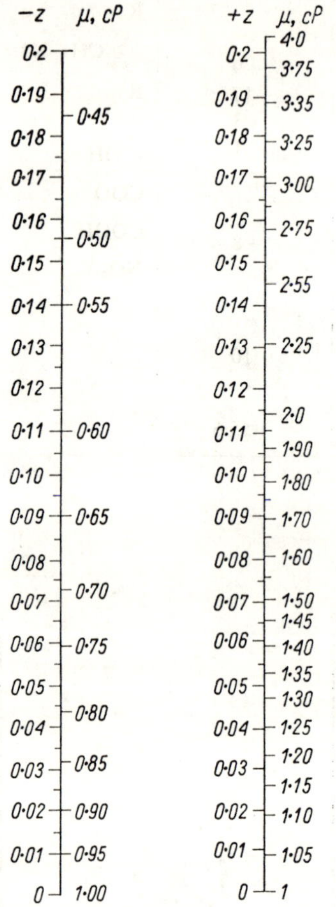

FIG. 4.14. Alignment chart for solving the Souders equation (4.21).

According to Reid and Sherwood,[306] this method is most accurate when applied to aromatic and alicyclic compounds, the average error then being less than $\pm 10\%$. Satisfactory results are also obtained for the viscosities of esters. The viscosities calculated for unsaturated or branched hydrocarbons are less accurate (average error $\pm 20\%$). Calculated viscosities for alcohols and compounds containing several halogen atoms in the molecule, are usually too high; those for acids are too low. Gambill[118] has carried out such calculations in 144 cases. He obtained an average error of 45%: the maximum error was 770%.

EXAMPLE 4.2. Estimate the viscosity of *o*-xylene, $C_6H_4(CH_3)_2$, at 20°C. The density of *o*-xylene at 20°C is 0·8812 g/cm³ and its molecular weight M is 106·16.

Solution

Calculation of the viscosity constant I:

8 contributions of C	8×50·2 =	401·6	
10 contributions of H	10×2·7 =	27·0	
3 double bonds	3×(−15·5) =		−46·5
1 6-member ring			−21·0
2 side groups CH_3	2×(−9) =		−18·0
1 *ortho*-position		+3·0	
		+431·6	−85·5
		I = 346·1	

$$z = \frac{346 \cdot 1}{106 \cdot 16} \times 0 \cdot 8812 - 2 \cdot 9 = 2 \cdot 87 - 2 \cdot 90 = -0 \cdot 03$$

According to the alignment chart in Fig. 4.14 (left-hand axis) the viscosity corresponding to $z = -0{\cdot}03$ is

$$\mu_{20} = 0{\cdot}86 \text{ cP}$$

The experimentally determined value (Table 4.1) is $\mu_{20} = 0{\cdot}807$ cP. The error of the calculated viscosity is +6·6%.

b. Thomas' Method

Thomas[366] has suggested a method of calculating liquid viscosities from the equation

$$\log \frac{8{\cdot}569\mu}{\rho\, 0{\cdot}5} = \theta\left(\frac{1}{T_r} - 1\right) \tag{4.22}$$

where μ is the liquid viscosity,
ρ is the liquid density,
T_r ($= T/T_c$) is the reduced temperature,
and θ is a coefficient calculated by summing group contributions.

Contributions to the coefficient θ arising from various groups and bonds are listed in Table 4.5. The equation may be employed in the temperature region $T_r < 0{\cdot}7$.

TABLE 4.5. GROUP CONTRIBUTIONS TO THE COEFFICIENT θ IN THE THOMAS EQ. (4.22)

Atoms, groups, bonds	Contribution	Atoms, groups, bonds	Contribution
C	−0·462	double bond	+0·478
H	+0·249	C_6H_5	+0·385
O	+0·054	S	+0·043
Cl	+0·340	CO (ketones, esters)	+0·105
Br	+0·326	CN (cyanides)	+0·381
I	+0·335		

The disadvantage of Thomas' method, when compared with Souders' method, is that it involves the use of the reduced temperature and so the critical temperature of the compound must be known.

Authors vary in their estimation of the accuracy of this method. Thomas found an average error of 39% for seven compounds exhibiting large deviations (benzene, halogenated alkylenes); for 108 other liquids, the average error amounted to 5%.

Reid and Sherwood[306] give calculations for 36 liquids at 128 various temperatures; the average error amounts to 21%. The worst errors which they report are for alcohols and acids and attain 90%. For monohalogenated aromatics (except benzene), hydrocarbons with large molecules (such as n-paraffins), and esters, the erroı is smaller than ±15%. Gambill[118] regards Thomas' method as the most universal and accurate method for the additive calculation of viscosities when no experimentally determined values are available.

EXAMPLE 4.3. Estimate the viscosity μ of ethylene dichloride, $C_2H_4Cl_2$, at 20°C by Thomas' method given that the density at 20°C is 1·2529 g/cm³. (From Table 1.31 $t_c = 288·4$°C, whence $T_c = 561·6$°K.)

Solution

Calculation of the value of the constant θ:

2 contributions of C	2×(−0·426) =	−0·852
4 contributions of H	4× (0·249) = 0·996	
2 contributions of Cl	2× (0·340) = 0·680	
	1·676	−0·852
	θ = 0·824	

Inserting θ into Eq. (4.22)

$$\log \frac{8·569\mu}{1·2529^{0·5}} = 0·824\left(\frac{561·6}{293·2} - 1\right)$$

we obtain

$$\mu = 0·744 \text{ cP}$$

It is seen from Table 4.1 that the experimental value is 0·833 cP. The error in the predicted value is thus −10·4%.

c. Calculation of Liquid Viscosities at the Normal Boiling Point from Rheochor Values

It has been seen in Chapter 1 (section 4d), that the rheochor can be calculated additively with fairly high accuracy by summing up the relevant contributions (Table 1.27):

$$R_{ch} = \frac{M(10\mu_b)^{0·125}}{\rho_{l,b} + 2\rho_n} \tag{4.23}$$

where R_{ch} is the rheochor,
M is the molecular weight,
μ_b is the liquid viscosity at the normal boiling temperature T_b, cP,
$\rho_{l,b}$ is the liquid density at the normal boiling temperature, g/cm³,
ρ_n is the density of the saturated vapour, g/cm³,
and T_b is the normal boiling temperature, °K.

The quantity R_{ch} is usually found to be approximately independent of temperature and variations with pressure are negligible for pressures up to 6000 atm.[118]

Very often the term $2\rho_n$ is sufficiently small in comparison with $\rho_{l,b}$ to be neglected and since $M/\rho_{l,b} = V_b$, Eq. (4.23) then reduces to

$$R_{ch} \approx (10\mu_b)^{0·125}\ V_b \tag{4.24}$$

(cf. Eq. (1.52))

whence

$$\mu_b \approx 0·1\left(\frac{R_{ch}}{V_b}\right)^8$$

where V_b is the molar volume at the normal boiling point. If the value of the liquid density $\rho_{l,b}$ at the normal boiling temperature is unknown, the value of the molal volume V_b (cm³/mole) at the normal boiling point may be calculated from the Le Bas contributions (Table 1.20).

The error in μ_b calculated by means of Eq. (4.24) usually amounts to 15–20%, though it frequently reaches 50%.[306] Gambill[118] has checked Eq. (4.24) against viscosity data for 35 liquids. He found that the average error in μ_b was $\pm 19\%$, and the maximum error 97%.

For non-associated liquids with known molal critical volume V_c the value of the rheochor R_{ch} can be calculated to within $\pm 5\%$ from the relationship

$$R_{ch} \approx 0{\cdot}426\ V_c \tag{4.25}$$

EXAMPLE 4.4. Estimate the viscosity μ_b of ethyl methyl ketone, $C_2H_5—CO—CH_3$, at its normal boiling point.

Solution

Calculation of the rheochor (for contributions cf. Table 1.20)

4 contributions of C	$4\times 12{\cdot}8 = 51{\cdot}2$
8 contributions of H	$8\times 5{\cdot}5 = 44{\cdot}0$
1 contribution of 0 (ketone)	$13{\cdot}2$
	$R_{ch} = 108{\cdot}4$

Calculation of the molal volume at the normal boiling point (Table 1.13)

4 contributions of C	$4\times 14{\cdot}8 = 59{\cdot}2$
8 contributions of H	$8\times 3{\cdot}7 = 29{\cdot}6$
1 contribution of 0	$7{\cdot}4$
	$V_b = 96{\cdot}2$

Substituting into Eq. (4.24)

$$\mu_b = 0{\cdot}1\left(\frac{108{\cdot}4}{96{\cdot}2}\right)^8 = 0{\cdot}244 \text{ cP}$$

This value agrees with the one determined experimentally.

6. METHODS OF CALCULATING THE LIQUID VISCOSITY WHICH ARE BASED ON THE THEOREM OF CORRESPONDING STATES

Direct application of the theory of corresponding states does not lead to generalizations of any value as far as liquid viscosity is concerned. In contrast to gases, for which the reduced viscosity $\mu_r = \mu/\mu_c$ may be represented as a function of the reduced temperature $T_r = T/T_c$ and reduced pressure $p_r = p/p_c$, the function being approximately identical for all gases (Ch. 3, section 6b), the viscosities of various liquids when represented in the reduced coordinate system give different non-coincident curves.

Substantial progress toward the development of a corresponding-states type of correlation for liquid viscosities has been made by Smith and Brown,[335] who find that the quantity to be compared is the ratio $\mu/\sqrt{M}$, where μ is the liquid viscosity, and M the molecular weight.

The authors state in their paper that the use of this ratio was, in part, suggested to them by some earlier work by Onnes.[264] (In 1894 Onnes had postulated that the group

$$\left(\frac{T_c}{M^3 p_c^4}\right)^{\frac{1}{6}} \tag{4.26}$$

should have the same value for all substances when in corresponding states.) As a result of their empirical study, Smith and Brown found that the equation

$$(\mu/\sqrt{M}) = f(T_r, p_r) \tag{4.27}$$

provides a good representation of the viscosity data for liquid normal paraffins, the function f in this equation being the same for all paraffins. When homologous series of associated liquids (such as ethers, acids and alcohols) are considered, Smith and Brown find that it is necessary to modify their relationship by replacing the molecular weight M by the viscosity constant I from Souders' equation (section 5a, Table 4.4). The equation

$$(\mu/\sqrt{I}) = f(T_r, p_r) \tag{4.28}$$

is then obtained. They find that the form of the function f depends on the homologous series considered. If, for example, $\mu^\circ/\sqrt{I}$ is plotted as a function of T_r, a set of separate curves is obtained, one for each homologous series considered (Fig. 4.15). The above finding enables the viscosity of a given liquid to be predicted if the viscosity of another liquid of the same homologous series is known as a function of temperature.

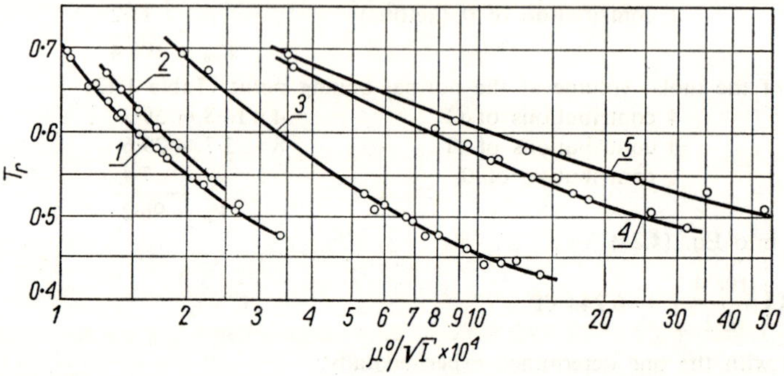

FIG. 4.15. Generalized graph of liquid viscosity vs. reduced temperature. I is the viscosity constant in Eq. (4.28); *1*—liquid hydrocarbons (n-pentane, n-hexane, n-heptane, n-octane, n-nonane, n-decane, n-undecane, n-dodecane), *2*—ethers (diethyl, dipropyl), *3*—acids (propionic, n-butyric, isobutyric, n-valeric), *4*—alcohols (ethyl, n-propyl, n-buthyl), 5—isoalcohols (isopropyl, isobutyl) (after Smith and Brown[335]).

Following the work of Bauer, Magat and Surdin, Pospekhov[293] claims that if the "reduced viscosity" μ/μ_m is plotted as a function of θ, where $\theta = (T - T_m)/(T_c - T_m)$, common curves are obtained for the members of certain homologous series, e.g. for normal paraffins with an even number of carbon atoms (butane, hexane, n-octane); the data for n-pentane and n-heptane lie on a different curve.†

By considering one of the molecular theories of viscosity,[31] Bondi,[29] came to the conclusion that the relationship

$$\mu_1^* \sim \exp(1/T_1^*) \tag{4.29}$$

should exist between the reduced viscosity

$$\mu_1^* = \mu r_0^2/(m\varepsilon)^{1/2} \tag{4.30}$$

and the reduced temperature

$$T_1^* = kT/\varepsilon \tag{4.31}$$

where r_0 is the intermolecular distance at which the intermolecular potential energy is a minimum (cm),

ε is the minimum value of this potential energy in ergs,

and m ($= M/N_A$) is the weight of a molecule, in g.

† Cf. section 8b. T_m is the melting temperature.

Bondi confirmed the validity of relationship (4.29) for several series of compounds (hydrocarbons: *n*-paraffinic, iso-paraffinic, naphthenic, aromatic; ethers, and others) and gave two formulae (one for "rigid" molecules and the other for "flexible" molecules), by means of which the viscosity of liquids belonging to the above-mentioned series of compounds may be calculated to within several per cent.

7. SOME SIMPLE RULES WHICH PROVIDE APPROXIMATE ESTIMATES OF LIQUID VISCOSITY

Gambill[118] has reviewed several rules for obtaining rough estimates of the liquid viscosity μ. According to one such rule the viscosities, μ_b, of pure liquids at the normal boiling point may be assumed to be given (on average) by

$$\mu_b \approx 0{\cdot}29 \text{ cP} \tag{4.32}$$

Gambill tested this equation on 33 liquids and found the average error to be 26·3%; the maximum error was 147%. According to another rule, the liquid viscosity at the melting point is given by

$$\mu_m \approx 2 \text{ cP} \tag{4.33}$$

This rule is approximately true for certain classes of compounds, for instance for normal paraffins with fewer than 10 carbon atoms per molecule, but for many other compounds the error may reach 700%.

In contrast to the above-mentioned rules, which are only of approximate value, a simple formula[282] given by Arrhenius in many cases provides a comparatively accurate estimate:

$$\mu_b = 0{\cdot}275\rho_{l,b}^{0{\cdot}5} \tag{4.34}$$

In this equation μ_b is the viscosity
and $\rho_{l,b}$ is the liquid density (both quantities being measured at the normal boiling point T_b).

Gambill[118] has checked Eq. (4.34) for 18 non-associated and associated liquids, and found the average error to be 21·2%, and the maximum error 51·6%. We see that, as far as accuracy is concerned, the Arrhenius equation is little inferior to other, more complicated methods (sections 5a,b,c).

Example 4.5. Estimate the liquid viscosity μ_b at the normal boiling point by means of the Arrhenius equation (4.34) for:

1. *p*-Dichlorobenzene, $C_6H_4Cl_2$, $\rho_{l,b} = 1{\cdot}11$

$$\mu_b = 0{\cdot}275\sqrt{1{\cdot}11} = 0{\cdot}29 \text{ cP}$$

The experimental $\mu_b = 0{\cdot}304$ cP; the error amounts to $-4{\cdot}5\%$.

2. Ethyl methyl ketone, C_2H_5—CO—CH_3, $\rho_{l,b} = 0{\cdot}748$

$$\mu_b = 0{\cdot}275\sqrt{0{\cdot}748} = 0{\cdot}238 \text{ cP}$$

The experimental $\mu_b = 0{\cdot}244$ cP; the error is $-2{\cdot}5\%$.

3. n-Butyl bromide, n-C_4H_9Br, $\rho_{l,b} = 1{\cdot}425$

$$\mu_b = 0{\cdot}275\sqrt{1{\cdot}425} = 0{\cdot}328 \text{ cP}$$

The experimental $\mu_b = 0{\cdot}305$ cP; the error is $+7{\cdot}5\%$.

8. THE EFFECTS OF TEMPERATURE AND PRESSURE ON LIQUID VISCOSITIES

(see also sections 1, 2 and 3)

a. The Effects of Temperature on Liquid Viscosities

The effect of temperature on liquid viscosity has been the object of many investigations; the literature reports more than 100 different equations. A review of these works is given by Partington.[282]

If the viscosity of a substance is known at two temperatures, viscosities at other temperatures may be obtained using the graphical methods considered in section 2.

If an analytical expression is required, Eq. (4.13) is usually sufficiently accurate for non-polar liquids

$$\mu = Ae^{B/T}$$

Taking logarithms

$$\log\mu = A' + B'/T \tag{4.35}$$

This equation contains two constants A and B, the determination of which requires a knowledge of the liquid viscosity at two different temperatures. Use may sometimes be made of the fact that for many organic liquids and their mixtures a plot of log A vs. B is linear.[392]

Equation (4.13) has been improved by Andrade, who introduced the specific volume v:

$$\mu = \frac{C}{v^{1/3}} e^{D/vT} \tag{4.36}$$

According to Andrade, this equation yields more accurate predictions than those obtained from Eq. (4.13), but Srinivasan and Prasad[339] did not find that this was so.

Girifalco[128] has shown that in the case of associated liquids, when Eq. (4.13) fails, good results are obtained by allowing the "constant" B to vary with temperature in the following manner:

$$B = \frac{C'}{T} + C''$$

He has also provided a theoretical interpretation for this fact. When his expression for B is combined with Eq. (4.35), the equation

$$\log\mu = \frac{\alpha}{T^2} + \frac{\beta}{T} + \gamma \tag{4.37}$$

is obtained, containing three constants α, β, γ. For non-associated liquids, with slightly polar molecules, $\alpha = 0$ and Eq. (4.37) reduces to Eq. (4.35). Several illustrative values of the constants α, β and γ are listed in Table 4.6. In order to determine the values of

TABLE 4.6. CONSTANTS IN EQ. (4.37)

Compound	$\alpha \times 10^{-4}$	$\beta \times 10^{-2}$	γ
Water	23·611	−6·8922	−0·3908
Methanol	3·428	2·750	−1·567
Ethanol	−1·145	7·807	−2·459
Acetic acid	−15·759	15·965	−3·521
Sulphuric acid	26·37	−4·690	−0·09771

the constants α, β and γ, values of the viscosity of the given liquid at three different temperatures are required. In the case of ethanol and sucrose solutions it has been found[19] that β varies linearly with γ. The average error of μ calculated from Eq. (4.37) for these solutions is less than 1·5%.

Kopylov[203] has checked an equation for associated liquids which was originally proposed by Tamman and Hesse:

$$\mu = Ae^{a/RT^3} \tag{4.38}$$

He finds that it gives good results only at temperatures below the normal boiling point. This author examined the temperature–viscosity relationships for n-butanol, methanol and carbon tetrachloride up to the critical region. He found that, over this range of conditions, the viscosities of these substances could be represented by the equation

$$\mu = A/(t+\alpha)^n \tag{4.39}$$

where t is the temperature
and A, α and n are constants.

Mean differences between calculated and experimental values are about 3%.

Cornelissen and Waterman[67] have shown that in many cases where the simple equation (4.13) fails (e.g. for associated liquids or oils of high viscosity), the dependence of the kinematic viscosity of liquids on the absolute temperature T is given with high accuracy by the equation

$$\log\nu = A/T^x + B \tag{4.40}$$

The exponent x is characteristic of a given liquid, for example for different hydrocarbons x was found to equal 1, 2 and 3; for mixtures of hydrocarbons (lubricating oils) $x = 3·5$. For sulphuric acid solutions $x = 3$, for mercury $x = 1·5$, etc.

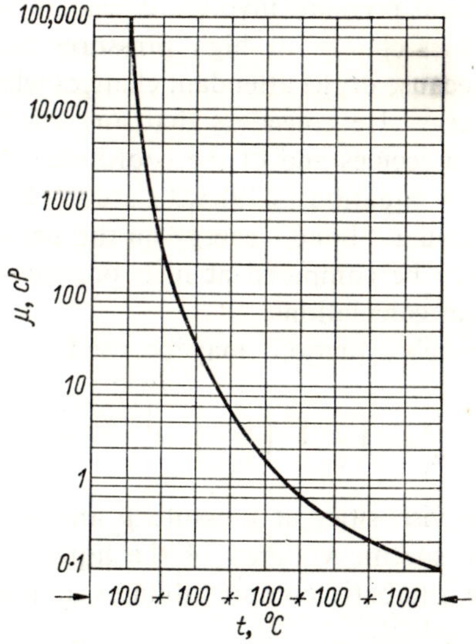

FIG. 4.16. Generalized viscosity vs. temperature curve for liquids.

If the viscosity of a substance is only known at one temperature (T_1), values at other temperatures may be obtained, using the graph of Lewis and Squires[221] (Fig. 4.16). In this graph temperature increments are shown on the abscissa, but no zero point for this scale is given. Liquid viscosity (in cP) is plotted on a logarithmic scale along the ordinate·

The graph is to be used as follows. The known value of viscosity μ (at temperature t) is marked on the ordinate, and from the point corresponding to the value of μ on this axis a straight line is drawn parallel to the abscissa to intersect with the curve. The projection of the intersection point on the abscissa determines the point corresponding to the temperature t. In this way the temperature axis is "calibrated", and the viscosity at any other temperature can be read. This graph cannot be used for associated liquids.

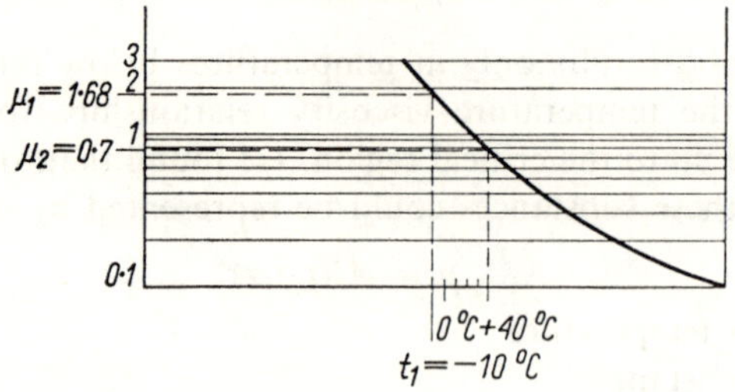

FIG. 4.17. Example of the application of the Lewis and Squires graph.

EXAMPLE 4.6. At temperature $t_1 = -10°C$ the viscosity (μ_1) of carbon tetrachloride is 1·68 cP. What is the viscosity (μ_2) of this compound at temperature $t_2 = 40°C$?

Solution

Fig. 4.17, which is an enlarged section of Fig. 4.16, shows the necessary construction lines (see above). The value of μ_2 read from the graph is 0·7 cP. The experimental value is 0·74 cP.

b. The Effects of Pressure on Liquid Viscosities

The calculation of the viscosities of liquids at high pressures from a knowledge of these viscosities at atmospheric pressure involves considerable difficulties. According to Panchenkov[278] liquid viscosities change at high pressures because of the reduction of the liquid volume and also because of the attendant changes which take place in the average intermolecular potential energy. He considers that any effects arising from changes in the effective volume of the molecules and of the coordination number may be neglected. Panchenkov supplemented his equation (4.19) with a complex equation, representing the dependence of the intermolecular "bond" energy on the pressure. The modified formula enables the liquid viscosity to be computed at high pressures, but it is too complicated to be useful for technological calculations.

Up to 2000 atm Andrade's equation[6] may be used

$$\frac{\mu_p}{\mu_1} = \left(\frac{v_1}{v_p}\right)^{\frac{1}{6}} \left(\frac{\beta_{1s}}{\beta_{ps}}\right)^{\frac{1}{2}} e^{\left(\frac{C(v_1-v_p)}{Tv_1v_p}\right)} \tag{4.41}$$

where μ_p and μ_1 are the viscosities at pressure p and at 1 atm,
v_p and v_1 are the specific volumes of the liquid at pressure p and at 1 atm,
β_{ps} and β_{1s} are the adiabatic compressibilities at pressure p and at 1 atm,
C is a constant,
and T is the absolute temperature.

The principal difficulty in using this formula is the lack of data concerning the adiabatic (i.e. isentropic) compressibility

$$\beta_s = -\frac{1}{V}\left(\frac{\partial V}{\partial p}\right)_s \tag{4.42}$$

In his original paper Andrade suggested that the ratio of the isothermal compressibilities could, with little error, be substituted for the ratio of the adiabatic compressibilities in Eq. (4.41). The isothermal compressibility is given by

$$\beta_T = -\frac{1}{V}\left(\frac{\partial V}{\partial p}\right)_T \tag{4.43}$$

Alternatively the value of β_s may be calculated from the molar velocity of sound U_u and this in turn may be predicted by summing up group contributions (Eqs. (1.56) and (1.58)). If the liquid densities ρ_1 and ρ_p (or v_1 and v_p) are known, the sound velocity u can be calculated from the molar velocity U_u, and β_s may be obtained from the equation

$$\beta_s = \frac{1}{u^2\rho} \tag{4.44}$$

(cf. Eq. (1.58)): Andrade's equation (4.41) may then be employed.

According to Gambill,[119] this is the most useful of all known methods for estimating the effect of pressure upon liquid viscosity. The method breaks down at pressures greater than 2000 atm, but this pressure range is of no practical importance.

Methods of extrapolating and interpolating viscosity data when viscosities are known at two or more different pressures, have already been discussed (cf. Fig. 4.7).

c. Viscosities of Saturated Liquids at Pressures Greater than Atmospheric

Reid and Sherwood[306] call attention to the fact that as the saturation curve is followed both temperature and pressure increase and that these factors have opposing effects: the increase in temperature lowers the liquid viscosity, whereas the increase in pressure raises it.

In this region use may be made of the alignment chart of Grunberg and Nissan.[137] This chart gives the viscosity ratio μ_p/μ_0 for gases, vapours or, at sufficiently elevated temperatures, liquids as a function of the reduced temperature T/T_c and reduced density ρ/ρ_c (Fig. 3.16). The authors of the nomograph have used it to estimate the viscosities of water and liquefied carbon dioxide at elevated temperatures and pressures, the error in the estimates amounting to 5–10%. This chart is only applicable at reduced densities less than 1·8. If $\rho/\rho_c > 1{\cdot}8$, the effect of temperature upon the viscosity of a boiling liquid may be estimated with sufficient accuracy from Eq. (4.13).

An extension of the method of Smith and Brown[335] (which is based on the law of corresponding states) may also be used to predict the viscosities of boiling liquids at temperatures between the normal boiling point and the critical temperature. The application of this method at pressures near atmospheric has already been described in section 6. Under the conditions considered in the present section Smith and Brown find that, in a given homologous series, the following quantities have the same value for each compound, when evaluated at the same reduced temperature:

Liquid Hydrocarbons

$$\frac{\mu p_r}{\sqrt{M}}\,\frac{p_c'}{p_c} \tag{4.45}$$

where μ is the viscosity of the boiling liquid at elevated pressure,
p_r is the reduced pressure,
p_c' is the critical pressure of pentane,
and p_c is the critical pressure of the liquid compound considered.

Alcohols and Formates

$$\frac{\mu p_r}{\sqrt{I}} \frac{p_c''}{p_c}$$

where I is the viscosity constant (Eq. (4.21) and Table 4.4),
and p_c'' is the critical pressure of the ethyl compound in the series.

The application of Smith and Brown's method is limited to those homologous series for which the viscosity–temperature relationship is accurately known for at least one compound over an extensive range of temperatures.

9. SELECTION OF THE BEST METHOD FOR ESTIMATING THE VISCOSITY OF A PURE LIQUID

The selection of the most appropriate method to use in any particular instance depends to a large extent on the amount of viscosity data and other information which is available, both for the liquid in question and for similar liquids.

a. Estimation of the Viscosity of a Liquid for which no Viscosity Data are Available

Very approximate predictions can be made using, for example, Souders' method (section 5a), Thomas' method (section 5b) or the theoretical formula of Eyring (Eq. (4.16c)).

If the viscosity of a liquid compound from the same homologous series is known over a wide range of temperatures and pressures, Smith and Brown's method (section 6) may be used.

If the viscosity is required at the normal boiling point, the rheochor method of viscosity calculation (section 5c) or the empirical formula of Arrhenius (Eq. (4.34)) may be used.

All the methods cited above are equally inaccurate, the average error of the estimates amounting to 20–30%.

Since no one method is notably more accurate than the others, it seems reasonable, in order to obtain as accurate a value of the liquid viscosity as possible, to employ several different methods, and to accept the arithmetic mean value.

b. Calculation of the Liquid Viscosity at Normal Pressure and a Temperature which Differs from that at which the Viscosity has been Determined Experimentally

If the viscosity is known at one temperature only Lewis and Squires's graph, shown in Fig. 4.14, may be employed. Viscosities estimated in this manner may be in error by 10 or 20%.

Another possible procedure is to apply Bachinskii's equation in the form of Eqs. (4.6) and (4.7). If both the liquid density–temperature relationship and the molal volume at the critical point are known, a knowledge of the viscosity at any one arbitrary temperature is sufficient to determine the value of the constant (the value of the other constant is calculated from Eq. (4.7)). This method is more accurate than that of Lewis and Squires. A very rapid method is to use the one viscosity which is available to locate on the point nomograph (Fig. 4.5) the point which corresponds to the given liquid; once this point is known the viscosity of this liquid may be read off the nomograph at all other temperatures.

When two experimental values of viscosity are available, μ_1 at temperature T_1 and μ_2 at temperature T_2, one of the graphical methods may be used. If an analytic expression is required, the exponential formula (Eq. (4.13)) containing two constants is most often recommended, though this equation is inaccurate for highly polar substances.

More accurate, but also more tedious, are the graphical methods: the Dühring plot (Ch. 1, section 7c) or Othmer plot (Fig. 4.8). In order to obtain good results using these methods it is necessary to select as reference compound a substance which has properties similar to those of the liquid under consideration.

The error in the estimates obtained by using these methods, does not usually exceed 1–2%.

If the liquid under consideration consists of highly polar molecules, Eq. (4.13) is not sufficiently accurate, and the use of Eq. (4.37) containing three constants is recommended. In order to apply this equation it is necessary to determine these three constants and for this purpose it is necessary to know the viscosity of the liquid at three different temperatures.

c. Estimation of Liquid Viscosity under High Pressures

It is recommended that Andrade's formula (Eq. (4.41)), should be used for this purpose. It should be noted, however, that the determination of the constant C in this equation involves a knowledge of the liquid viscosity μ_1 under normal pressure and also of the viscosity at one high pressure.

A simpler method of representing the effect of pressure upon viscosity, is to construct a Dühring plot. For this purpose the viscosity of the liquid under consideration is required at two pressures and complete viscosity–pressure data are required for some reference liquid. This method has been discussed in section 2 (see Fig. 4.7).

10. VISCOSITIES OF SOLUTIONS AND LIQUID MIXTURES

In technical design work, it is not the viscosities of pure compounds which are usually lacking but rather those of liquid mixtures, solutions of solid compounds (electrolytes or non-electrolytes) in liquids, of suspensions, emulsions, etc. The prediction of these viscosities from molecular considerations would involve an understanding of a separate and very complicated field of phenomena; present knowledge is inadequate to enable the viscosities of even the simpler mixtures of compounds to be obtained by direct calculation.

Only a few prediction methods are available for mixtures, and most of these have little or no theoretical basis.

a. Viscosities of Liquid Mixtures: Results of Experiments

The viscosities of aqueous solutions of non-electrolytes are found to increase with increase in concentration and to decrease with increase of temperature. Fig. 4.18 shows the variation of the viscosity of a 60% saccharose solution with temperature.[145]

The viscosities of solutions of solid organic compounds in organic solvents usually rise with increase in the solute concentration, though the reverse behaviour is also known (for instance the solution of *p*-nitrotoluene in ethanol). Figure 4.19 shows the viscosities of some typical solutions as functions of concentration.

The viscosities of aqueous solutions of electrolytes either rise with increase in the concentration of the dissolved compound, as for instance in the case of solutions of sodium hydroxide (Fig. 4.20) or else they vary in an irregular manner. In this latter case the viscosity may show a minimum at some point of the concentration range. This occurs for solutions of some ammonium salts (Fig. 4.21).[145]

Mixtures of organic liquids may be divided, with respect to the behaviour of their viscosity, into several types.

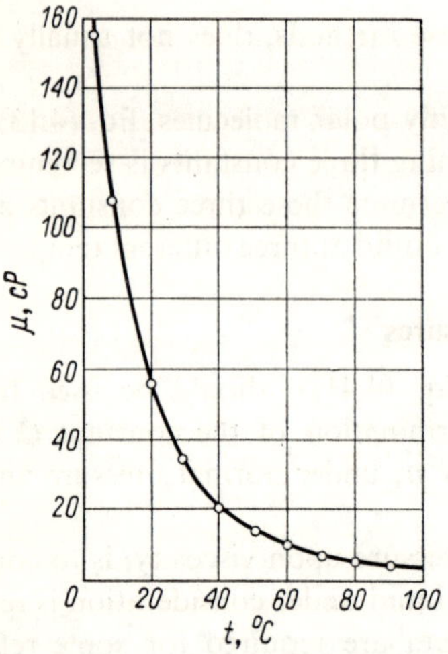

FIG. 4.18. Viscosity of an aqueous saccharose solution (60%) vs. temperature (after Bingham and Jacksos).

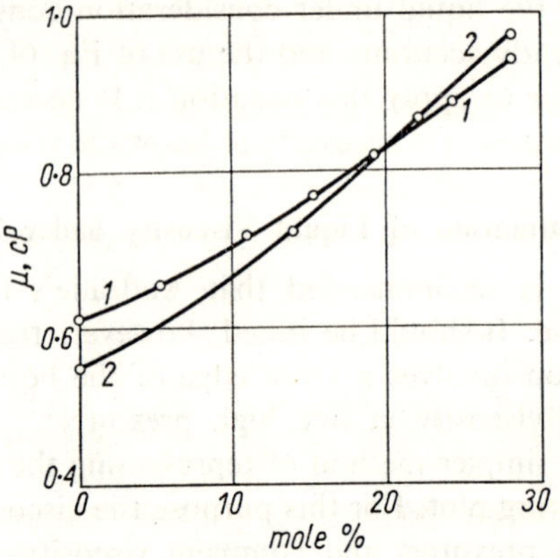

FIG. 4.19. Effect of concentration on the viscosities of solutions; *1*—naphthalene in benzene, *2*—diphenyl in toluene (after Kendall and Monroe[194]).

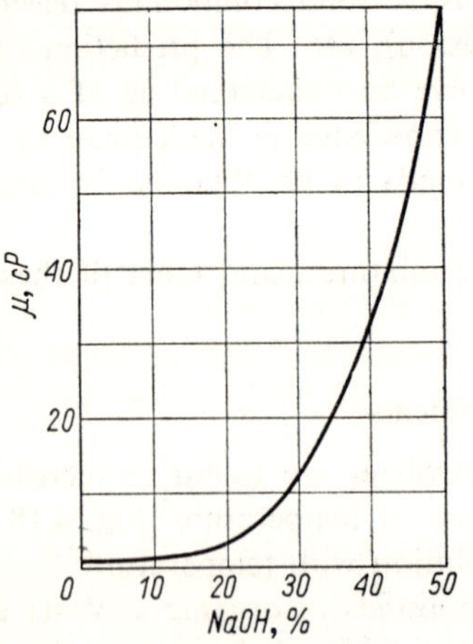

FIG. 4.20. Effect of concentration on the viscosities of aqueous sodium hydroxide solutions.

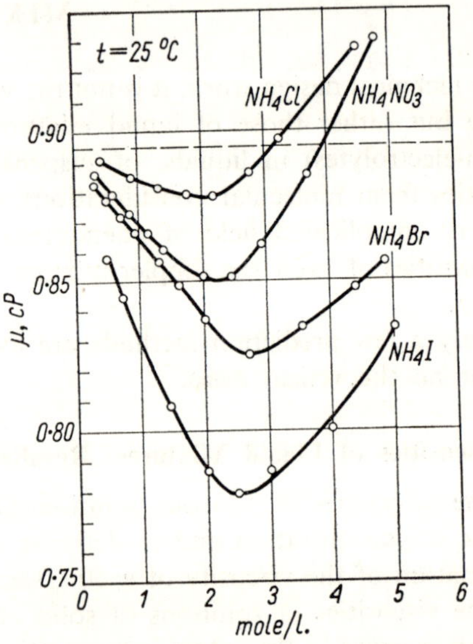

FIG. 4.21. Viscosities of aqueous solutions of ammonium salts as a function of concentration (after Getman).

Kendall and Monroe[194] have investigated the viscosities of *ideal mixtures* of organic liquids, i.e. mixtures for which neither heat effects nor changes of volume occur when they are made up from their pure components. The results of these investigations are presented in Fig. 4.22. Although viscosity does not vary linearly with the composition in these cases,

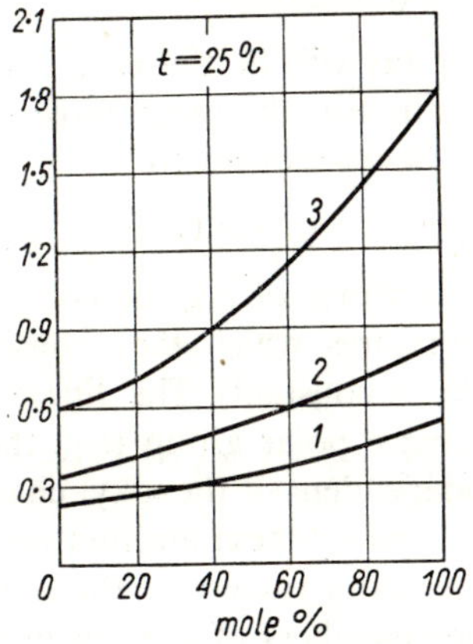

FIG. 4.22. Relationships between the viscosities of perfect mixtures of organic liquids and their compositions (mole % of the first component) at 25°C; *1*—toluene–diethyl ether, *2*—decane–hexane, *3*—nitrobenzene–benzene.

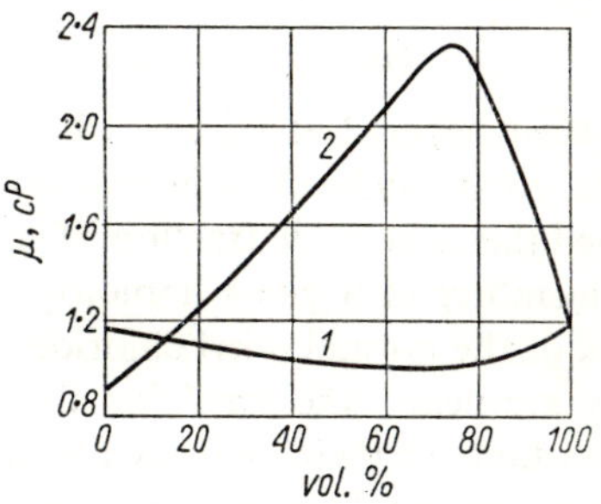

FIG. 4.23. Relationships between viscosities of liquid mixtures and their compositions (vol. % of first component) at 25°C; *1*—ethanol–phenetole (Baker), *2*—acetic acid–water (Dunstan).

the deviations from linearity are much smaller than those which arise in many non-ideal liquid mixtures. These latter deviations may be either positive or negative, as illustrated in Fig. 4.23.[145] The viscosity–composition curve for the ethanol–phenetole system (curve *1*) shows a minimum, while that for the acetic acid–water system exhibits a maximum (curve *2*).

b. Influence of the Molecular Properties of the Components of a Mixture upon its Viscosity

The viscosity of a liquid mixture depends on the molecular properties of its components, i.e. as discussed in section 3, on properties such as molecular volume v_m, shape factor ϕ, molecular mass m, polarizability α, and dipole moment $\bar{\mu}$.

The relationship developed by Lutskii[230] for a pure compound has the form

$$(\mu)_{p,T} = \mathrm{m}^{0\cdot5}\alpha^{1\cdot5}(A+B\bar{\mu}) \tag{4.46}$$

where A and B are coefficients, the value of which depends on the shape of the molecules (cf. Eq. (4.11) and Fig. 4.12). For binary mixtures (subscripts 1 and 2 denote the components) Lutskii and Obukhova[233] assume that

$$(\mu_{1,2})_{p,T} = m_{1,2}^{0\cdot5}\alpha_{1,2}^{1\cdot5}(A_{1,2}+B_{1,2}\,\mu_{1,2}) \tag{4.47}$$

The quantities $m_{1,2}$, $\alpha_{1,2}$ and $\mu_{1,2}$ are taken to be proportional to the mole fractions N_1 and N_2 $(=1-N_1)$ of the components so

$$(\mu_{1,2})_{p,T} = [N_1 m_1+(1-N_1)m_2]^{0\cdot5}[N_1\alpha_1+(1-N_1)\alpha_2]^{1\cdot5}\times[N_1(A_1'+B_1'\bar{\mu}_1)+ +(1-N_1)(A_2'+B_2'\bar{\mu}_2)] \tag{4.48}$$

Equation (4.48) may be considerably simplified in certain cases. For example, if mixtures are considered for which the components have the same shape factor, so that $\phi_1 = \phi_2$ then $A_1 = A_2$ and $B_1 = B_2$. If furthermore the components have the same dipole moment ($\bar{\mu}_1 = \bar{\mu}_2$) and the same polarizability ($\alpha_1 = \alpha_2$), Eq. (4.48) reduces to

$$(\mu_{1,2})_{p,T} = (N_1 m_1 + (1-N_1)m_2)^{0.5} \times (\text{constant}) \tag{4.49}$$

If the mixtures considered have the same shape factor (as will be the case if they have some dominant group, such as a benzene ring, in common), the same molecular mass ($m_1 = m_2$) and the same polarizability ($\alpha_1 = \alpha_2$), Eq. (4.48) reduces to

$$(\mu_{1,2})_{p,T} = (\text{constant})'(N_1\bar{\mu}_1 + (1-N_1)\bar{\mu}_2) + (\text{constant})'' \tag{4.50}$$

In their paper, Lutskii and Obukhova considered mixtures of a given component (1) with the compounds of a homologous series. The mixtures which they considered all contained the same mole fraction of component 1 (so N_1 = constant). The dipole moments of the members of a given homologous series containing a polar group (e.g. the normal alcohols or the normal alkyl chlorides) are virtually independent of the alkyl chain length, i.e. they are nearly the same for all members of the series so these authors assume that $\bar{\mu}_2$ = constant. In their analysis Lutskii and Obukhova also assume that the shape factor ϕ_2 may be taken to be the same for each of the higher members of a given homologous series. This assumption can hardly be rigidly valid though it may be justifiable in the present context: it implies that within a given homologous series, chain length is comparatively unimportant in determining viscosity behaviour except in so far as it dictates M and R_D. On writing $\bar{\mu}_2$ = constant, ϕ_2 = constant and N_1 = constant equation (4.48) reduces to

$$(\mu_{1,2})_{p,T,1,N_1,\phi_2,\bar{\mu}_2} = \text{constant} \times M_{1,2}^{0.5} R_{D1,2}^{1.5} \tag{4.51}$$

where $M_{1,2} = N_1 M_1 + N_2 M_2$ and $R_{D1,2} = N_1 R_{D_1} + N_2 R_{D_2}$. M is here the molecular weight. The molar refraction R_D has been used in Eq. (4.51) rather than α (to which R_D is roughly proportional) since it is a more readily available quantity.

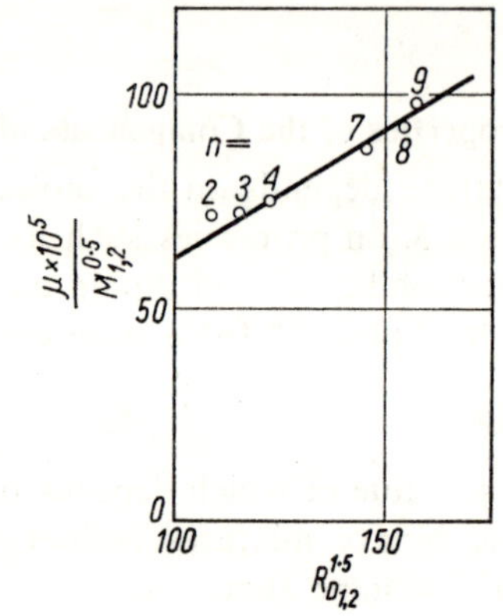

FIG. 4.24. ($\mu \times 10^5/M_{1,2}^{0.5}$) vs. $R_{D1,2}^{1.5}$ for mixtures of benzene with alcohols of the series n-$C_nH_{2n+1}OH$ at t = 25°C, N_2 = 0·2 (after Lutskii and Obukhova[233]).

According to the arguments given above, Eq. (4.51) should be applicable to a set of mixtures containing a given mole fraction N_1 of a given component 1 and mole fractions $(1-N_1)$ of second components drawn from a specified homologous series. Figure 4.24 shows a test of this relationship for binary mixtures composed of benzene and one of the alcohols of the series n-$C_nH_{2n+1}OH$ ($n = 2, 3, 4, 7, 8, 9$). The quantity $(\mu_{1,2} \times 10^5/M_{1,2}^{0\cdot5})$ is shown plotted as a function of $R_{D_{1,2}}^{1\cdot5}$. It is seen that the points fall close to a straight line.

Lutskii and Obukhova also considered mixtures of a given component (1) with a series of substances each member of which had the same shape factor but for which the molecular mass, polarizability and dipole moments were all different. They found that if for mixtures containing the same mole fraction of component 1, they plotted the group $\left(\frac{\mu_{1,2}}{M_{1,2}^{0\cdot5} R_{D_{1,2}}^{0\cdot5}}\right)$ as a function of the dipole moment $\bar{\mu}_2$ of component 2 most of the points lay on a single straight line. This finding is exemplified for mixtures of acetone with compounds of the type C_6H_5X in Fig. 4.25. The substances considered in this figure would be expected to

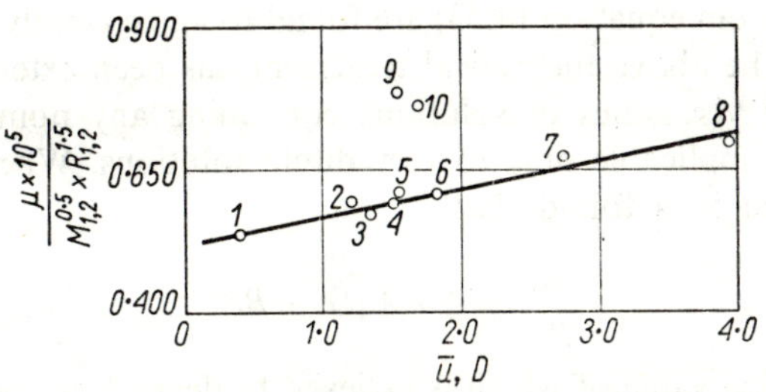

FIG. 4.25. The viscosities of acetone–C_6H_5X mixtures containing 80 mole % acetone shown as a function of the dipole moment $\bar{\mu}$(D) of the C_6H_5X component (X denotes: *1*—C_2H_5, *2*—OC_2H_5, *3*—I, *4*—Br, *5*—Cl, *6*—$COOC_2H_5$, *7*—CHO, *8*—NO_2, *9*—NH_2, *10*—CH_2OH; $t = 25°C$ and $N_2 = 0\cdot2$) (after Lutskii and Obukhova[233]).

have similar molecular shapes since they all have a dominant group (the benzene ring) in common. The only two points which are badly off line are for substances with an NH_2 group or an OH group. Since pure substances containing these groups do not correlate well (section 3), these discrepancies might have been anticipated. In conclusion it would appear from the above examples that the correlations proposed by Lutskii and Obukhova are reasonably accurate and may be of value for prediction purposes.

c. Estimation of the Viscosities of Solutions of Solid Compounds in Liquids

Viscosities of Electrolyte Solutions: Theoretical Investigations

Falkenhagen[95] was the first worker to develop a successful theory for the viscosities of strong binary electrolytes.† For very dilute solutions he deduced that

$$\frac{\mu_m}{\mu_0} = 1 + A\sqrt{c} \tag{4.52}$$

where μ_m is the viscosity of the solution,
μ_0 is the viscosity of the solvent,
c is the concentration,
and A is a constant (always greater than zero), the value of which depends on the electrostatic forces between the ions.

† i.e. electrolytes which dissociate to give two ions per molecule.

An empirical equation of this form (in which A was an arbitrary constant) was proposed by Jones and Dole[179] at about the same time.

Falkenhagen's theory enables the value of the constant A to be calculated. The equation for A, although complicated in the general case, assumes a fairly simple form if the ionic valencies are equal (so that $z_1 = z_2 = z$) as also are the mobilities of the ions at infinite dilution (so that $l_1^\infty = l_2^\infty = l^\infty$). In this case

$$A = \frac{0{\cdot}517\, z^2}{l^\infty \mu_0 \sqrt{\varepsilon_0 T}} \tag{4.53}$$

where ε_0 is the dielectric constant of the solvent,
T is the absolute temperature,
and l^∞ is the ionic mobility at infinite dilution.

Values of A calculated from equation (4.53) are found to agree closely with values obtained from viscosity data.† The above theoretical treatment has been extended by Onsager and Fuoss[265] to include the viscosities of solutions containing any number of ionic species.

Equation (4.52) is applicable only to very dilute solutions. When more concentrated solutions are considered it is found that[179]

$$\frac{\mu_m}{\mu_0} = 1 + A\sqrt{c} + Bc \tag{4.54}$$

where B is a constant, the value of which is believed to depend on the interaction between the ions and molecules of the solvent.

In the case of still higher concentrations (up to one mole per litre), the three constant equation

$$\frac{\mu_m}{\mu_0} = 1 + A\sqrt{c} + Bc + Dc^2 \tag{4.55}$$

is required to represent viscosity data. Illustrative values of the constants in this equation are listed in Table 4.7. It may be seen from this table that the constant B may be greater or less than zero; in the latter case the viscosity of the solution can decrease with increase in the concentration of the dissolved compound.

TABLE 4.7. CONSTANTS IN EQ. (4.55) FOR THE VISCOSITIES OF AQUEOUS SOLUTIONS AT 25°C (AFTER KAMINSKY[187])

Compound	Valencies	A (mole/l.)$^{-1/2}$	B (mole/l.)$^{-1}$	D (mole/l.)$^{-2}$
NaCl	1–1	0·0062	0·0793	0·0080
KCl	1–1	0·0052	−0·0140	0·001
KI	1–1	0·0047	−0·0755	0
K_2SO_4	1–2	0·0135	0·1937	0·032
$MgCl_2$	2–1	0·0165	0·3712	0
$MgSO_4$	2–2	0·0230	0·5937	0·02
$CeCl_3$	3–1	0·0310	0·0555	0·11

Values of the constant B may be calculated additively by summing the contributions of the cation and the anion.[68] Since the mobilities and the temperature coefficients of the

† See *The Physical Chemistry of Electrolytic Solutions* by Harned and Owen, 3rd ed., p. 241, Reinhold, 1957.

mobilities of the K^+ and Cl^- ions are similar, the values of B for these ions would also be expected to be similar.[139] On the basis of this assumption, the following table of ionic contributions has been compiled (Table 4.8).

TABLE 4.8. CONSTANT B IN EQ. (4.54) FOR AQUEOUS SOLUTIONS OF ELECTROLYTES AT 25°C (AFTER KAMINSKY[188])

Cation	B	Cation	B	Anion	B
Li^+	0·1495	Be^{2+}	0·3923	Cl^-	−0·0070
Na^+	0·0863	Mg^{2+}	0·3852	Br^-	−0·042
K^+	−0·0070	Ca^{2+}	0·285	I^-	−0·0685
Rb^+	−0·030	Sr^{2+}	0·265	OH^-	0·12
Cs^+	−0·045	Ba^{2+}	0·220	ClO_3^-	0·024
NH_4^+	−0·0074	Fe^{2+}	0·416	NO_3^-	−0·046
Ag^+	0·091	La^{3+}	0·588	MnO_4^-	−0·059
H^+	0·069	Ce^{3+}	0·5765	SO_4^{2-}	0·2085

Equations (4.52) to (4.55) should be applicable not only to solutions of electrolytes in water but also to solutions of electrolytes in non aqueous solvents such as methanol, ethanol or nitrobenzene. The limited evidence so far available suggests that this is so.

EXAMPLE 4.7. Using the general method of calculating the values of additive quantities, estimate the value of constant B for magnesium sulphate at 25°C, using: (a) the data in Table 4.7 and (b) Table 4.8.

Solution

(a) $B_{MgSO_4} = B_{MgCl_2}+B_{K_2SO_4}-2B_{KCl} = 0{\cdot}3712+0{\cdot}1937-2\times 0{\cdot}0140 = 0{\cdot}5929$

(b) $B_{MgSO_4} = B_{Mg^{2+}}+B_{SO_4^{2-}} = 0{\cdot}3852+0{\cdot}2085 = 0{\cdot}5937$

The value determined experimentally

$B_{MgSO_4} = 0{\cdot}5937$

Error in the estimate: (a) −0·13%,
(b) 0%.

Empirical Methods for Estimating the Viscosities of Aqueous Solutions of Electrolytes and of Solid Non-electrolytes

Empirical methods of predicting the viscosities of aqueous solutions of salts or solid non-electrolytes when no viscosity data are already available are very unreliable. Mention may be made of a "rough and ready" rule proposed by Kern.[195] When designing heat exchangers, he suggests that, in order to prevent underdesign, a viscosity twice that of water be used for "salts in water where the concentration does not exceed 30 per cent and where it is known that a syrup type of solution does not result. ... A solution of sodium hydroxide in water even under very low concentrations should be considered syrupy and cannot be estimated". It is an implication of this rule that

$$(\mu_{solution} < 2\mu_{water})_{c<30\%} \tag{4.56}$$

However, this formula is very crude and can by no means be recommended. Anderson[8] has attempted to formulate more accurate relationships.

If a limited number of viscosity determinations have been made on the aqueous system of interest, it is possible to use empirical methods to predict viscosities under conditions for which they have not been determined, with rather greater accuracy. Davis[74], for example, has developed a curious, though convenient, graphical method for extending viscosity data for solutions of hydroxides and inorganic salts. His method is based on the finding that a plot of log(sin 10μ) against salt concentration is linear for solutions of a substantial number of inorganic electrolytes in water (see Fig. 4.26); μ is expressed in cP and the concentration in moles/1000 g water. This fact greatly facilitates the interpolation and extrapolation of viscosity data for solutions of electrolytes, since for the determination of the entire isotherm it is necessary to know the viscosity of the solution at only two concentrations at a given temperature. To avoid the troublesome procedure of finding the value of the logarithm of a trigonometric sine function, Davis has devised a special

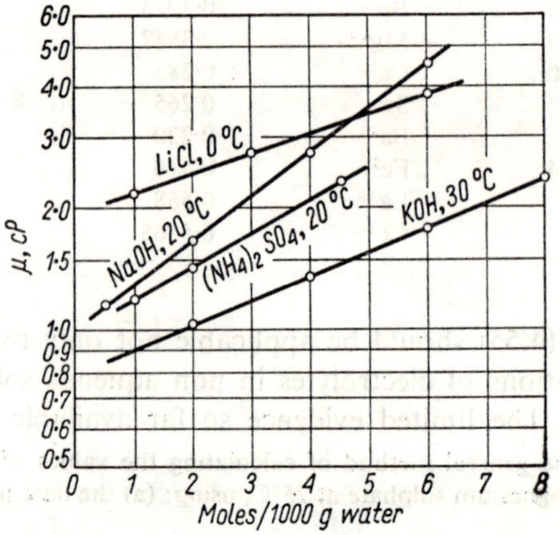

FIG. 4.26. Davis' viscosity/concentration plots.

type of graph paper with a net suitable for this purpose. Such a net with a few straight lines for the solutions of some electrolytes is presented as an example in Fig. 4.26. To construct a similar line for any other salt solution, values of the viscosity of the solution at two different concentrations at the same temperature are required.

For the purpose of interpolating or extrapolating viscosity data for electrolyte solutions as a function of temperature the Dühring plot is very convenient. This method, which has been described in section 2 of this chapter, requires complete viscosity data for a reference liquid, and a knowledge of the viscosity of the solution considered at two or more different temperatures. Examples of Dühring plots for salt solutions of specified composition are given in Fig. 4.27 for 10% and 25% solutions of sodium nitrate, and in Fig. 4.28, which has been constructed for a series of aqueous saccharose solutions.

If solvation or association occur, straight lines will not necessarily be obtained on the Dühring plot. Cornelissen and Waterman[67] have shown that the effect of temperature on the kinematic viscosity of solutions of this type is well represented by the equation

$$\log \nu = A/T^x + B \tag{4.57}$$

where A, B and x are constants for a given solution. They find for example that plots of log ν vs. A/T^3 are linear for sulphuric acid solutions. These authors also applied their equation successfully to many other mixtures and single liquids, both associated and non-associated.

The viscosities of aqueous electrolyte solutions containing two salts (or hydroxides) which do not react with each other, may be estimated using Zdanovskii's formula[407] (Eq. 4.58). This formula has been checked on the systems $NaCl–NH_4Cl–H_2O$, $NaCl–KCl–H_2O$, $NaOH–Na_2CO_3–H_2O$, and $NaCl–MgCl_2–H_2O$, the ternary mixtures being made up by the author by mixing two binary solutions, each of them containing one of the salts dissolved in water. According to Zdanovskii the viscosity μ_m of a ternary mixture can be calculated, if the viscosities μ_1 and μ_2 of the binary solutions from which it was constituted as well as their densities ρ_1 and ρ_2 are known. For this purpose he proposes that the equation

$$\mu_m = \frac{x_1\rho_1 + x_2\rho_2}{x_1(\rho_1/\mu_1) + x_2(\rho_2/\mu_2)} \tag{4.58}$$

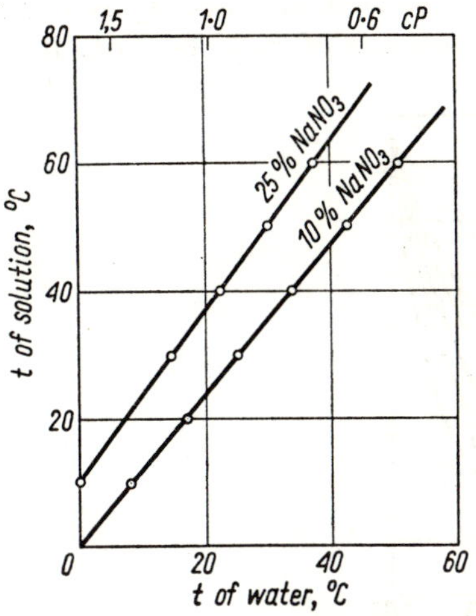

FIG. 4.27. Dühring plots giving the viscosities of aqueous sodium nitrate solutions as a function of temperature (water as reference liquid).

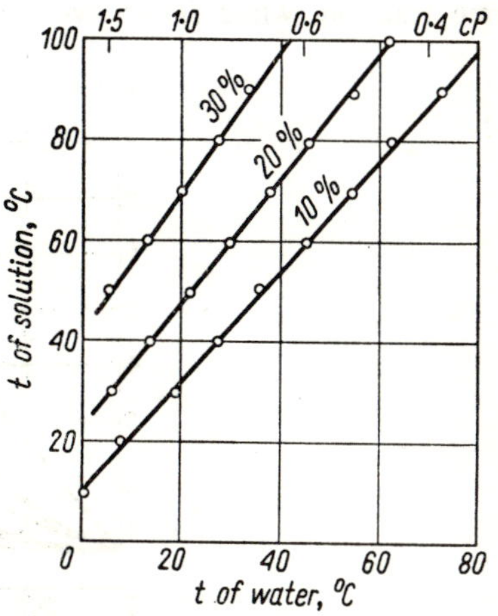

FIG. 4.28. Dühring plots giving the viscosities of aqueous saccharose solutions as a function of temperature (water as reference liquid).

should be used, in which x_1 and x_2 denote the volume fractions of solutions 1 and 2 from which the ternary mixture was constituted. Equation (4.58) may be rewritten in terms of the specific volumes v_1 and v_2 of the solutions 1 and 2 giving

$$\mu_m = \frac{v_1 + v_2}{(v_1/\mu_2) + (v_2/\mu_1)} \tag{4.59}$$

The solutions investigated behaved approximately as if they were ideal mixtures. The calculated viscosities were found to be in good agreement with the experimental data; the average error of these estimates was of the order of 0·5%, the maximum error did not exceed 1%. A subsequent investigation of the viscosities of solutions containing water, cadmium chloride and one of the compounds HCl, LiCl, NaCl, NH_4Cl or KCl showed[110] that the Zdanovskii formula gave deviations of 6–9%. The Kendall and Monroe formula (Eq. (4.60)), which is discussed below, gave deviations of 10–12%, and the Arrhenius equation gave deviations of 8–10%.

d. Viscosities of Mixtures Made up from Two Liquids

This class of mixture is often of great importance in technological calculations. As has already been mentioned (section 10a), the law of additivity is inapplicable even in the case of ideal mixtures. In illustration of this, mention may be made of the work of Kendall and Monroe.[194] These authors investigated the viscosities of mixtures composed of benzene and benzyl benzoate, which are approximately ideal. They compared the experimentally determined viscosities with values calculated by assuming:

A a linear dependence of the viscosity μ on composition,

B a linear dependence of the fluidity φ $(=1/\mu)$ on composition,

C a linear dependence of the logarithm of viscosity, $\log \mu$, on the composition of the mixture.

The experimental results are shown in Fig. 4.29. Three curves are shown. The first curve (*I*) shows viscosity as a function of mass fraction, curve *II* shows viscosity as a function of volume fraction, while *III* shows viscosity as a function of mole fraction. It is seen

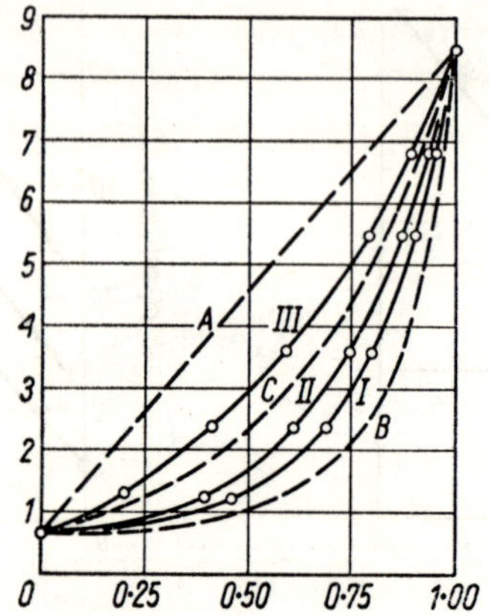

FIG. 4.29. Dependence of the viscosities of benzyl benzoate–benzene mixtures on composition (dashed lines calculated, continuous lines experimental data) (after Hatschek[145]).

from the graph that none of these curves approaches the straight line *A* which corresponds to the law of additivity. On the basis of their investigations Kendall and Monroe found that an additivity rule could be applied to the cube roots of the viscosities of the components. Their empirical equation

$$\mu_m^{1/3} = x_1 \mu_1^{1/3} + x_2 \mu_2^{1/3} \tag{4.60}$$

(where x_1, x_2 denote the mole fractions in the mixture) holds well in many cases of non-associated non-polar similar liquids. A different, but apparently equally accurate, empirical equation has been recommended by Arrhenius,[15] who assumed that the logarithm of the viscosity was an additive quantity so that

$$\log \mu_m = x_1 \log \mu_1 + x_2 \log \mu_2 \tag{4.61}$$

where x_1 and x_2 are the mole fractions of the components. Arrhenius originally recommended the use of mass fractions for x_1 and x_2, but Kendall has shown that the introduction of mole fractions gives a better agreement with experiment. Equation (4.61) is found to fit the data for the phenetole–diphenyl ether system with an average error of 0·2%. Pospekhov[294] has checked the formula on several other binary systems and finds that the predicted viscosities for the mixtures agree quite well with the experimentally determined ones.

For the empirical equations (4.60) and (4.61), to give good results it is necessary either for the liquids constituting the mixture to be similar in structure, non-polar and non-associated or for one component to be present in dominant amount in the mixture. Furthermore, the difference between the viscosities of the pure components, making up the mixture, should be small (below 15 cP).

Fairly good results can also be achieved by applying the Zdanovskii equations (4.58) and (4.59). These may be used to estimate the viscosities of binary liquid mixtures, the subscripts 1 and 2 in this case referring to pure components 1 and 2. For systems such as ethyl ether–benzene, nitrobenzene–benzene and methyl propyl ketone–acetone, the Zdanovskii formula is found to express the viscosity–composition relationship much more closely than does the Arrhenius formula (Eq. (4.61)).[408] On the other hand, in several other cases Deich[76] found that the Arrhenius formula gives good results, for instance for the benzene–toluene system the maximum error amounts to 0·6%, for benzene–*p*-xylene 0·7%, benzene–cumene 0·3% and toluene–ethylbenzene 0·1%.

The utility of the equations considered above lies in the fact that they may be used to predict the viscosities of organic mixtures from a knowledge only of the viscosities μ_1 and μ_2 of the pure components.† More accurate estimates may be made if at least one viscosity determination has been made on a mixture of components 1 and 2. Such a determination enables the constant in an equation such as that of Tamura and Kuvata (4.66) to be determined. Other more accurate equations exist but these contain more constants, and this requires more extensive viscosity data for the mixtures. For example, very good interpolations may be made using the transformed Panchenkov equation.[79,278,279] In contrast to most of the formulae considered in this section, the Panchenkov equation has been derived on the basis of theoretical considerations of the mechanism of flow of a viscous liquid (section 4). According to his equation

$$\mu_m = A_s \rho^{4/3} T^{1/2} (e^{\varepsilon_m/RT} - 1) \tag{4.62}$$

where μ_m is the viscosity of the liquid mixture,
A_s is a coefficient,
ρ is the density of the mixture,
T is the absolute temperature,
ε_m is the energy of the "bonds" between the molecules of the mixture,
and R is the gas constant.

The Panchenkov equation contains two quantities A_s and ε_m which must be evaluated from viscosity data. This requires knowledge of the viscosities of at least two mixtures. The average error of estimates made using Eq. (4.62) is smaller than 1%, and the maximum errors amount to only a few per cent.

There is a superficial resemblance between the behaviour of vapour–liquid saturation pressures and fluidities. For example, when considering the maxima or minima which appear on the viscosity–composition curves for multi-component systems, one cannot help thinking of an analogy with the phenomena of azeotropy. This analogy has been analysed by Faust[98] in terms of the results obtained by Zawidzki,[406] who discovered a relationship between the phenomena considered. Persuing this analogy further, Reik[307] has recently introduced the concept of a fictitious *partial viscosity*. According to this author the partial viscosity for component i in a mixture is given by

$$\mu_i = x_i \mu_{i0} / \gamma_i^{1/3} \tag{4.63}$$

† Many of the empirical equations proposed by various authors for estimating the viscosities of mixtures have been reviewed by Partington[282] and by Zdanovskii.[407]

where x_i is the mole fraction of component i in the liquid mixture,
μ_{i0} is the viscosity of pure component i,
and γ_i is the activity coefficient of component i in the mixture.

This formula is analogous to the equation for the partial pressure of a constituent of a liquid mixture. According to Reik, the viscosity of a mixture may be estimated as the sum of the partial viscosities. His equation is, however, difficult to apply because the values of the activity coefficients of the constituents of the mixture must be known.

One-constant Equations for Viscosity as a Function of Composition

Both the Panchenkov and the Reik equations are rather cumbersome for practical use in technological calculations and the usual procedure is to employ an equation with only one constant, which can be evaluated from a knowledge of the viscosity of one mixture only. Such formulae have been devised by several authors. For example Grunberg and Nissan[136] recommend the equation

$$\log\mu_m = x_1\log\mu_1+x_2\log\mu_2+x_1x_2d \tag{4.64}$$

where μ_m, μ_1 and μ_2 are the viscosities of the mixture and of components 1, 2, respectively,
x_1 and x_2 are mole fractions,
and d is an interaction coefficient which is dependent on temperature but independent of composition.

Hind, McLaughlin and Ubbelohde[151] investigated the viscosity–temperature relationship for camphor–pyrene mixtures, and applied Eq. (4.13):

$$\mu_m = A\,e^{B/RT}$$

They found that the quantity B is approximately independent of the composition of the mixture in contrast to A, which varies considerably with the mixture composition.

Van der Wyk[378] has obtained the relationship

$$\ln\mu_m = x_1^2\ln\frac{\mu_1\mu_2}{\mu_{1,2}^2}+2x_1\ln\frac{\mu_{1,2}}{\mu_2}+\ln\mu_2 \tag{4.65}$$

where $\mu_{1,2}$ is a coefficient of interaction, and the other symbols are the same as in Eq. (4.64).

Tamura and Kurata[357] have proposed the equation

$$\log\mu_m = x_1\mu_1\phi_1+x_2\mu_2\phi_2+2\mu_{1,2}(x_1x_2\phi_1\phi_2)^{0\cdot5} \tag{4.66}$$

In this equation the notation is as in Eqs. (4.63) and (4.64) except that ϕ_1 and ϕ_2 are the volume fractions of the components 1 and 2 in the mixture. The interaction coefficient $\mu_{1,2}$ is constant at a fixed temperature; if, however, the temperature is changed, this coefficient varies in the same way as the viscosity of a pure compound (i.e. approximately in accordance with Eq. (4.13)). The value of the constant $\mu_{1,2}$ may be determined from one measured value of μ_m, i.e. the viscosity of one mixture. If more than one value of μ_m is known, $\mu_{1,2}$ should be calculated from each value and the average value of $\mu_{1,2}$ should be used. Eq. (4.66) provides a surprisingly accurate representation of viscosities as a function of composition. The viscosities of nonpolar–polar, polar–polar, and nonpolar–nonpolar mixtures (including mixtures showing maxima or minima on the viscosity–composition curves) are represented with errors of less than 5–7%.

Prediction of the Viscosities of Completely Miscible Liquid Mixtures from Molecular Structure

If neither the viscosities of the pure components nor the viscosity of any mixture are known, the methods based on the principle of additivity, which have been described in sections 5 a, b, c may be employed.

Souders' equation (4.21) has been modified by Lima,[225] so as to be suitable for the estimation of the viscosities of binary liquid mixtures. The modified equation is

$$\log(\log 10\mu_m) = \rho_m \frac{x_1 I_1 + x_2 I_2}{x_1 M_1 + x_2 M_2} - 2{\cdot}9 \tag{4.67}$$

where μ_m is the viscosity of the mixture,
ρ_m is the density of the mixture,
x_1 and x_2 are the mole fractions of the components 1 and 2,
I_1 and I_2 are the viscosity constants of these components, calculated either on the basis of the contributions given in Table 4.4 or from the nomograph in Fig. 4.13,
and M_1 and M_2 are the molecular weights of the components.

This formula may be applied to mixtures of nonpolar or slightly polar liquids which are similar in structure. Viscosities estimated in this way are only approximate, the average error being of the order of 12%.

An alternative method for estimating the viscosity of a liquid mixture with neither μ_1, μ_2 nor μ_m known, for which good results have been claimed, has been suggested by Shukla and Bhatnagar.[327] These authors have found that the molar average of the rheochors of the pure components of a mixture can be used to estimate the viscosity of this mixture using the equation

$$\mu_m^{1/8} = \frac{(x_1 R_{ch1} + x_2 R_{ch2} + \dots + x_n R_{chn})\rho_m}{M_m} \tag{4.68}$$

where μ_m is the viscosity of the mixture,
$x_1, x_2 \dots x_n$ are the mole fractions of the components,
$R_{ch1}, R_{ch2} \dots R_{chn}$ are the rheochors of the pure components,
ρ_m is the density of the mixture,
and M_m is the mean molecular weight of the mixture.

The values of the rheochors $R_{ch1}, R_{ch2} \dots$ for the pure components may be calculated by summing group contributions (section 4d). The authors have checked the formula on a dozen or so mixtures including, among others, toluene–acetic acid, benzene–acetic acid, chloroform–acetic acid, toluene–acetone, toluene–methanol–benzene, ethanol–methanol–acetic acid and acetone–ethanol–methanol. The average error in the estimated viscosity was found to be 0·5–1%; the maximum error was 4·4%.

All the formulae for estimating the viscosities of liquid mixtures, so far considered in this section, are empirical in character. This accounts for their large number and also for the fact that a given formula may only be applicable over a rather restricted range of conditions. This situation has been explained to some extent by McAllister,[243] who based his arguments on Eyring's viscosity model. McAllister considered the factors which would be expected to influence the energy of activation for the viscous flow of a two component liquid mixture. The magnitude of this energy of activation would be expected to depend on the environment in which the molecule under consideration moves. In the simplest case this molecule may move between two other molecules (like or unlike); this is the basis of the so called *three-body model.* As a refinement, each molecule may be considered to move between three other molecules, as in the *four-body model.* McAllister came to the conclusion that the energy of activation for the viscous flow of a binary liquid mixture should be calculable using the principle of additivity (concentration being expressed in mole fractions x_1 and x_2). On the basis of the "three-body model" he arrived at the following formula for the kinematic viscosity ν_m of a binary mixture:

$$\begin{aligned}\ln\nu_m = {} & x_1^3\ln\nu_1+3x_1^2x_2\ln\nu_{1,2}+3x_1x_2^2\ln\nu_{2,1}+x_2^3\ln\nu_2- \\ & -\ln[x_1+x_2M_2/M_1]+3x_1^2x_2\ln[(2+M_2/M_1)/3]+ \\ & +3x_1x_2^2\ln[(1+2M_2/M_1)/3]+x_2^3\ln(M_2/M_1)\end{aligned} \tag{4.69}$$

If the ratio of molecular weights $M_1/M_2 = 1$, the last four terms vanish. Some empirical formulae, e.g. Eq. (4.61), represent particular cases of Eq. (4.69). This equation contains two constants $\nu_{1,2}$ and $\nu_{2,1}$, which can be determined if the viscosities of mixtures of two

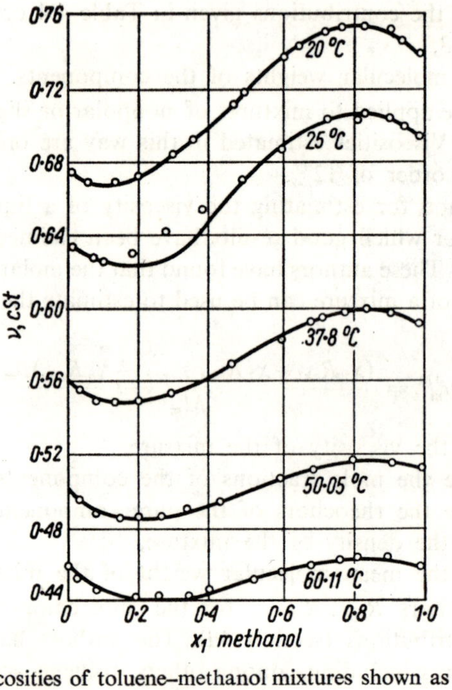

FIG. 4.30. Kinematic viscosities of toluene–methanol mixtures shown as functions of composition (after McAllister[243]).

different compositions are known. Equation (4.69) provides a good fit for the viscosity–composition curves for numerous binary mixtures, e.g. methanol–toluene, benzene–toluene and cyclohexane–heptane. Fig. 4.30 shows the kinematic viscosity ν for the system toluene–methanol as a function of composition. The experimental results are marked as open circles and the variation of ν with composition as predicted by Eq. (4.69) is shown by a continuous line for each temperature. The agreement between the estimated and the experimental results is very good.

McAllister also showed that each of the quantities linear for ν_1, $\nu_{1,2}$, $\nu_{2,1}$ and ν_2 should vary linearly with the reciprocal of the absolute temperature. If values of these quantities are known at two temperatures it should therefore be possible to read off values for other temperatures from a suitable graph.

In some cases Eq. (4.69) is an over simplification: thus in the case of acetone–water mixtures it is necessary to employ a more complicated formula derived from the "four-body model" and containing three constants: $\nu_{1,1,1,2}$, $\nu_{1,1,2,2}$ and $\nu_{2,2,2,1}$.

e. Estimation of the Viscosities of Heterogenous Mixtures (Suspensions, Colloidal Solutions, Emulsions)

Einstein[83] was the first to formulate the relationship between the viscosity μ of a system consisting of perfectly rigid spheres suspended in liquid and the volumetric concentration φ of the solid phase. On theoretical grounds he arrived at the relationship

$$\mu_m = \mu_1(1+2{\cdot}5\varphi) \tag{4.70}$$

where μ_1 denotes the viscosity of the pure liquid. This formula was subsequently checked experimentally and the coefficient of φ was found to be 2·9 rather than 2·5.

Later investigations have shown that, owing to the occurrence of complicating phenomena such as the formation of adsorbed layers on the surface of the particles of the suspended phase and the existence of electric charges on the particles, the Einstein equation quite frequently fails to predict the viscosities of suspensions accurately.

Various authors have recommended other formulae for calculating the viscosities of suspensions and some of these equations are listed in Table 4.9. Most of the equations in this table are only accurate when applied to rather disperse suspensions ($\phi < 0{\cdot}1$). These equations are not suitable for colloidal solutions of lyophiles. Gambill recommends equation (2) in Table 4.9, presented by Kunitz, as being the most reliable. Steinour's equation (8) affords approximate results.

TABLE 4.9. FORMULAE FOR CALCULATING THE VISCOSITIES OF SUSPENSIONS OF SOLIDS IN LIQUIDS (AFTER GAMBILL[120])

No.	Equation	Range of application	References
1	$\mu_m/\mu_l = (1+0{\cdot}5\phi)/(1-\phi)^2$ (if $\phi < 0{\cdot}04$, $\mu_m = \mu_l(1+2{\cdot}5\phi)$)	$\phi < \approx 0{\cdot}2$	(83)
2	$\mu_m/\mu_l = (1+0{\cdot}5\phi)(1-\phi)^4$	$\phi < 0{\cdot}1$	(209)
3	$\mu_m/\mu_l = 1+4{\cdot}5\phi$	$\phi < 0{\cdot}4$	(143)
4	$\mu_m/\mu_l = 1/(1-\phi^{1/3})$	$0{\cdot}5 < \phi < 0{\cdot}9$	(144)
5	$\mu_m/\mu_l = 1+2{\cdot}5\phi+7{\cdot}17\phi^2+16{\cdot}2\phi^3$	$\phi < 0{\cdot}35$	(379)
6	$\mu_m/\mu_l = 1+2{\cdot}5\phi+10{\cdot}05\phi^2$	$\phi < 0{\cdot}3$	(234)
7	$\mu_m/\mu_l = \left[1+\dfrac{2{\cdot}5\phi}{2(1-1{\cdot}35\phi)}\right]^2$	$\phi < 0{\cdot}5$	(84)
8	$\mu_m/\mu_l = 10^{1{\cdot}82\phi/(1-\phi)}$	$\phi < \approx 0{\cdot}4$	(344)
9	$\mu_m/\mu_l = e^{[2{\cdot}5\phi/(1-0{\cdot}61\phi)]}$		(379)

For a rapid, approximate estimation of the viscosity of solid suspensions in water use may be made of the alignment chart prepared by Davis.[75] This nomograph which provides a rapid graphical solution to Hatschek's equation (4), is presented in Fig. 4.31. When using the nomograph the temperature of the suspension (°C) is marked off on the left-hand axis and the volume fraction ϕ of the dry solid in the suspension is marked off on the right-hand axis. The straight line connecting these points, cuts the middle axis at a point corresponding to the viscosity of the suspension of the solid in water (cP).

In addition to the equations listed in Table 4.9, mention may be made of an equation proposed by Kurgaev[213] for the viscosity of a suspension of a solid in a liquid. He defines the concentration C, of the solid phase

$$C = n^3 \frac{\pi d^3}{6} \tag{4.71}$$

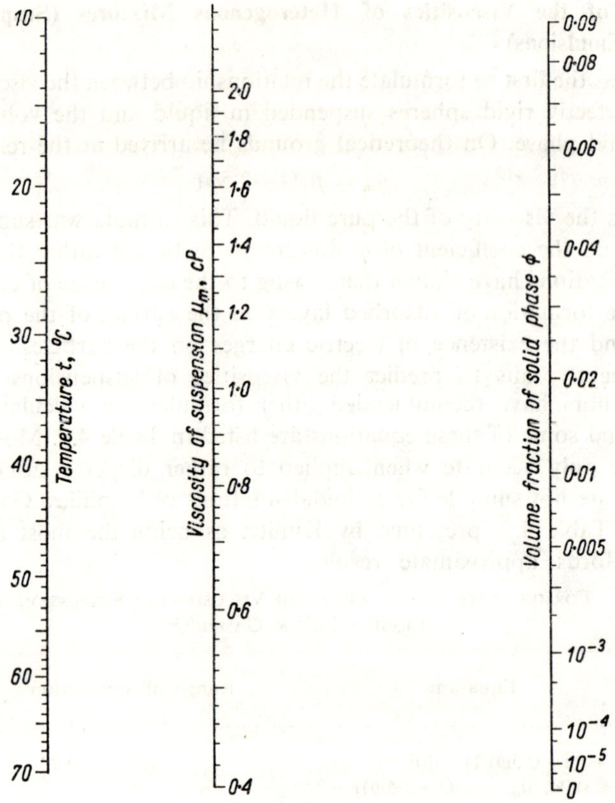

FIG. 4.31. Alignment chart for calculating the viscosity of a suspension of a solid (volume fraction ϕ) in water (after Davis[75]).

where d is the diameter of the solid particles,
and n is the number of particles per unit length of liquid. According to Kurgaev

$$\mu_m = \mu_l \left[1 + \frac{2C}{(1-1{\cdot}2C^{2/3})^2}\right] \tag{4.72}$$

where μ_m is the viscosity of the suspension,
and μ_l is the viscosity of the pure liquid.

For concentrations $C < 0{\cdot}25$ this formula gives good results as may be seen from the following comparison:

C	0·05	0·10	0·15	0·20	0·25	0·30
μ_m/μ_l, calculated by Eq. (4.72)	1·14	1·36	1·68	2·12	2·85	3·85
μ_m/μ_l, experimental	1·17	1·35	1·67	2·10	2·8	4·1

For the calculation of the viscosity of a mixture of two immiscible liquids, Gambill[120] recommends the Taylor formula[362]

$$\mu_m/\mu_l = 1+2{\cdot}5\phi\left(\frac{\mu_d+0{\cdot}4\mu_l}{\mu_d+\mu_l}\right) \tag{4.73}$$

where μ_m is the viscosity of the mixture (emulsion),
μ_l is the viscosity of the liquid appearing as the continuous phase,
μ_d is the viscosity of the liquid appearing as the disperse phase,
and ϕ is the volume fraction of the disperse phase.†

Neogy and Chosh[255] have investigated aqueous emulsions of xylene, stabilized by various emulsifiers and have compared their results with values predicted using equations recommended by various authors. They find that Richardson's modified equation

$$\mu_m/\mu_l = e^{(k\varphi+c)} \tag{4.74}$$

in which φ denotes the volume ratio of the phases, is the most satisfactory, the error in the predicted viscosities amounting to 5%. The value of the constant C depends on the nature of the emulsifier used.

f. Selection of the Best Methods for Estimating the Viscosities of Liquid Mixtures, Solutions and Suspensions

When interpolating or extrapolating experimental data for electrolytic solutions as functions of composition or temperature the Davis graph (Fig. 4.26) or plots of the Dühring type are most suitable. In the case of solutions containing two dissolved electrolytes, the Zdanovskii equations (4.58) and (4.59) may sometimes be used.

The choice of the best method for calculating the viscosity of a mixture of two organic compounds depends on the data available for these liquids. If the structural formulae of the liquids making up the mixture are known, either the rule of the additivity of the rheochor, i.e. Eq. (4.68), or the Souders equation modified by Lima, i.e. Eq. (4.67) may be used, the latter being less accurate.

If the viscosities of the pure compounds forming the mixture (at the temperature of the mixture) are known, Eq. (4.60) or (4.61) is often employed; the estimates obtained from Eq. (4.61) usually seem to be the more accurate.

The viscosity of a mixture of two liquids may be estimated with good accuracy, if the viscosity of a neighbouring mixture (constituted from the same pure components but of different composition) has been determined experimentally, provided that viscosities have also been determined for the pure components. For this purpose the equation of Tamura and Kurata (4.66) is recommended.

If values of the viscosity have been determined experimentally for two or three mixtures, the use of McAllister's formula (Eq. (4.69)) is recommended.

If a quick, but not highly accurate estimation of the viscosity of a suspension is required, the Davis nomograph may be used (Fig. 4.31). More accurate results may be obtained from Eq. (2) in Table 4.9 or from the Kurgaev equation (4.71). The methods discussed above are not applicable to lyophylic colloids.

† In the derivation of this formula, the disperse phase was assumed to be in the form of spherical particles, and $\phi \leqslant 0{\cdot}03$.

CHAPTER 5

THERMAL CONDUCTIVITIES OF GASES

1. INTRODUCTION AND SOME SIMPLE CONCLUSIONS FROM THE KINETIC THEORY

The rate of heat conduction in a homogeneous medium is given by Fourier's equation, according to which the quantity Q (kcal) of heat conducted through the medium in time τ (hr) between two parallel planes a distance $|x|$ (m) apart is proportional to the cross-sectional area A (m^2) through which the heat is being transmitted and to the temperature difference Δt (°C) between the planes, i.e.

$$Q = \lambda A \cdot \left(\frac{\Delta t}{|x|}\right) \tau \tag{5.1}$$

This equation may be represented in the differential form:

$$dQ = -\lambda A \cdot \left(\frac{dt}{dx}\right) d\tau \tag{5.2}$$

where λ is the thermal conductivity, kcal/m.hr.°C,
and dt/dx is the temperature gradient.

The phenomena of heat conduction in gases have been interpreted in terms of the kinetic theory by Maxwell, Clausius, Stefan, von Lang, Boltzmann, Meyer and others.† Heat conduction in a gas involves the transference of energy from molecule to molecule and hence, like viscosity and diffusion, it constitutes one of the transport processes. Simplified considerations lead to the conclusion that, like the other transport phenomena, gaseous heat conduction should depend on the mean free path $\bar{l}$ of the molecules, the mean molecular speed $\bar{w}$, the gas density ρ and the heat capacity c_v of unit mass of the gas at constant volume. The expression obtained is

$$\lambda^\circ = \tfrac{1}{3} c_v \rho \bar{w} \bar{l} \tag{5.3}$$

where λ° denotes the thermal conductivity of a gas at a moderate pressure. Now it has been seen that an analogous simplified treatment of the phenomenon of viscosity in gases (i.e. of the process of momentum transfer between colliding molecules) leads to the equation (3.15)

$$\mu^\circ = \tfrac{1}{3} \rho \bar{w} \bar{l} \tag{5.4}$$

where μ° denotes the dynamic gas viscosity at moderate pressures. On combining Eqs. (5.4) and (5.3), we obtain

$$\lambda^\circ = \mu^\circ c_v$$

In practice this simple equation is not found to be accurately obeyed and it is convenient to define a quantity K, known as Maxwell's modulus[241], by the equation

$$\lambda^\circ = K \mu^\circ c_v \tag{5.5}$$

† This problem is considered in greater detail by Partington.[282]

For a given gas, the value of K is almost constant over a fairly wide range of temperatures and pressures, but it depends on the number of atoms in the gas molecule. For gases containing more than three atoms in one molecule, the constant is approximately $K \approx 1{\cdot}30$, for gases with triatomic molecules $K \approx 1{\cdot}70$, and for gases with diatomic molecules $K \approx 1{\cdot}90$. A more rigorous development of the classical kinetic theory of gases shows that in the case of monatomic gases the value of K should be approximately equal to 2·5. This prediction has been substantiated experimentally. For example, for helium $K = 2{\cdot}503$ and for argon $K = 2{\cdot}506$ (Ref. 263).

According to Jeans,[172] the relationship $\lambda = c_v\mu$, obtained from simplified considerations, refers to the energy of rotational motion. On the other hand, the relation $\lambda = 2{\cdot}5\, c_v\mu$, found by Maxwell, refers to the energy of translational motion of the molecules.

For gases with more than one atom per molecule, C_v is larger than for monatomic gases being given approximately by $C_v = \frac{1}{2}nR$. C_p for a polyatomic gas is given by $C_p = \frac{1}{2}(n+2)R$ so $(C_p/C_v) = (n+2)/n$, where n denotes the total number of degrees of freedom available to the molecule (for a monatomic gas $n = 3$). Now Maxwell's modulus K for a polyatomic gas is given by[306]

$$K = \frac{\lambda^\circ}{\mu^\circ c_v} = \frac{[3\times\frac{5}{2}+(n-3)]}{n} = \frac{9(n+2)-5n}{4n}$$

On substituting $\varkappa = C_p/C_v$ Eucken's equation[90] is obtained

$$\frac{\lambda^\circ}{\mu^\circ} = \frac{9\varkappa-5}{4}c_v \quad \text{or} \quad \frac{\lambda^\circ}{\mu^\circ} = \frac{9-\dfrac{5}{\varkappa}}{4}c_p \tag{5.6}$$

These equations may be used to estimate λ°/μ° though the estimated values may be in error by as much as 13% when substances with strongly polar molecules are considered (cf. Ref. 306).

By considering the analogy between diffusivity and thermal conductivity, Andrussov,[9] (cf. also Ref. 5) has found the following relationship

$$\lambda_p = \varkappa\mu c_p \tag{5.7}$$

where λ_p is the thermal conductivity at constant pressure,
$\varkappa = c_p/c_v$
and μ is the viscosity.

The variation of λ_p with temperature is given according to Andrussov, by the formula

$$\lambda_p = \lambda_{p0}\left(\frac{T}{273{\cdot}16}\right)^{n_\lambda} \tag{5.8}$$

In this formula λ_{p0} denotes the thermal conductivity at constant pressure and a temperature of 273·16°K. The exponent n_λ, which depends on temperature t(°C), is calculated from the equation

$$n_\lambda = A - B\times10^{-4}t + C\times10^{-7}t^2 \tag{5.9}$$

The values of the constants A, B and C for several gases are compiled in Table 5.1. The agreement between the values calculated from Eq. (5.7) with those determined experimentally is quite good, the average error in the estimated values being approximately 3%.

TABLE 5.1. VALUES OF THE COEFFICIENTS IN EQ. (5.9)

Gas	A	B	C	Gas	A	B	C
H_2	0·674	−0·336	—	CO_2	1·192	2·38	0·56
D_2	0·678	−0·33	—	N_2O	1·21	2·88	0·8
N_2	0·773	0·57	0·4	C_3H_8	1·75	4·98	—
CO	0·792	0·51	0·5	H_2S	1·13	1·74	—
O_2	0·830	0·91	0·64	SO_2	1·31	3·2	0·9
NO	0·82	1·64	—	CH_4	1·33	0·70	—
HCl	1·05	1·66	—	C_2H_4	1·52	2·82	—
H_2O	1·122	1·60	—	C_2H_6	1·49	0·98	—

EXAMPLE 5.1. Estimate the thermal conductivity of carbon dioxide at 300°C.
At 0°C, $c_p/c_v = 1{\cdot}301$, $c_p = 0{\cdot}199$ cal/g.°C, $\mu_0^\circ = 137{\cdot}5\times10^{-6}$ P,
$A = 1{\cdot}192$, $B = 2{\cdot}38$, $C = 0{\cdot}56$.

Solution

According to Eq. (5.7) the thermal conductivity of carbon dioxide at 0°C is:

$$\lambda_{p0} = 1{\cdot}301\times137{\cdot}5\times10^{-6}\times0{\cdot}199 = 35{\cdot}6\times10^{-6} \text{ cal/cm.sec.°C}$$

(the value of λ_{p0} determined experimentally lies, according to various authors, in the range 33·7 to 36·1 × 10⁻⁶).

Calculation of the value of the exponent n_λ:

$$n_\lambda = 1{\cdot}192-2{\cdot}38\times10^{-4}\times300+0{\cdot}56\times10^{-7}\times300^2 = 1{\cdot}126$$

$$\lambda_{\text{calc}} = 35{\cdot}6\times10^{-6}\times\left(\frac{573}{273}\right)^{1{\cdot}126} = 81{\cdot}9\times10^{-6} \text{ cal/cm.sec.°C}$$

The value found experimentally is $\lambda_{\text{exp}} = 84\times10^{-6}$
The error in the estimated value is thus −2·5%.

Schaefer and Thodos[318] have correlated the thermal conductivities of diatomic gases (with the exception of hydrogen) as a function of reduced temperature, $T_r = T/T_c$. They find that, at moderate pressures (where the thermal conductivity λ° does not depend on pressure) λ° can be calculated at the critical temperature T_c from the equation:

$$(\lambda^\circ)_{T_c} = 3{\cdot}05\times10^{-5}\frac{p_c^{9/16}}{M^{1/3}T_c^{1/5}} \tag{5.10}$$

where p_c is the critical pressure,
and M is the molecular weight. They also find that, for diatomic gases, the following relations are obeyed:

$$\lambda^\circ/(\lambda^\circ)_{T_c} = T_r \quad \text{if } T_r \leqslant 1 \tag{5.11}$$

$$\lambda^\circ/(\lambda^\circ)_{T_c} = 1{\cdot}207\, T_r^{0{\cdot}729} \quad \text{if } T_r \geqslant 2{\cdot}5 \tag{5.12}$$

EXAMPLE 5.2. Estimate the thermal conductivity of nitric oxide at a pressure of 1 atm and a temperature of 573°K, and of carbon monoxide at the same pressure and a temperature of 123°K. The data available (from Ref. 318) are

	NO	CO
T_c, °K	180	133
$10^{-5}(\lambda^\circ)_{T_c}$, cal/cm.sec.°C	3·83	2·79

a. Thermal conductivity of nitric oxide at 573°K:

$$\lambda^\circ = 1{\cdot}207\left(\frac{573}{180}\right)^{0{\cdot}729}\times3{\cdot}83\times10^{-5} = 1{\cdot}075\times10^{-4} \text{ cal/cm.sec.°C}$$

The experimental value is $1{\cdot}07\times10^{-4}$.

b. Thermal conductivity of carbon monoxide:

$$\lambda^\circ = \frac{123}{133} \times 2{\cdot}79 \times 10^{-5} = 0{\cdot}258 \times 10^{-4} \text{ cal/cm.sec.}^\circ\text{C}$$

The experimental value is $0{\cdot}26 \times 10^{-4}$.

In both cases the agreement between the calculated and experimental values is very good.

Misic and Thodos[253] investigated the thermal conductivity of gaseous hydrocarbons. For methane and naphthenic and aromatic hydrocarbons they found that the relationship

$$(\lambda^\circ \Lambda) C_p = 0{\cdot}445 \times 10^{-5} T_r \text{ was obeyed for } T_r < 1{\cdot}0 \quad (5.13)$$

where

$$\Lambda = M^{1/2} T_c^{1/6} / p_c^{2/3} \quad (5.14)$$

λ° is the thermal conductivity at moderate pressures, $0{\cdot}2 < p < 5$ atm,
C_p is the molal heat capacity,
and T_r $(= T/T_c)$ is the reduced temperature.

For $3{\cdot}0 > T_r > 1{\cdot}0$, they found that

$$(\lambda^\circ \Lambda) C_p = (14{\cdot}52\, T_r - 5{\cdot}14)^{2/3} \times 10^{-6} \quad (5.15)$$

For all the other gaseous hydrocarbons which they studied, Misic and Thodos found that Eq. (5.15) was obeyed for $0{\cdot}6 < T_r < 3{\cdot}0$.

The authors report that Eqs. (5.13) and (5.15) represent the data with average errors of 1·8 and 2·6%, respectively.

Provided the heat capacity c_p of the gas and its viscosity μ° are known, the nomograph shown in Fig. 5.1 may be used for a rapid calculation of the value of the thermal conductivity λ°. This nomograph, which is based on Eucken's equation (5.6), has been developed by Johnson and Huang.[177] The nomograph contains three reference lines (the oblique lower lines): one for water vapour and hydrogen sulphide, the second for gases with more than 4 atoms in the molecule ($n > 4$), and the third for gases with $n < 4$. To determine the thermal conductivity of a gas, a horizontal line is drawn from the point corresponding to the heat capacity c_p of this gas (left-hand axis) to intersect with the oblique line representing the viscosity μ° of the gas. Next, a line is drawn vertically downwards from the intersection point until it intersects that reference line which corresponds to the value of n for the gas. From the second intersection point a horizontal line is drawn to intersect the right-hand axis. The desired value of the thermal conductivity λ° is then read from this axis.

EXAMPLE 5.3. At 0°C the viscosity of chlorine, μ°, is 0·012 cP and the heat capacity, c_v, is 0·114 cal.g.°C. Find the thermal conductivity λ°.

Solution

The necessary construction lines connecting the appropriate points on the nomograph are indicated by dashed lines in Fig. 5.1.

The value determined experimentally is $\lambda^\circ = 18 \times 10^{-6}$ cal/cm.sec.

The value read from the nomograph is $\lambda^\circ = 19 \times 10^{-6}$ cal/cm.sec.

Bromley[47] has proposed a more accurate method for calculating the value of λ°/μ°. He arrived at his final equations by considering individually the contributions to the total energy of the gas arising from the translational, rotational and vibrational motions of the molecules. These considerations led the author to the following equations:

(a) Monatomic gases

$$\frac{\lambda^\circ}{\mu^\circ} = 2{\cdot}5 C_v \frac{1}{M} \quad (5.16)$$

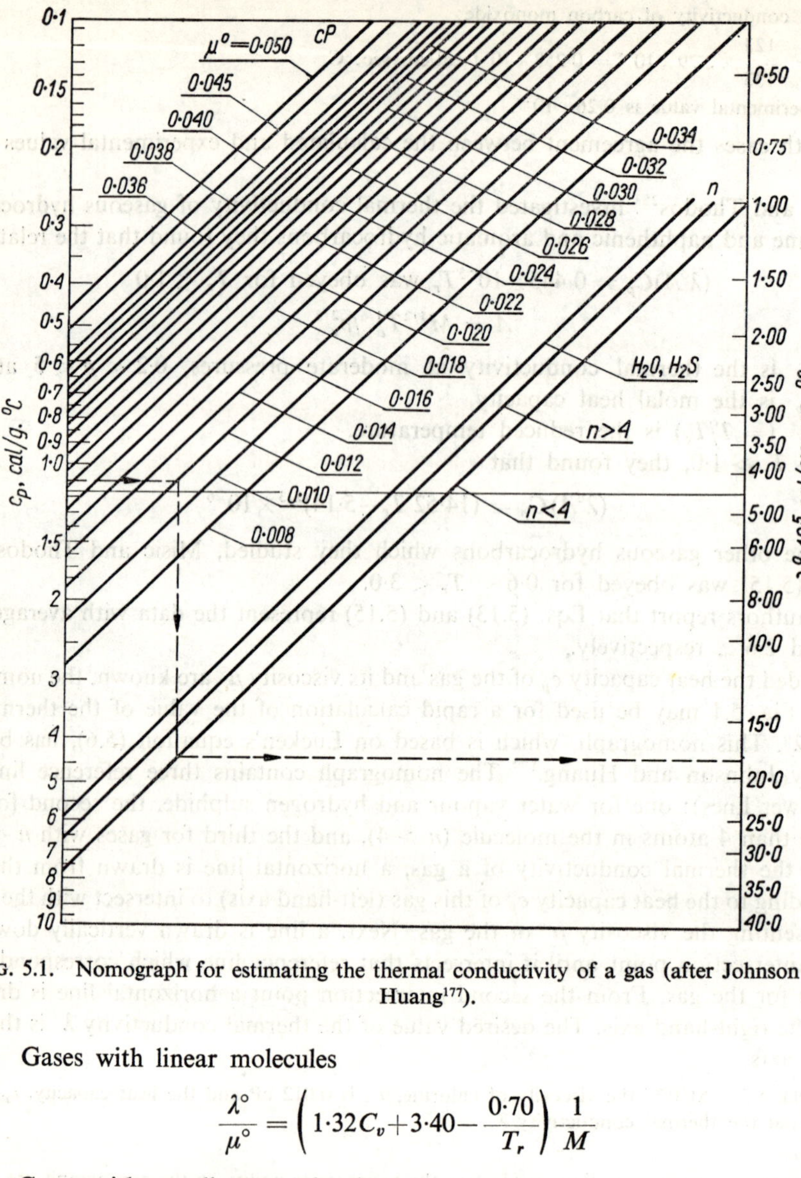

FIG. 5.1. Nomograph for estimating the thermal conductivity of a gas (after Johnson and Huang[177]).

(b) Gases with linear molecules

$$\frac{\lambda^\circ}{\mu^\circ} = \left(1{\cdot}32C_v + 3{\cdot}40 - \frac{0{\cdot}70}{T_r}\right)\frac{1}{M} \tag{5.17}$$

(c) Gases with non-linear molecules

$$\frac{\lambda^\circ}{\mu^\circ} = \left(1{\cdot}30C_v + 3{\cdot}60 - 0{\cdot}3C_{\text{int}} - \frac{0{\cdot}69}{T_r} - 3\alpha\right)\frac{1}{M} \tag{5.18}$$

In the above equations,

λ° is the thermal conductivity of the gas at low pressure, cal/cm.sec.°C,

μ° is the "dynamic" viscosity, poises,

C_v is the molal heat capacity at constant volume, cal/mole.°C,

M is the molecular weight,

T_r ($= T/Tc$) is the reduced temperature,

α is a coefficient, the value of which is computed from Eq. (5.19) (this coefficient characterizes the interaction of colliding polar molecules),

and C_{int} is the "internal" heat capacity (i.e. the contribution to the molal heat capacity arising from internal rotations within the molecules), cal/mole. °C.

The equation for calculating the coefficient α is as follows:

$$\alpha = \frac{3\rho_b}{M}(\Delta S_{vb} - 8{\cdot}75 - R \ln T_b) \tag{5.19}$$

where ρ_b is the liquid density at the normal boiling temperature (T_b), g/cm³,

$\Delta S_{vb}\left(=\frac{L}{T_b}\right)$ is the entropy of vaporization at the temperature T_b, cal/mole. °K,

$R =$ 1·985 cal/mole. °K,

and T_b is the normal boiling point, °K.

Reid and Sherwood[306] give the following values of α:

Compound	H_2O	NH_3	CH_3OH	n-C_3H_7OH	CH_3—CO—CH_3	C_6H_6
Value of α	0·90	0·51	0·33	0·24	0·06	0·00

Using the group contributions listed in Table 5.2, it is possible to predict values of C_{int}. The contributions of the groups R—CH= have been calculated in two ways: first from the properties of propylene (†) and then from those of *cis*-2-butene (‡). The values of

TABLE 5.2. INTERNAL-ROTATIONAL HEAT CAPACITY CONTRIBUTIONS C_{int} IN EQ. (5.18)

Group	Contribution of the group at a temperature of (°K)								
	200	298	400	500	600	700	800	900	1000
—CH_2—CH_2—	2·64	3·18	2·74	2·39	2·12	1·90	1·73	1·60	1·51
CH_3—CH_2—	1·74	2·10	2·17	2·09	1·96	1·82	1·69	1·59	1·51
R—C—H	1·76	2·12	2·19	2·12	2·00	1·85	1·71	1·62	1·55
R—C—	1·60	1·99	2·18	2·23	2·18	2·08	1·96	1·85	1·76
R—CH=†	2·02	2·02	1·82	1·65	1·50	1·40	1·32	1·26	1·21
R—CH=‡	1·48	1·26	1·16	1·11	1·07	1·05	1·04	1·03	1·02
=CH—CH=	2·09	3·02	3·25	2·96	2·67	2·44	2·24	2·09	1·96
≡C—C≡	0·81	0·92	0·76	0·60	0·48	0·39	0·35	0·32	0·31
R—C= (with bond below)	1·91	2·15	2·10	1·95	1·79	1·65	1·54	1·46	1·38
CH_3—C=O (with R below)	1·79	1·52	1·34	1·24	1·18	1·10	1·09	1·08	1·06
CH_3—OH	1·4	1·35	1·25	1·20	1·18	1·16	1·15	1·14	1·13
RCH_2—O—CH_2R	2·24	2·33	2·26	1·09	1·91	1·71	1·66	1·55	1·47

† Calculated from the properties of propylene.
‡ Calculated from the properties of *cis*-2-butene.

the contributions are those given by Reid and Sherwood[306] on the basis of the work of Souders, Matthews and Hurd,[338] and Bromley.[47] The average errors incurred when Bromley's equations are used are about 50% smaller than those involved in using Eucken's equation (5.6).[306]

The equations discussed so far in this chapter are based on the classical kinetic theory of gases. The modern kinetic theory of gases takes the forces of molecular interaction into consideration. As in the case of dynamic viscosity, values of the thermal conductivity may be estimated theoretically, provided values of the force constants σ and ε in Lennard-Jones equation (1.29) are known. (These force constants may either be taken from Table 1.16 or estimated using Eqs. (1.33)–(1.37).)

For a pure monatomic gas (the molecules of which may be regarded as perfectly spherically symmetrical with no internal degrees of freedom) Bird, Hirschfelder and Curtiss[25] arrive, as a first approximation, at the following equation, which should be valid at moderate pressures,

$$\lambda^\circ \times 10^7 = 1989{\cdot}1 \frac{\sqrt{T/M}}{\sigma^2 \Omega^{(2,2)*}} \tag{5.20}$$

An analogous relationship exists which gives, as a first approximation, the dynamic viscosity μ° of a monatomic gas in terms of the intermolecular force constants (Ch. 3). On combining these relationships the equation

$$\lambda^\circ = \frac{15}{4}\frac{R}{M}\mu^\circ = 2{\cdot}5 c_v \mu^\circ \tag{5.21}$$

is obtained relating the thermal conductivity to the dynamic viscosity. According to this equation the constant K in Eq. (5.5) should be 2·5 for a monatomic gas. This is in agreement with the results of classical kinetic theory. The notation used in Eqs. (5.20) and (5.21) is as follows:

λ° is the thermal conductivity, cal/cm.sec.°C,
T is the absolute temperature, °K,
M is the molecular weight, g/mole,
σ and ε are the force constants in Lennard-Jones' equation (1.29),
$\Omega^{(2,2)*}$ is a function, the value of which is taken from Table 3.1 for a given value of $T^* = kT/\varepsilon$,
R is the gas constant,
μ° is the dynamic viscosity, g/cm.sec,
and c_v is the heat capacity at constant volume, cal/g.°C.

Equation (5.20) is only true in the first approximation. A more accurate value for the thermal conductivity λ' of a monatomic gas may be obtained by multiplying the value λ°, found from Eq. (5.20), by the correction factor f_λ. Values of this correction factor are listed as a function of T^* ($= kT/\varepsilon$) in Table 3.2.

$$\lambda' = \lambda^\circ f_\lambda \tag{5.22}$$

To obtain an approximate value of the thermal conductivity of a pure polyatomic gas, we must multiply the value of the thermal conductivity, calculated for a monatomic gas, by Eucken's factor (5.6)

$$\lambda^\circ = \frac{15}{4}\frac{R}{M}\mu\left(\frac{4}{15}\frac{C_v}{R}+\frac{3}{5}\right) \tag{5.23}$$

where C_v denotes the molal heat capacity of a polyatomic gas.

The viscosity of a gas with non-polar molecules may be calculated at moderate pressures by applying either Eq. (3.22) (Hirschfelder) or (3.30) (Bromley and Wilke). Substituting the value of viscosity, calculated in such a way, in Eucken's equation (5.6), we obtain (after Reid and Sherwood[306])

$$\lambda^\circ = 266{\cdot}9 \times 10^{-7} \frac{(T/M)^{0{\cdot}5}}{\sigma^2 \Omega^{(2,2)*}}(C_v + 4{\cdot}47) \tag{5.24}$$

$$\lambda^\circ = 231\times 10^{-7}\left(\frac{T_c}{M}\right)^{0\cdot5}\frac{(C_v+4\cdot47)}{\sigma^2}f_1(1\cdot33T_r) \tag{5.25}$$

$$\lambda^\circ = 333\times 10^{-7}\left(\frac{T_c}{M}\right)^{0\cdot5}V_c^{-0\cdot67}f_1(1\cdot33T_r)(C_v+4\cdot47) \tag{5.26}$$

where λ° is the thermal conductivity of the gas at a "moderate" pressure, cal/cm. sec. °K,
T is the absolute temperature, °K,
M is the molecular weight, g/mole,
σ is the Lennard-Jones force constant, Å,
$\Omega^{(2,2)*}$ is a function of the variable $T^* = kT/\varepsilon$ (its value is given in Table 3.1),
C_v is the molal heat capacity at constant volume, cal/mole. °K,
T_c is the critical temperature of the gas, °K,
V_c is the molal critical volume, cm³/mole,
and f_1 is a function of the reduced temperature, the value of which may be read from Table 3.2.

The use of Eqs. (5.24), (5.25) and (5.26) requires a knowledge of the force constants σ and ε or alternatively of the critical constants and the specific heat. All this information may not always be available, a fact which restricts the use of these formulae for technological calculations. If all that is required is to predict the viscosity at one temperature when the viscosity is already known at some other temperature, the force constants are not required, as may be seen as follows: at the critical temperature $T = T_c$, $T_r = 1$, and $f_1(1\cdot33T_r) = f_1(1\cdot33) = 0\cdot833$ (Table 3.2). By comparing the values of λ° calculated for T (λ_T°) and for the critical temperature, the value of λ_c° is found from Eq. (5.25)

$$\frac{\lambda_T^\circ}{\lambda_c^\circ} = 1\cdot20 f_1(1\cdot33T_r)\frac{(C_v+4\cdot47)_T}{(C_v+4\cdot47)_c} \tag{5.27}$$

The values of the molal heat capacity C_v refer to temperatures T and T_c, respectively. Equation (5.27) may be used for estimating the value of $\lambda_{T_2}^\circ$, if the value of $\lambda_{T_1}^\circ$ is known. Substituting and dividing we obtain

$$\frac{\lambda_{T_2}^\circ}{\lambda_{T_1}^\circ} = \frac{f_1(1\cdot33T_{r2})\,(C_v+4\cdot47)_{T_2}}{f_1(1\cdot33T_{r1})\,(C_v+4\cdot47)_{T_1}} \tag{5.28}$$

where $T_{r1} = T_1/T_c$, $T_{r2} = T_2T_c$. When applying this method we need not know the values of λ_c and $(C_v)_c$ at the critical point, but the knowledge of the effect of temperature upon the molal heat capacity of the gas is indispensable. According to Reid and Sherwood,[306] the average error of estimation is $\pm 2\%$, though for CO_2 and H_2O, the calculated values of λ_T° are subject to a larger error of the order of $\pm 5\%$ (maximum 20%).

Recent work on the thermal conductivity of gases has been reviewed by Liley (cf. Ref. 5).

Variation of Thermal Conductivity of Gases with Pressure and Temperature

As has been shown in Chapter 3 the value of the dynamic gas viscosity μ in the region of moderate pressures (from a few millimetres of mercury to several atmospheres) does not depend on pressure (Eq. (5.4)). Since the ratio of the thermal conductivity of a gas to its viscosity is, according to Eq. (5.5), approximately constant, the thermal conductivity of the gas at moderate pressures should not depend on pressure, either. This conclusion, formulated for the first time by Stefan,[342] is in agreement with experiment.

The reason for the independence of the thermal conductivity of a gas on pressure is evident from a study of Eq. (5.3): the heat capacity c_p and mean molecular speed $\overline{w}$ of the molecules are approximately independent of pressure; the mean free path of the molecule varies in inverse proportion with pressure, the density varies proportionally to pressure, and hence the product ρl is approximately constant. It may be assumed for practical purposes that the thermal conductivity of a gas does not depend on pressure, provided that[217]

$$p_r \leqslant \frac{2}{3} T_r \quad \text{and} \quad T_r > 1$$

where p_r $(= p/p_c)$ is the reduced pressure
and T_r $(= T/T_c)$ is the reduced temperature.

The variation of thermal conductivity with pressure is greatest in the vicinity of the critical point; at very high pressures the thermal conductivity of a gas does not depend so strongly on pressure (Fig. 5.10).

Enskog's work leads to an explicit equation for the effect of pressure upon thermal conductivity. Suppose that the thermal conductivity of a gas at temperature T and pressure p, is λ_{Tp} and that the value of the thermal conductivity at temperature T and atmospheric pressure (i.e. at a moderate pressure) is $\lambda_{T,1}$. Then employing Enskog's equation,[140] we obtain

$$\lambda_{T,p} = \lambda_{T,1}\left(\frac{b_0}{V}\right)\left(\frac{1}{y}+1{\cdot}2+0{\cdot}755\,y\right)$$

where V is the molal volume and the quantities b_0 and y are defined by Eqs. (3.55)–(3.60). On substituting $b_0 = \frac{2}{3}\pi N\sigma^3$ (where σ is the diameter of the molecule and N is Avogadro's number), the equation

$$\lambda_{T,p} = \lambda_{T,1}\left(\frac{2\pi N\sigma^3}{3V}\right)\left(\frac{1}{y}+1{\cdot}2+0{\cdot}755\,y\right) \tag{5.29}$$

is obtained. This equation is analogous to Enskog's Eq. (3.58), which is used for the determination of the viscosity of a fluid. In practice, Enskog's equations are rarely used for technological calculations because they require a knowledge of quantities characterizing the compound (σ, y), which are difficult to estimate. When applied to monatomic gases, however, they give satisfactory results.

Predvoditelev,[299] while investigating the mechanism of heat conduction in fluids, came to the conclusion that the following relationship should exist between the thermal conductivity λ_p of a gas under elevated pressure, the thermal conductivity λ_1 at moderate pressures and the specific gravity γ,

$$\lambda_p = \lambda_1 + B\gamma$$

Experimentally, Vargaftig[381] finds that the conductivities of many compounds obey the relationship

$$\lambda_p = \lambda_1 + B\gamma^n \tag{5.30}$$

where B and n are constants characteristic of the given gas,
γ is the specific gravity, kg (force)/m^3,
and λ is the thermal conductivity, kcal/m.hr.°C.

Tsederberg[373] has checked this equation for oxygen, hydrogen, methane, carbon dioxide, nitrogen, benzene and water vapour. He has shown that the value of the exponent n depends on the molecular weight M of the compound, viz.,

$$n = 1{\cdot}16 + 0{\cdot}0024\ M$$

To calculate the value of the constant B in Eq. (5.30), at least one experimentally determined value of λ_p is required. Several values of B and n are listed below:

Gas	O_2	H_2	CH_4	CO_2	N_2
B	$1{\cdot}246\times10^{-5}$	$2{\cdot}950\times10^{-4}$	$4{\cdot}914\times10^{-4}$	$1{\cdot}380\times10^{-5}$	$1{\cdot}662\times10^{-5}$
n	1·24	1·16	1·20	1·26	1·23

Eq. (5.30) is usually obeyed to within 3–4%.

Since the ratio of the thermal conductivity of a gas to its viscosity is approximately constant, it follows that the thermal conductivity of a gas depends on temperature in the same way as does the gas viscosity. Because of this Sutherland's equation (3.20) may be used to calculate the thermal conductivities of gases, being applied in the form

$$\lambda = \lambda_0 \frac{(273+C)}{(T+C)}\left(\frac{T}{273}\right)^{3/2} \tag{5.31}$$

where λ_0 is the thermal conductivity at 0°C,
and C is Sutherland's constant (for H_2, $C = 94$; for NO, $C = 195$; for N_2, $C = 114$; for O_2, $C = 144$; for C_6H_6, $C = 114$; for air, $C = 125$; for CO, $C = 156$).

The value of the constant C can be estimated approximately from Vogel's formula,[386] $C \approx 1{\cdot}5T_b$ (where T_b is the normal boiling temperature). Combining the constants in Eq. (5.31) into one coefficient, we obtain

$$\lambda = \frac{B\sqrt{T}}{1+C/T}$$

It has been found experimentally (Jakob[91]) that for some gases the coefficient B is not constant, but varies linearly with temperature. The general formula, representing the variation of the thermal conductivity of a gas with temperature, is in these cases more complex

$$\lambda = \frac{D(1+ET)\sqrt{T}}{1+C/T} \tag{5.32}$$

Values of the constants in Eq. (5.32) are given below for several gases (after Eucken and Jakob[91]):

Gas	Ar	CO_2	Air	CH_4	Hg vapour	H_2O vapour	H_2
C, °K	142	274	123·6	198	960	673	83
$E\times10^6$, °K^{-1}	0	1026	0	153	0	0 (?)	247
$D\times10^6$, kcal/m.hr.°C$^{1\cdot5}$	1292	1162	1818	2391	864	2888	10,750

If $E = 0$, Eq. (5.32) becomes identical with Eq. (5.31). When Eq. (5.32) is used, the average error of estimation is ±3%, and the maximum error *ca.* 6%.

2. EXPERIMENTAL DATA FOR THE THERMAL CONDUCTIVITY OF GASES

Experimental data for the thermal conductivities of gases are rather limited, and many of the values which have been determined are very inaccurate. Table 5.3 lists values of the thermal conductivity of about three score of gaseous compounds (or vapours) at a pressure of 1 atm and at various temperatures.[11]

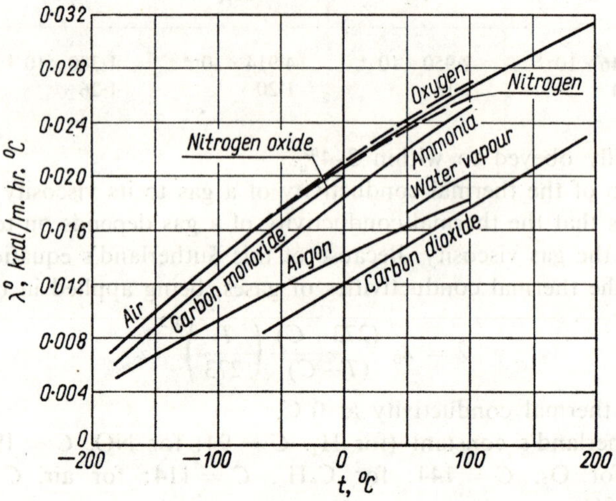

FIG. 5.2. Thermal conductivities of gases (after Eucken and Jakob[91]).

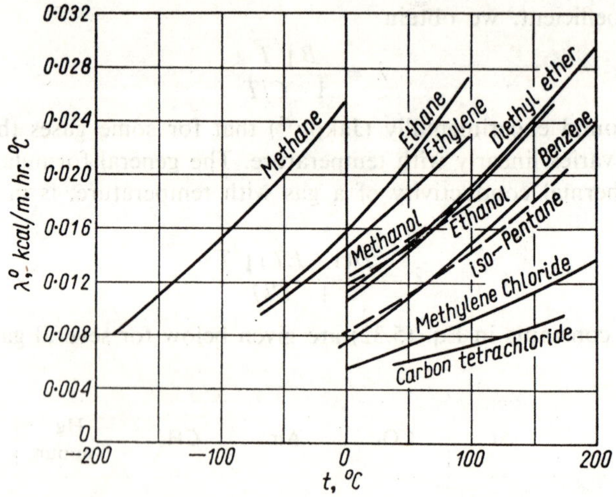

FIG. 5.3. Thermal conductivities of gaseous organic compounds (after Eucken and Jakob[91]).

The thermal conductivities of several gases and vapours are shown as functions of temperature in Figs. 5.2 and 5.3. All the data refer to the region of moderate pressures, in which the value of $\lambda°$ does not depend on pressure.

TABLE 5.3. THERMAL CONDUCTIVITIES OF GASES, $\lambda^{\circ} \times 10^4$ cal/cm.sec.°C

Compound	Temperature, °C									
	−180	−150	−100	−50	0	20	50	100	200	300
Helium	1·63	1·95	2·46	2·96	3·43	3·61	3·83	4·08	—	—
Neon	0·49	—	0·82	—	1·09	—	—	1·33	—	—
Argon	0·14	0·19	0·26	0·32	0·39	0·42	0·45	0·52	—	—
Krypton	—	—	—	—	0·21	—	—	—	—	—
Xenon	—	—	—	—	0·12	—	—	0·16	—	—
Hydrogen	—	2·20	2·78	3·50	4·19	4·45	—	5·47	6·34	7·39
Deuterium	—	—	—	—	3·06	—	—	3·77	—	—
Chlorine	—	—	—	—	0·19	—	—	—	—	—
Oxygen	0·20	0·28	0·39	0·49	0·58	0·62	0·68	0·76	—	—
Nitrogen	0·21	0·29	0·39	0·49	0·57	0·61	0·66	0·73	0·85	0·95
Air	0·21	0·37	0·39	0·49	0·576	0·61	0·64	0·74	0·88	—
Mercury vapour	—	—	—	—	—	—	—	—	0·18	—
Water vapour	—	—	—	—	—	—	—	0·60	0·78	1·03
Hydrogen sulphide	—	—	—	—	0·30	—	—	—	—	—
Sulphur dioxide	—	—	—	—	0·20	—	—	—	—	—
Carbon disulphide	—	—	—	—	0·16	—	—	—	—	—
Ammonia	—	—	—	0·41	0·52	—	—	0·81	1·17	(1·58)
Nitrous oxide	—	—	—	0·30	0·36	—	—	0·50	—	—
Nitric oxide	—	—	—	0·47	0·56	—	0·54	0·74	0·91	1·07
Carbon monoxide	0·19	0·26	0·36	0·46	0·53	—	—	—	—	—
Carbon dioxide	—	—	—	—	0·34	0·38	—	0·50	0·68	0·84
Compound	**−180**	**−150**	**−100**	**−50**	**0**	**6·5**	**20**	**50**	**100**	**200**
Methane	—	—	—	0·58	0·73	—	0·79	—	1·11	1·52
Ethane	—	—	—	0·32	0·43	—	0·49	0·60	0·77	1·13
Ethylene	—	—	—	0·31	0·40	—	—	—	0·64	—
Acetylene	—	—	—	—	0·44	—	—	—	0·72	1·01
Propane	—	—	—	—	0·36	—	0·41	—	0·63	0·96
n-Butane	—	—	—	—	0·32	—	0·37	—	—	—
iso-Butane	—	—	—	—	0·33	—	—	—	—	—
n-Pentane	—	—	—	—	0·31	0·32	0·34	—	0·52	0·81
iso-Pentane	—	—	—	—	0·30	0·31	0·34	0·40	0·52	0·82
n-Hexane	—	—	—	—	0·30	—	—	—	0·47	0·75
Hexylene	—	—	—	—	0·24	0·26	0·29	—	0·44	—
Cyclohexane	—	—	—	—	—	—	—	—	0·42	—
Benzene	—	—	—	—	0·21	0·22	0·25	0·30	0·41	0·68
Methyl chloride	—	—	—	—	0·22	0·23	0·25	0·30	0·38	0·57
Methyl bromide	—	—	—	—	0·15	0·15	0·17	0·20	0·25	—
Methyl iodide	—	—	—	—	0·11	0·12	0·12	0·15	0·18	—
Methylene chloride	—	—	—	—	0·16	0·16	0·17	0·20	0·25	0·37
Chloroform	—	—	—	—	0·15	0·16	0·17	0·19	0·24	0·32
Carbon tetrachloride	—	—	—	—	0·14	0·15	0·15	0·17	0·21	0·27
Ethyl chloride	—	—	—	—	0·22	0·24	0·26	0·31	0·39	0·59
Ethyl bromide	—	—	—	—	0·17	—	—	—	—	—
Ethyl iodide	—	—	—	—	0·14	0·15	—	—	—	—
Methanol	—	—	—	—	0·34	0·35	—	—	0·52	0·75
Ethanol	—	—	—	—	0·33	—	0·36	0·42	0·50	0·82
Acetone	—	—	—	—	0·23	0·24	0·26	0·31	0·40	0·65
Methyl acetate	—	—	—	—	0·24	0·25	0·28	—	—	—
Diethyl ether	—	—	—	—	0·31	0·33	0·35	0·42	0·54	0·82
Ethyl acetate	—	—	—	—	0·22	—	0·25	0·29	0·39	0·58

Continued overleaf

TABLE 5.3 (*cont.*)

Compound	Temperature, °C									
	−180	−150	−100	−50	0	6·5	20	50	100	200
Methylamine	—	—	—	—	—	0·38	—	—	—	—
Ethylamine	—	—	—	—	—	0·34	—	—	—	—
Dimethylamine	—	—	—	—	—	0·35	—	—	—	—
Propylamine	—	—	—	—	—	0·30	—	—	—	—
Trimethylamine	—	—	—	—	—	0·33	—	—	—	—
Butylamine	—	—	—	—	—	0·30	—	—	—	—
Diethylamine	—	—	—	—	—	0·30	—	—	—	—
Amylamine	—	—	—	—	—	0·28	—	—	—	—
Dipropylamine	—	—	—	—	—	0·26	—	—	—	—
Triethylamine	—	—	—	—	—	0·27	—	—	—	—

It is a consequence of Eq. (5.5) that the value of Maxwell's modulus, i.e. the ratio of the thermal conductivity $\lambda°$ to the product of viscosity $\mu°$ and heat capacity c_v at constant volume, should be approximately constant for a given gas, being independent of temperature in the region of moderate pressures

$$K = \frac{\lambda°}{\mu° c_v} \tag{5.33}$$

According to Eucken's equation (5.6), the value of K should decrease as the temperature is increased, because the quantity c_v rises with increase in temperature. The experimental results, however, do not support these conclusions. This may be seen from Table 5.4 where some values of K, listed by Reid and Sherwood,[306] are given. It may be seen from this table that K for some gases increases as the temperature rises, whereas in other cases it decreases or remains virtually constant.

The thermal conductivity of a gas, like the gas viscosity, is independent of pressure only at moderate pressures; on the other hand, at very low pressures (below several millimetres of mercury) as well as at elevated pressures (above a dozen or more atmospheres) the thermal conductivity of a gas rises with increase in pressure.

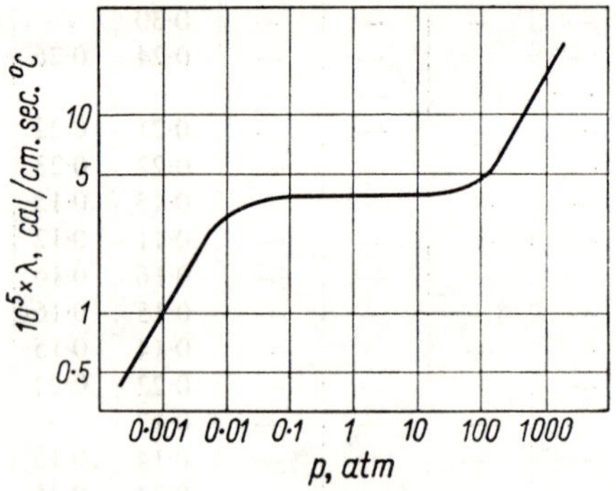

FIG. 5.4. Variation of the thermal conductivity of argon with pressure at 0°C (after Franck[108]).

By way of illustration, the effect of pressure p on the thermal conductivity (λ) of argon is presented in Fig. 5.4. Owing to the use of a log–log scale, a very wide pressure range has been presented in the graph. It is seen that λ is approximately independent of pressure p over the range 0·01 to 100 atm, and fairly accurately so over the range 0·01 to 10 atm.

TABLE 5.4. EFFECT OF TEMPERATURE UPON $K = \lambda^\circ/\mu^\circ c_v$

Compound	T, °K	$\lambda^\circ/\mu^\circ c_v$
C_2H_2	198	1·63
	273	1·51
	323	1·52
	373	1·52
CO_2	195	1·80
	200	1·67
	273	1·67
	380	1·75
	473	1·56
	598	1·55
	631	1·79
	1273	1·58
CS_2	273	1·58
CO	100	1·76
	200	1·85
	273	1·86
	380	1·94
Cl_2	273	1·74
H_2	91	2·35
	200	2·08
	273	2·03
	473	2·12
	673	2·19
NO	100	1·63
	200	1·76
	273	1·83
	380	1·92
N_2	273	1·95
	500	1·86
	700	1·74
	873	1·83
	1273	1·73
O_2	100	1·81
	200	1·91
	273	1·92
	380	2·00
CH_3—CO—CH_3	273	1·28
	373	1·25
	457	1·30
CH_2Cl_2	273	1·42
	373	1·40
	485	1·40
n-C_5H_{12}	273	1·32
C_6H_6	273	1·51
	373	1·48
	486	1·49
n-C_4H_{10}	273	1·35
	373	1·31
i-C_4H_{10}	273	1·37
	373	1·31
CCl_4	273	1·38
	373	1·30
	457	1·28
$CHCl_3$	273	1·48
	358	1·45
	457	1·42
C_6H_{12} (cyclohexane)	339	1·40
	393	1·36
	433	1·38
C_2H_6	203	1·57
	273	1·57
	373	1·57
CH_3COO—C_2H_5	319	1·24
	373	1·24
	457	1·27
C_2H_5—O—C_2H_5	273	1·35
	358	1·46
	486	1·45
C_2H_4	273	1·57
	323	1·53
	373	1·46
C_6H_{14}	273	1·21
	358	1·28
	433	1·37
CH_4	90	1·68
	140	1·73
	200	1·79
	273	1·78
	380	1·78
CH_3Cl	273	1·55
	373	1·57
	486	1·59
C_3H_8	273	1·47
	373	1·42
	413	1·46

Continued overleaf

TABLE 5.4 (*cont.*)

Compound	T, °K	$\lambda°/\mu°c_v$	Compound	T, °K	$\lambda°/\mu°c_v$
CH_3CHO	313	1·26	CH_3OH	273	1·45
	353	1·29		358	1·37
	393	1·31		400	1·39
				433	1·37
CH_3CN	353	1·22	n-C_3H_7OH	304	1·46
	393	1·23		378	1·39
				402	1·36
NH_3	213	1·51	i-C_3H_7OH	304	1·54
	273	1·54		376	1·44
	373	1·42		400	1·41
			SO_2	273	1·48
C_2H_5OH	293	1·44			
	401	1·35	H_2O	373	1·41
				473	1·40
				573	1·41
H_2S	273	1·53		673	1·42

The effect of pressure on the thermal conductivity of gases composed of polyatomic molecules, is very considerable. The thermal conductivity of methyl alcohol at 110°C, increases by about 4%, as the pressure is raised by 1 atm.[140] A similar effect of pressure on conductivity has been observed for most vapours of organic compounds. In the case of polar gases, the increase in conductivity is probably caused by the formation of associated molecules (Schäfer[316]).

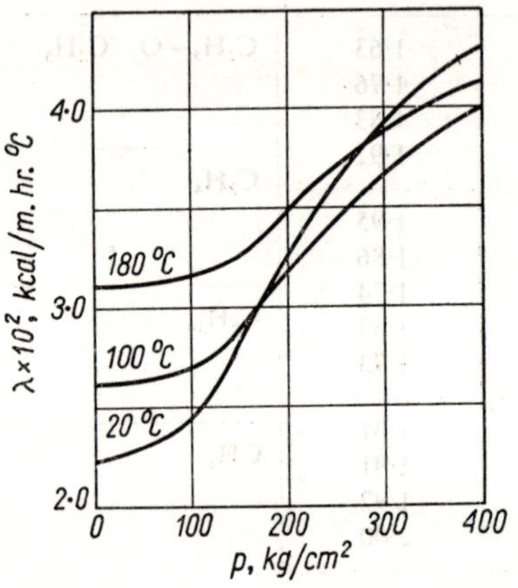

FIG. 5.5. Variation of the thermal conductivity of air with pressure and temperature (after Stolarov[349]).

The form of the variation of thermal conductivity (λ) of a gas with pressure (p) depends on the temperature, the dependence of λ on p being usually more pronounced at lower temperatures. Curves representing the relationship between λ and p for air are given in Fig. 5.5. Since a linear scale has been used for the pressure axis on this figure, the behaviour of λ in the region of very low pressures is not visible.

It should be mentioned that in the range of reduced pressures from about 0·5 to about 2, the form of the variation of thermal conductivity with pressure is not the same for all gases. Within this range, the value of λ usually rises very rapidly with increase in pressure, though cases are known where the thermal conductivity shows a maximum or minimum as, for instance, in the case of ethane.[306] In Figs. 5.6 and 5.7, isobars are shown giving the thermal conductivities of N_2 and H_2O in the vicinity of the saturation curve.

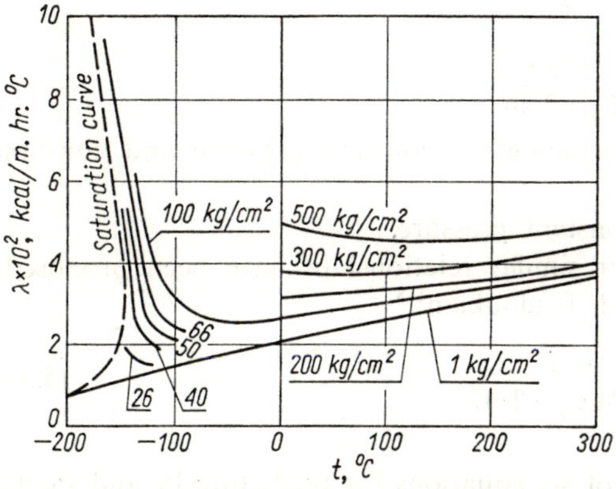

FIG. 5.6. Variation of the thermal conductivity of nitrogen with pressure and temperature. The saturated vapour curve is shown on the left (after Franck[107]).

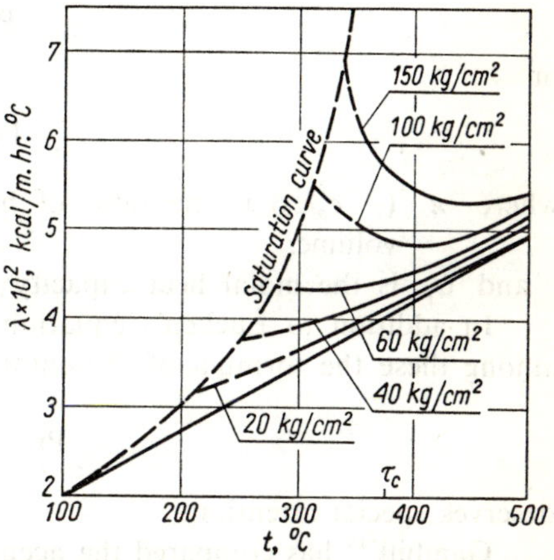

FIG. 5.7. Thermal conductivity of saturated and superheated water vapour (after Franck[107]).

Vargaftig and Tarzimanov[383] have tabulated the results of measurements of the thermal conductivity of water vapour in the temperature region of 100–800°C and in the pressure region of 1–600 kg/cm^2.

Vargaftig and Zaitseva[384] determined the thermal conductivity of D_2O in the vapour state and obtained the following values

t, °C	100	200	300	400	500
$\lambda_{D_2O}/\lambda_{H_2O}$	0·970	0·996	1·017	1·032	1·050

where λ_{D_2O} and λ_{H_2O} are the thermal conductivities of the vapours of D_2O and H_2O, respectively.

3. PRANDTL NUMBER FOR GASES

It is seen from Table 5.4, that the quantity

$$K = \frac{\lambda^\circ}{\mu^\circ c_v}$$

varies only slightly with temperature. The Prandtl number behaves similarly:

$$\mathrm{Pr} = \frac{c_p \mu^\circ}{\lambda^\circ} \tag{5.34}$$

A knowledge of methods of calculating the Prandtl number is of great importance not only for chemical engineering calculations but also as a means for estimating the thermal conductivities of gases. Most often the Prandtl number of gases is represented as a function of heat capacity or molal heat capacity of these gases. Eucken's formula was one of the first to be used and was obtained by transformation of Eqs. (5.6)

$$\mathrm{Pr} = \frac{9 - \frac{5}{\varkappa}}{4} \tag{5.35}$$

or

$$\mathrm{Pr} = \frac{C_p}{C_p + 2{\cdot}46}$$

where $\varkappa$ $(= c_p/c_v)$ is the ratio of heat capacities at constant pressure and constant volume,

and C_p is the molal heat capacity at constant pressure.

In addition to Eucken's equation other similar relationships have been proposed; among these the formula of Hilsenrath and Touloukian[158]

$$\mathrm{Pr} = \frac{C_p}{1{\cdot}204\, C_p + 1{\cdot}47} \tag{5.36}$$

deserves special attention.

Gambill[115] has compared the accuracy of six equations for predicting Pr and shown that when the molar heat capacity, C_p, of a gas is smaller than 15, best results are obtained by using Eucken's equation (5.35). If $C_p > 15$, more accurate results may be obtained by using Eq. (5.36). The maximum error of estimation in both cases is of the order of $\pm 15\%$.

According to Bromley[47] the value of the Prandtl number depends on the structure and properties of the gas molecules:

(a) For monatomic gases, e.g. Ar, He, Ne, Xe,

$$\mathrm{Pr} = 0{\cdot}67 \text{ (maximum error } \pm 5\%)$$

(b) For non-polar, "linear" molecules, e.g. O_2, Cl_2, CO, C_2H_2, CS_2, CO_2, N_2O, NO,

$$\mathrm{Pr} = 0{\cdot}73 \text{ (maximum error } \pm 15\%)$$

(c) For non-polar or almost non-polar, non-linear molecules, e.g. CH_4, CH_3—CO—CH_3, C_6H_6, C_2H_6, CCl_4, CH_2Cl_2,

$$\mathrm{Pr} = 0{\cdot}79 \text{ (maximum error } \pm 15\%)$$

(d) For strongly polar molecules, e.g. CH_3OH, C_2H_5OH, SO_2, H_2S, HCl,

$$\mathrm{Pr} = 0{\cdot}86 \text{ (maximum error } \pm 8\%)$$

Gambill[115] is of the opinion that the use of Eqs. (5.35) (when $C_p > 15$) and (5.36) (when $C_p < 15$) is about equally accurate with the use of Bromley's indications (a)—(d) for predicting the Prandtl number. The values of the Prandtl number, listed in Table 5.5, have been calculated on the basis of measured values of $\lambda°$, $\mu°$, c_p and on the basis of Eucken's equation (5.35).

As already stated, the value of the Prandtl number depends only slightly on temperature. An illustration of this is provided in Table 5.6 where Pr numbers calculated for air[249] are shown as a function of temperature.

TABLE 5.5. VALUES OF THE PRANDTL NUMBER DETERMINED EXPERIMENTALLY AND CALCULATED FROM EUCKEN'S EQUATION (5.35) (IN THE REGION OF MODERATE PRESSURES)

Gas	H_2	O_2	N_2	SO_2	NH_3	H_2S	CH_4	C_2H_4	CO_2	CO	H_2O
Pr exp.	0·74	0·74	0·74	0·80	0·78	0·77	0·79	0·83	0·80	0·74	0·78
Pr calc.	0·735	0·735	0·738	0·794	0·787	0·768	0·787	0·811	0·784	0·738	0·764

TABLE 5.6. VARIATION OF THE PRANDTL NUMBER WITH TEMPERATURE FOR AIR (AT $p = 1$ kg/cm²)

t, °C	−100	0	100	200	400	600	800	1000	1400	1800
Pr	0·742	0·723	0·722	0·722	0·722	0·723	0·725	0·727	0·736	0·744

The values of the Prandtl number for saturated and superheated vapours depend strongly on temperature and pressure; the values found for steam by Rubin[313] are listed in Table 5.7.

TABLE 5.7. PRANDTL NUMBER FOR STEAM†

Pressure kg/cm²	Saturated steam	Temperature, °C		
		316	427	538
1	1·063	1·003	1·002	0·995
34	1·90	1·40	1·19	1·12
68	2·67	2·06	1·44	1·27
102	3·81	3·53	1·73	1·42
136	5·27	—	2·09	1·60

† According to the Bureau of Standards[115] and Mikheev,[249] the values of Rubin are 30% too high.

The effect of pressure upon the Prandtl number for gases may be estimated by the use of Enskog's Eqs. (3.56) and (5.29) for the variation of the thermal conductivity and viscosity of a gas with pressure,

$$\frac{(\mathrm{Pr})_{T,p}}{(\mathrm{Pr})_{T,1}} = \frac{(c_p)_{T,p}}{(c_p)_{T,1}} \frac{\left(\frac{1}{y}+0{\cdot}80+0{\cdot}761\,y\right)}{\left(\frac{1}{y}+1{\cdot}20+0{\cdot}755\,y\right)} \tag{5.37}$$

where Pr is the Prandtl number,
c_p is the heat capacity at constant pressure,
$y = z-1$ where z is the compressibility of the real gas according to Eqs. (3.57)–(3.60),
T, p are subscripts which denote that the quantity is given for temperature T and pressure p,
and $T, 1$ are subscripts which denote that the quantity is given for temperature T and a pressure of 1 kg/cm².

Equation (5.37) gives fairly accurate results. It is necessary, however, to know the p–V–T relations of the real gas under consideration in order to calculate the value of y (or z).

McLaughlin (cf. Ref. 5), starting from Enskog's equations, found for nitrogen at 50°C the following relationship between the thermal conductivity and pressure:

Pressure, atm	12·2	116·5	300	521	783	889	1080
$\lambda_{calc}\times 10^6$, cal/cm.sec.°C	72·0	85·3	115	150	189	204	228
$\lambda_{exp}\times 10^6$, cal/cm.sec.°C	67·6	82·8	115	152	189	203	222

4. APPLICATION OF THE THEOREM OF CORRESPONDING STATES TO THE ESTIMATION OF THE THERMAL CONDUCTIVITIES OF GASES

As has been shown in Chapter 3, section 6b both the reduced viscosity $\mu_r = \mu/\mu_c$ of a gas and the ratio of gas viscosity at an elevated pressure to the viscosity in the region of moderate pressures μ_p/μ° are functions of the reduced parameters: $T_r = T/T_c$ and $p_r = p/p_c$. Since the viscosities and the thermal conductivities of gases are interrelated (Eq. (5.5)) it might be anticipated that similar relationships should exist between the thermal conductivity and the reduced parameters.

Experiments confirm the validity of this presumption, and various authors have presented numerical or graphical methods for estimating the thermal conductivities of gases on the basis of the theory of corresponding states.

For monatomic gases (Ne, Ar, Kr, Xe), the reduced conductivity at moderate pressures is a unique function of the reduced temperature in the temperature range from −183°C to 1100°C (deviations are about 4·5%). The dependence of λ_r on T_r for reduced temperatures in the range 0·8 to 8 is accurately given by the equation:[358]

$$\lambda_r = \frac{4{\cdot}24\times 10^{-6}}{1+1{\cdot}00/T_r} \tag{5.38}$$

(average error *ca.* 1%).

The modern kinetic theory of gases, which takes account of the intermolecular forces also leads to a generalized relationship between λ_r° and T_r (Eq. (5.27)).

a. Reduced Thermal Conductivity of Gases

Gamson[123] has examined the variation of the reduced thermal conductivity with reduced temperature and reduced pressure and has shown that the following general relationship, which is approximately valid for gases, exists

$$\lambda_r = F(T_r, p_r) \tag{5.39}$$

In this equation

$$\lambda_r = \lambda_p/\lambda_c \tag{5.40}$$

where λ_r is the reduced thermal conductivity of the gas,
λ_p is the thermal conductivity of the gas at pressure p,
λ_c is the thermal conductivity of the gas at the critical point,
T_r ($= T/T_c$) is the reduced temperature,

T_c is the critical temperature,
p_r ($= p/p_c$) is the reduced pressure,
and p_c is the critical pressure.

Gamson has constructed a chart from which, knowing the critical parameters T_r and p_r of a gas, it is possible to determine the value of the reduced thermal conductivity λ_r, corresponding to these parameters (Fig. 5.8). An analogous chart for monatomic gases has been prepared by Owens and Thodos.[273]

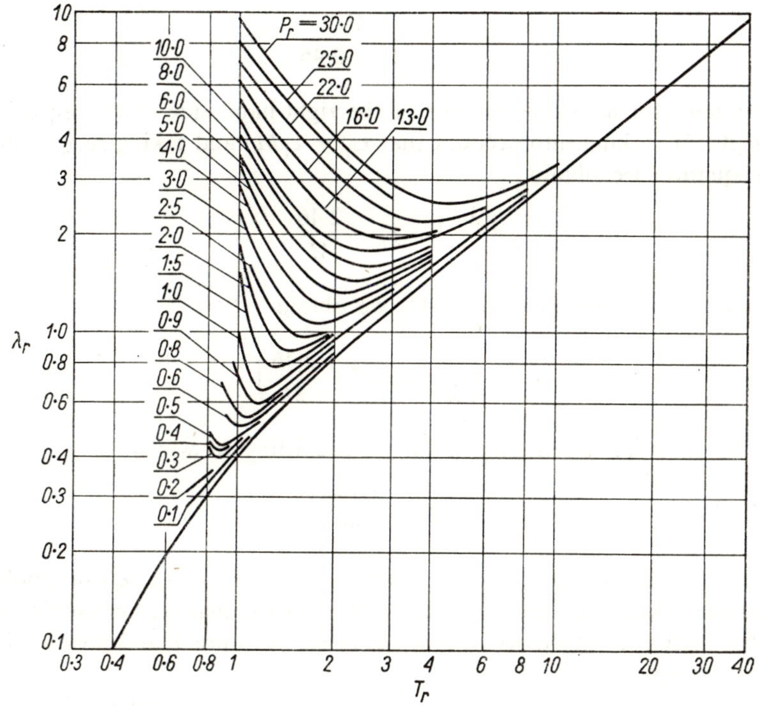

FIG. 5.8. Generalized relationship between the reduced conductivity $\lambda_r = \lambda/\lambda_c$ of a gas and reduced temperature $T_r = T/T_c$ and reduced pressure $p_r = p/p_c$ (after Gamson[123]).

Having estimated λ_r, values of λ_p can be derived from the equation:

$$\lambda_p = \lambda_r \lambda_c \tag{5.41}$$

It is obvious that in order to apply Gamson's method the value of the thermal conductivity λ_c at the critical point is required. Gamson has shown that in the case of 25 different gases the ratio of the reduced thermal conductivity λ_r° to the reduced viscosity μ_r° (both quantities being measured at moderate pressures where λ and μ do not depend on pressure), is a unique function of the reduced temperature

$$\lambda_r^\circ/\mu_r^\circ = f(T_r) \tag{5.42}$$

The values of this function are listed in Table 5.8. After introducing the values of λ_r° and μ_r° into Eq. (5.42), we obtain

$$\left(\frac{\lambda_r^\circ}{\mu_r^\circ}\right) = \frac{\lambda^\circ}{\lambda_c}\frac{\mu_c}{\mu^\circ} \tag{5.43}$$

TABLE 5.8. VALUES OF THE FUNCTION $\lambda_r^\circ/\mu_r^\circ = f(T_r)$

T_r	0·4	0·5	1·0	1·4	2·0	3·0	5·0	10	40
$\lambda_r^\circ/\mu_r^\circ$	0·565	0·655	0·855	0·931	0·970	1·018	1·11	1·192	1·395

Now on combining Eqs. (5.34) and (5.35) it is found that

$$\frac{c_p^\circ \mu^\circ}{\lambda^\circ} = \frac{C_p^\circ}{C_p^\circ + 2{\cdot}46}$$

Dividing the numerator and denominator of the right-hand side of this equation by the molecular weight M (taking into account that $C_p^\circ = c_p^\circ M$) and taking reciprocals of both sides of the equation we obtain

$$\left(\frac{\lambda^\circ}{\mu^\circ}\right) = c_p^\circ + \frac{2{\cdot}46}{M} \tag{5.44}$$

After substituting this value into Eq. (5.43) we obtain

$$\lambda_c = \left(\frac{\mu_r^\circ}{\lambda_r^\circ}\right)\mu_c\left(c_p^\circ + \frac{2{\cdot}46}{M}\right) \tag{5.45}$$

An estimate for μ_c is provided by Eq. (3.66), according to which

$$\mu_c = 7{\cdot}70\frac{M^{1/2}p_c^{2/3}}{T_c^{1/6}} \tag{5.46}$$

λ_c may be calculated from Eq. (5.45) on inserting the value of $(\mu_r^\circ/\lambda_r^\circ)$ interpolated from Table 5.8 and the value of μ_c given by Eq. (5.46). To summarize, Gamson's method for calculating the thermal conductivity λ_p of a gas at temperature T and pressure p proceeds in the following steps:

(a) Find in the tables the critical constants T_c and p_c for the gas and also its heat capacity c_p° at low pressure and temperature T.
(b) Calculate the reduced parameters T_r and p_r (Eq. 5.40).
(c) For the calculated value of T_r interpolate from Table 5.8 the value of $\lambda_r^\circ/\mu_r^\circ$ (T_r corresponds to the same temperature for which c_p° is calculated).
(d) Calculate the value of the viscosity μ_c at the critical point from Eq. (5.46).
(e) Substitute in Eq. (5.45) the values of $\lambda_r^\circ/\mu_r^\circ$ and μ_c calculated in steps (c) and (d), and thus find the thermal conductivity λ_c at the critical point,
(f) Read from Fig. 5.8 the value of the reduced thermal conductivity λ_r for the values of T_r and p_r calculated in step (b).
(g) Insert the values of λ_c and λ_r into Eq. (5.41), and hence calculate the required value of the thermal conductivity λ_p at pressure p and temperature T.

It should be mentioned that alternative methods exist for estimating the thermal conductivity λ_c of a gas at the critical point. One such method utilizes the following relationship which is approximately true at the critical temperature and moderate pressures:

$$\left(\frac{\mu_r^\circ}{\lambda_r^\circ}\right)_{T_c} \approx 1{\cdot}17 \tag{5.47}$$

Now it may be seen from Uyehara and Watson's graph (Fig. 3.14), that at the critical temperature and a "moderate" pressure

$$(\mu_r^\circ)_{T_c} = 0{\cdot}45$$

hence,

$$(\lambda_r^\circ)_{T_c} = \frac{(\mu_r^\circ)_{T_c}}{1{\cdot}17} = \frac{0{\cdot}45}{1{\cdot}17} = 0{\cdot}385 \tag{5.48}$$

or,

$$(\lambda_r^\circ)_{T_c} = \frac{(\lambda^\circ)_{T_c}}{\lambda_c} = 0{\cdot}385$$

and thus

$$\lambda_c = 2{\cdot}60\,(\lambda^\circ)_{T_c} \tag{5.49}$$

A modification to this method has been proposal by Schaefer and Thodos[318] who, on the basis of data for N_2, O_2 and H_2, found the relationship:

$$\lambda_c = \frac{0{\cdot}8980\,(\lambda^\circ)_{T_c}}{z_c} \tag{5.50}$$

where z_c ($= p_cV_c/RT_c$) is the compressibility factor in the critical state. This equation makes it possible to estimate the thermal conductivity λ_c of a gas at the critical point, if the thermal conductivity of this gas at the critical temperature, but under a moderate pressure, is known.

An alternative method for calculating λ_c, when the thermal conductivity λ_p of the gas at some pressure p and temperature T is known, is to use Gamson's graph. The value of λ_r is read from the graph for the appropriate T_r and p_r, and the value of λ_c is calculated from Eq. (5.41).

The above two methods, which utilize Eqs. (5.49) and (5.41), respectively, as a means of calculating λ_c from a known value of λ_p, are much more reliable and accurate than those involving the use of Eq. (5.45), in which no experimental value of λ_p is required. Values of λ_c for a number of compounds are shown in Table 5.9.

The accuracy of Gamson's method is rather low; the author estimates the average error to be $\pm 15\%$; in some cases calculations based on an estimated value of the thermal conductivity λ_c at the critical point are very uncertain and the error may then be much greater than that quoted above. More recently other graphs showing the relationship between λ_r and T_r and p_r have been published. These graphs are somewhat similar to that of Gamson but are based on the work of Owens and Thodos,[273] on the thermal conductivities of the noble gases as well as on the work of Schaefer and Thodos[317] on the conductivity of hydrogen. Attempts at employing the Owens and Thodos graph for the calculation of the thermal conductivities of di-, tri- and polyatomic gases have shown that in the case of diatomic gases the average error is of the order of 4–5%, while the maximum errors are about 37%. In the case of triatomic gases the errors are about 50 per cent greater, and for polyatomic gases the method (which is based on data for monatomic gases) fails completely. In the case of ethane, for instance, the average error is 32%, and the maximum error 119%.

Stiele and Thodos (cf. Ref. 5) present several graphs showing the variation of λ_r with T_r and p_r for different gases. A graph of this type for ethylene has been published by Owens and Thodos.[274] The average deviation of the values of the thermal conductivity for ethylene, read from the graph, from the values determined experimentally, is equal

TABLE 5.9. THERMAL CONDUCTIVITIES λ_c (kcal/m.hr.°C) OF VAPOURS AT THE CRITICAL POINT (AFTER GAMSON[123] AND THODOS (REF. 5))

Compound	λ_c	Compound	λ_c
Acetone	0·0682	Helium	0·0128
Acetylene	0·0473	Heptane	0·0457
Ammonia	0·0793	Hydrogen	0·0572
Argon	0·0220	Hydrogen bromide	0·0296
Benzene	0·0909	Hydrogen chloride	0·0404
Bromine	0·0244	Hydrogen fluoride	0·0936
Carbon dioxide	0·0387	Iodine	0·0228
Carbon monoxide	0·0310	Krypton	0·0139
Carbon tetrachloride	0·0309	Methane	0·0426
Chlorine	0·0346	Neon	0·0185
Chloroform	0·0365	Nitrogen	0·0308
Dichlorodifluoromethane	0·0326	Nitrogen oxide	0·0425
Diethyl ether	0·0754	Nitrous oxide	0·0394
Ethane	0·0484	Oxygen	0·0374
Ethyl acetate	0·0714	Sulphur dioxide	0·0309
Ethyl chloride	0·0534	Water	0·1161
Ethylene	0·0426	Xenon	0·0123
Fluorine	0·0343		

to 1·87%. These authors used their graph to estimate the thermal conductivities of binary mixtures of ethylene with nitrogen and carbon dioxide, using pseudocritical parameters estimated from Kay's rule.[193] The errors in the predicted values ranged from 1·4% to 6·2%. This graph should not be used to estimate the thermal conductivities of *liquids* other than liquefied ethylene. On the other hand, this graph may be used to predict the thermal conductivities $\lambda°$ of gases other than ethylene ($\lambda°$ is the thermal conductivity of the gas at a moderate pressure, i.e. at a pressure in the range for which the conductivity is independent of pressure). Owens and Thodos use a log–log scale to represent $\lambda_r°$ as a function of T_r and obtain straight line plots indicating that the following relationship is obeyed:

$$\lambda_r° = 0{\cdot}364 T_r^{1{\cdot}786} \tag{5.51}$$

where $\lambda_r°$ is the reduced thermal conductivity of the gas at the reduced temperature T_r ($\lambda°$ refers to a moderate pressure). For gases other than ethylene

$$\lambda_r° = \lambda°/\lambda_c^* \tag{5.52}$$

where λ_c^*, the quasicritical thermal conductivity, may be calculated from Eq. (5.52), if one experimentally determined value of conductivity $\lambda°$ of the given compound is known. If no value of $\lambda°$ is known, the following equation may be employed:

$$\lambda_c^* = 62{\cdot}2 \times 10^{-5} \frac{M^{1/2} p_c^{2/3}}{T_c^{1/6}} \tag{5.53}$$

Having evaluated λ_c^*, thermal conductivities may be calculated from the equation

$$\lambda° = 0{\cdot}364 T_r^{1{\cdot}786} \lambda_c^* \tag{5.54}$$

which is obtained by combining Eqs. (5.51) and (5.52). Stiele and Thodos have checked these equations on many different series of compounds and for reduced temperatures in the range 0·5 to 1·0, they obtained the following mean deviations from the measured values: for aliphatic hydrocarbons −1·7%, for alcohols −0·9%, for ethers −1·3%, for halogenated compounds −2·2%, for aldehydes −1·1%, for esters −1·8%, for ketones −1·5%, for amines −1·2%, for cyclic hydrocarbons −3·2%. The authors give values of λ_c^* for several score of compounds. A few of their values of λ_c^*, expressed in kcal/hr.m.°K, are presented below:

Acetone	0·0720	Ethyl acetate	0·0770
Carbon tetrachloride	0·0382	Ethyl chloride	0·0584
Chloroform	0·0461	Ethyl ether	0·0785
Ethane	0·0525		

When these values are compared with the figures for λ_c shown in Table 5.9 it is evident that the values of λ_c^* and λ_c are not identical.

EXAMPLE 5.4 (after Owens and Thodos[274]). Estimate the thermal conductivity λ_1° of ethyl ether at a pressure of 1 atm and a temperature (t_1) of 210°C.

At a temperature (t_2) of 100°C and a pressure of 1 atm, the thermal conductivity of ethyl ether (λ_2°) is $5{\cdot}30\times10^{-5}$ cal/(cm.sec.°K). The critical temperature of ethyl ether (t_c) is 193·8°C.

Solution

The reduced temperature (T_{r2}) at the temperature t_2 for which a value of λ° is known is given by

$$T_{r,2} = \frac{100+273{\cdot}2}{193{\cdot}8+273{\cdot}2} = 0{\cdot}798$$

From Eq. (5.51), the reduced thermal conductivity at this temperature is

$$\lambda_r^\circ = 0{\cdot}364\times0{\cdot}798^{1{\cdot}786} = 0{\cdot}243.$$

Hence

$$\lambda_c^* = (5{\cdot}30\times10^{-5})/0{\cdot}243 = 21{\cdot}8\times10^{-5} \text{ cal/(cm.sec.°K)}$$

or

$$\lambda_c^* = 21{\cdot}8\times10^{-5}\times3{\cdot}6\times10^{2} = 0{\cdot}0785 \text{ kcal/(m.hr.°K)}$$

which is in agreement with the value given in the table above.

Calculation of λ_1° at t_1:

$$T_{r,1} = \frac{210+273{\cdot}2}{193{\cdot}8+273{\cdot}2} = 1{\cdot}04$$

$$\lambda_1^\circ = 0{\cdot}364\times1{\cdot}04^{1{\cdot}786}\times21{\cdot}8\times10^{-5} = 8{\cdot}50\times10^{-5} \text{ cal/cm.sec.°K}$$

The calculated value of λ_1° agrees well with the experimental one, which is $8{\cdot}55\times10^{-5}$ cal/(cm.sec.°K).

Tsederberg[371] has investigated the variation of the reduced thermal conductivity λ_r° with the reduced temperature T_r and finds that

$$\log\frac{\lambda^\circ}{(\lambda^\circ)_{T_c}} = \log\lambda_r^\circ = a\log T_r + b\log T_r^2 + c\log T_r^3 \qquad (5.55)$$

where λ° denotes the thermal conductivity of the gas at temperature T and a moderate pressure. The values of the coefficients a, b and c in the equation depend on the number n of atoms in the molecule, different sets of coefficients being recommended for $T_r > 1$ and for $T_r < 1$. For the case when $T_r < 1$, the values of the coefficients may be taken from Fig. 5.9. For di- and triatomic gases, the values of the coefficients a, b, and c are presented below:

		a	b	c
Diatomic gases,	$T_r < 1$	1·000	0	0
Triatomic gases,	$T_r < 1$	1·325	−0·6311	−1·540
Diatomic gases,	$T_r > 1$	+0·9340	−0·0323	−0·0511
Triatomic gases,	$T_r > 1$	+1·440	+0·0115	+0·2141

If Eq. (5.55) is to be used for prediction purposes, one experimentally determined value of $\lambda°$ must be available. (This value may then be substituted into Eq. (5.55) to give $\lambda^{\circ}_{T_c}$.)

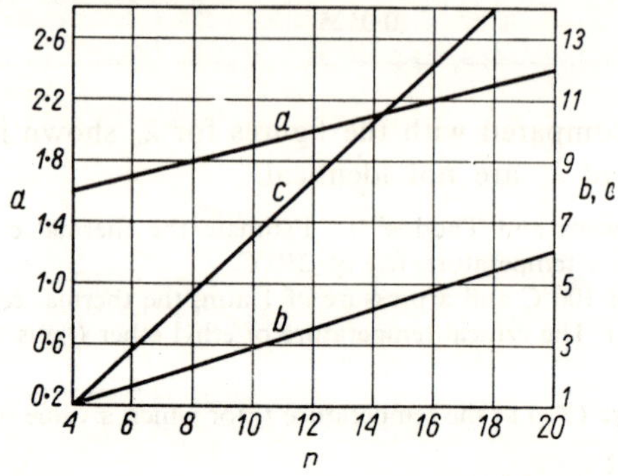

FIG. 5.9. Dependence of the coefficients a, b, c on the number of atoms in the molecule, when $T_r < 1$.

When this method is used, the average error of estimation is ±4·7%, and the maximum error ±11%.

b. Relationship Between the Ratio ($\lambda_p/\lambda°$) and the Reduced Parameters T_r and p_r (λ_p is the thermal conductivity of the gas at an elevated pressure and $\lambda°$ is the conductivity at the same temperature, but at a "moderate" pressure)

In order to use the method described in the previous section (4a), a knowledge is required of the thermal conductivity of the gas at the critical point. This quantity usually has to be predicted, a procedure which is likely to be subject to appreciable error. More accurate predictions are possible if all that is required is to calculate the value of the thermal conductivity λ_p of a gas under a high pressure, if the value $\lambda°$ of this coefficient at the same temperature but under a moderate pressure is known.

As Comings and Mayland,[65] Comings and Nathan[66] as well as Leng and Comings[66] have shown, the ratio $(\lambda_p/\lambda°)_T$ is nearly a unique function of the reduced temperature ($T_r = T/T_c$) and of the reduced pressure ($p_r = p/p_c$), i.e.

$$(\lambda_p/\lambda°)_T = F(T_r, p_r) \tag{5.56}$$

On the basis of this finding, generalized graphs of $(\lambda_p/\lambda°)_T$ vs. T_r and p_r have been constructed.[220] These graphs are presented in Figs. 5.10 and 5.11 for the regions of high and low pressures, respectively. The use of these graphs is very simple.

(a) The values of the reduced parameters $T_r = T/T_c$ and $p_r = p/p_c$ are calculated.

(b) λ°_T for the gas under consideration is obtained from a tabulation at the temperature T ($\lambda°$ does not depend on pressure since it refers to a moderate pressure).

(c) The value of the ratio $(\lambda_p/\lambda^\circ)_T$ is read from the graph in Fig. 5.10 or Fig. 5.11.
(d) The value of λ_p is calculated from the equation

$$\lambda_p = \left(\frac{\lambda_p}{\lambda^\circ}\right)_T \lambda_T^\circ \tag{5.57}$$

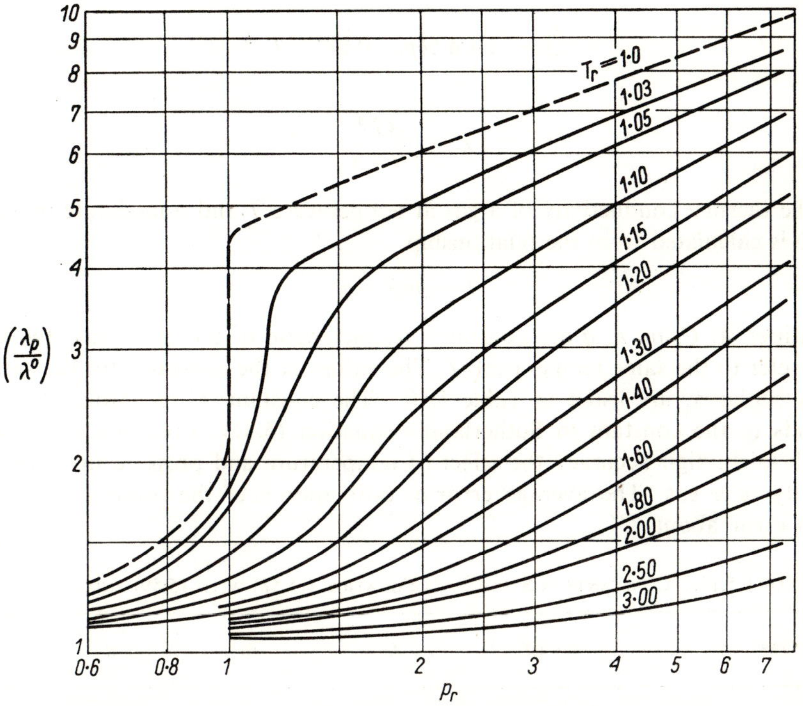

FIG. 5.10. Generalized relationship between the ratio λ_p/λ° (thermal conductivity of a gas at pressure p and temperature T to that at moderate pressure and temperature T) and reduced temperature T_r and reduced pressure p_r for the region of high pressures (after Lenoir[220], and Reid and Sherwood[306]).

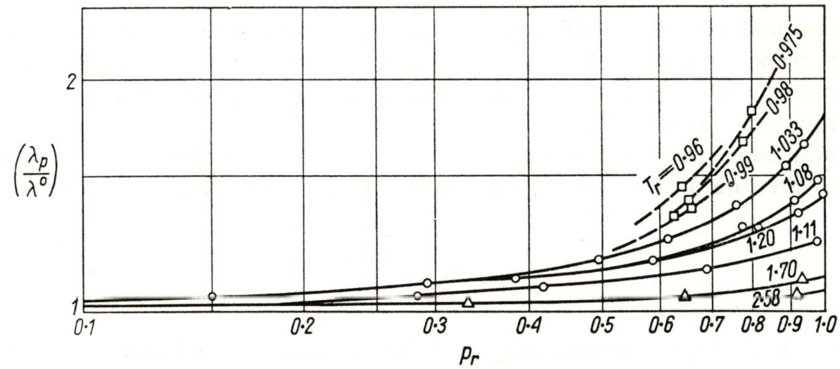

FIG. 5.11. Generalized relationship as in Fig. 12.10 for the region of pressures p_r from 0·1 to 1·0 (after Lenoir[220], and Reid and Sherwood[306]).

According to Reid and Sherwood,[306] this method is unreliable in the range of reduced pressures from 0·5 to 2·0; it may give results which are subject to large errors, also if it is applied to gases composed of polyatomic molecules with a complex structure or to strongly polar molecules.

Stolarov[350] has shown that the variation of the ratio of $(\lambda_p/\lambda^\circ)_T$ with T_r and p_r may be represented by the equation

$$\frac{\lambda_p}{\lambda^\circ} = 1+4{\cdot}5(p_r-0{\cdot}8)^{0{\cdot}145}T_r^{-b} \tag{5.58}$$

where

$$\lambda^\circ = \frac{AT^{3/2}}{T+C} \tag{5.59}$$

denotes the thermal conductivity of a gas at temperature T and a moderate pressure. The value of b is calculated from the relationship

$$b = mp_r^n+q \tag{5.60}$$

The quantities A, C, m, n, q are constants characteristic of the given gas; the values of λ_p and λ° refer to the same temperature T. The values of the constants for the gases investigated by Stolarov, are listed in Table 5.10. The constant C in Stolarov's equation corresponds to the constant in Sutherland's equation (5.31). Stolarov's equation reproduces with fairly high accuracy the effect of temperature and pressure upon the thermal conductivity of a gas. The average error of estimation is of the order of $\pm 3\%$, and the maximum error about 7%.

TABLE 5.10. CONSTANTS IN STOLAROV'S EQUATION ((5.58) WITH (5.59) AND (5.60))

Gas	$A\times 10^6$	C	m	n	q
Hydrogen	32·1	89·1	13·5	−1·57	2
Nitrogen	4·74	102·9	13·5	−1·57	2
Air	5·63	165	13·5	−1·57	2
Methane	20·0	980	10	−1·60	3
Carbon dioxide	35·5	4337·0	13·5	−1·60	3

One essential conclusion which may be drawn from Stolarov's equation is that since the values of the constants A, C, m, n and q are different for different gases, the ratio $(\lambda_p/\lambda^\circ)_T$ cannot strictly be a unique function of T_r and p_r. Equation (5.56) cannot therefore be rigidly accurate.

By examining the available data for the variation of the thermal conductivities of gases with pressure at a constant temperature, Franck[107] has obtained the relationship:

$$\frac{\lambda_p}{\lambda^\circ} = \frac{1}{1-\dfrac{v_c}{3v}}+F(C_p-C_p^\circ) \tag{5.61}$$

where λ_p is the thermal conductivity of the gas at temperature T and pressure p,
λ° is the thermal conductivity of a gas at temperature T and a moderate pressure,
v_c is the specific volume of the gas at the critical point,

v is the specific volume of the gas at temperature T and pressure p,
C_p is the molal heat capacity of a gas at a constant pressure and temperature T,
and C_p° is the molal heat capacity of a gas at temperature T and a constant moderate pressure.

The value of the coefficient F is:

0·071 for diatomic gases, and
0·035 for polyatomic gases.

Franck's equation is reasonably accurate, the average error in the predicted conductivities being only a few per cent. A practical disadvantage in the use of this equation is that it requires a knowledge of v and C_p at the temperature T and pressure p.

5. SELECTION OF THE BEST METHOD FOR ESTIMATING THE THERMAL CONDUCTIVITY OF A GAS

A summary of the different methods, listing the data required for the calculation, and indicating the probable accuracy, is given in Table 5.11.

a. Conductivities at Moderate Pressures

All known methods for estimating the thermal conductivities of gases from other properties are rather inaccurate. The starting point for such calculations is the determination or estimation either of Maxwell's modulus K (Eq. (5.5)) or of the Prandtl number Pr (Eq. (5.34)) using the methods described earlier in this chapter. If either K or Pr are known, conductivities may be calculated from viscosity and heat capacity data using the equation

$$\lambda = \frac{c_p \mu}{\mathrm{Pr}} \quad \text{or} \quad \lambda = K \mu c_v \tag{5.62}$$

Methods for predicting heat capacities are discussed in textbooks on thermodynamics (cf. Ref. 171) and methods for predicting the viscosities of gases have been described in Chapter 3. Maxwell's modulus may be predicted using Eucken's equation (5.6) or Bromley's equations (5.16), (5.17) and (5.18); the Prandtl number can be predicted either from Eucken's equation (5.35), or from Hilsenrath and Touloukian's equation or from the list of the average values given by Bromley (cf. section 3).

The most accurate method of predicting conductivities from viscosities is to calculate the ratio λ°/μ° from Bromley's equations (5.16), (5.17) and (5.18). This method cannot always be applied, however, because the necessary values of the group contributions for computing C_{int} are not always available. In this case either Eucken's equation (if $C_p < 15$) or Hilsenrath and Touloukian's equation (if $C_p > 15$) may be applied.

The accuracy of predictions based on the evaluation of Maxwell's modulus is approximately the same as that of predictions based on the evaluation of the Prandtl number. If none of the above methods are applicable then Eq. (5.26), proposed by Reid and Sherwood, can be used.

b. Effect of Temperature on the Thermal Conductivities of Gases

Although relatively inaccurate, Sutherland's formula (5.31) is often employed. The value of the constant C in this equation depends upon the gas considered and this must be known as well as the value of λ_0, the thermal conductivity at a "moderate" pressure and 0°C.

TABLE 5.11. COMPARISON OF METHODS FOR PREDICTING THE THERMAL CONDUCTIVITY OF A GAS

Author	Equation	Data necessary for the calculations	Average error	Remarks
(a) *Calculation of the thermal conductivity of a gas at moderate pressures when no experimental data are available*				
Eucken (calculation of Maxwell's modulus)	(5.6)	μ°, $\varkappa$, c_v (or c_p)	±8%	
Bromley (calculation of Maxwell's modulus)	(5.16) – (5.19)	μ°, C_v, T_c, ρ_b, T_b, L	±5%	contributions known only for some groups
Bromley (calculation of the Prandtl number)	data from literature (Ref. 47)	μ°, c_p	±8%	
Eucken (calculation of the Prandtl number)	(5.35)	C_p, μ°	±8%	for $M < 15$
Hilsenrath, Touloukian (calculation of the Prandtl number)	(5.36)	C_p, μ°	±8%	for $M > 15$
Hirschfelder, Bird, Curtiss	(5.20), (5.22)	σ, ε		for monatomic gases
Reid, Sherwood	(5.26)	T_c, V_c, C_v		μ° need not be known
(b) *Effect of temperature on thermal conductivity* λ°				
Sutherland	(5.31)	C, λ_0		if the constant C is unknown, we must know a value of λ_1° (beside λ_0)
Reid, Sherwood	(5.28)	T_c, C_v, value λ_1°	±2%	
Vargaftig	(5.30)	two values: λ_1°, λ_2°	±3% }	the value λ° must be known for calculating the constants
Jakob	(5.32)	three values: λ_1°, λ_2°, λ_3°	±3% }	
Tsederberg	(5.55)	T_c, one value λ_1°	±4%	
(c) *Effect of pressure* p *and temperature* T *on thermal conductivity*				
Gamson	Graph in Fig. 5.8	T_c, p_c, $(c_p^\circ)_T$ or T_c, p_c, $(c_p^\circ)_T$, one value λ°, or T_c, p_c, $(c_p^\circ)_T$, $(\lambda^\circ)_{T_c}$	±15% to ±20%	the maximum error may exceed 100%
Comings, Nathan	Graphs in Figs. 5.10, 5.11	T_c, p_c one value λ°	±15% to ±20%	
Stolarov	(5.58)–(5.60)	five constants	±3%	five values of λ (different p, T) are required for calculating the constants
Franck	(5.61)	$(\lambda^\circ)_T$, v_c, v, C_p, C_p°	±3%	

Good results are obtained when Eq. (5.28) and Table 3.2 are used. In order to use this equation, it is sufficient to know one value of λ_1° (at any temperature), T_c and C_v. However, the effect of temperature upon C_v must be known. The average error of calculation is of the order of $\pm 2\%$; for CO_2 and H_2O, however, it is larger (*ca.* $\pm 5\%$).

Quite accurate results are also obtained by using Tsederberg's equation (5.55); in this case, it is sufficient to know the conductivity at one temperature and the critical temperature T_c of the compound.

c. Effect of Pressure and Temperature on the Thermal Conductivity of a Gas

The most universally applicable method is that of Gamson. This method, which involves the use of a generalized chart (Fig. 5.8), does not require the knowledge of any experimentally determined values of the conductivity. In the most general case it is sufficient to know the critical constants of the compound (T_c, p_c) and its heat capacity $(c_p^\circ)_T$ at the temperature T, at which the thermal conductivity λ_p is to be calculated. Unfortunately, calculations using Gamson's method are frequently inaccurate. Somewhat better results are obtained when the value of the conductivity λ_c at the critical point (necessary for the use of Gamson's graph) is calculated from one value of thermal conductivity λ_1, measured under arbitrary conditions. In this case the use of the generalized graph of Comings and Nathan (Figs. 5.10 and 5.11) is less tedious.

The accuracy of both methods is approximately the same, the average error being 15–20%.

A more accurate, but more tedious, method of estimating the thermal conductivity λ_p of a gas at T_2, p, when the conductivity $\lambda_{T_1}^\circ$ at temperature T_1 and under moderate pressure is known, consists in:

(a) calculation of the thermal conductivity $\lambda_{T_2}^\circ$ under moderate pressure, but at temperature T_2 (e.g. by means of Eq. (5.28)),

(b) calculation of λ_p under pressure p, on the basis of the conductivity $\lambda_{T_2}^\circ$, by employing Franck's equation (5.61).

In this case to compute the specific volume of the gas at temperature T_2 and pressure p as well as v_c at the critical point we must know one value of the conductivity $\lambda_{T_1}^\circ$, the relationship between the molal heat capacity of the gas and pressure and temperature, and the equation of state for the gas.

EXAMPLE 5.5. Estimate the thermal conductivity of nitric oxide using different methods, at temperature $t = 300°C$ and pressure 1 atm (region of moderate pressures). According to the data presented in Table 5.3, the experimental value is $\lambda^\circ = 1{\cdot}07 \times 10^{-4}$ cal/cm.sec.°C.

Solution

Physico-chemical constants of nitric oxide:
Molecular weight $M = 30{\cdot}01$
"Dynamic" viscosity at 0°C $\mu_0^\circ = 0{\cdot}01777$ cP (Table 3.4)
Sutherland's constant $C = 128$ (Table 3.4).
At 300°C ($T = 573°K$) (Eq. (3.20))

$$\mu^\circ = 17{\cdot}77 \times 10^{-5} \times \frac{(273+128)}{(573+128)} \times \frac{(573)^{3/2}}{(273)} = 32{\cdot}7 \times 10^{-5} \text{ poises}$$

Molal heat capacity

$$C_p = 6{\cdot}461 + 2{\cdot}358 \times 10^{-3} T = 6{\cdot}461 + 2{\cdot}358 \times 573 \times 10^{-3} = 7{\cdot}811 \text{ cal/mole.°C}$$
$$C_v = 7{\cdot}811 - 1{\cdot}985 = 5{\cdot}826 \text{ cal/mole.°C}$$

Heat capacity

$$c_p = 7{\cdot}811/30{\cdot}01 = 0{\cdot}260 \text{ cal/g.°C}$$
$$c_v = 5{\cdot}826/30{\cdot}01 = 0{\cdot}194 \text{ cal/g.°C}$$
$$\varkappa = C_p/C_v = 1{\cdot}340$$

From Table 1.31

$$t_c = -94°C, \quad T_c = 179{\cdot}2°K, \quad \rho_c = 0{\cdot}52, \quad V_c = \frac{30{\cdot}01}{0{\cdot}52} = 57{\cdot}7 \text{ cm}^3/\text{mole}$$

$$T = 300+273{\cdot}2 = 573{\cdot}2°K, \quad T_r = \frac{573{\cdot}2}{179{\cdot}2} = 3{\cdot}20$$

a. Eucken's equation (Maxwell's modulus) (5.6)

$$\lambda° = \frac{(9\times 1{\cdot}328-5)}{4}\times 0{\cdot}194\times 32{\cdot}7\times 10^{-5} = 1{\cdot}105\times 10^{-4} \text{ cal/cm.sec.°C}$$

The error in this estimate is $+3{\cdot}3\%$.

b. Eucken's equation (Prandtl number) (5.35)

$$\text{Pr} = \frac{C_p}{C_p+2{\cdot}46} = \frac{7{\cdot}811}{7{\cdot}811+2{\cdot}46} = 0{\cdot}762$$

$$\lambda° = \frac{0{\cdot}260\times 32{\cdot}7\times 10^{-5}}{0{\cdot}762} = 1{\cdot}115\times 10^{-4} \text{ cal/cm.sec.°C}$$

The error in this estimate is $+4{\cdot}2\%$.

c. Equation of Hilsenrath and Touloukian (Prandtl number) (5.36)

$$\text{Pr} = \frac{C_p}{1{\cdot}204C_p+1{\cdot}47} = \frac{7{\cdot}811}{1{\cdot}204\times 7{\cdot}811+1{\cdot}47} = 0{\cdot}719$$

$$\lambda° = \frac{0{\cdot}260\times 32{\cdot}7\times 10^{-5}}{0{\cdot}719} = 1{\cdot}181\times 10^{-4} \text{ cal/cm.sec.°C}$$

The error in the estimate is $+9{\cdot}4\%$.

d. Bromley's equation (Maxwell's modulus) (5.17)

$$\lambda° = \left(1{\cdot}32\times 5{\cdot}826+3{\cdot}40-\frac{0{\cdot}70}{3{\cdot}20}\right)\times\frac{32{\cdot}7\times 10^{-5}}{30{\cdot}01} = 1{\cdot}185\times 10^{-4} \text{ cal/cm.sec.°C}$$

The error in the estimate is $+10{\cdot}7\%$.

e. When the gas viscosity $\mu°$ is unknown, Eq. (5.26) is used.
The data are taken from Table 3.2.

$$f_1(1{\cdot}33T_r) = f_1(1{\cdot}33\times 3{\cdot}20) = f_1(4{\cdot}26) = 2{\cdot}167$$

$$\lambda° = 333\times 10^{-7}\times\left(\frac{179{\cdot}2}{30{\cdot}1}\right)^{1/2}(57{\cdot}7)^{-2/3}\times 2{\cdot}167\times(5{\cdot}826+4{\cdot}47) = 1{\cdot}214\times 10^{-4} \text{ cal/cm.sec.°C}$$

The error in this prediction is $+13{\cdot}5\%$.

f. According to the nomograph shown in Fig. 5.1
For $c_p = 0{\cdot}260$, $\mu° = 0{\cdot}0327$ cP, $n = 2$

$$\lambda° = 11{\cdot}5\times 10^{-5} \text{ cal/cm.sec.°C}$$

error: $+7{\cdot}5\%$.

g. The experimentally determined value of the thermal conductivity of nitric oxide at 0°C is $\lambda_0° = 0{\cdot}56\times\times 10^{-4}$ cal/cm.sec.°C
Equation (5.28) is used: $T_{r_2} = 3{\cdot}20$

$$C_{v2} = 5{\cdot}826 \text{ cal/mole.°C, for 0°C} \qquad T_{r_1} = \frac{273{\cdot}2}{179{\cdot}2} = 1{\cdot}525$$

$$C_{v1} = 5{\cdot}120 \text{ cal/mole.°C.}$$

In Table 3.2 we find

$$f_1(1{\cdot}33\times 1{\cdot}525) = f_1(2{\cdot}022) = 1{\cdot}215$$

$$f_1(1{\cdot}33\times 3{\cdot}20) = 2{\cdot}167$$

$$\lambda_{300}° = 0{\cdot}56\times 10^{-4}\times\frac{2{\cdot}167}{1{\cdot}215}\times\frac{(5{\cdot}826+4{\cdot}47)}{(5{\cdot}120+4{\cdot}47)} = 1{\cdot}072\times 10^{-4} \text{ cal/cm.sec.°C}$$

error: $0{\cdot}0\%$.

EXAMPLE 5.6. Estimate by Bromley's method the thermal conductivity $\lambda°$ of the vapour of n-propanol CH_3—CH_2—CH_2OH at a pressure of 1 atm and temperature of 100°C.

In the calculation, Bromley's equation (5.18) for gases with non-linear molecules should be employed.

a. Available data:
Molecular weight $M = 60{\cdot}09$

Viscosity of the vapour of n-propanol at a pressure of 1 atm and temperature of 0°C $\mu_0^\circ = 68{\cdot}2\times 10^{-4}$ cP
Sutherland's constant in equation (3.20) $C = 515{\cdot}6$
Normal boiling temperature $t_b = 97{\cdot}2$°C, i.e. $T_b = 370{\cdot}4$°K
Liquid density at 20°C $\rho_{20} = 0{\cdot}8044$ g/cm³
Critical temperature $t_c = 263{\cdot}7$°C, i.e. $T_c = 536{\cdot}9$°K
Critical pressure $p_c = 50$ atm
Temperature variation of the molal heat capacity at constant pressure (Ref. 36) $C_p = -0{\cdot}62+74{\cdot}67\times 10^{-3}T+25{\cdot}22\times 10^{-6}T^2$
Ratio $C_p/C_v = 1{\cdot}27$
Molal heat of vaporization at 20°C: $L_{20} = 10{\cdot}82$ kcal/mole.

Solution

Calculation of the values of the necessary physico-chemical constants.
Viscosity of the vapour at 100°C:

$$\mu_{100}^\circ = 68{\cdot}2\times 10^{-4}\left(\frac{273{\cdot}2+515{\cdot}6}{373+515{\cdot}6}\right)\left(\frac{373}{273}\right)^{3/2} = 96{\cdot}5\times 10^{-4} \text{ cP}$$

Molal heat capacity at 100°C:

$$C_p = -0{\cdot}62+74{\cdot}67\times 10^{-3}\times 373+25{\cdot}22\times 10^{-6}\times 373^2 = 30{\cdot}74 \text{ cal/mole.°C}$$

$$C_v = C_0(1{\cdot}27 = 30{\cdot}74)\ 1{\cdot}27 = 24{\cdot}21 \text{ cal/mole.°C}$$

Reduced temperatures:

$$T_{r20} = \frac{293}{536{\cdot}9} = 0{\cdot}546 \qquad T_{rb} = \frac{370{\cdot}4}{536{\cdot}9} = 0{\cdot}690$$

$$T_{r100} = \frac{373}{536{\cdot}9} = 0{\cdot}695$$

The density ρ_b of liquid n-propanol at the normal boiling point may be estimated using Watson's equation (Refs. 160, 393):

$$\frac{0{\cdot}8044}{\rho_b} = \frac{0{\cdot}1745-0{\cdot}0838\times 0{\cdot}546}{0{\cdot}1745-0{\cdot}0838\times 0{\cdot}690} \quad \text{whence} \quad \rho_b = 0{\cdot}729 \text{ g/cm}^3$$

The molal heat of vaporization at the normal boiling point is given by Watson's equation (Ref. 393):

$$\frac{L_b}{10{\cdot}82} = \left(\frac{1-0{\cdot}690}{1-0{\cdot}546}\right)^{0{\cdot}38} \quad \text{or} \quad L_b \approx 9600 \text{ cal/mole}$$

Hence the entropy of vaporization is:

$$\Delta S_b = \frac{L_b}{T_b} = \frac{9600}{370{\cdot}4} = 26{\cdot}0 \text{ cal/mole.°C}$$

From Eq. (5.19), the coefficient α is given by

$$\alpha = \frac{3\times 0{\cdot}729}{60{\cdot}1}(26{\cdot}0-8{\cdot}75-4{\cdot}57\log 370{\cdot}4) = 0{\cdot}20$$

Before Eq. (5.18) can be used, the internal-rotational contributions $C_{i,r}$ must be calculated. According to Table 5.2, the values of these contributions for the molecule of n-propanol are

	CH_3—CH_2—	2·15
	—CH_2—CH_2—	2·85
	—CH_2—OH	1·28
Hence	$C_{i,r} =$	6·28

Substituting into equation (5.18),

$$\lambda_{100}^\circ = \frac{96{\cdot}5\times 10^{-6}}{60{\cdot}09}\left(1{\cdot}30\times 24{\cdot}21+3{\cdot}60-0{\cdot}3\times 6{\cdot}28-\frac{0{\cdot}69}{0{\cdot}695}-3\times 0{\cdot}20\right) = 50{\cdot}7\times 10^{-6} \text{ cal/cm.sec.°C}$$

EXAMPLE 5.7. Estimate the thermal conductivity λ_p of methane at a temperature of $t = 100$°C and pressure of $p = 400$ kg/cm² ($p = 387$ atm) by various methods.

Solution

a. The physico-chemical data for methane are as follows:
Molecular weight $M = 16{\cdot}04$
"Dynamic" viscosity at 0°C and pressure of 1 atm
$\mu_0^\circ = 1022\times 10^{-5}$ cP (Table 3.4)
Sutherland's constant $C = 164$ (Table 3.4)

Critical constants (Table 1.31)

$$t_c = -82{\cdot}5°C, \quad T_c = 190{\cdot}7°K, \quad p_c = 45{\cdot}8 \text{ atm},$$

$$\rho_c = 0{\cdot}162 \text{ g/cm}^3, \qquad V_c = \frac{16{\cdot}04}{0{\cdot}162} = 98{\cdot}8 \text{ cm}^3/\text{mole}$$

Molal heat capacity

$$C_p = 4{\cdot}75+12\times10^{-3}T+3{\cdot}03\times10^{-6}T^2-2{\cdot}63\times10^{-9}T^3 = 8{\cdot}29 \text{ cal/mole.°C}$$

Thermal conductivity at 0°C and 1 atm

$$\lambda_0^\circ = 0{\cdot}73\times10^{-4} \text{ cal/cm.sec.°C (Table 5.3)}$$

b. The following parameters may now be calculated:

Absolute temperature $T = 100+273{\cdot}2 = 373{\cdot}2°K$

Reduced parameters

$$T_r = \frac{100+273{\cdot}2}{190{\cdot}7} = 1{\cdot}955 \qquad p_r = \frac{387}{45{\cdot}8} = 8{\cdot}46$$

The "dynamic" viscosity at 100°C and 1 atm is given by Sutherlands' equation to be $\mu° = 10{\cdot}22\times$

$$\times10^{-5}\frac{(273+164)}{(373+164)}\times\left(\frac{373}{273}\right)^{3/2} = 13{\cdot}31\times10^{-5} \text{ poises}$$

Molal heat capacities at 373°K

$$C_p = 4{\cdot}75+12\times10^{-3}\times373+3{\cdot}03\times10^{-6}\times373^2-2{\cdot}63\times10^{-9}\times373^3 = 9{\cdot}516 \text{ cal/mole.°C}$$

$$C_v = 9{\cdot}516-1{\cdot}985 = 7{\cdot}531 \text{ cal/mole.°C}$$

Heat capacities at 373°K

$$c_p = \frac{9{\cdot}516}{16{\cdot}04} = 0{\cdot}594 \text{ cal/g.°C} \qquad c_v = \frac{7{\cdot}531}{16{\cdot}04} = 0{\cdot}471 \text{ cal/g.°C}$$

$$\varkappa = 9{\cdot}516/7{\cdot}531 = 1{\cdot}26$$

c. The thermal conductivity $\lambda°$ at 100°C and 1 atm may be calculated from these parameters using either Sutherland's (5.31) or Jakob's (5.32) equation.

According to Eq. (5.31)

$$\lambda° = 0{\cdot}73\times10^{-4}\frac{(273+198)}{(373+198)}\left(\frac{373}{273}\right)^{3/2} = 0{\cdot}96\times10^{-4} \text{ cal/cm.sec.°C}$$

According to Eq. (5.32)

$$\lambda° = \frac{23{\cdot}91\times10^{-4}\times(1+0{\cdot}453\times10^{-3}\times373)\times\sqrt{373}}{1+\dfrac{198}{373}} = 352{\cdot}1\times10^{-4} \text{ kcal/m.hr.°C.}$$

On changing the units,

$$\lambda° = 0{\cdot}98\times10^{-4} \text{ cal/cm.sec.°C}$$

Alternatively $\lambda°$ at 100°C may be estimated from λ_0° using Eq. (5.28)

Since $\quad T_{r1} = \dfrac{273{\cdot}2}{190{\cdot}7} = 1{\cdot}432$

it follows from Eq. (5.28) that

$$\frac{\lambda°}{\lambda_0^\circ} = \frac{f_1(1{\cdot}33\times1{\cdot}955)}{f_1(1{\cdot}33\times1{\cdot}432)}\times\frac{(7{\cdot}531+4{\cdot}47)}{(8{\cdot}29-1{\cdot}985+4{\cdot}47)} = \frac{1{\cdot}4962}{1{\cdot}1581}\times\frac{12{\cdot}00}{10{\cdot}77} = 1{\cdot}44$$

so $\lambda° = 0{\cdot}73\times10^{-4}\times1{\cdot}44 = 1{\cdot}05\times10^{-4}$ cal/cm.sec.°C.

The experimentally determined value of $\lambda°$ is $1{\cdot}11\times10^{-4}$ cal/cm.sec.°C so the last method gives the most accurate result.

d. Calculation of the value of the thermal conductivity $\lambda_{T_c}^\circ$ at the critical temperature and under moderate pressure

$$T_{r2} = T_c/T_c = 1 \qquad T_{r1} = 273{\cdot}2/190{\cdot}7 = 1{\cdot}432$$

In order to use Eq. (5.28), to calculate $\lambda_{T_c}^\circ$ it is necessary to know the molal heat capacity C_p of the compound at the critical temperature T_c and under moderate pressure. Under such conditions the experimentally determined heat capacity of methane at a constant pressure is

$$c_p = 0{\cdot}498 \text{ cal/g.°C} \qquad \text{(Ref. 275)}$$

(The value of c_p may also be calculated on the basis of the differences of enthalpy taken from the entropy diagram for methane)

$$C_p = 0{\cdot}498\times16{\cdot}04 = 7{\cdot}97 \text{ cal/mole.°C} \qquad C_v = 7{\cdot}97-1{\cdot}99 = 5{\cdot}98$$

As a rule, the formula determining C_p of methane is valid for temperatures above 0°C. Extrapolating to the temperature of −82°C we get

$$C_p = 4{\cdot}75 + 12\times10^{-3}\times190 + 3{\cdot}03\times10^{-6}\times190^2 - 2{\cdot}63\times10^{-9}\times190^3 = 8{\cdot}13 \text{ cal/cm.sec.°C}$$

the result is quite close to the experimentally determined value (+1·7%).

According to Eq. (5.28) (assuming $C_{v,T_c} = 5{\cdot}98$ and $C_{v,0} = 6{\cdot}30$) we obtain

$$\frac{\lambda^\circ_{T_c}}{\lambda^\circ_0} = \frac{f_1(1{\cdot}33\times1)}{f_1(1{\cdot}33\times1{\cdot}432)}\times\frac{(5{\cdot}98+4{\cdot}47)}{(6{\cdot}30+4{\cdot}47)} = \frac{0{\cdot}8330}{1{\cdot}1555}\times\frac{10{\cdot}45}{10{\cdot}77} = 0{\cdot}700$$

hence $\lambda^\circ_{T_c} = 0{\cdot}73\times10^{-4}\times0{\cdot}700 = 0{\cdot}51\times10^{-4}$ cal/cm.sec.°C

e. Calculation of the value of the thermal conductivity λ_c at the critical point.

From Eq. (5.49) we find

$$\lambda_c = 2{\cdot}60\ \lambda^\circ_{T_c} = 2{\cdot}60\times0{\cdot}51\times10^{-4} = 1{\cdot}32\times10^{-4} \text{ cal/cm.sec.°C}$$

Alternatively λ_c may be estimated using Fig. 5.8 and the value of λ° at 0°C, i.e. with recourse to a knowledge of $\lambda^\circ_{T_c}$. The procedure is as follows:

$\lambda^\circ_0 = 0{\cdot}73\times10^{-4}$ cal/cm.sec.°C at 273°K. Now at 0°C and atmospheric pressure, $T_r = 1{\cdot}432$ and $p_r = 1/45{\cdot}8 = 0{\cdot}0218$. From Fig. 5.8, we read the value of $\lambda_r = 0{\cdot}602$, hence

$$\lambda_c = \lambda^\circ_0/\lambda_r = 0{\cdot}73\times10^{-4}/0{\cdot}602 = 1{\cdot}21\times10^{-4} \text{ cal/cm.sec.°C}$$

If no experimental value of the thermal conductivity λ is known, λ_c may be estimated as follows: from Eqs. (5.46) and (5.45)

$$\mu_c = 7{\cdot}70\times\frac{16{\cdot}04^{1/2}\times45{\cdot}8^{2/3}}{190{\cdot}7^{1/6}} = 164\ \mu\text{P}$$

Now according to Table 5.8, the interpolated value of

$(\lambda^\circ_r/\mu^\circ_r)$ for $T = 373$, or $T_r = 1{\cdot}955$, is equal to $\lambda^\circ_r/\mu^\circ_r = 0{\cdot}968$.

It has been shown under (b) that $c^\circ_p = 0{\cdot}594$ cal/g.°C.

Substituting these values in Eq. (5.45), we obtain

$$\lambda_c = 0{\cdot}968\times164\times10^{-6}\times\left(0{\cdot}594+\frac{2{\cdot}46}{16{\cdot}04}\right) = 1{\cdot}18\times10^{-4} \text{ cal/cm.sec.°C}$$

For methane, Gamson (Table 5.9) gives the value

$\lambda_c = 0{\cdot}0426$ kcal/m.hr.°C, i.e. $\lambda_c = 1{\cdot}18\times10^{-4}$ cal/cm.sec.°C

Solution

Calculation of the value of λ_p for methane at a temperature of 100°C and a pressure of 387 atm.

Comings' method

It was found above that

$T_r = 1{\cdot}955 \qquad p_r = 8{\cdot}46$

At a temperature of 100°C and under a pressure of $p = 1$ atm $\lambda^\circ = 1{\cdot}05\times10^{-4}$ cal/cm.sec.°C. From Fig. 5.10 we obtained by extrapolation

$\lambda_p/\lambda^\circ = 1{\cdot}9$

From Eq. (5·57)

$$\lambda_p = \lambda^\circ\times1{\cdot}9 = 1{\cdot}9\times1{\cdot}05\times10^{-4} = 2{\cdot}0\times10^{-4} \text{ cal/cm.sec.°C}$$

Gamson's method

It was found above that

$T_r = 1{\cdot}955, \qquad p_r = 8{\cdot}46, \qquad \lambda_c = 1{\cdot}18\times10^{-4}$ cal/cm.sec.°C

From Fig. 5.8

$\lambda_r = 1{\cdot}7$

According to Eq. (5.41)

$$\lambda_p = 1{\cdot}7\times1{\cdot}18\times10^{-4} = 2{\cdot}0\times10^{-4} \text{ cal/cm.sec.°C}$$

Franck's method

It was found above that

$T_r = 1{\cdot}955, \qquad p_r = 8{\cdot}46, \qquad \lambda^\circ = 1{\cdot}05\times10^{-4}$ cal/cm.sec.°C

By extrapolating from the graph given by Hougen and Watson[160] we find

$(C_p - C^\circ_p) = 3{\cdot}2$

$$v_c = \frac{1}{\rho_c} = \frac{1}{0{\cdot}162} = 6{\cdot}18 \text{ cm}^3/\text{g}$$

Using the equation of state $pV = zRT$, we calculate the specific volume v of methane at a temperature of 100°C and a pressure of 100 atm. The compressibility factor z is read from Hougen and Watson's graph[160]

$$z = 1{\cdot}10, \quad R = 82{\cdot}06 \text{ cm}^3.\text{atm/mole.°K}$$

$$V = \frac{1{\cdot}10 \times 82{\cdot}06 \times 373}{387} = 87{\cdot}1 \text{ cm}^3\text{/mole}, \quad v = \frac{87{\cdot}1}{16{\cdot}04} = 5{\cdot}44 \text{ cm}^3\text{/g}$$

Substituting the value of v in Franck's equation (5.61), we obtain

$$\lambda_p = 1{\cdot}05 \times 10^{-4} \left(\frac{1}{1 - 6{\cdot}18/(3 \times 5{\cdot}44)} + 0{\cdot}035 \times 3{\cdot}2 \right) = 1{\cdot}81 \times 10^{-4} \text{ cal/cm.sec.°C}$$

The value determined experimentally by Lenoir and Comings[219] is

$$\lambda_p = 1{\cdot}801 \times 10^{-4} \text{ cal/cm.sec.°C}$$

The errors in these estimates are as follows:

Comings' method:	+11·1%
Gamson's method:	+11·1%
Franck's formula:	0·0%

6. THERMAL CONDUCTIVITIES OF GAS MIXTURES

The phenomenon of heat conduction through gaseous mixtures has not yet been sufficiently examined. Experimental information is rather scanty, and because of the complex character of the phenomenon all the known methods of estimating the thermal conductivity of gas mixtures are often unsatisfactory for the solution of chemical engineering problems.

a. Experimental Data

In general the thermal conductivity of gas mixtures is not an additive function and cannot therefore be calculated directly from the proportions of the components and their conductivities.

Lehmann[217] has investigated the relationship between thermal conductivity and composition for a number of binary gas mixtures. He has found that deviations from additivity occur when the constituents of the mixture differ considerably either in molecular weight, M, or in polarity as measured by the dipole moments, μ, of the molecules. Deviations also occur when the constituents of a mixture differ considerably in molecular diameter.

Lehmann has classified binary mixtures into three groups:

1. *Mixtures of constituents which differ greatly in molecular weight* $((M_1/M_2) > 4)$. Such mixtures, which usually contain hydrogen or helium, exhibit negative deviations from additivity. Their actual conductivity is usually smaller than that calculated on the basis of linear proportionality, and the curve representing the variation of the thermal conductivity λ_m of the mixture with composition is of a concave shape. Hydrogen–carbon dioxide is an example of such a system (Fig. 5.12). In this case $M_1/M_2 = 44/2 = 22$.

For mixtures containing equal volumes of the constituents (equimolal mixtures) at 0°C Lehmann has shown the following negative deviations from additivity to occur: H_2–CO_2, $\Delta = -39{\cdot}6\%$; H_2–N_2, $\Delta = -36\%$; H_2–Ar, $\Delta = -29{\cdot}9\%$; H_2–CO, $\Delta = -30{\cdot}5\%$.

2. *Mixtures of constituents showing no great differences in molecular weight* $((M_1/M_2) \leq 4)$ *or dipole moment.* The thermal conductivities λ_m of such mixtures approximately obey the law of additivity. Figure 5.13 shows the effect of the composition on λ_m for mixtures of two non-polar gases, argon and nitrogen. The thermal conductivities

λ_m of mixtures of polar gases with similar dipole moments, chloroform ($\bar{\mu}_1 = 1{\cdot}15$ D) and diethyl ether ($\bar{\mu}_2 = 1{\cdot}14$ D), are shown in Fig. 5.14. In both cases the relationship is almost exactly rectilinear.

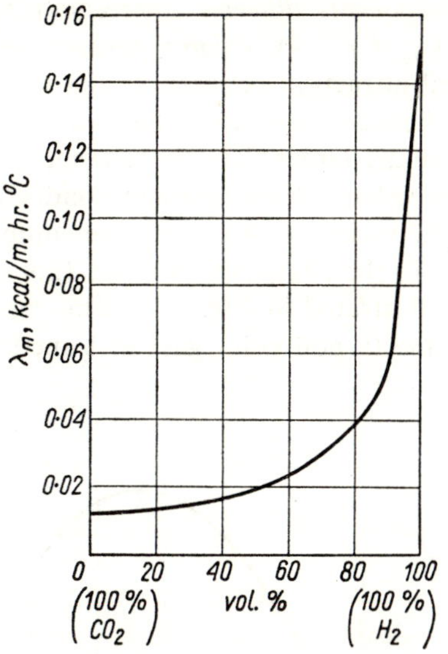

FIG. 5.12. Variation of the thermal conductivity λ_m of hydrogen–carbon dioxide mixture with composition (after Lehmann[217]).

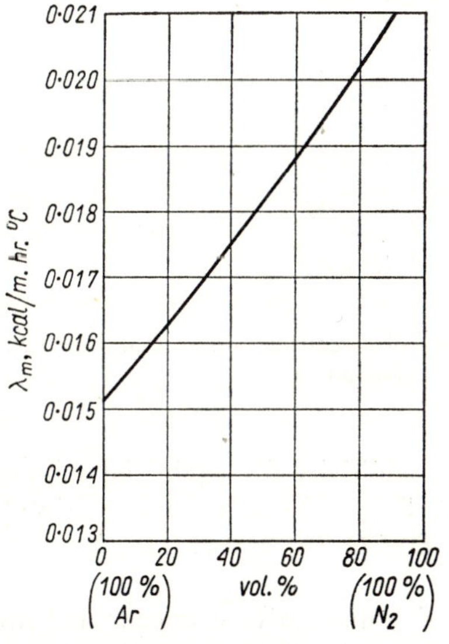

FIG. 5.13. Variation of the thermal conductivity λ_m of the nitrogen–argon system with composition (after Lehmann[217]).

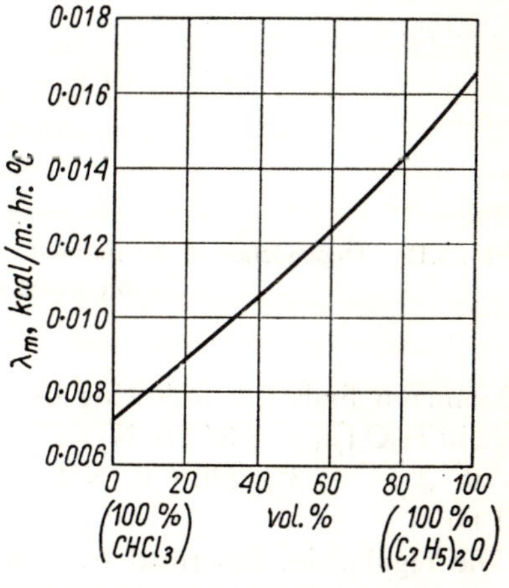

FIG. 5.14. Dependence of the thermal conductivities λ_m of gaseous chloroform–diethyl ether mixtures on composition (after Lehmann[217]).

Lehmann has calculated the deviations from additivity exhibited by the thermal conductivities λ_m of many gaseous mixtures, composed either of non-polar molecules or of molecules with similar dipole moments. For equi-molal mixtures the values of the deviations Δ are: H_2–D_2, Δ = −3·6% (0°C); CH_4–air, Δ = 0% (20°C). For the polar system $(C_2H_5)_2NH/(C_2H_5)_2O$, Δ = 0·1%. The dipole moments of these components are 0·92 D and 1·14 D, respectively.

3. *Gaseous mixtures of components differing appreciably in polarity but whose molecular weights are similar* $((M_1/M_2) < 4)$, *exhibit positive deviations from additivity.* In this case, the forces of mutual attraction between unlike molecules are smaller than the arithmetic mean of the forces between like molecular pairs and hence the mean free paths of the molecules in the mixture are larger than the average of the mean free paths in the pure components.[138] Thus the values of the thermal conductivity λ_m are higher.

The curves representing the variation of λ_m with the composition of the mixture, may exhibit maxima at some point in the composition range (not necessarily at the equimolal composition). Such a case is illustrated in Fig. 5.15. This figure shows the thermal conductivities of mixtures of hexane (a non-polar gas) with methyl alcohol (a polar gas) for which $\bar{\mu} = 1{\cdot}67$ D.

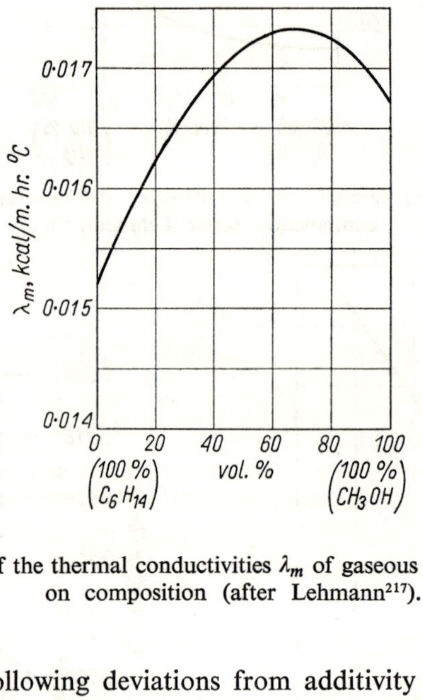

FIG. 5.15. Dependence of the thermal conductivities λ_m of gaseous hexane–methanol mixtures on composition (after Lehmann[217]).

Lehmann finds the following deviations from additivity for equimolal mixtures of this class: H_2O ($\bar{\mu}_1 = 1{\cdot}84$ D)–N_2 ($\bar{\mu}_2 = 0$), Δ = +9·4% (+65°C); NH_3 ($\bar{\mu}_1 = 0{\cdot}589$ D)-air ($\bar{\mu}_2 = 0$, Δ = +7·3% (20°C); NH_3 ($\bar{\mu}_1 = 0{\cdot}589$ D)-CO ($\bar{\mu}_2 = 0{\cdot}11$ D), Δ = +6·6% (+22°C).

The regularities described by Lehmann are of a qualitative character and are not rigorously obeyed in all cases. Keyes[197] has found that the thermal conductivity of a mixture of equal volumes of carbon dioxide and nitrogen at 50°C is lower than that calculated additively (Δ = −3·3%), whereas at 350°C the same mixture shows a value of λ_m which is higher than that calculated additively (Δ = +3%).

The thermal conductivities of gaseous mixtures increase with rise of temperature. Values of the thermal conductivity λ_m found for several binary mixtures containing equal volumes of the pure components are listed in Table 5.12.

TABLE 5.12. VARIATION OF THE THERMAL CONDUCTIVITIES OF GAS MIXTURES WITH TEMPERATURE (VOL. %) (AFTER LEHMANN[217])

Components	50% N_2 + 50% CO_2		50% H_2O + 50% N_2		50% NH_3 + 50% air		50% $(CH_3)_2CO$ + 50% C_6H_6	
Temperature, °C	50	350	65	330	20	80	76·7	125·1
λ_m, kcal/m.hr.°C	0·191	0·405	0·238	0·447	0·222	0·272	0·132	0·173

All these considerations refer to the thermal conductivity of gas mixtures at moderate pressures, where the conductivity does not depend on pressure. After exceeding a certain pressure limit (of the order of several or a few dozen atmospheres) the thermal conductivities of mixtures rise with increase in pressure, as in the case of pure gases.

Experimental data on the thermal conductivities λ_m of gas mixtures at high pressures is meagre. In Table 5.13 the results of the investigations of Junk and Comings[183] on the effect of pressure p upon λ_m for a mixture of ethylene and nitrogen are listed.

TABLE 5.13. EFFECT OF PRESSURE p (atm) ON THE THERMAL CONDUCTIVITIES ($\lambda_m \times 10^6$, cal/cm.sec.°C) OF ETHYLENE–NITROGEN MIXTURES AT 42°C

C_2H_4, mole % N_2, mole %	79·2 20·8	67·5 32·5	38·8 61·2	20·0 80·0
$p = 200$ atm	156	139	104	94·3
$p = 150$ atm	130	115	91·9	84·8
$p = 100$ atm	98·5	88·6	79·4	75·3
$p = 50$ atm	67·4	67·0	67·4	67·4

If a gaseous mixture consists of gases which react with each other then, after the equilibrium state has been established, the thermal conductivity λ_m may be many times as high as the thermal conductivity of mixtures of non-reacting gases (e.g. the mixture $N_2O_4 \rightleftarrows 2NO_2$ exhibits unusually high values of λ_m[44,45,52,168,207]).

b. Estimation of the Thermal Conductivities of Gas Mixtures from the Kinetic Theory of Gases

By considering the analogy between the transport of momentum and energy in gases, and using Sutherland's equation (3.80) for the viscosity of gaseous mixtures, Wassiliewa[394] has derived the following form for the relationship between λ_m of a mixture of gases 1, 2, ..., etc. and composition ($x_1, x_2, \ldots$, etc. denoting mole fractions of the components):

$$\lambda_m = \frac{\lambda_1}{1+A_{1,2}\dfrac{x_2}{x_1}+A_{1,3}\dfrac{x_3}{x_1}+\ldots} + \frac{\lambda_2}{1+A_{2,1}\dfrac{x_1}{x_2}+A_{2,3}\dfrac{x_3}{x_2}+\ldots} + \ldots \quad (5.63)$$

where $A_{1,2}$, $A_{1,3}$, $A_{2,1}$ ... are coefficients dependent on the nature of the constituent gases. Although the above-mentioned analogy is not exact (Bock[26]), Eq. (5.63) has been the basis for several other more accurate formulations.

The values of the coefficients A have been calculated by Sutherland[354] on the basis of the simplified kinetic theory of non ideal gases. Lindsay and Bromley[226] have corrected these relationships, and obtained:

$$A_{1,2} = \frac{1}{4}\left\{1+\left[\frac{\mu_1}{\mu_2}\left(\frac{M_2}{M_1}\right)^{0\cdot 75}\frac{\left(1+\frac{C_1}{T}\right)}{\left(1+\frac{C_2}{T}\right)}\right]^{0\cdot 5}\right\}^2 \frac{\left(1+\frac{C_{1,2}}{T}\right)}{\left(1+\frac{C_1}{T}\right)} \tag{5.64}$$

where μ_1 and μ_2 are the viscosities of the pure gaseous compounds,
M_1 and M_2 are the molecular weights,
C_1 and C_2 are constants in Sutherland's equation (5.31) (we can assume $C \approx 1\cdot 5\ T_b$, where T_b denotes the normal boiling temperature of the compound),
and $C_{1,2} = f\sqrt{C_1 C_2}$, where $f = 1$ when both gases 1 and 2 are either polar or non-polar, and $f \approx 0\cdot 733$ when one gas consists of non-polar molecules and the other gas of strongly polar molecules. An expression for the constant $A_{2,1}$ in Eq. (5.63) can be obtained from (5.64) by interchanging the subscripts 1 and 2.

These authors have shown that despite the large errors (of the order of 20%) involved in calculating the constants C, the estimation of λ_m is quite satisfactory, a typical error being about 1%.

The ratio μ_1/μ_2 in Eq. (5.64) may be calculated, provided the values of thermal conductivities λ_1, λ_2 of pure compounds are known. Using Eucken's equation, we obtain:

$$\frac{\mu_1}{\mu_2} = \frac{\lambda_1}{\lambda_2}\frac{(C_{p2}+2\cdot 47)}{(C_{p1}+2\cdot 47)} \tag{5.65}$$

where C_{p1} and C_{p2} denote the molal heat capacities.

Lindsay and Bromley have checked the accuracy of Eqs. (5.63) and (5.64), and found the average error to be $\pm 1\cdot 9$%; but at high temperatures, the error may be very high (Keyes[197]).

If the value of the kinematic diffusivity $D_{1,2}$ is known, we may use the simplified equation of Lindsay and Bromley[226] to calculate the thermal conductivity of a mixture of two gases 1 and 2. They obtained their equation by analogy with the Buddenberg and Wilke equation for the viscosities of gaseous mixtures (Eq. (3.88)), and so their treatment rests on the analogy between the kinematic viscosity $\nu = \mu/\rho$ and the thermal diffusivity $\alpha = \lambda/(\rho c_p)$. Accordingly

$$\lambda_m = \frac{\lambda_1}{1+\frac{1\cdot 114\lambda_1}{D_{1,2}\rho_1 c_{p1}}\frac{x_2}{x_1}} + \frac{\lambda_2}{1+\frac{1\cdot 114\lambda_2}{D_{2,1}\rho_2 c_{p2}}\frac{x_1}{x_2}} \tag{5.66}$$

where λ_1 and λ_2 are the thermal conductivities of gases 1, 2,
ρ_1 and ρ_2 are the densities of gases 1, 2,
c_{p1} and c_{p2} are the heat capacities of gases 1, 2,
and x_1 and x_2 are the mole fractions of gases 1, 2.

The average error of estimation by this equation is $\pm 3\cdot 5$%, the maximum error $\pm 11\cdot 7$%.

Cheung, Bromley and Wilke[60] have derived an equation by means of which the thermal conductivity of a gaseous multi-component mixture may be calculated with great accuracy:

$$\lambda_k = \sum_{i=1}^{n} \frac{\lambda_{ki}}{1+\sum\limits_{\substack{j=1\\ i\neq j}}^{n} \frac{D_{ii}}{D_{ij}}\left(\frac{M_{ij}}{M_i}\right)^{1/8}\frac{x_j}{x_i}} + \sum_{i=1}^{n} \frac{\lambda_{di}}{1+\sum\limits_{\substack{j=1\\ i\neq j}}^{n} \frac{D_{ii}}{D_{ij}}\frac{x_j}{x_i}} \tag{5.67}$$

The quantities appearing in this equation are calculated from the following formulae: λ_{ki} (the collisional conductivity of the i-th component)

$$\lambda_{ki} = \frac{2{\cdot}5c_{vt}+1{\cdot}0c_r}{2{\cdot}5c_{vt}+1{\cdot}0c_r+1{\cdot}32(c_v+c_{ir}+\ldots)}\lambda_i \tag{5.68}$$

λ_{di} (the diffusional conductivity of the i-th component)

$$\lambda_{di} = \lambda_i - \lambda_{ki} \tag{5.69}$$

$$M_{ij} = \frac{M_i+M_j}{2} \tag{5.70}$$

The ratio D_{ii}/D_{ij} is calculated from Arnold's equation:[13]

$$\frac{D_{ii}}{D_{ij}} = \left(\frac{M_i}{M_{ij}}\right)^{\frac{1}{2}}\left\{\frac{1}{2}\left[1+\left(\frac{V_j}{V_i}\right)^{\frac{1}{3}}\right]\right\}^2\left[\frac{T+0{\cdot}733F\times 1{\cdot}5\sqrt{T_{bi}T_{bj}}}{T+1{\cdot}5T_{bi}}\right] \tag{5.71}$$

In the case of mixtures of non-polar gases, D_{ii}/D_{ij} may be replaced by the identical value ϕ_{ij} given by Bromley and Wilke in Figs. 3.18 and 3.19, provided the viscosities μ_1 and μ_2 of pure components are known.

The symbols used in Eqs. (5.67)–(5.71) are as follows:

c_{ir} is the internal rotational heat capacity,
c_r is the rotational heat capacity,
c_{vt} is the translational heat capacity at constant volume,
c_v is the vibrational heat capacity,
D_{ii} is the coefficient of self-diffusion,
D_{ij} is the coefficient of binary diffusion,
F is a numerical factor which is a function of the ratio V_j/V_i,
M is the molecular weight,
T_{bi} is the normal boiling temperature of component i, °K,
V_i is the molecular volume of the i-th component in the liquid state at the boiling point,
x_i is the mole fraction of the i-th component,
λ is the thermal conductivity,
and λ_k is the conductivity of a gas mixture from conduction only.

The authors of Eq. (5.67) have used it to calculate the thermal conductivities of 16 ternary mixtures and found the average error to be 2·2%. The maximum error did not exceed 6%. However, the use of this formula is tedious.

Bird, Hirschfelder and Curtiss[25] have presented a method for calculating the thermal conductivities of gaseous mixtures, which takes into consideration the forces of interaction between the molecules of the gases and represents a development of the classical works of Chapman and Cowling.[59] The calculation of the thermal conductivity for a mixture of two monatomic gases involves tedious calculations, as in the case of calculating the viscosity of a gas mixture. First, we must find the value of the thermal conductivity of a hypothetical gas with a molecular weight

$$M_{1,2}=\frac{M_1+M_2}{M_1 M_2}$$

and the force constants $\sigma_{1,2}$ and $\varepsilon_{1,2}$. This is given by

$$\lambda_{1,2}\times 10^7=\frac{1989{\cdot}1}{\sigma_{1,2}^2\Omega^{(2,2)*}}\sqrt{\frac{T(M_1+M_2)}{2M_1 M_2}} \tag{5.72}$$

where M_1 is the molecular weight of gas 1,
M_2 is the molecular weight of gas 2,
$\varepsilon_{1,2}$ and $\sigma_{1,2}$ are the force constants calculated for a mixture of gases 1 and 2 from Eq. (1.39),
and $\Omega^{2,2*}$ is a function according to the data given in Table 3.1 for $T^*=kT/\varepsilon_{1,2}$.

Alternatively, if the value of $D_{1,2}$ is known, the value of $\lambda_{1,2}$ may be calculated from the relationship:

$$\lambda_{1,2}=\frac{25}{8}\frac{pD_{1,2}}{\bar{A}_{1,2}^*T} \tag{5.73}$$

where p is the pressure (atm),
and $\bar{A}_{1,2}^*$ is a function, the value of which is read from Table 3.3 for $T^*=kT/\varepsilon_{1,2}$.

Knowing the thermal conductivities, λ_1 and λ_2, for the pure gases 1 and 2, as well as $\lambda_{1,2}$ for a hypothetical gas, we can find the thermal conductivity λ_m of the mixture according to the equation

$$\frac{1}{\lambda_m}=\frac{X_\lambda+Y_\lambda}{1+Z_\lambda} \tag{5.74}$$

The quantities X_λ, Y_λ and Z_λ are calculated from the following equations, in which x_1 and x_2 denote the mole fractions of gases 1 and 2 in the mixture

$$X_\lambda=\frac{x_1^2}{\lambda_1}+\frac{2x_1x_2}{\lambda_{1,2}}+\frac{x_2^2}{\lambda_2}$$

$$Y_\lambda=\frac{x_1^2}{\lambda_1}U^{(1)}+\frac{2x_1x_2}{\lambda_{1,2}}U^{(y)}+\frac{x_2^2}{\lambda_2}U^{(2)}$$

$$Z_\lambda=x_1^2U^{(1)}+2x_1x_2U^{(z)}+x_2^2U^{(2)}$$

$$U^{(1)}=\frac{4}{15}\bar{A}_{1,2}^*-\frac{1}{12}\left(\frac{12}{5}\bar{B}_{1,2}^*+1\right)\frac{M_1}{M_2}+\frac{1}{2}\frac{(M_1-M_2)^2}{M_1M_2}$$

$U^{(2)}$ is given by a similar expression with M_1 and M_2 interchanged.

$$U^{(y)}=\frac{4}{15}\bar{A}_{1,2}^*\frac{(M_1+M_2)^2}{4M_1M_2}\frac{\lambda_{1,2}^2}{\lambda_1\lambda_2}-\frac{1}{12}\left(\frac{12}{5}\bar{B}_{1,2}^*+1\right)-\frac{5}{32\bar{A}_{1,2}^*}\left(\frac{12}{5}\bar{B}_{1,2}^*-5\right)\frac{(M_1-M_2)^2}{M_1M_2}$$

$$U^{(z)}=\frac{4}{15}\bar{A}_{1,2}^*\left[\frac{(M_1+M_2)^2}{4M_1M_2}\left(\frac{\lambda_{1,2}}{\lambda_1}+\frac{\lambda_{1,2}}{\lambda_2}\right)-1\right]-\frac{1}{12}\left(\frac{12}{5}\bar{B}_{1,2}^*+1\right)$$

The quantities $\bar{A}^*_{1,2}$ and $\bar{B}^*_{1,2}$, appearing in the equation for U, may be found in Table 3.3 knowing the value of $T^* = kT/\varepsilon_{1,2}$.

Bird, Hirschfelder and Curtiss,[25] recommend the following empirical method for use when the gases constituting the mixture are polyatomic. The thermal conductivities of the pure components λ_1 and λ_2 are assumed to be known.

First, we assume the gases are monatomic and calculate hypothetical values of the thermal conductivities λ'_1 and λ'_2 by means of Eq. (5.20) or (5.21). Then with the help of Eq. (5.74) we get λ'_m of the mixture. If the mole fractions of the gases in the mixture are x_1 and x_2, the desired thermal conductivity λ_m of the mixture is found from the equation

$$\lambda_m = \lambda'_m \left(x_1 \frac{\lambda_1}{\lambda'_1} + x_2 \frac{\lambda_2}{\lambda'_2} \right) \tag{5.75}$$

These methods involve, however, long and troublesome calculations and demand a knowledge of certain physico-chemical constants for the gaseous components of the mixtures. For these reasons they have only limited application.

Brokaw,[43] has developed a modified method and, using a first approximation equation, he reports an average error of $+2{\cdot}6\%$, and a maximum error of $6{\cdot}7\%$ for 19 non-polar binary and ternary mixtures.

By applying a second approximation he diminishes the average error to $0{\cdot}5\%$, and the maximum error to $2{\cdot}1\%$.

In his later works, Hirschfelder[155] extended the method of calculating the thermal conductivity to the mixtures of polyatomic gases. His equation for calculation of the thermal conductivity of multi-component mixtures has the following form:

$$\lambda_m = \lambda'_n + \sum_{i=1}^{n} (\lambda_i - \lambda'_i) \left[1 + \sum_{\substack{k=1 \\ k \neq i}}^{n} \frac{D_{ii}}{D_{ik}} \frac{x_k}{x_i} \right]^{-1} \tag{5.76}$$

where D_{ii} is the self-diffusion coefficient of pure component i,
D_{ik} is the binary diffusion coefficient,
x_i is the mole fraction of the i-th component,
λ is the true coefficient of thermal conductivity,
and λ' is the coefficient that would be obtained if all internal degrees of freedom were "frozen".

For a mixture of two polyatomic gases the above formula takes the form

$$\lambda_m = \lambda'_m + (\lambda_1 - \lambda'_1) \left(1 + \frac{x_2}{x_1} \frac{D_{11}}{D_{12}} \right)^{-1} + (\lambda_2 - \lambda'_2) \left(1 + \frac{x_1}{x_2} \frac{D_{22}}{D_{12}} \right)^{-1} \tag{5.77}$$

The value of λ' is obtained from the Eucken equation, as modified by Hirschfelder. According to this equation,

$$\lambda = (\mu/M) \left[\frac{5}{2} C_{v\,\mathrm{trans}} + \left(\frac{\rho D}{\mu} \right) C_{v\,\mathrm{int}} \right] \tag{5.78}$$

where $C_{v\,\mathrm{trans}}$ and $C_{v\,\mathrm{int}}$ are the heat capacities at constant volume corresponding to the translational and internal degrees of freedom,
D is the coefficient of self-diffusion,
ρ is the density,
and μ is the viscosity.

Assuming that the intermolecular potential can be represented by the Lennard-Jones potential or the Buckingham equation it is possible to show that:[155]

$$\lambda/\lambda' = 0{\cdot}354(C_v/R)+0{\cdot}469 = 0{\cdot}354(C_p/R)+0{\cdot}115 \tag{5.79}$$

where C_p and C_v are the heat capacities at constant pressure and constant volume, respectively,
and R is the gas constant.

This method is rather tedious to employ and Mason and Saxena[237,238] have derived a simplified equation for the calculation of the thermal conductivities of mixtures of gases with polyatomic molecules

$$\lambda_m = \sum_{i=1}^{n} \lambda_i \left[1+\sum_{\substack{j=1\\ j\neq i}}^{n} G_{ij}(x_j/x_i)^{-1}\right] \tag{5.80}$$

where the notation used is the same as above, and

$$G_{ij} = (1{\cdot}065/2\sqrt{2})[1+(M_i/M_j)]^{-1/2}[1+(\lambda_i'/\lambda_j')^{1/2}(M_i/M_j)^{1/4}]^2 \tag{5.81}$$

If the viscosities μ_i and μ_j of the gases are known, the value of λ_i'/λ_j' can be calculated from the relationship

$$\lambda_i'/\lambda_j' = (\mu_i M_i)/(\mu_j M_j) \tag{5.82}$$

If this is not the case, Eq. (5.79) is employed.

The theoretical equations of Hirschfelder and of Mason and Saxena, presented here are, in general, more accurate than the empirical equations, (5.66) and (5.75), given earlier. They have been checked by a number of authors.[238,239,240]

Whether the empirical equations of Lindsay and Bromley ((5.63) and (5.64)) or those of Mason and Saxena ((5.80), (5.81), (5.82)) are used, the average deviations of the estimates from the measured values are of the same order. On the other hand, the Hirschfelder equation (5.76) is more accurate. At elevated temperatures, the deviations of the estimates calculated by the equation of Mason and Saxena from the experimental values are, in general, larger.

In the case of mixtures of monatomic gases, the average deviation is 1·2%, except for He–Ar mixtures for which it is found to be 5·5%. For seven mixtures of polyatomic gases the average deviation was 3·8%, and the maximum deviation 15·1%. Estimations carried out using Lindsay and Bromley's equations, (5.63) and (5.64), gave somewhat better results, the average deviation being 2·6%, and the maximum deviation 10·0%. Particularly large deviations were found in the case of the H_2–N_2 system.

c. Empirical Equations for Estimating the Thermal Conductivities of Gas Mixtures

If it is assumed that the thermal conductivity λ_m of a gas mixture is an additive quantity:

$$\lambda_m = \lambda_1 x_1+\lambda_2 x_2+\lambda_3 x_3+ \dots \tag{5.83}$$

where λ_1, λ_2 etc. are the conductivities of components 1, 2 ... etc.,
and x_1, x_2 ..., etc. are the mole fractions of components 1, 2 ..., etc. ($x_1+x_2+\dots = 1$).

Various authors have suggested other empirical equations, in which the principle of additivity is applied to the logarithms of the thermal conductivities or to the reciprocals of the conductivities:

$$\frac{1}{\lambda_m} = \frac{x_1}{\lambda_1} + \frac{x_2}{\lambda_2} + \frac{x_3}{\lambda_3} \cdots \tag{5.84}$$

In some cases the best results are obtained with Brokaw's equation,[42] which is based on the combination of the principles of the additivity of the conductivities themselves and of the reciprocals of conductivity. Brokaw has observed that in the case of mixtures containing hydrogen or helium (i.e. if $M_2/M_1 > 4$) Eq. (5.83) gives too high and Eq. (5.84) too low results. According to Brokaw we obtain the best results by employing the empirical equation

$$\lambda_m = \underline{a}\lambda'_m + (1-\underline{a})\lambda''_m \tag{5.85}$$

where λ'_m is the value calculated from Eq. (5.83),
λ''_m is the value calculated from Eq. (5.84),
and $\underline{a}$ is a fraction, the value of which depends on x_1, the mole fraction of the lighter component of the mixture (H_2 oı He).

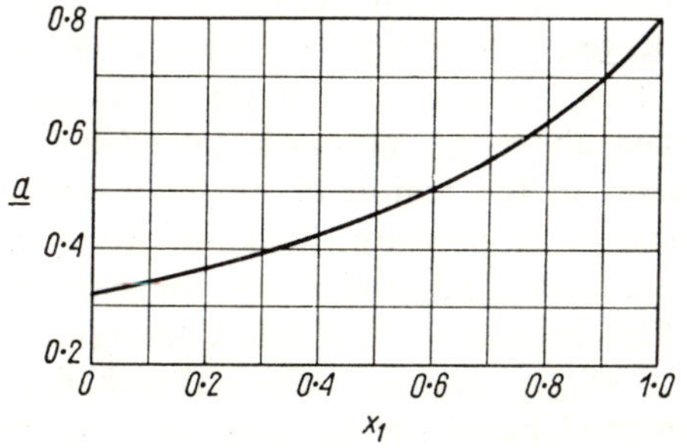

FIG. 5.16. Dependence of the coefficient $\underline{a}$ in Eq. (5.85) on the mole fraction x_1 of the lighter component (H_2 or He) in the mixture (after Brokaw[42]).

The value of the factor $\underline{a}$ is taken either from the empirical graph (Fig. 5.16) or from Table 5.14.

TABLE 5.14. VALUES OF $\underline{a}$ IN EQ. (5.85) (AFTER BROKAW[42])

x_1 (H_2 or He)	0	0·1	0·2	0·3	0·4	0·5	0·6	0·7	0·8	0·9	0·95	1·0
$\underline{a}$	0·32	0·34	0·37	0·39	0·42	0·46	0·50	0·55	0·61	0·69	0·74	0·80

For multi-component mixtures, Brokaw recommends the use of the value $\underline{a}$ — 0·5, whereas according to Lehmann[217] the value of $\underline{a}$ should be taken from Fig. 5.16.

Brokaw has shown that the error of the estimation by using Eq. (5.85) is on the average equal to ±2·6%, the maximum error being 11·4%. This method cannot be used for highly polar gases.

In order to calculate the thermal conductivity λ_m of a gas mixture containing hydrogen or helium, we may use, according to Lehmann,[217] the following empirical equation with fairly good results

$$\lambda_m = \frac{x_1\lambda_1\sqrt[3]{M_1}+x_2\lambda_2\sqrt[3]{M_2}+\dots}{x_1\sqrt[3]{M_1}+x_2\sqrt[3]{M_2}+\dots} \tag{5.86}$$

(The notation used here is the same as in the former equations).

Finally, for the case when a gaseous mixture contains non-polar and polar molecules, Lehmann[217] has suggested the equation

$$\lambda_m = (x_1\lambda_1+x_2\lambda_2+x_3\lambda_3+\dots)\left(1+\frac{x_p-x_p^2}{3\cdot 5}\right) \tag{5.87}$$

where x_p denotes the sum of mole fractions of the polar constituents of the mixture; the remaining notation is the same as in the previous equations.

d. Estimation of the Variations of Thermal Conductivities of Gaseous Mixtures with Temperature and Pressure

The thermal conductivities of gas mixtures at any temperature may be calculated by means of the equations discussed above provided that the relevant pure component data, such as heat capacity, density, thermal conductivity, etc. is available for the temperature of interest. According to Keyes,[197] Bromley's equation (5.66) is not very satisfactory at high temperatures and Brokaw recommends the use of Hirschfelder's equations.

The equations discussed above are primarily intended for calculating the thermal conductivities of gas mixtures in the region of moderate pressures for which the conductivity is almost independent of pressure. Reid and Sherwood[306] have tried to employ Lindsay and Bromley's equations (5.63) and (5.64) for estimating the thermal conductivity of gas mixtures at high pressures; Junk and Comings[183] have compared the values of the thermal conductivity of gas mixtures, which they had measured, with the values calculated by Comings' method based on the theorem of corresponding states (section 4b), using pseudocritical parameters.

In order to apply Comings' method we must estimate first the value of the thermal conductivity λ_m° of the mixture at a moderate pressure and the same temperature T for which the conductivity $\lambda_{m,p}$ of a mixture under an elevated pressure p is to be calculated. Next, when the critical parameters p_{c1}, T_{c1} and p_{c2}, T_{c2} of the pure components are known, we calculate the pseudocritical parameters of the mixture

$$T_{pcm} = x_1 T_{c1}+x_2 T_{c2} \tag{5.88}$$

$$p_{pcm} = x_1 p_{c1}+x_2 p_{c2}$$

Having evaluated the parameters $T_{pr} = T/T_{pcm}$ and $p_{pr} = p/p_{pcm}$ we read the value of $(\lambda_{m,p}/\lambda_m^\circ)$ from Comings' graph (Figs. 5.10 and 5.11) and employ Eq. (5.57)

$$\lambda_{m,p} = \left(\frac{\lambda_{m,p}}{\lambda_m^\circ}\right)\lambda_m^\circ \tag{5.89}$$

According to Reid and Sherwood,[306] the methods of Lindsay and Bromley as well as of Comings are almost equally inaccurate; in the cases examined by these authors, the error of estimation reached 50%, while the average error was of the order of ±20%.

7. CHOICE OF METHOD FOR THE ESTIMATION OF THE THERMAL CONDUCTIVITIES OF GAS MIXTURES

Lehmann[217] has made a critical review of various methods for estimating the thermal conductivities λ_m of gaseous mixtures. The properties of the molecules constituting the mixture are decisive in selecting the best method. Lehmann classifies mixtures of gases into three groups as follows:

1. *Mixtures composed of gases differing greatly in molecular weight.* These are usually mixtures of hydrogen or helium with compounds of higher molecular weights ($M_2/M_1 > 4$). The actual values of the conductivity λ_m for such mixtures are lower than the values calculated additively by means of Eq. (5.83).

In this case it is recommended that the quantity λ_m be estimated from Brokaw's equation (5.85), which may be written in the form

$$\lambda_m = \underline{a}\sum x_i\lambda_i+(1-a)\frac{1}{\sum \frac{x_i}{\lambda_i}} \tag{5.90}$$

where x_i is the mole fraction of the component i in the mixture,
and λ_i is the thermal conductivity of pure component i.

The value of $\underline{a}$ for a given content of H_2 or He is either read from Fig. 5.16 or interpolated from the data in Table 5.14. For equimolar mixtures, Lehmann has found the average error of the estimation to be $\pm 4{\cdot}3\%$, the maximum error being $9{\cdot}0\%$.

Somewhat less accurate results can be obtained in this case by using the empirical equation (5.86). The average error of estimating the thermal conductivity of equimolar mixtures is then $\pm 7{\cdot}3\%$, the maximum error being $14{\cdot}3\%$.[217]

If the molecular weights of the constituents of a mixture are close to each other (the mixture contains neither H_2 nor He), the selection of the proper equation depends on the polarity of the molecules making up the mixture.

2. *Mixtures containing molecules of an (approximately) equal degree of polarity.* These are, for example, non-polar molecules or molecules with very small or very large dipole moments.

In this case the simple rule of additivity is applicable, and Eq. (5.83) should be used. The average error of estimation is $\pm 2\%$, the maximum error being about 8%.

3. *Mixtures composed of non-polar and polar molecules.* In this case, Lehmann recommends the use of Eq. (5.87). The thermal conductivities of mixtures of this type are greater than given by the additivity rule (Fig. 5.15).

For equimolar mixtures Eq. (5.87) gives a value of λ_m which is higher by about 7·1% than the value calculated additively. Actually, this difference fluctuates for various mixtures from 3·2 to 11·9%.

It should be mentioned that after considering the application of various equations to different mixtures as well as comparing his calculations with the results of his measurements, Lehmann recommends only empirical equations and warns against the use of equations based on the kinetic theory of gases, such as the equations of Lindsay and Bromley, and the equations of Hirschfelder *et al.* The former equations give very good results in many cases, whereas the latter are too complicated to be suitable for rapid technological calculations.

If a gaseous mixture contains a component with polar molecules, Reid and Sherwood[306] recommend the use of Lindsay and Bromley's equations (5.63), (5.64). In this case the equations of Mason and Saxena ((5.80)–(5.82)) may also be used.

For the estimation of the thermal conductivity of gaseous mixtures at high pressures a corresponding states treatment combined with the use of pseudocritical parameters is recommended. This method involves the use of Kay and Comings' graphs (Figs. 5.10 and 5.11) and Gamson's equation (5.41).

EXAMPLE 5.8. Estimate by various methods the thermal conductivity λ_m of a mixture containing 60% (vol.) of H_2 and 40% of CO at a temperature of 180°C and pressure p of 1 atm (it may be assumed approximately that volume percentage = mole percentage; hence, hydrogen $x_1 = 0{\cdot}6$, carbon monoxide $x_2 = 0{\cdot}4$, $T = 180+273 = 453$°K).

Data

Physico-chemical constants read from the tables:

Conductivity (0°C, 1 atm) from Table 5.3

H_2: $\lambda_1^\circ = 4{\cdot}19\times10^{-4}$ cal/cm.sec.°C

CO: $\lambda_2^\circ = 0{\cdot}53\times10^{-4}$ cal/cm.sec.°C

Sutherland's constants in Eq. (5.31) from section 1

H_2: $C_1 = 94$

CO: $C_2 = 156$

Molecular weights

H_2: $M_1 = 2{\cdot}02$

CO: $M_2 = 28{\cdot}01$

Dynamic viscosity from Table 3.4 (values interpolated to a temperature of 180°C, and pressure of 1 atm):

H_2: $\mu_1 = 1177\times10^{-5}$ cP

CO: $\mu_2 = 2388\times10^{-5}$ cP

Dipole moment

H_2: $\bar{\mu}_1 = 0$ D

CO: $\bar{\mu}_2 = 0{\cdot}11$ D

Diffusivity $D_{1,2} = 1{\cdot}58$ cm²/sec (180°C, 1 atm).

The heat capacities at 180°C (453°K) are given by:

H_2: $C_{p1} = 6{\cdot}952-0{\cdot}4576\times10^{-3}\times453+0{\cdot}9563\times10^{-6}\times453^2-0{\cdot}2079\times10^{-9}\times453^3$
$= 6{\cdot}923$ cal/mole.°C

$c_{p1} = 6{\cdot}923/2{\cdot}02 = 3{\cdot}461$ cal/g.°C

CO: $C_{p2} = 6{\cdot}726+0{\cdot}4001\times10^{-3}\times453+1{\cdot}283\times10^{-6}\times453^2-0{\cdot}5307\times10^{-9}\times453^3$
$= 7{\cdot}075$ cal/mole.°C

$c_{p2} = 7{\cdot}075/28{\cdot}01 = 0{\cdot}2525$ cal/g.°C

The thermal conductivities (180°C, 1 atm) are, according to Eq. (5.31)

$$H_2:\quad \lambda_1 = 4{\cdot}19\times10^{-4}\left(\frac{273+94}{453+94}\right)\left(\frac{453}{273}\right)^{3/2} = 6{\cdot}02\times10^{-4}\ \text{cal/cm.sec.°C}$$

$$CO:\quad \lambda_2 = 0{\cdot}53\times10^{-4}\left(\frac{273+156}{453+156}\right)\left(\frac{453}{273}\right)^{3/2} = 0{\cdot}80\times10^{-4}\ \text{cal/cm.sec.°C}$$

Sutherland's constant $C_{1,2}$ for the mixture

$$C_{1,2} = \sqrt{94\times156} = 121$$

Densities (180°C, 1 atm) are:

$$H_2:\quad \rho_1 = \frac{2{\cdot}02\times273}{22416\times453} = 0{\cdot}0000537 = 5{\cdot}37\times10^{-5}\ \text{g/cm}^3$$

$$CO:\quad \rho_2 = \frac{28{\cdot}01\times273}{22416\times453} = 0{\cdot}000751 = 75{\cdot}1\times10^{-5}\ \text{g/cm}^3$$

Solution

a. Estimation of λ_m from Eq. (5.83)

$$\lambda_m = 0{\cdot}6\times 6{\cdot}02\times 10^{-4}+0{\cdot}4\times 0{\cdot}80\times 10^{-4} = 3{\cdot}93\times 10^{-4} \text{ cal/cm.sec.°C}$$

b. Estimation of λ_m from Brokaw's equation (5.85). The content of the lighter component (H_2) is $x_1 = 0{\cdot}6$. According to the data presented in Table 5.14 $\underline{a} = 0{\cdot}50$, hence

$$\lambda'_m = 3{\cdot}93\times 10^{-4} \text{ (cf. } (a))$$

$$\frac{1}{\lambda''_m} = \frac{0{\cdot}6}{6{\cdot}02\times 10^{-4}} + \frac{0{\cdot}4}{0{\cdot}80\times 10^{-4}} = 0{\cdot}60\times 10^4 \qquad \lambda''_m = 1{\cdot}67\times 10^{-4}$$

$$\lambda_m = 0{\cdot}5\times 3{\cdot}93\times 10^{-4}+(1-0{\cdot}5)\times 1{\cdot}67\times 10^{-4} = 2{\cdot}80\times 10^{-4} \text{ cal/cm.sec.°C.}$$

c. Estimation of λ_m by means of the empirical equation (5.86)

$$\lambda_m = \frac{0{\cdot}6\times 6{\cdot}02\times 10^{-4}\times \sqrt[3]{2}+0{\cdot}4\times 0{\cdot}80\times 10^{-4}\times \sqrt[3]{28}}{0{\cdot}6\times \sqrt[3]{2}+0{\cdot}4\times \sqrt[3]{28}} = 2{\cdot}81\times 10^{-4} \text{ cal/cm.sec.°C}$$

d. Estimation of λ_m by means of the first equation of Lindsay and Bromley, (5.63) and (5.64)

$$A_{1,2} = \tfrac{1}{4}\left\{1+\left[\frac{1177}{2388}\left(\frac{28}{2}\right)^{0{\cdot}75}\frac{1+\dfrac{94}{453}}{1+\dfrac{156}{453}}\right]^{0{\cdot}5}\right\}^2 \frac{1+\dfrac{121}{453}}{1+\dfrac{94}{453}} = 2{\cdot}05$$

$$A_{2,1} = \tfrac{1}{4}\left\{1+\left[\frac{2388}{1177}\left(\frac{2}{28}\right)^{0{\cdot}75}\frac{1+\dfrac{156}{453}}{1+\dfrac{94}{453}}\right]^{0{\cdot}5}\right\}^2 \frac{1+\dfrac{121}{453}}{1+\dfrac{156}{453}} = 0{\cdot}581$$

$$\lambda_m = \frac{6{\cdot}02\times 10^{-4}}{1+2{\cdot}05\times\dfrac{0{\cdot}4}{0{\cdot}6}} + \frac{0{\cdot}80\times 10^{-4}}{1+0{\cdot}581\times\dfrac{0{\cdot}6}{0{\cdot}4}} = 2{\cdot}97\times 10^{-4} \text{ cal/cm.sec.°C}$$

e. Estimation of λ_m by means of the second equation of Lindsay and Bromley (5.66)

$$\lambda_m = \frac{6{\cdot}02\times 10^{-4}}{1+\dfrac{1{\cdot}114\times 6{\cdot}02\times 10^{-4}}{1{\cdot}58\times 5{\cdot}37\times 10^{-5}\times 3{\cdot}461}\times\dfrac{0{\cdot}4}{0{\cdot}6}} + \frac{0{\cdot}80\times 10^{-4}}{1+\dfrac{1{\cdot}114\times 0{\cdot}80\times 10^{-4}}{1{\cdot}58\times 75{\cdot}1\times 10^{-5}\times 0{\cdot}2525}\times\dfrac{0{\cdot}6}{0{\cdot}4}} = 2{\cdot}94\times 10^{-4} \text{ cal/cm.sec.°C}$$

Comparison of the estimates

	λ_m, cal/cm.sec.°C	Error in the estimate
(a) Additive estimation	$3{\cdot}93\times 10^{-4}$	$+44{\cdot}5\%$
(b) Brokaw's equation (5.85)	$2{\cdot}80\times 10^{-4}$	$+2{\cdot}9\%$
(c) Empirical equation (5.86)	$2{\cdot}81\times 10^{-4}$	$+3{\cdot}3\%$
(d) Lindsay and Bromley's equations (5.63) and (5.64)	$2{\cdot}97\times 10^{-4}$	$+9{\cdot}2\%$
(e) Lindsay and Bromley's equation (5.66)	$2{\cdot}94\times 10^{-4}$	$+8{\cdot}1\%$
Experimental value of λ_m	$2{\cdot}72\times 10^{-4}$	

EXAMPLE 5.9. Estimate the thermal conductivity $\lambda_{m,p}$ of a mixture consisting of 40% (mole) of ethylene and 60% of nitrogen at a temperature of 42°C and a pressure p of 200 atm

$$x_1 = 0{\cdot}4, \qquad x_2 = 0{\cdot}6, \qquad T = 42+273 = 315\text{°K}$$

Data

Physico-chemical constants taken from the tables:

Thermal conductivity (0°C, 1 atm) from Table 5.3

C_2H_4: $\lambda_1^\circ = 0{\cdot}40\times 10^{-4}$ cal/cm.sec.°C

N_2: $\lambda_2^\circ = 0{\cdot}57\times 10^{-4}$ cal/cm.sec.°C

Sutherland's constants

C_2H_4: $C_1 = 241$

N_2: $C_2 = 114$ (Section 1)

Dynamic viscosity from Table 3.4 (values interpolated to a temperature of 42°C and pressure of 1 atm)

C_2H_4: $\mu_1 = 1078\times10^{-5}$ cP

N_2: $\mu_2 = 1852\times10^{-5}$ cP

Molecular weights

C_2H_4: $M_1 = 28{\cdot}05$

N_2: $M_2 = 28{\cdot}02$

Critical constants

C_2H_4: $t_{c1} = 9{\cdot}7$°C, $T_{c1} = 282{\cdot}9$°K, $p_{c1} = 50{\cdot}9$ atm

N_2: $t_{c2} = -147{\cdot}1$°C, $T_{c2} = 126{\cdot}1$°K, $p_{c2} = 33{\cdot}5$ atm

Thermal conductivity at the critical point from Table 5.9

C_2H_4: $\lambda_{c1} = 0{\cdot}0426$ kcal/m.hr.°C $= 1{\cdot}18\times10^{-4}$ cal/cm.sec.°C

N_2: $\lambda_{c2} = 0{\cdot}0283$ kcal/m.hr.°C $= 0{\cdot}785\times10^{-4}$ cal/cm.sec.°C

Dipole moments

C_2H_4: $\bar{\mu}_1 = 0$

N_2: $\bar{\mu} = 0$

Thermal conductivities (315°K, 1 atm) are given by:

$$C_2H_4:\quad \lambda^{\circ}_{1T} = 0{\cdot}40\times10^{-4}\times\frac{273+241}{315+241}\times\left(\frac{315}{273}\right)^{3/2} = 0{\cdot}45\times10^{-4}\ \text{cal/cm.sec.°C}$$

$$N_2:\quad \lambda^{\circ}_{2T} = 0{\cdot}57\times10^{-4}\times\frac{273+114}{315+114}\times\left(\frac{315}{273}\right)^{3/2} = 0{\cdot}63\times10^{-4}\ \text{cal/cm.sec.°C}$$

Pseudocritical and reduced parameters of the mixture are:

$T_{pc} = 0{\cdot}4\times282{\cdot}9+0{\cdot}6\times126{\cdot}1 = 189$°K

$p_{pc} = 0{\cdot}4\times50{\cdot}9+0{\cdot}6\times33{\cdot}5 = 40{\cdot}4$ atm

$T_{rm} = 315/189 = 1{\cdot}67$

$p_{rm} = 200/40{\cdot}4 = 4{\cdot}95$

The thermal conductivity of the mixture at moderate pressure (315°K, 1 atm)

Both gases are non-polar, and so it is reasonable to use the rule:

$\lambda^{\circ}_{mT} = 0{\cdot}4\times0{\cdot}45\times10^{-4}+0{\cdot}6\times0{\cdot}63\times10^{-4} = 0{\cdot}56\times10^{-4}$ cal/cm.sec.°C

Solution

a. Estimation of the value of $\lambda_{m,p}$ by Comings' method and Eq. (5.57)

For $T_{rm} = 1{\cdot}67$ and $p_{rm} = 4{\cdot}95$, we read from the graph in Fig. 5.10 the value of $\lambda_p/\lambda_0 = 1{\cdot}95$

$\lambda_{m,p} = 1{\cdot}95\times0{\cdot}56\times10^{-4} = 1{\cdot}09\times10^{-4}$ cal/cm.sec.°C

b. Estimation of the value of $\lambda_{m,p}$ by the Gamson method and Eq. (5.41)

For $T_{rm} = 1{\cdot}67$ and $p_{rm} = 4{\cdot}95$, the value read from Fig. 5.8 is $\lambda_{rm} = 1{\cdot}3$

$\lambda_{cm} = 0{\cdot}4\times1{\cdot}18\times10^{-4}+0{\cdot}6\times0{\cdot}785\times10^{-4} = 0{\cdot}94\times10^{-4}$ cal/cm.sec.°C

$\lambda_{m,p} = 0{\cdot}94\times10^{-4}\times1{\cdot}3 = 1{\cdot}22\times10^{-4}$ cal/cm.sec.°C

Comparison of the results of estimation

	λ_m, cal/cm.sec.°C	Error in the estimate
(a) Comings' method	$1{\cdot}09\times10^{-4}$	−4%
(b) Gamson's method	$1{\cdot}22\times10^{-4}$	+10·9%
Experimental value of λ_m	$1{\cdot}1\times10^{-4}$	

CHAPTER 6

THERMAL CONDUCTIVITIES OF LIQUIDS

1. BASIC CONCEPTS: EXPERIMENTAL DATA

If a liquid is placed between two planes at different temperatures, heat will be conducted from the surface at the higher temperature to the surface with the lower temperature. In contradistinction to the convection process (in which the energy is transferred by large groups of molecules moving together in the same definite direction), the process of heat conduction involves the transmission of energy from one stationary layer of molecules to neighbouring layers without any large scale motions occurring within the body of the fluid.

A knowledge of the values of the thermal conductivities of liquids is of practical and theoretical importance. For example, engineering calculations of heat transfer rates from a liquid to a solid or vice versa are often based on the hypothesis of *laminar films* or *boundary layers*. According to this hypothesis the main resistance to the transfer of heat across the interface is caused by the existence of a relatively immobile liquid layer at the wall. Heat is transferred through this layer by conduction, and not by convection.

The thermal conductivities of liquids are much greater than those of gases and in contrast to the latter, the thermal conductivities of most liquids decrease with increasing temperature. This difference is paralleled by an analogous difference between the variation of the viscosities of gases and of liquids with temperature. However, the effect of temperature on the thermal conductivities of liquids is comparatively insignificant, whereas the viscosities of liquids are quite sensitive to changes in temperature. The effect of temperature on thermal conductivity is illustrated for several liquids in Fig. 6.1. In addition Tables 6.1 and 6.2 list the thermal conductivities λ of a number of organic liquids at different temperatures.[11,147,306,311] (The units used in Tables 6.1 and 6.2 are cal/cm.sec.°C and kcal/m.hr.°C, respectively.) Table 6.2 also lists approximate values of λ for some industrial oils.

As may be seen from Fig. 6.1 the thermal conductivities of most liquids fall very slightly with increase in temperature, a rise of 10°C in the temperature usually producing a fall of 0·5% to 2% in λ. An average figure is close to 1%, but the exact value of the temperature coefficient depends on the temperature range considered. Over a sufficiently narrow range of temperatures, say from T_0 to T, the variation of λ may be taken to be linear so that

$$\lambda = \lambda_0(1+a\Delta T) \tag{6.1}$$

where λ and λ_0 are the thermal conductivities at temperatures T and T_0, respectively,
$\Delta T = T-T_0$
and a is the temperature coefficient of λ (usually 0·0005 to 0·002). For water a is +0·001, for castor oil it is −0·0005, while for paraffin oil it is −0·0005.[282]

TABLE 6.1. THERMAL CONDUCTIVITIES OF LIQUIDS ($\lambda \times 10^6$, cal/cm.sec.°C)

Compound	Temperature, °C								
	0	12	20	40	50	60	75	80	100
Acetaldehyde	—	—	400	—	—	—	—	—	—
Acetic acid	—	470	—	—	—	—	—	—	—
Acetone	406	398	386	359	—	—	—	—	—
Amyl alcohol	400	—	390	—	380	—	380	—	370
iso-Amyl alcohol	360	—	360	—	360	—	350	—	350
Amyl bromide	—	240	—	—	—	—	—	—	—
Amyl chloride	—	280	—	—	—	—	—	—	—
Aniline	410	—	410	—	410	—	400	—	—
Benzene	—	—	353	—	328	—	—	—	—
Bromobenzene	—	—	267	—	—	—	—	241	—
n-Butyl alcohol	—	—	—	364	—	349	—	334	—
iso-Butyl alcohol	370	340	—	—	—	—	—	—	—
Butyric acid	—	360	—	—	—	—	—	—	—
iso-Butyric acid	—	340	—	—	—	—	—	—	—
Carbon disulphide	376	—	—	350	—	—	—	—	—
Carbon tetrachloride	—	—	247	237	234	—	—	—	—
Chlorobenzene	317	—	308	298	—	290	—	—	—
Chloroform	—	290	—	—	—	—	—	—	—
m-Cresol	—	—	359	—	—	—	—	348	—
Decane	360	—	—	—	—	—	340	—	—
Diethyl ether	340	—	310	—	—	—	—	—	—
Dimethylaniline	—	—	341	—	—	—	—	316	—
Ethyl acetate	—	—	—	348	—	315	—	—	—
Ethyl alcohol	450	—	430	403	—	386	—	—	—
Ethyl bromide	300	—	290	—	—	—	—	—	—
Ethyl butyrate	—	320	—	—	—	—	—	—	—
Ethyl formate	—	380	—	—	—	—	—	—	—
Ethyl iodide	270	—	270	—	260	—	260	—	260
Ethyl valerate	—	310	—	—	—	—	—	—	—
Ethylbenzene	—	—	316	—	—	—	—	280	—
Fluorobenzene	—	—	320	(30°C)	—	—	—	—	—
Formic acid	—	650	—	—	—	—	—	—	—
Glycerol	680	—	680	—	680	—	680	—	680
Glycol	610	—	620	—	—	620	—	—	640
Heptane	340	—	330	—	—	—	—	—	—
Heptyl alcohol	400	—	390	—	380	—	370	—	370
Hexane	340	—	330	—	—	—	—	—	—
Hexyl alcohol	390	—	390	—	380	—	370	—	370
Iodobenzene	250	—	244	238	232	—	—	—	—
Methyl alcohol	510	—	500	487	—	461	—	—	—
Methyl butyrate	—	340	—	—	—	—	—	—	—
Methyl valerate	—	320	—	—	—	—	—	—	—
Nitrobenzene	—	380	—	—	—	—	370	—	—
n-Octane	350	—	350	—	—	—	330	—	330
n-Pentane	330	—	320	—	—	—	—	—	—

TABLE 6.1 (*cont.*)

Compound	Temperature, °C								
	0	12	20	40	50	60	75	80	100
Propionic acid	—	390	—	—	—	—	—	—	—
Propyl alcohol	410	—	400	376	—	362	—	—	—
iso-Propyl alcohol	370	—	—	335	—	322	—	—	—
Propyl bromide	—	260	—	—	—	—	—	—	—
Propyl chloride	—	280	—	—	—	—	—	—	—
Toluene	370	—	360	—	350	—	320	—	—
Valeric acid	—	330	—	—	—	—	—	—	—
Water	1350	—	1430	1500	—	1560	—	1600	1625
o-Xylene	331	—	321	310	—	—	—	—	—
m-Xylene	340	—	—	—	—	—	—	—	—

TABLE 6.2. THERMAL CONDUCTIVITIES OF SOME LIQUIDS (kcal/m.hr.°C) (Ref/s 147, 311)

t, °C	λ
Diethylene glycol $(C_2H_4OH)_2O$	
0	0·174
20	0·176
100	0·184
Methyl acetate	
12	0·139
Propyl acetate	
12	0·118
Propyl chloride	
12	0·102
Methyl chloride	
−20	0·168
−10	0·161
0	0·154
20	0·140
Trichloroethylene	
−60	0·117
20	0·100
Petrol	
0	0·12
White spirit	
0	0·115
50	0·111
100	0·107
Kerosene, refined	
0	0·134
20	0·130
50	0·125
100	0·115
Paraffin oil	
0	0·108
50	0·108
100	0·108
200	0·107
Ammonia	
0	0·464
20	0·425
100	0·269
Carbon dioxide $p = 60$ kg/cm²	
10	0·087
20	0·076
$p = 90$ kg/cm²	
10	0·092
20	0·081
30	0·071
Sulphur dioxide	
−20	0·192
−10	0·187
0	0·182
20	0·171
Methylene chloride	
−20	0·139
−10	0·137
0	0·136
20	0·133
Ethylene chloride	
−20	0·125
20	0·116
80	0·103
Spindle oil	
0	0·124
50	0·122
100	0·120
150	0·145
Cylinder oil	
0	0·133
50	0·130
100	0·128
200	0·122
Sulphuric acid	
20	0·27

Palmer[276] has observed that associated liquids have high thermal conductivities, this being due to the formation of hydrogen bonds (as, for example, in water, mono- and polyhydric alcohols, aniline, etc.).

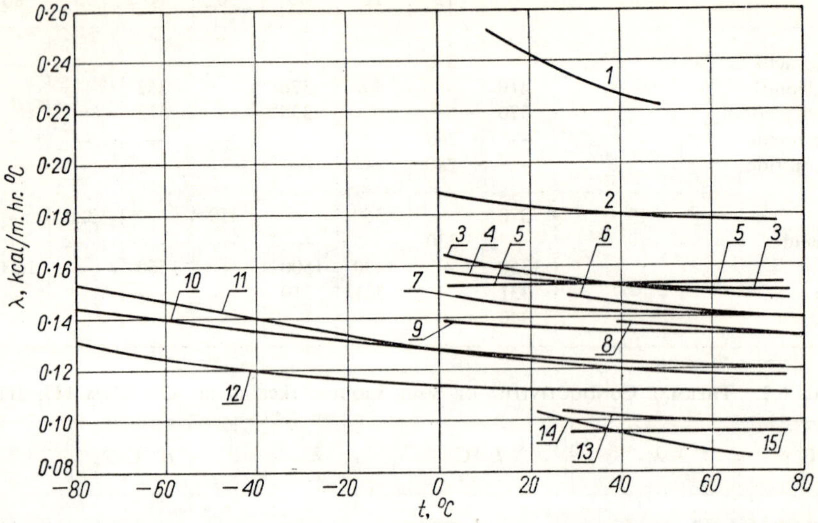

FIG. 6.1. The thermal conductivities of some typical liquids as functions of temperature (after Eucken[91]); *1*—glycerol, *2*—methanol, *3*—ethanol, *4*—acetone, *5*—aniline, *6*—n-nonyl alcohol, *7*—olive oil, *8*—n-hexyl alcohol, *9*—carbon disulphide, *10*—toluene, *11*—ethyl ether, *12*—pentane, *13*—ethyl bromide, *14*—carbon tetrachloride, *15*—ethyl iodide.

It appears from the investigations of Abas-Zade[1] that although the thermal conductivities of liquids usually decrease with increase in temperature up to the critical point, further rise of temperature produces an increase in λ. Abas-Zade found that the thermal conductivities of saturated organic liquids could be represented by the equation

$$\lambda - \lambda^\circ = \alpha\rho^n \tag{6.2}$$

where λ is the thermal conductivity of the liquid at its vapour pressure at temperature t,

λ° is the thermal conductivity of the liquid at the same temperature but at atmospheric pressure,

ρ is the density,

and α and n are constants characteristic of the liquid.

When a liquid–vapour system is heated in the vicinity of the critical point, the conductivity of the vapour increases to approach that of the liquid. This phenomenon is illustrated for carbon dioxide, for which the critical temperature (t_c) is 31·1°C, in Table 6.3.

TABLE 6.3. THERMAL CONDUCTIVITY OF SATURATED CARBON DIOXIDE ($\lambda \times 10^4$, cal/cm.sec.°C) (Ref. 363)

Temperature, °C	20	22	24	25	26	27	28	29	30	31
λ for liquid	2·05	2·01	1·96	1·95	1·88	1·80	1·77	1·63	1·53	1·15
λ for vapour	0·68	0·70	0·758	0·773	0·805	0·833	0·870	0·912	0·968	1·09

The thermal conductivity λ of water and some other compounds (e.g. certain forms of sulphur) shows an increase, instead of a decrease, with increase in temperature. Palmer explains this for water as being due to association and formation of hydrogen bonds. The thermal conductivity of water reaches a maximum at about 120°C (Table 6.5 and Fig. 6.2).

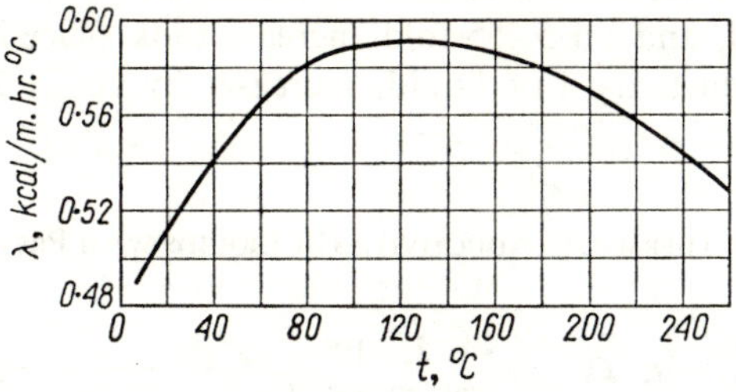

FIG. 6.2. Variation of the thermal conductivity of water with temperature (below 100°C the data are for a pressure of 1 atm; above this temperature they are at a pressure equal to the vapour pressure) (after Eucken[91]).

Most thermal conductivity measurements on liquids have been carried out at temperatures below the normal boiling temperature. However, Scheffy and Johnson[319] have measured the conductivities of some liquids at temperatures between the boiling temperature T_b and the critical temperature T_c.

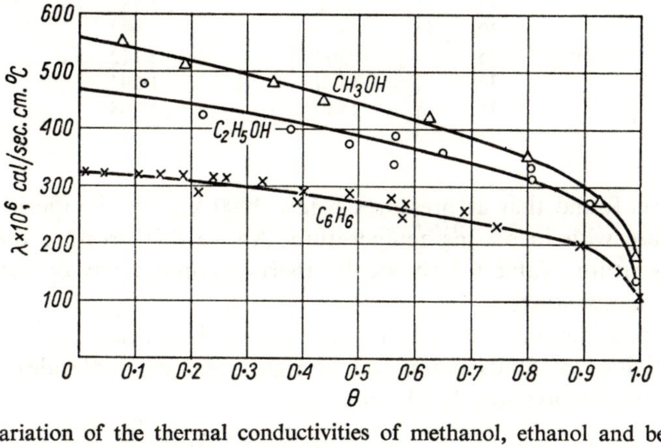

FIG. 6.3. Variation of the thermal conductivities of methanol, ethanol and benzene with a reduced temperature θ given by $\theta = (T-T_m)/(T_c-T_m)$.

Figure 6.3 shows the thermal conductivities of methanol, ethanol and benzene as functions of a reduced temperature θ defined by

$$\theta = \frac{T-T_m}{T_c-T_m}$$

where T_m is the melting temperature. The curves for the liquids shown are all similar in shape. In practice thermal conductivities are usually required at values of θ less than about 0·5. It may be seen from Fig. 6.3 that in this range the relationship between λ and θ is almost linear, whereas for $\theta > 0{\cdot}7$ a pronounced curvature appears. At the melting temperature the liquid thermal conductivity λ_m is approximately three times as high as at the critical point:

$$\lambda_m \approx 3\lambda_c \qquad (6.3)$$

The thermal conductivities of liquids increase with pressure. This is illustrated in Table 6.4, the figures in which are taken from the work at Bridgman. In this table λ_0 denotes the thermal conductivity at zero pressure. The effect of pressure on thermal conductivity is greater the more compressible is the liquid, and hence is greater at higher temperatures. In the experiments, the results of which are listed in Table 6.4, very high pressures were employed, and a 1·5–2·5-fold increase took place in the conductivity. On average the thermal conductivity λ of liquids increases by about 20% when the pressure is raised by 700 kg/cm^2.

TABLE 6.4. VARIATION OF THE THERMAL CONDUCTIVITIES OF LIQUIDS WITH PRESSURE (AFTER BRIDGMAN[37,38])

Compound	t, °C	$\lambda_0 \times 10^6$, cal/cm.sec.°C	λ_p/λ_0	
			$p = 6000$ kg/cm^2	$p = 12{,}000$ kg/cm^2
Methanol	30	505	1·724	2·097
Methanol	75	493	1·785	2·191
Ethanol	30	430	1·744	2·122
Ethanol	75	416	1·845	2·278
Diethyl ether	30	329	2·009	2·451
Diethyl ether	75	322	2·043	2·537
Carbon disulphide	30	382	1·663	1·962
Carbon disulphide	75	362	1·789	2·154
n-Pentane	30	322	1·987	2·481
n-Pentane	75	307	2.112	2·740
Water	30	1440	1·293	ice
Water	75	1540	1·308	1·506

Bridgman has found that at pressures below 3000 kg/cm^2 the thermal conductivities of liquids decrease with increasing temperature. Above this pressure they increase with increase in temperature. Table 6.5 shows the thermal conductivity of water as a function of pressure and temperature.

Theoretical considerations based on Enskog's work indicate that the viscosities and thermal conductivities of fluids composed of rigid spherical molecules should be interrelated, as functions of pressure, by the equation

$$\lambda_p/\lambda^\circ = \mu_p/\mu^\circ$$

where λ_p and μ_p are the conductivity and viscosity at an elevated pressure p,
and λ° and μ° are the conductivity and viscosity at a moderate pressure.

In practice this equation is not accurately obeyed by real fluids at very high pressures or if the molecules are complex. Bridgman,[38] for example, obtains the following values for the ratios (μ_p/μ_0) and (λ_p/λ_0) for methanol at 30°C:

Pressure, kg/cm^2	1	968	1936	3871	5807	7742	11614
μ_p/μ_0	1·00	1·47	1·93	2·96	4·13	5·62	9·95
λ_p/λ_0	1·00	1·20	1·34	1·56	1·72	1·86	2·10

TABLE 6.5. VARIATION OF THE THERMAL CONDUCTIVITY OF WATER WITH TEMPERATURE AND PRESSURE (AFTER CHIRKIN[61])

t, °C	Thermal conductivity λ, kcal/m.hr.°C				
	$p = 1$ kg/cm²	100 kg/cm²	200 kg/cm²	300 kg/cm²	400 kg/cm²
0	0·474	0·477	0·480	0·484	0·488
20	0·515	0·519	0·523	0·528	0·533
40	0·545	0·549	0·554	0·559	0·564
60	0·567	0·571	0·575	0·580	0·585
80	0·580	0·584	0·589	0·594	0·599
100	0·587	0·593	0·598	0·603	0·608
120	0·590	0·596	0·602	0·608	0·614
140	0·589	0·596	0·602	0·608	0·615
160	0·587	0·593	0·599	0·606	0·613
180	0·580	0·587	0·593	0·600	0·609
200	0·570	0·578	0·584	0·593	0·602
220	0·555	0·564	0·572	0·582	0·591
240	0·540	0·547	0·557	0·567	0·577
260	0·520	0·526	0·536	0·547	0·558
280	0·484	0·500	0·512	0·526	0·538
300	0·464	0·466	0·482	0·499	0·513
320	0·425	—	0·444	0·463	0·482
340	0·376	—	0·392	0·422	0·447
360	0·306	—	0·314	0·364	0·401

It may be seen from the above table that at very high pressures the effect of pressure upon the fluid viscosity is much higher than that upon the thermal conductivity. However, at more moderate pressures, the parallelism between the variations in viscosity and thermal conductivity is approximately obeyed for liquefied carbon dioxide.[64]

Temperature, °C	10	20	20	30	30
Pressure, kg/cm²	50	70	80	80	90
μ_0/μ_p	6·35	5·2	5·6	3·7	4·2
λ_0/λ_p	6·4	5·8	6·0	5·1	5·2

It is found that the deviations from the parallelism between the variations in viscosity and thermal conductivity are the more marked the more complex is the molecular structure of the substance considered. The thermal conductivities of fused salts and molten metals are rather different from those of other liquids. Those two types of liquid are sometimes used as heating media and, therefore, a knowledge of their thermal conductivities is of practical importance.

The thermal conductivities of fused salts have been measured in a few cases only, since these measurements are difficult to carry out. It is common practice to study eutectic mixtures in order to depress the freezing temperature of the melt as far as possible. Table 6.6 gives the thermal conductivities of several mixtures of fused salts. (The compositions of those mixtures are given in mole per cent in parentheses.) However, the accuracy of these data is rather low and they may be in error by as much as ±25%.

TABLE 6.6. THERMAL CONDUCTIVITIES OF FUSED SALTS (AFTER GAMBILL[121])

Composition of the mixture, mole per cent	T, °K	λ, kcal/m.hr.°C
LiF–RbF (43–57)	748	1·8
RbF–ZrF_4–UF_4 (48–48–4)	698	1·5
NaF-ZrF_4–UF_4 (53·5–40–6·5)	813	1·8
NaF–BeF_2 (57–43)	633	3·6
NaF–ZrF_4–UF_4 (50–46–4)	793	1·9
NaF–KF–LiF–UF_4 (10·9–43·5–44·5–1·1)	725	3·4
NaF–KF–LiF (11·5–42–46·5)	727	3·8
KNO_3–$NaNO_2$–$NaNO_3$ (44·15–48·9–6·95)	415·4	0·49
NaOH (97·6 wt.%)	591·6	0·75

TABLE 6.7. THERMAL CONDUCTIVITIES OF MOLTEN METALS (AFTER BREBACH[34])

Molten metal	t, °C	λ, cal/cm.sec.°C	Molten metal	t, °C	λ, cal/cm.sec.°C
Bismuth	300	0·042	Cadmium	349	0·107
	450	0·0378			
	600	0·0373	Mercury	20	0·0200
				100	0·0200
Lead	403	0·0390			
	600	0·0360	Sodium	100	0·203
				210	0·191
Tin	301	0·0835			

A few typical data for molten metals are given in Table 6.7. According to Ewing, Walker, Grand and Miller[93] the thermal conductivity of a molten metal can be predicted from the equation

$$\lambda = 2{\cdot}61\times10^{-8}(T/r) - \frac{2\times10^{-17}(T/r)^2}{c_p\rho} + \frac{97c_p\rho^2}{MT} \tag{6.4}$$

where λ is the thermal conductivity, w/cm.°C,
T is the temperature, °K,
r is the electric resistivity of the metal, ohm.cm,
ρ is the density, g/cm^3,
and M is the molecular (or atomic) weight.

The average error in the predicted conductivity is claimed to be 5–10%.

It should be noted that the thermal conductivities of fused salts and molten metals are many times greater than those of liquid organic compounds.

2. THEORETICAL CONSIDERATIONS

Although it is believed that the process of momentum transfer leads to the phenomenon of internal friction (viscosity) and that this is related to the process of energy transfer (heat conduction), attempts to develop an exact theory for transport phenomena in liquids have, so far, been hindered by insufficient knowledge of the liquid state. Whereas the viscosities of liquids at ambient temperature vary very widely (from about 10^{-3} to about 10^4 P, i.e. by several orders of magnitude), the thermal conductivities of different organic liquids are rather similar, the variations being no greater than fivefold.[33] It would appear, therefore, that the mechanisms of momentum and energy transfer may be different. This differing behaviour may be responsible in part for the wide variations in the Prandtl numbers of different liquids exemplified in Table 6.8. Recent attempts to develop a theory

TABLE 6.8. PRANDTL NUMBERS FOR SOME LIQUIDS (REFS. 33, 78)

Liquid	*t*, °C	Pr	Liquid	*t*, °C	Pr
Water	0	11·3	Ethylene glycol	25	141
	20	7·3			
	40	4·38	Glycerol	20	12500
	60	3·01		30	5370
	80	2·15			
			Pentane	30	2·40
Ethyl alcohol	30	13·9	Hexane	30	3·26
	70	8·7	Heptane	30	4·16
			Octane	30	5·42
n-Amyl alcohol	30	46·7	Nonane	30	6·19

for heat conduction in liquids have been of two types: one based on the exact principles of statistical mechanics, and the other involving attempts to relate the thermal conductivity to the *free volume* in liquids. Bearman and Vaidhyanathan[21] have shown from statistical mechanical considerations that, the product of the viscosity μ and the thermal conductivity λ should be constant, at given temperature and pressure for systems forming regular solutions. They have checked this result against data for the system benzene–carbon tetrachloride at 30°C and give the following table of values.

Mole fraction of benzene	0·0	0·2	0·4	0·8	0·9	1·0
$\lambda \times \mu$ (kcal.cP/mole.hr.deg)	0·073	0·070	0·068	0·067	0·068	0·069

Most of the earlier theories for the mechanism of heat conduction in liquids are based on a model developed by Bridgman.[37] According to this model, the liquid molecules may be regarded as oscillating about equilibrium positions centred on a cubic lattice, the distance between adjacent molecules being equal to l. Since a molecule of a monatomic gas under a moderate pressure has an average energy of (3/2) kT, i.e. of $2{\cdot}02\times10^{-16}T$ ergs (where T is the absolute temperature), distributed among three degrees of freedom, the energy of a liquid molecule, oscillating at the lattice point, will be equal to $4{\cdot}04\times10^{-16}\,T$ ergs with due allowance for the three degrees of freedom of potential energy and the three degrees of freedom of kinetic energy. The difference between the energy of two adjacent molecules in the direction of heat transfer should be equal to $4{\cdot}04\times10^{-16}\,l\,(\partial T/\partial x)$. According to Bridgman energy is transmitted from molecules in hot layers to adjacent molecules in layers with a lower temperature by these molecular vibrations. The number of impacts per unit time experienced by a given molecule as it oscillates is given by (u/l) where u is the sonic velocity in the liquid. Each such impact leads to a transfer of energy of

$4 \cdot 04 \times 10^{-16}\, l\, (\partial T/\partial x)$ ergs and the rate of transfer of energy through a given molecule is thus equal to $4 \cdot 04 \times 10^{-16}\, u(\partial T/\partial x)$. Now in any lattice plane there will be $(1/l^2)$ molecules per unit cross-section so the amount of energy transferred through the unit cross-section per unit time is $4 \cdot 04 \times 10^{-16}\, (u/l^2)(\partial T/\partial x)$. The thermal conductivity λ of the liquid is given by $\lambda =$ (rate of energy transfer per unit area)/$(\partial T/\partial x)$. According to Bridgman's model, therefore

$$\lambda = 4 \cdot 04 \times 10^{-16} \frac{u}{l^2} \tag{6.5}$$

where u is the sound velocity in the liquid, cm/sec,
and l is the distance between two adjacent molecules, cm.

One mole of a liquid occupies the volume $V = M/\rho$ (where M is the molecular weight and ρ is the density) and contains N molecules (where N is Avogadro's number).

If the molecules are taken to be arranged on a cubic lattice, the average length of the edge of the elementary cube is:

$$l = \sqrt[3]{\frac{M}{N\rho}} \tag{6.6}$$

Equations (6.5) and (6.6) enable λ to be predicted from u and ρ. As seen from Table 6.9, the predicted values for λ agree fairly closely with the measured values. Brillouin[40] and Eyring,[199,295] have developed expressions for λ using arguments which are rather similar to those of Bridgman. Brillouin's formula contains an empirical coefficient characterizing the given liquid, and is unsuitable for engineering use.

TABLE 6.9. COMPARISON OF LIQUID THERMAL CONDUCTIVITIES CALCULATED FROM BRIDGMAN'S EQUATIONS (6.5) AND (6.6) WITH EXPERIMENTAL VALUES (AFTER JELLINEK[174])

Compound	$\mu \times 10^{-5}$, cm/sec	$l^{-2} \times 10^{-14}$, cm^{-2}	$\lambda \times 10^{-3}$ CGS units	
			calculated	measured
Methyl alcohol	1·13	6·00	27·4	21·1
Ethyl alcohol	1·14	4·74	21·8	18·0
Propyl alcohol	1·24	3·94	19·7	15·4
Butyl alcohol	1·05	3·49	14·9	16·7
Diethyl ether	0·92	3·19	11·9	13·7
Acetone	1·14	4·00	18·5	17·9
Carbon disulphide	1·18	4·61	21·9	15·9
Water	1·50	10·4	63·0	60·1

Using the cell theory of liquids, Fourier[106] has derived the equation

$$\lambda = \frac{c_v^2 l^2 T}{V^2 \mu}$$

where c_v is the heat capacity at constant volume,
l is the mean distance between adjacent molecules,
T is the absolute temperature,
and μ is the dynamic viscosity.

3. SEMI-EMPIRICAL AND EMPIRICAL EQUATIONS FOR ESTIMATING THE THERMAL CONDUCTIVITIES OF LIQUIDS

Denbigh,[78] noting that according to the *hole theory of liquids* both the viscosity and the thermal conductivity should be related to the free volume[130] and hence to the latent heat of vaporization (L), examined whether a correlation could be found between the Prandtl number ($\mathrm{Pr} = c_p\mu/\lambda$) and the latent heat. He found that

$$\log \mathrm{Pr} = \frac{aL}{RT} + b \tag{6.7}$$

where the constants a and b are approximately 0·2 and $-1{\cdot}8$, respectively. This equation can be written in the alternative form:

$$\log \mathrm{Pr} = 0{\cdot}10 \frac{L}{T_b}\frac{T_b}{T} - 1{\cdot}8 \tag{6.8}$$

where L/T_b is the entropy of vaporization at the normal boiling temperature. Unfortunately, values of λ calculated from Denbigh's equation are not very accurate.

Various other methods for estimating the thermal conductivities of liquids have been developed which are based either on Bridgman's considerations Eqs. (6.4) and (6.5), or on the principle of additivity in conjunction with theory of corresponding states. In addition, there are several purely empirical relationships which are sometimes used.

a. Empirical Relationships

One of the earliest equations used for predicting the thermal conductivities λ of liquids is the Weber equation[395]

$$\lambda = 1{\cdot}29 c_p \rho V^{-1/3} \tag{6.9}$$

where c_p is the heat capacity at constant pressure, kcal/kg.°C,
ρ is the density, kg/l.,
$V(= M/\rho)$ is the molar volume, l./kg-mole,
and M is the molecular weight.

Equation (6.9) is usually used in the form

$$\lambda = 1{\cdot}29 c_p \rho \sqrt[3]{\frac{\rho}{M}} \tag{6.10}$$

λ in equations (6.9) and (6.10) is expressed in kcal/m.hr.°C. In order to obtain λ in cal/cm.sec.°C, the coefficient 1·29 must be replaced by $3{\cdot}59\times10^{-3}$; c_p should then be expressed in cal/g.°C and ρ in g/cm^3. If the conductivity is to be expressed in Btu/(hr)(ft^2)(°F/ft), the appropriate coefficient is 0·869. Estimates of λ obtained from the Weber equations are of a very approximate character, the average error being $\pm13\%$. The maximum error is about 60% (Ref. 306). Kuong[212] has constructed a nomograph for estimating the conductivities of liquids. This nomograph (Fig. 6.4), is based on Weber's equation (6.10).

In 1948 Predvoditelev[299] provided a theoretical interpretation for the original Weber formula, and showed that the numerical coefficient I in the equation

$$\lambda = I c_p \rho^{4/3} M^{-1/3} \tag{6.11}$$

should be a function of temperature. At 30°C, $I = 4{\cdot}28 \times 10^{-3}$, if λ is expressed in cal/cm.sec.°C, or 1·54, if λ is given in kcal/m.hr.°C. Since the product Ic_p is approximately independent of temperature, λ can be calculated at any temperature, provided c_p for the liquid is known as a function of temperature and the value of I at one temperature (say at 30°C) is known.

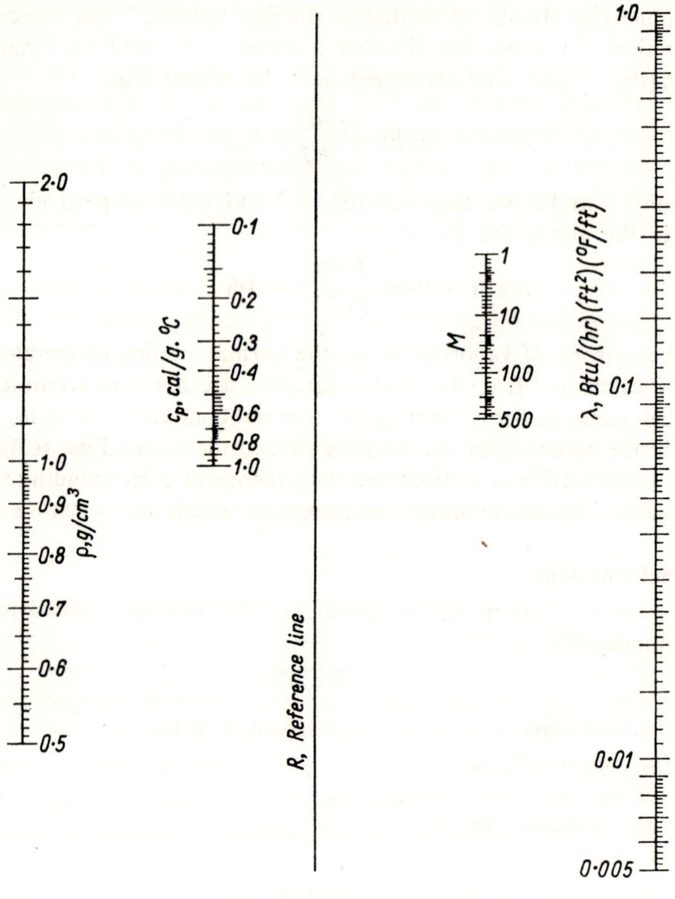

FIG. 6.4. Nomograph based on Weber's equation for estimating the thermal conductivities of liquids (after Kuong[212]).

Many attempts have been made to improve the Weber equation. Smith[333] presents several empirical formulae which in his opinion are more accurate than the Weber equation. The first of his formulae is identical in form with the Weber equation but the coefficient is changed to 0·0043 so that

$$\lambda = 4{\cdot}3 \times 10^{-3} c_p \rho \sqrt[3]{\frac{\rho}{M}} \tag{6.12}$$

where λ is expressed in cal/cm.sec.°C, c_p in cal/g.°C and ρ in g/cm³. According to the second of Smith's empirical formulae,

$$\lambda = \frac{0{\cdot}293 \rho^{2\cdot15} c_p^{\,1\cdot55} M^{0\cdot192}}{\mu^{0\cdot12}} \tag{6.13}$$

where λ is expressed in kcal/m.hr.°C, ρ in kg/l., c_p in kcal/kg.°C, M in kg/kg-mole, and μ in cP. The third of Smith's empirical formulae is as follows:

$$\lambda\times 10^6 = 11+6450(c_p-0{\cdot}45^3)+1250\left(\frac{\rho}{M}\right)^{1/3}+100\left(\frac{\mu}{\rho}\right)0{\cdot}111 \tag{6.14}$$

where λ is expressed in cal/cm.sec.°C, c_p in cal/g.°C, ρ in g/cm^3 and μ in cP. According to Perry, values of λ calculated from this equation are subject to an average error of $\pm 6{\cdot}7\%$.

It has been seen above that the thermal conductivities of liquids are influenced by the degree of molecular association. This fact has led several authors (Vargaftig[380], Palmer[276], and others) to introduce factors into the Weber formula to allow for this effect. Both Vargaftig and Palmer suggest that the right-hand side of Weber's equation (6.10) should be multiplied by the factor $1/\alpha$, where α is the degree of association of the liquid (for non-associated liquids $\alpha = 1$). An estimate of the degree of association may be obtained from the molar entropy of vaporization of the liquid (L_b/T_b). For non-associated liquids this quantity is, according to Trouton's rule, approximately equal to 21. However, for associated liquids this equality is not fulfilled and the ratio

$$\frac{L_b/T_b}{21}$$

(which is ≈ 1 for non-associated liquids) has been used as a factor for correcting calculated thermal conductivities for the effects of association (see, for example, Eq. (6.16)).

Predvoditelev's[299] equation takes the form

$$\lambda = 1{\cdot}29\rho c_p\left(\frac{\rho_0}{M}\right)^{1/3}\frac{e^{-n/T}}{\varepsilon} \tag{6.15}$$

where ρ_0 is the density at 0°C,
ρ is the density at temperature T,
and c_p is the heat capacity at constant pressure and temperature T.

In this formula, the association of the liquid and the variation of the numerical coefficient l in Eq. (6.11) with temperature are allowed for by the factors ε and $e^{-n/T}$, respectively. The constants ε and n can be determined if conductivity is known at two different temperatures. Equation (6.15) may than be used to predict thermal conductivities at other temperatures.

Palmer[276] recommends the following formula:

$$\lambda = \frac{94{,}700\times 10^{-6}\rho c_p(\rho/M)^{1/3}}{L_b/T_b} \tag{6.16}$$

In this equation the entropy of vaporization, L_b/T_b, is expressed in cal/mole.°C, and the thermal conductivity λ in cal/cm.sec.°C.

According to Reid and Sherwood, the equations of Smith and of Palmer are, in practice, less reliable than the original Weber formula (6.9). When they tested these equations against data for about thirty liquids they found that:

Weber's equations (6.9) and (6.10) gave an average error of $\pm 13\%$,
Smith's equation (6.12) gave an average error of $\pm 18\%$,
Smith's equation (6.14) gave an average error of $\pm 20\%$,
and Palmer's equation (6.16) gave an average error of $\pm 18\%$.

When applied to non-associated liquids, the most accurate of the equations considered above is that of Predvoditelev, the average error being about $\pm 5\%$.

b. Estimation of Liquid Thermal Conductivities Using the Principle of Group Contributions

It follows from Bridgman's theory of heat conduction in liquids (Eqs. (6.5) and (6.6)) that the thermal conductivity λ should depend on the velocity of sound u and the molal volume V of the liquid. Both these quantities are additive and may be estimated from a knowledge of the molecular structure by summing group contributions. The molar sound velocity may be calculated from Eq. (1.55) using the contributions compiled in Table 1.29, while the molal volume may be calculated by the methods described in Chapter 1, section 4a.

Since both u and V are additive quantities and Bridgman's equation is obeyed with reasonable accuracy, it would appear that it should be possible to calculate conductivities using a method based on group contributions. Such a method has been devised by Sakiadis and Coates.[314] The reasoning adopted by these authors is similar to that of Bridgman[37] except that the energy transfer during a collision is expressed in terms of the heat capacity of the compound and that the distance L between the peripheries of adjacent molecules is smaller than the distance Δ between their molecular centres (i.e. than the quantity l of Bridgman) by the diameter d of the molecule.

$$L = \Delta - d \tag{6.17}$$

Denoting the sonic velocity in the liquid by u, the density of the liquid by ρ and its heat capacity by c_p, the above authors obtained the relationship

$$\lambda = c_p \rho u L \tag{6.18}$$

The sonic velocity u may be calculated from the molar sonic velocity U_u, introduced by Rao, using the equation

$$u = \left(\frac{U_u \rho}{M}\right)^3 \tag{6.19}$$

(cf. Eq. (1.55)). In equation (6.19)

M is the molecular weight,

and U_u is the molar sonic velocity which may be calculated additively from the group contributions in Table 1.30.

The second quantity, which is required for the application of Eq. (6.18) is the distance L between peripheries of adjacent molecules. This may be calculated from Eq. (6.17), provided Δ is known; according to Sakiadis and Coates, this is an additive quantity which may be calculated using the method of group contributions. Appropriate contributions for estimating Δ at a temperature of 19°C are compiled in Table 6.10. In this table Δ is expressed in Å (10^{-8} cm). $\Delta(\Delta)$ for a given group is the increment in Δ, which arises when that group is substituted in the molecule.

Applying Cailletet and Mathias' rule for extrapolating the density to the temperature of 0°K at which the molecules touch one another ($L = 0$), as well as to the critical temperature at which the densities of liquid and gas are equal, Sakiadis and Coates have obtained the following formula:

$$L = \Delta\left[1 - \left(\frac{\rho_f - \rho_r T_{f,r}}{\rho_{f,r} - 2T_{f,r}}\right)^{\frac{1}{2}}\right] \tag{6.20}$$

TABLE 6.10. CONTRIBUTIONS FOR CALCULATING THE QUANTITY Δ IN THE SAKIADIS AND COATES EQUATION

I. Linear molecules

Number of carbon atoms in the molecule	C_1	C_2	C_3	C_4	C_5	C_6 †
Acids	4·10	4·70	5·00	5·20	—	→
Alcohols	4·30	4·60	5·09	—	—	→
Aldehydes	...	4·45	5·01	5·20	—	→
Amides	4·13	4·80	5·10			
Amines (mono-, di-)	...	...	5·03	5·20	—	→
Esters	...	...	4·91	5·03	—	→
Ethers	...	...	...	5·29	—	→
Halogenated hydrocarbons	...	4·70 ‡				
Hydrocarbons	...	...	...	...	...	5·40 →
Ketones	...	...	5·10	—	—	→
Mercaptans	...	...	4·96	5·01	—	→
Nitrated alkanes	...	4·60 §	5·01			
Nitriles	...	4·43	4·84			

II. Cyclic compounds
(Values of t, the thickness of the disk-shaped ring, in Å).

	t, Å		t, Å
Benzene, pyridine	5·30	Cyclohexane, piperidine	5·73
Naphthalene	5·81	Cycloheptane	5·86
Cyclopentane	5·50		

III. Contribution due to branching in linear and cyclic compounds. Substitution of a CH_3 group for hydrogen $\Delta(\Delta) = 0{\cdot}42n$ §§ (for ring compounds the contributions $\Delta(\Delta)$ are added to the dimension t)

1. Chain compounds

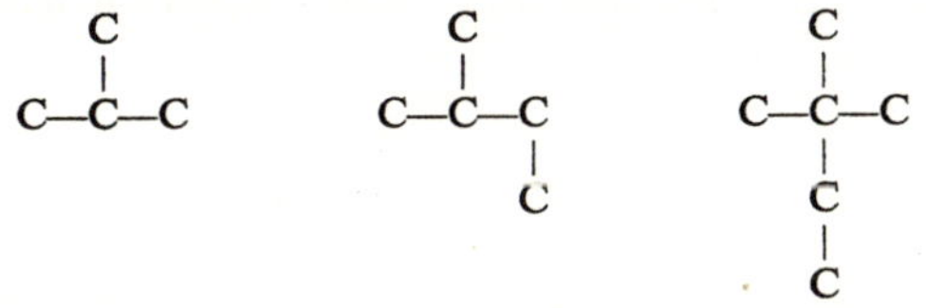

$n = 1$ $n = 1$ $n = 2$

2. Cyclic compounds

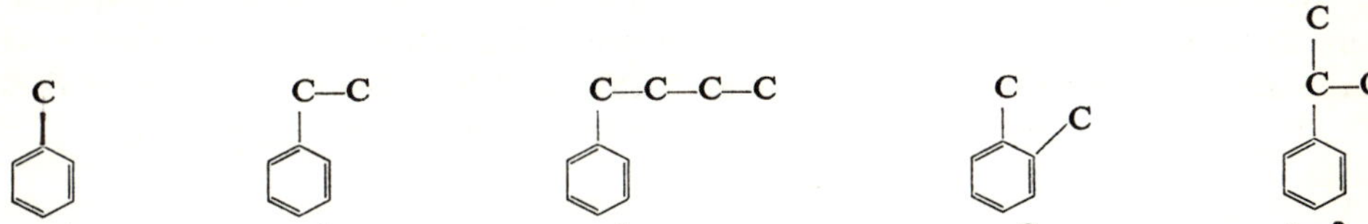

$n=1$ $n=1$ $n=1$ $n=2$ $n=2$

† The arrows indicate that the contribution is effectively the same for all larger molecules.

‡ Methyl iodide.

§ Nitromethane.

§§ n refers to the number of $—CH_3$ groups substituted as shown in the formulae. For example, if two $—CH_3$ groups are substituted in the 2,4-positions (as in 2,4-dimethylpentane), the effective n is 1, since the molecule is presumed to rotate and the swept volume is the same, whether one or two $—CH_3$ groups have been substituted. The same applies to the case of 2,2,4-trimethylpentane.

Continued overleaf

TABLE 6.10 (*cont.*)

IV. Substitution† of hydrogen by

	$\Delta(\Delta)$, Å ‡	$\Delta(\Delta)$, Å §
1. NH_2, CHO	0·10	0·80
2. F, I	0·10	
3. OH	0	0·20
4. CN, O (ketones)	0	
5. NO_2	0·50	0·75

† For ring compounds the contributions should be added to the *t* dimension.

‡ Single substitution. When substitution takes place in combination with Cl or Br, the position of the substitution is immaterial. When the substitution takes place in combination with —CHO, —NH_2 it takes place in the *p*-position.

§ Substitutions in the *o*- and *m*-positions in combination with —CHO, —NH_2, —OH.

where L is the distance between adjacent molecules, Å,
Δ is the distance between the centres of adjacent molecules, Å,
ρ_r $(=\rho/\rho_c)$ is the reduced density,
ρ_f is the density at freezing point,
$\rho_{f,r}$ $(=\rho_f/\rho_c)$ is the reduced density at freezing point,
T_f is the freezing temperature, °K,
$T_{f,r}$ $(=T_f/T_c)$ is the reduced temperature at freezing point,
and T_c is the critical temperature.

Instead of the quantities at the freezing point, one may substitute quantities corresponding to the normal boiling point though the first method gives better results.

If the liquid density at a temperature T, the freezing temperature T_f and the critical temperature T_c of a compound are known, the liquid density ρ_f at the freezing point may be estimated by means of the Sakiadis and Coates equation based on the Cailletet and Mathias rule:

$$\rho_f = \rho \frac{1-0{\cdot}48T_f/T_c}{1-0{\cdot}48T/T_c}$$

Having thus found the sonic velocity u and the value of Δ, and hence also of L, by Eq. (6.20), we can find the required value of λ by means of Eq. (6.18).

Reid and Sherwood remark[306] that although the above method has, perhaps, rather more theoretical basis than some of the other prediction methods, it is in practice rather inaccurate. Typical errors are in the range 5–15%, and in a few cases the error rises to 57%. It seems, therefore, that this method is not as useful as the simple empirical formulae.

c. Estimation of the Thermal Conductivities of Liquids Using Group Contributions and the Law of Corresponding States

Sakiadis and Coates suggest another method for estimating the thermal conductivities of organic liquids. This method is based on the law of corresponding states. Reid and Sherwood[306] consider that this method may be summarized by the equation

$$\lambda = 2{\cdot}26\theta_1 - 1{\cdot}26\theta_2 - 2{\cdot}10(\theta_1-\theta_2)T_r \tag{6.21}$$

where λ is the thermal conductivity at the reduced temperature T_r, cal/cm.sec.°C,
T_r $(= T/T_c)$ is the reduced temperature,
T_c is the critical temperature,
and θ is given by

$$\theta_1 = \lambda_{0\cdot6} + A \tag{6.22}$$

where A is an additive constant for functional groups listed in Table 6.11,
and $\lambda_{0\cdot6}$ is the thermal conductivity of the liquid at the reduced temperature $T_r = 0{\cdot}6$.

TABLE 6.11. CONTRIBUTIONS FOR CALCULATING THE TERM A IN EQ. (6.22)

Group	$A\times10^6$, cal/cm.sec.°C
—COO—	−29
—OH	+29
—Br	−103
—I	−128
—NO_2	0
Isomers: 1-CH_3	−25
2-CH_3	−43
3-CH_3	−59
Double bond	+4

Now, according to Sakiadis and Coates,

$$\lambda_{0\cdot6} = f(X) \tag{6.23}$$

where $X =$ (number of carbon atoms in the molecule)$+n_c$. The value of n_c is read from Table 6.12 and having calculated X, the corresponding value of $\lambda_{0\cdot6}$ is read from Table 6.13. θ_2 is a constant for a given type of compound; some typical values are listed in Table 6.12. (This method is unsuitable for predicting the conductivities of compounds belonging to other series than those mentioned in this table.) In order to use this method it is necessary to know the structural formula of the compound as well as its critical temperature.

TABLE 6.12. DATA FOR CALCULATING THE VALUE OF X IN EQ. (6.23) AND ALSO THE CONSTANT θ_2 IN EQ. (6.21)

Series	n_c	$\theta_2\times10^6$, cal/cm.sec.°C
Aliphatic saturated and unsaturated hydrocarbons	0	+41
Esters	−2	+21
Nitroparaffins	0	+91
Aliphatic alcohols	0	+153
Alkyl halides (mono)	...	−41
Chloro-	0	
Bromo-	1	
Iodo-	2	

TABLE 6.13. VALUES OF THE FUNCTION X FOR CALCULATING THE VALUE OF $\lambda_{0.6}$ ACCORDING TO EQ. (6.23) ($f(X)$, cal/cm.sec.°C)

X	1	2	3	4	5	6	7	8	9	10
$f(X)\times 10^6$	424	368	339	321	306	294	290	286	284	281

The estimation is performed in the following order:

1. The reduced temperature T_r is calculated.
2. A is found from Table 6.11.
3. The value of n_c is found from Table 6.12.
4. The number n of carbon atoms in the molecule is added to n_c, as $n+n_c = X$.
5. The value $f(X) = \lambda_{0.6}$ corresponding to the value of X is found in Table 6.13.
6. The values $\lambda_{0.6}$ and A are summed to find the value of θ_1 (Eq. (6.22)).
7. The constant θ_2 is found from Table 6.12.
8. The values θ_1 and θ_2 are substituted in Eq. (6.21).

Although this method is sometimes unreliable (as for example when applied to alcohols), Reid and Sherwood,[306] consider that it is somewhat more accurate than the other methods described above. When checked against data for 20 liquids, the average error in the calculated conductivities was found to be $\pm 12\%$; the maximum error was about 20%.

d. Estimation of Liquid Thermal Conductivities by Bondi's Method[28]

This method is suitable for non-associating compounds composed of large, polyatomic molecules. Earlier work by Bondi and others had shown that the conductivities of substances with large molecules (e.g. of long-chain compounds) were higher when the molecules were *flexible* (i.e. when the constituent groups could change their position with respect to one another) than when the molecules were rigid. Bondi's method is applicable both to substances with *rigid* molecules and to substances with *flexible molecules.* As has been seen in Chapter 1 (section 3) the forces of interaction between two spherical molecules may be described by two "force constants" σ and ε. σ is the distance between the molecular centres for which the intermolecular potential is zero and ε is the minimum potential energy of the two-molecule system. The molecules of the substances considered by Bondi cannot be considered to be spherically symmetrical. The approach adopted by Bondi was to divorce himself from any specific molecular model and simply to take the standard energy of vaporization as a measure of the intermolecular force (in place of ε) and then to find a suitable measure of the molecular dimensions to take the place of σ. The quantities which he finally chose to characterize the molecules were:

1. The standard energy of vaporization E° given by $E^\circ = \Delta H_V - RT$, where ΔH_V is the heat of vaporization, calculated at that temperature for which $V/V_w = 1{\cdot}70$. V is the molal volume and V_w the Van der Waals volume.[27]

2. The average minimum distance $\bar{d}_w$ between the centres of peripheral atoms on neighbouring molecules in contact at 0°K.

If each molecule has n outer atoms, then

$$\bar{d}_w = \frac{1}{n}\sum_{n=1}^{n=i} d_{w(i)}$$

where

$$d_{w(i)} = 2r_{w(i)}$$

is the Van der Waals' diameter of an external atom of type *i*. $d_{w(i)}$ may be calculated using the Van der Waals' radii r_w for carbon and hydrogen given in the following table:[27]

Bond type	Carbon $r_{w(C)}$	Hydrogen $r_{w(H)}$
Aliphatic, single	1·70	1·20
Olefinic, double	1·70	1·20
Aromatic ring	1·77	1·00

Exceptions are methyl and methylene groups (for which the value $d_{w(CH_3)} = d_{w(CH_2)} = 4{\cdot}00$ Å is recommended) and aromatic rings for which $d_{w(CH)} = 3{\cdot}73$ Å and $d_{w(\overset{\|}{C})} = 3{\cdot}54$ Å.

The formula finally arrived at by Bondi for estimating the thermal conductivity of a liquid is:

$$\lambda = \frac{R}{N_A d_w^2}\left(\frac{JE^\circ}{M}\right)^{\frac{1}{2}}\left(3{\cdot}22-\frac{8{\cdot}80cRT}{E^\circ}\right)\left(1+\frac{3_c-6}{N_s-1}\right) \tag{6.24}$$

where λ is the thermal conductivity, cal/cm.sec.°K,
R is the gas constant ($R = 1{\cdot}986$ cal/mole.°K),
N_A is Avogadro's number ($N_A = 0{\cdot}602\times10^{24}$),

TABLE 6.14. CONTRIBUTIONS FOR CALCULATING THE STANDARD ENERGY OF VAPORIZATION E°, kcal/mole (AFTER BONDI[28])

A. Hydrocarbons (N_B number of carbon atoms in the alkyl chain)

Group	Contribution
(CH_3)—	1·97
(CH_2)=	1·07
(C_2H_5)—	3·13
(C_2H_4)=	2·14
n-(C_3H_7)—	4·20
n-(C_3H_6)=(1,2)	3·01
i-(C_3H_7)—	3·93
i-(C_3H_6)=(1,1)	3·07
$(CH{\equiv}C)$—	3·75
$(CH_3C{\equiv}C)$—	4·71
Cyclopentyl	
(C_5H_9)—	$6{\cdot}10-0{\cdot}2\sum_{2}^{N_B}\frac{1}{N_B-1}$
(C_5H_8)=	4·8
Phenyl	
(C_6H_5)—	$7{\cdot}27-0{\cdot}3\sum_{2}^{N_B}\frac{1}{N_B-1}$
Anthryl	
$(C_{14}H_9)$	14·1

Group	Contribution
n-(C_4H_9)—	5·35
s-(C_4H_9)—	5·03
i-(C_4H_9)—	4·99
t-(C_4H_9)—	4·62
$(CH_2{=}CH)$—	3·22
(C_2H_2)=	2·30
$(CH_3{-}CH{=}CH)$—	4·16
$(CH_3{-}C{=}CH_2)$—	4·26
Cyclohexyl	
(C_6H_{11})—	$6{\cdot}85-0{\cdot}2\sum_{2}^{N_B}\frac{1}{N_B-1}$
(C_6H_{10})=	5·5
Decalyl	
$(C_{10}H_{17})$—	9·60
Naphthyl	
$(C_{10}H_7)$—	10·65
$(C_{10}H_6)$=	$-0{\cdot}3\sum_{2}^{N_B}\frac{1}{N_B-1}$

Continued overleaf

TABLE 6.14 (*cont.*)

B. Organic compounds containing nitrogen
(N_c number of carbon atoms in the molecule)

Nitrile		
aliphatic mono (R)		
	—(C=N)	$5{\cdot}35+\dfrac{1{\cdot}00}{N_c}$
aliphatic di(α, ω)		(mono 0·40)
aromatic (φ)	—(CN)	5·0 (?)
Pyridine	=N— †	2·30
		8·03
Nitro		
aliphatic primary		
	—NO_2	$5{\cdot}60+\dfrac{1{\cdot}60}{N_c}$
aliphatic secondary		(prim. 0·3)
aromatic (φ)		4·87
Amine		
aliphatic ‡	=N—	(0·40)
	=NH	3·26
	—NH_2	4·33
heteroaliphatic	=NH	2·9
aromatic (φ)	—NH_2	4·24
heteroaromatic §	=NH	3·20
Nitrate		
aliphatic primary		$5{\cdot}20+\dfrac{1{\cdot}40}{N_c}$
	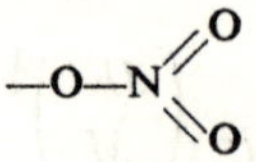	
aliphatic secondary		(prim. 0·2)

† $x = \Delta E_v^{\circ}$ (cpd)—5/6φH
‡ Reliable data available only for methyl derivatives
§ $x = \Delta E_v^{\circ}$ (cpd)—4/5 cyclo C_5 or equivalent

C. Organic compounds containing oxygen
(N_c number of carbon atoms in the molecule)

Ethers		
aliphatic (R_2) †	—O—	$1{\cdot}60+\dfrac{0{\cdot}40}{N_c}$
hetero-aliphatic (six-membered rings)	—O—	1·75
aromatic (φ_2)	—O—	1·0
heteroaromatic	—O—	1·50
Aldehydes		
aliphatic (R)	—(CHO)	5·20
aromatic (φ)	—(CHO)	5·00
Ketones		
aliphatic (R_2) †	=CO	$2{\cdot}40+\dfrac{3{\cdot}80}{N_c}$
aromatic	=CO	5·1
Esters		
aliphatic R_2	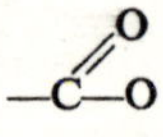	$3{\cdot}43+\dfrac{0{\cdot}20}{N_c}$
aromatic (φ_2)	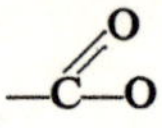	3·30

† n-alkyls

D. Organic compounds containing sulphur
(N_c number of carbon atoms in the molecule)

Sulphhydryl (Thiol) —SH		
aliphatic (n-alkyl)		$3{\cdot}36+\dfrac{0{\cdot}95}{N_c}$
s-alkyl,		(n-alkyl 0·15)
t-alkyl,		(n-alkyl 1·1)
aromatic		3·50
Disulphides —S—S—		
aliphatic (n-alkyl)$_2$		$4{\cdot}40+\dfrac{1{\cdot}90}{N_c}$
Thioether —S—		
aliphatic (n-alkyl)$_2$		$3{\cdot}36+\dfrac{0{\cdot}95}{N_c}$
cyclic (thiocycloalkane)		$2{\cdot}50+\dfrac{2{\cdot}00}{N_c}$
heteroaromatic (thiophene)	—S=	2·90

d_w is the Van der Waals' diameter,
$J = 4{\cdot}186\times10^7$ erg/cal,
$E°$ is the standard energy of vaporization (which may be calculated by summing the contributions listed in Table 6.14),
M is the molecular weight,
c is one third of the number of degrees of freedom per molecule, contributing to thermal conduction,
T is the absolute temperature, °K,
and N_s is the number of skeletal atoms per molecule.

TABLE 6.15. FORMULAE FOR CALCULATING THE CONSTANT 3c (AFTER BONDI[28]) (N_c TOTAL NUMBER OF CARBON ATOMS PER MOLECULE)

(1) n-Paraffins	$3c = 6+0{\cdot}43\ (N_S-1)\Big/\left(1+\frac{5{\cdot}0}{N_S-1}\right)$
Isoparaffins: the effect of widely spaced methyl groups is usually negligible. Enough data are generally available to calculate c for highly branched compounds when the need arises.	
(2) n-Alkylcyclohexanes	$3c = 6+0{\cdot}44\ (N_c-6)$
(3) n-Alkyl benzenes	$3c = 6+0{\cdot}43\ (N_S-1)\Big/\left(1+\frac{2{\cdot}5}{N_S-1}\right)$
(4) n-Alkyl cyanides	$3c = 6+0{\cdot}43\ (N_S-1)\Big/\left(1+\frac{5{\cdot}3}{N_S-1}\right)$
(5) α, ω-n-Alkyl dicyanides	$3c = 6+0{\cdot}45\ (N_S-5)$
(6) Aryl monocyanides	$3c = 6{\cdot}6+0{\cdot}4\ (N_S-1)$
(7) Mononitroparaffins	$3c = 6+0{\cdot}43\ (N_S-1)\Big/\left(1+\frac{5}{N_S-1}\right)$
(8) Mononitrobenzenes	$3c = 6+0{\cdot}4\ (N_S-1)$
(9) n-Alkyl mononitrates	$3c = 6+0{\cdot}43\ (N_S-1)\Big/\left(1+\frac{5{\cdot}3}{N_S-1}\right)$
(10) Alkyl polynitrates	$3c = 6+0{\cdot}4\ (N_S-1)\Big/\left(1+\frac{5}{N_S-1}\right)$
(11) Straight chain aliphatic ethers and acetals, etc.	$3c \approx 5{\cdot}8+0{\cdot}42\ (N_S-3)$
(12) Heterocycloaliphatic ethers	$3c = 6{\cdot}0+0{\cdot}4\ (N_S-3)$ for n-alkyl ethylene oxides $3c = 4{\cdot}8+0{\cdot}4\ (N_S-5)$ for n-alkyl tetrahydrofurans $3c = 4{\cdot}2+0{\cdot}4\ (N_S-6)$ for n-alkyl dioxanes
(13) Aliphatic monocarboxylate esters (for $N_c > 10$)	same as n-paraffins
(14) Aliphatic α, ω dicarboxylate esters	$3c = 6{\cdot}0+0{\cdot}47\ (N_S-5)$

N_S total number of skeletal units per molecule; here the ether oxygen, the carbonyl group and the NO_2 group are counted as units equivalent to the methyl or methylene groups.

The number of degrees of freedom $3c$ is calculated from the formulae compiled in Table 6.15. For a compound made up of *rigid* molecules, $3c \approx 6$, and the last term in Eq. (6.24) becomes unity.

It would appear from the data given in his paper that Bondi's method is more accurate than other known methods for predicting liquid thermal conductivities, the average error in the predicted values being as low as 8%. When applied to substances with large flexible molecules, the method of Bondi is clearly superior to other methods.

4. CHOICE OF BEST PREDICTION METHOD AND EXAMPLES OF THE PREDICTION OF THE THERMAL CONDUCTIVITIES OF PURE LIQUIDS BY VARIOUS METHODS

EXAMPLE 6.1. Estimate the thermal conductivity λ of n-butanol $CH_3—CH_2—CH_2—CH_2—OH$ at a temperature $t = 20°C$ and a pressure $p = 1$ atm.

Physico-Chemical Data required

The following physico-chemical data for n-butanol were taken from standard tables:

Molecular weight	$M = 74{\cdot}12$
Critical temperature (Table 1.31)	$T_c = 560{\cdot}2°K$
Critical pressure (Table 1.31)	$p_c = 48{\cdot}4$ atm
Freezing point	$T_f = 193{\cdot}3°K$
Normal boiling point	$T_b = 390{\cdot}2°K$
Dynamic viscosity (Table 4.1)	$\mu = 2{\cdot}95$ cP at 20°C
Density	$\rho = 0{\cdot}8098$ g/cm³ at 20°C
Molal heat capacity	$C_p = 43{\cdot}8$ cal/mole.°C at 20°C
Heat of vaporization at normal boiling temperature	$L_b = 10470$ cal/mole

The following quantities may be calculated from the above data:

Reduced temperature $T_r = 293{\cdot}2/560{\cdot}2 = 0{\cdot}523$

Heat capacity at constant pressure $c_p = \dfrac{43{\cdot}8}{74{\cdot}12} = 0{\cdot}590$ cal/mole

Density at the freezing point $\rho_f = 0{\cdot}8098 \times \dfrac{1-0{\cdot}48(193{\cdot}3/560{\cdot}2)}{1-0{\cdot}48(293{\cdot}2/560{\cdot}2)} = 0{\cdot}902$ g/cm³

The density ρ_c at the critical point may be estimated using Watson's equation, for which purpose the expansion factor ω is required

$$\omega = 0{\cdot}1745-0{\cdot}0838\times 0{\cdot}523 = 0{\cdot}1307$$

$$\rho_c = \frac{0{\cdot}8098\times 0{\cdot}044}{0{\cdot}1307} = 0{\cdot}273 \text{ g/cm}^3$$

The sonic velocity u

The molar sound velocity U_u is first calculated using the data given in Table 1.30. According to this table U_u for a substance with the structure

$$CH_3—CH_2—CH_2—CH_2—OH$$

is given by

$$U_u = 1850+872+872+872+137 = 4603$$

By Eq. (6.19),

$$u = \left(\frac{4603\times 0{\cdot}8098}{74{\cdot}12}\right)^2 = 1{\cdot}286\times 10^5 \text{ cm/sec}$$

The value of u, determined experimentally, is $1{\cdot}315\times 10^5$ cm/sec.

Calculation of the thermal conductivity of n-butanol

a. Bridgman's method (Eqs. (6.5) and (6.6))

The mean distance l between the molecules is given by

$$l = \sqrt[3]{\frac{74{\cdot}12}{6{\cdot}02\times 10^{23}\times 0{\cdot}8098}} = 5{\cdot}33\times 10^{-8} \text{ cm}$$

Hence

$$\lambda = \frac{4{\cdot}04\times 10^{-16}}{4{\cdot}18\times 10^7}\times\frac{1{\cdot}286\times 10^5}{(5{\cdot}33\times 10^{-8})^2} = 437\times 10^{-6} \text{ cal/cm.sec.°C}$$

(The factor $4{\cdot}18\times 10^7$, is necessary to convert ergs to calories).

b. Denbigh's method (Eq. (6.7))

$$\log \mathrm{Pr} = 0{\cdot}10\times\frac{10470}{390{\cdot}2}\times\frac{390{\cdot}2}{293{\cdot}2}-1{\cdot}8 = 1{\cdot}766$$

$$\mathrm{Pr} = \frac{c_p\mu}{\lambda} = 58{\cdot}4$$

$$\lambda = \frac{c_p\mu}{\mathrm{Pr}} = \frac{0{\cdot}590\times 0{\cdot}0295}{58{\cdot}4} = 300\times 10^{-6}\ \text{cal/cm.sec.°C}$$

c. Weber's method (Eq. (6.10))

$$\lambda = 1{\cdot}29\times 0{\cdot}590\times 0{\cdot}8098\times\sqrt[3]{\frac{0{\cdot}8098}{74{\cdot}12}} = 0{\cdot}138\ \text{kcal/m.hr.°C}$$

or

$$\lambda = 382\times 10^{-6}\ \text{cal/cm.sec.°C}$$

d. 1st formula of Smith (Eq. (6.12))

$$\lambda = 4{\cdot}3\times 10^{-3}\times 0{\cdot}590\times 0{\cdot}8098\sqrt[3]{\frac{0{\cdot}8098}{74{\cdot}12}} = 459{\cdot}10^{-6}\ \text{cal/cm.sec.°C}$$

e. 2nd formula of Smith (Eq. (6.13))

$$\lambda = \frac{0{\cdot}293\times 0{\cdot}8098^{2\cdot 15}\times 0{\cdot}59^{1\cdot 55}\times 74{\cdot}12^{0\cdot 192}}{2{\cdot}95^{0\cdot 12}}$$

$$\log\lambda = \log 0{\cdot}293+2{\cdot}15\log 0{\cdot}8098+1{\cdot}55\log 0{\cdot}59+0{\cdot}192\log 74{\cdot}12-0{\cdot}12\log 2{\cdot}95 = -0{\cdot}786$$

$$\lambda = 0{\cdot}163\ \text{kcal/m.hr.°C or}$$

$$\lambda = 452\times 10^{-6}\ \text{cal/cm.sec.°C}$$

f. 3rd formula of Smith (Eq. (6.14))

$$\lambda\times 10^6 = 11+6450(0{\cdot}59-0{\cdot}45)^3+1250\left(\frac{0{\cdot}8098}{74{\cdot}12}\right)^{0\cdot 333}+100\left(\frac{2{\cdot}95}{0{\cdot}8098}\right)^{0\cdot 111}$$

$$= 423\times 10^{-6}\ \text{cal/cm.sec.°C}$$

g. Palmer's formula (Eq. (6.16))

$$\lambda = \frac{94700\times 10^{-6}\times 0{\cdot}8098\times 0{\cdot}59\times(0{\cdot}8098/74{\cdot}12)^{0\cdot 33}}{(10470/390{\cdot}2)} = 377\times 10^{-6}\ \text{cal/cm.sec.°C}$$

h. 1st method of Sakiadis and Coates (section 3b)

Reduced temperature $T_r = 0{\cdot}523$

Reduced freezing temperature $T_{f,r} = 193{\cdot}3/560{\cdot}2 = 0{\cdot}344$

Reduced density ρ_r:

(i) at a temperature of 20°C $\rho_r = \dfrac{0{\cdot}8098}{0{\cdot}273} = 2{\cdot}962$

(ii) at the freezing temperature $\rho_{f,r} = \dfrac{0{\cdot}902}{0{\cdot}273} = 3{\cdot}306$

The distance Δ between the molecules may be estimated from the data presented in Table 6.10. For C_4 alcohols: $\Delta = 5{\cdot}09\times 10^{-8}$.

According to Eq. (6.20)

$$L = 5{\cdot}09\times 10^{-8}\left[1-\left(\frac{2{\cdot}962-2{\cdot}962\times 0{\cdot}344}{3{\cdot}306-2\times 0{\cdot}344}\right)^{0\cdot 5}\right] = 0{\cdot}727\times 10^{-8}\ \text{cm}$$

whence from Eq. (6.18)

$$\lambda = 0{\cdot}590\times 0{\cdot}8098\times 1{\cdot}286\times 10^5\times 0{\cdot}727\times 10^{-8} = 446\times 10^{-6}\ \text{cal/cm.sec.°C}$$

i. 2nd method of Sakiadis and Coates (section 3c) (method based on theorem of corresponding states)

θ_1 is calculated from the data given in Table 6.11:

$A = +29$ (1 group —OH).

According to the data in Table 6.12 for alcohols $n_c = 0$

The molecule of butanol includes 4 C atoms

$X = 4+0 = 4$

According to the data in Table 6.13, $f(X) = 321\times 10^{-6}$.

By substitution in Eq. (6.22)

$$\theta_1 = 321\times 10^{-6}+29\times 10^{-6} = 350\times 10^{-6}$$

According to the data in Table 6.12

$$\theta_2 = +153\times10^{-6}$$

On the basis of Eq. (6.21)

$$\lambda = 2{\cdot}26\times350\times10^{-6}-1{\cdot}26\times153\times10^{-6}-2{\cdot}1\times(350-153)\times10^{-6}\times0{\cdot}523$$
$$= 381\times10^{-6}\ \text{cal/cm.sec.°C}$$

Comparison of the results obtained by the above methods with experimental values

Experimental values of the thermal conductivity of n-butanol at 80, 60 and 40°C are listed in Table 6.1. These values may be extrapolated to 20°C using Eq. (6.1), the value obtained being 379×10^{-6} cal /cm.sec.°C.

Method of estimation	Calculated value	Average error
	$\lambda\times10^6$ cal/cm.sec.°C	
(a) Bridgman's method (Eqs. (6.5) and (6.6))	437	+15·4%
(b) Denbigh's method (Eq. (6.7))	300	−20·8%
(c) Weber's method (Eq. (6.10))	382	+0·8%
(d) 1st formula of Smith (Eq. (6.12))	459	+21·0%
(e) 2nd formula of Smith (Eq. (6.13))	452	+19·2%
(f) 3rd formula of Smith (Eq. (6.14))	423	+11·6%
(g) Palmer's formula (Eq. (6.16))	377	−0·5%
(h) 1st method of Sakiadis and Coates (section 3b)	446	+17·6%
(i) 2nd method of Sakiadis and Coates (section 3c)	381	+0·5%

EXAMPLE 6.2 (Ref. 28). Use the Bondi method (Eq. (6.24)) to estimate the thermal conductivity of n-octadecane, n-$C_{18}H_{38}$, at 40°C (313°K).

Solution

Estimation of $E°$ (Table 6.14)

$$2[(C_4H_9)-]+10[-(CH_2)-] = 2\times5{\cdot}35+10\times1{\cdot}07 = 21{\cdot}40\ \text{kcal/mole}$$
$$E° = 21{,}400\ \text{cal/mole}$$

Estimation of d_w

Since the contributions of the (CH_3)— and —(CH_2)— groups are the same and equal to 4·00 Å,

$$d_w = \frac{1}{18}\times18\times4{\cdot}00 = 4{\cdot}00\ \text{Å}$$

Estimation of 3c (Table 6.15)

$$N_S = 18;\ 3c = 6+0{\cdot}43(18-1)/\left(1+\frac{5{\cdot}0}{18-1}\right) = 6+7{\cdot}31/1{\cdot}294 = 11{\cdot}65$$

$$c = 3{\cdot}883$$

The molecular weight M is given by

$$M = 18\times12+38\times1 = 254$$

on substituting the above values in Bondi's equation (6.24), it is found that

$$\lambda = \frac{1{\cdot}987}{6{\cdot}02\times10^{23}\times4^2\times10^{-16}}\times\left(\frac{4{\cdot}186\times21{\cdot}4\times10^7\times10^3}{254}\right)^{1/2}\times$$
$$\times\left(3{\cdot}22-\frac{8{\cdot}80\times3{\cdot}883\times1{\cdot}987\times313}{21400}\right)\times\left(1+\frac{5{\cdot}65}{18-1}\right) = 3{\cdot}63\times10^{-4}\ \text{cal/cm.sec.°K}$$

The value of λ determined experimentally is $3{\cdot}64\times10^{-4}$, so in this case the calculated and experimental values of λ are in very good agreement.

Table 6.16 lists the prediction methods which have been discussed above and indicates the accuracy of each method and the data required for its use. The examples solved above do not necessarily provide a reliable indication of the accuracy of the methods used since the number of calculations given is too small. However, the results of a much larger number of similar calculations have been compiled by Reid and Sherwood;[306] it is from

TABLE 6.16. COMPARISON OF METHODS FOR ESTIMATING THE THERMAL CONDUCTIVITY OF LIQUIDS

Authors	Equation	Data necessary for the estimation	Average error	Remarks
Bridgman	(6.5)	ρ	± 25	Method theoretically justified
Denbigh	(6.7)	L_b, μ, c_p	± 30	
Weber	(6.9)	ρ, c_p	± 13	
Predvoditelev	(6.11)	ρ, c_p	± 5(?)	Can be used only for non-associated liquids
Smith (1st)	(6.12)	ρ, c_p	± 18	
Smith (3rd)	(6.14)	ρ, c_p, μ	± 20	
Palmer	(6.16)	ρ, c_p, L_b, T_b	± 18	
Sakiadis and Coates (1st)	(6.18)	ρ, c_p, T_f, T_c	± 16	Method theoretically justified
Sakiadis and Coates (2nd)	(6.21)	T_c	± 12	Can be used only for some series of compounds
Bondi	(6.24)	structural formula only	± 8	Used for non-associated liquids. Suitable for compounds with large molecules

their work that the estimates given here of the relative accuracies of the various prediction methods are drawn.

In general methods which are based on simplified theoretical concepts for the mechanism of heat conduction tend to be less accurate than the purely empirical ones. The most simple and expeditious method consists in the application of Weber's equation (6.9) or, in the case of non-associated liquids, its modification as suggested by Predvoditelev (6.11). The accuracy achieved is not great (the result may well be in error by 20–30%), but is almost on a par with that achieved using more complex methods.

It appears that the most accurate of the methods is that of Bondi (Eq. (6.24)). This method is particularly suitable if the liquid is composed of large flexible and non-associated molecules. However, this method can only be applied to those substances for which the required values of E° and $3c$ can be calculated from Tables 6.14 and 6.15. As usual, when employing empirical equations, it is recommended that the estimation be carried out by several methods (for example by those of Weber or Predvoditelev, and Sakiadis and Coates) and that the results should then be compared.

The rate at which the thermal conductivity of a liquid varies with temperature may be estimated from Predvoditelev's equation, if c_p is known as a function of temperature, since the product Ic_p does not depend on the temperature (section 3a).

The effect of pressure on the thermal conductivities of liquids is insignificant at pressures below about fifty atmospheres and is usually neglected.

5. THE THERMAL CONDUCTIVITIES OF LIQUID MIXTURES

Comparatively little progress has yet been made towards providing satisfactory methods for estimating the thermal conductivities of homogenous solutions and heterogenous mixtures (such as emulsions and suspensions). Since experimental data are rather scanty, the general rules for estimating the thermal conductivity may be given only for certain cases.

a. Experimental Results

In general the thermal conductivity of a liquid mixture is not a simple function of the conductivities and concentrations of the components, and simple additivity rules cannot generally be applied.

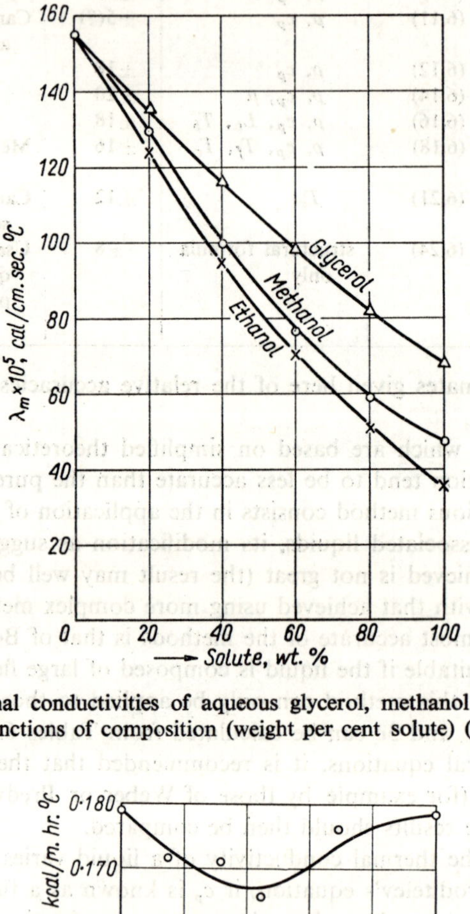

FIG. 6.5. The thermal conductivities of aqueous glycerol, methanol and ethanol solutions at 50°C as functions of composition (weight per cent solute) (after D'Ans[11]).

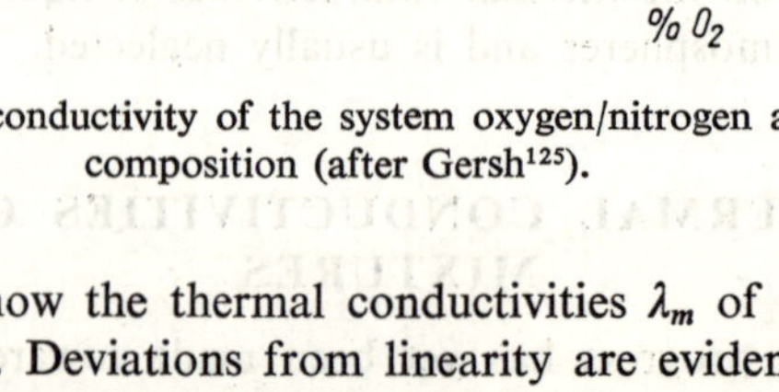

FIG. 6.6. The thermal conductivity of the system oxygen/nitrogen at 73°K as a function of composition (after Gersh[125]).

Figures 6.5 to 6.7 show the thermal conductivities λ_m of several binary systems as functions of composition. Deviations from linearity are evident in all the cases shown. The system nitrogen/oxygen (Fig. 6.6) is of particular interest: the thermal conductivities of both pure components are virtually the same (0·179 kcal/m.hr.°C), but the thermal conductivities of all the mixtures are less than that of the pure components, so the system exhibits a minimum.

Tsederberg[372] has attempted to correlate deviations from the simple additive law in terms of the molecular properties of the constituents. He expressed the additive law in the form

$$\lambda_m = \lambda_1 m_1 + \lambda_2 m_2 \tag{6.25}$$

where λ_m, λ_1 and λ_2 are the thermal conductivities of the mixture, component 1, and component 2, respectively,

and m_1 and m_2 are the mass fractions of components 1 and 2 in the mixture.

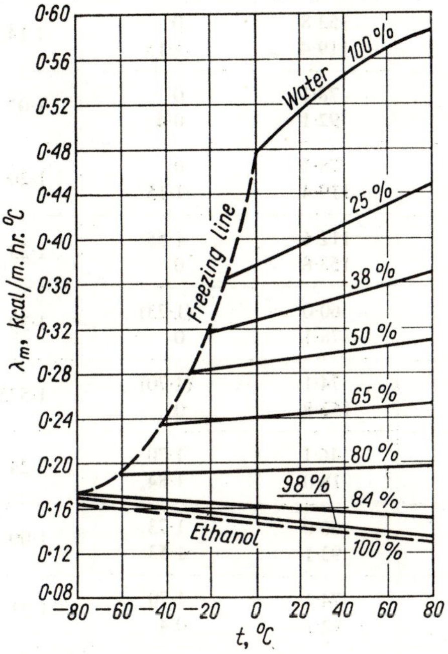

FIG. 6.7. The thermal conductivities of ethanol/water mixtures as functions of temperature with composition as parameter (after Tsederberg[372]).

Table 6.17 lists data for several organic liquid systems as well as values of the maximum deviations from the linear law as expressed by Eq. (6.25).

On the basis of this comparison, Tsederberg stated that the simple law of additivity (Eq. (6.25)), may be applied when either both kinds of molecules of a binary mixture are non-polar or one kind is non-polar and the other only slightly polar. If the dipole moments of the molecules of either or both the components are large (e.g. methyl and ethyl formates, $\mu = 1{\cdot}7$ D), the deviations from additivity can be considerable.

The effect of temperature upon thermal conductivity of a liquid mixture depends on whether the conductivities of the pure components increase or decrease with the rise of temperature. In the majority of cases the conductivities of the components, and hence the conductivities of the mixtures of these components decrease with increase in temperature. If one of the components is water, the thermal conductivity of which increases with the rise of temperature, the conductivity of the mixture may also increase; if the

TABLE 6.17. THERMAL CONDUCTIVITIES OF MIXTURES OF ORGANIC LIQUIDS: DEVIATIONS FROM THE ADDITIVE LAW (EQ. 6.26) (AFTER TSEDERBERG[372])

Components of the mixture (1/2)	Molecular weight M	Dipole moment $\bar{\mu}\times 10^{18}$	$\frac{\lambda_2}{\lambda_1}$	Temperature, °C	Maximum deviation Δ, %
Benzene Carbon tetrachloride	78·1 153·8	0 0	1·42	30	+4·4
Benzene Ethyl ether	78·1 74·1	0 1·14	1·07	15	+2·03
Carbon tetrachloride Chloroform	153·8 119·4	0 1·15	1·14	15	+2·01
Benzene Toluene	78·1 92·1	0 0·4	1·07	30	+2·53
Benzene Chloroform	78·1 119·4	0 1·15	1·20	15	+3·08
Chlorobenzene Carbon tetrachloride	112·6 153·8	1·55 0	1·23	30	+3·62
Methyl formate Benzene	60·0 78·1	(1·73) 0	1·30	15	5·73
Ethyl formate Carbon tetrachloride	74·1 153·8	(1·70) 0	1·575	15	+9·13
Ethanol Water	46·1 18·0	1·70 1·84	3·24	0 60	+20·00 +27·60
Acetic acid Aniline	60·0 93·1	1·73 1·53	1·00	25 35	−6·35 −9·97
Ethyl formate Toluene	74·1 92·1	1·70 0·4	1·23	15	+3·28
Ethyl ether Chloroform	74·1 119·4	1·14 1·15	1·13	15	+10·21
Methyl formate Chloroform	60·0 119·4	(1·73) 1·15	1·57	15	+8·30

water content in the mixture is reduced, the thermal conductivity λ may be independent of temperature, and at small water contents it will decrease with the rise of temperature. This is illustrated in Fig. 6.7 which shows the thermal conductivities of mixtures of water and ethanol as functions of temperature.

Considerable efforts have been devoted to the investigation of the thermal conductivities of aqueous solutions of electrolytes on account of their practical importance. In many cases it has been found that a linear (or almost linear) relationship exists between the thermal conductivity and the concentration of the dissolved electrolyte, when this is expressed as a molal concentration (molality) and not in weight per cent.[190] The results of some measurements of Kapustinskii and Ruzavin[190] as well as of Riedel are shown in Figs. 6.8 and 6.9. These authors measured the thermal conductivities of electrolytic solutions containing a variety of ions of differing radii. It is usually found that as the concentration of the electrolyte increases, the thermal conductivity decreases, though exceptions are known (e.g. solutions of NaOH).

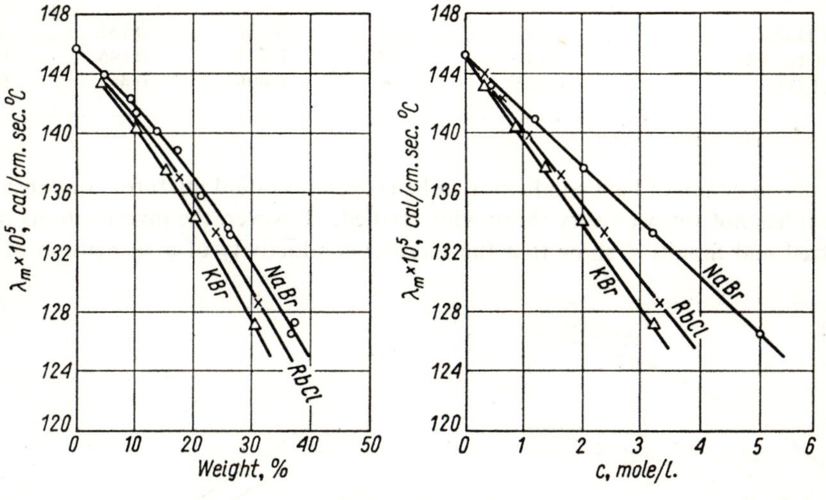

FIG. 6.8. The thermal conductivities of several electrolyte solutions as functions of concentration.

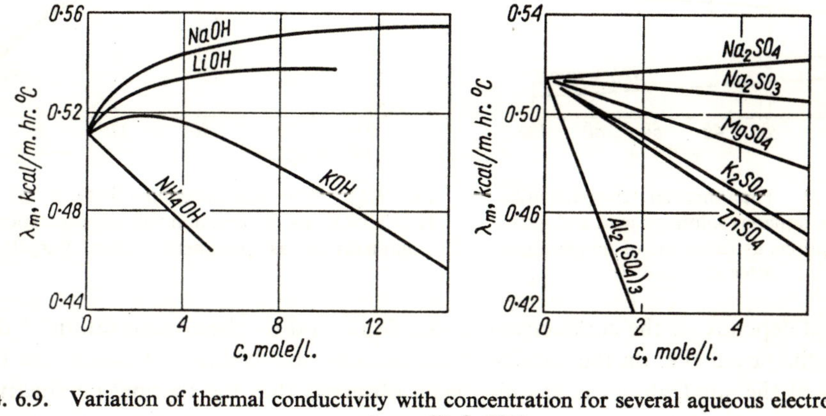

FIG. 6.9. Variation of thermal conductivity with concentration for several aqueous electrolyte systems (Ref. 310).

The thermal conductivities of dilute aqueous electrolytic solutions usually increase with rise of temperature, but in the case of concentrated solutions the reverse phenomenon may sometimes be observed as in the case of concentrated solutions of HNO_3 as shown in Figs. 6.10 and 6.11.

The thermal conductivities of aqueous solutions of a number of electrolytes are listed as functions of concentration in Table 6.18.

The thermal conductivity of a non-aqueous solution of an electrolyte is much lower than the conductivity of the same electrolyte dissolved in water. This is illustrated in the following table which is due to Eldarov.[85] This shows the thermal conductivity λ (kcal/m.hr.°C) of zinc chloride solutions in methanol, acetone and water at 25°C.

Concentration	0%	5%	10%	15%	20%
$ZnCl_2+CH_3OH$	0·181	0·181	0·181	0·181	0·181
$ZnCl_2+(CH_3)_2CO$	0·142	0·144	0·145	0·146	0·147
$ZnCl_2+H_2O$†	0·515	0·506	0·496	0·485	0·474

† After Riedel (Table 6.18), at 20°C.

The phenomenon of heat conduction in heterogenous liquid mixtures (suspensions and emulsions) has not yet been very thoroughly studied. However, the investigations by Jeferson, Witzell and Sibbitt[173] show that the thermal conductivity of a suspension of a solid

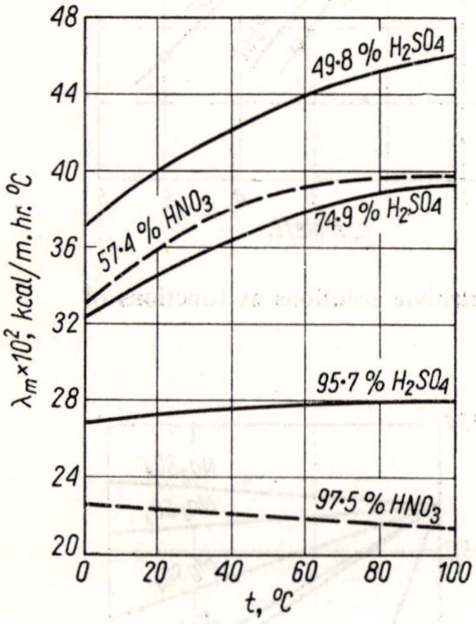

FIG. 6.10. The thermal conductivities of several aqueous sulphuric acid and nitric acid solutions as functions of temperature (after Osminin[267]).

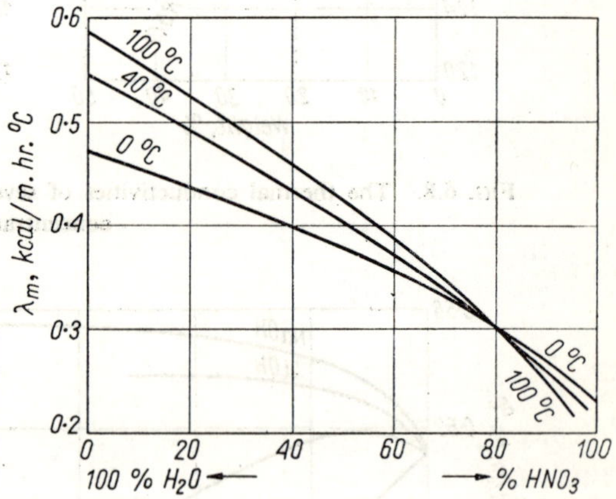

FIG. 6.11. The thermal conductivity of the system H_2O/HNO_3 as a function of concentration with temperature as parameter (after Vargaftig[382]).

in a liquid depends on the conductivity of the liquid and on the concentration of the solid phase in the liquid (i.e. on the density of the suspension): it does not depend on the conductivity of the solid phase. It has also been observed that the thermal conductivities of emulsions of organic compounds in water are lower than the conductivity of pure water.[386]

b. Methods for Estimating Thermal Conductivities of Mixtures

A number of empirical methods for estimating the thermal conductivities of liquid mixtures are known. The most appropriate method to use in any specific case depends on the nature of the mixture, as indicated below.

Homogenous Liquid Mixtures which Contain no Ionized Molecules

This group includes mixtures of organic liquids, aqueous solutions of organic non-ionized compounds (e.g. alcohols), mixtures of liquefied gases, etc.

TABLE 6.18. THERMAL CONDUCTIVITIES λ, kcal/m.hr.°C, OF AQUEOUS SOLUTIONS OF ELECTROLYTES AT 20°C (AFTER RIEDEL[310])

Compound	Weight per cent						
	5	10	20	30	40	50	†
LiOH	0·529	0·534	—	—	—	—	
NaOH	0·528	0·539	0·550	0·555	—	0·556	b
KOH	0·517	0·519	0·515	0·502	0·485	0·461	b
HCl	0·498	0·480	0·439	0·398	—	—	
HNO_3	0·505	0·494	0·471	0·448	0·428	—	a
H_2SO_4	0·507	0·499	0·480	0·459	0·435	0·408	c
H_2CrO_4	0·508	0·500	0·483	0·465	0·445	0·420	
H_3PO_4	0·505	0·498	0·479	0·458	0·438	0·418	c
LiCl	0·506	0·496	0·476	0·463	—	—	
LiBr	0·504	0·492	0·466	0·436	0·405	—	
LiI	0·504	0·492	0·462	0·435	0·403	0·367	a
Li_2SO_4	0·513	0·510	0·505	—	—	—	
NaCl	0·511	0·507	0·497	—	—	—	
NaBr	0·508	0·499	0·480	0·459	0·433	—	a
NaI	0·507	0·498	0·478	0·454	0·425	0·389	a
$NaNO_2$	0·512	0·509	0·502	0·494	0·482	—	
$NaNO_3$	0·512	0·508	0·499	0·489	0·478	—	
Na_2SO_3	0·514	0·513	0·510	—	—	—	
Na_2SO_4	0·515	0·516	—	—	—	—	
$Na_2S_2O_3$	0·513	0·510	0·503	0·490	0·468	—	
Na_2CO_3	0·518	0·522	—	—	—	—	
Na_2SiO_3	0·518	0·522	0·531	—	—	—	
Na_3PO_4	0·521	0·527	—	—	—	—	
$NaClO_3$	0·509	0·503	0·489	0·473	0·455	—	
$NaClO_4$	0·508	0·501	0·487	0·470	0·450	0·428	a
$NaBrO_3$	0·510	0·506	0·496	—	—	—	
$Na_2Cr_2O_7$	0·512	0·510	0·503	0·496	0·488	0·480	
NaOOC—CH_3	0·503	0·490	0·462	0·432	—	—	
KF	0·511	0·506	0·492	0·469	—	—	
KCl	0·507	0·499	0·481	—	—	—	
KBr	0·506	0·495	0·473	0·446	0·416	—	
KI	0·506	0·495	0·473	0·446	0·414	0·375	
KNO_2	0·509	0·502	0·487	0·470	0·453	0·437	a
KNO_3	0·509	0·502	0·487	—	—	—	
K_2SO_4	0·511	0·507	—	—	—	—	
K_2CO_3	0·512	0·509	0·501	0·485	0·464	0·438	a
$K_2(COO)_2$	0·510	0·505	0·493	—	—	—	a
$K_4Fe(CN)_6$	0·509	0·502	0·488	—	—	—	a
$MgCl_2$	0·504	0·493	0·470	0·444	—	—	
$MgBr_2$	0·504	0·493	0·466	0·434	0·395	0·350	a
$MgNO_3$	0·508	0·501	0·485	0·466	—	—	
$MgSO_4$	0·512	0·509	0·501	—	—	—	
$CaCl_2$	0·510	0·505	0·495	0·482	0·469	—	
$CaBr_2$	0·507	0·498	0·478	0·455	0·426	0·390	a
$Ca(NO_3)_2$	0·511	0·507	0·497	0·486	0·473	0·458	a

Continued overleaf

TABLE 6.18 (*cont.*)

Compound	Weight per cent						
	5	10	20	30	40	50	†
$SrCl_2$	0·511	0·506	0·495	0·483	—	—	a
$SrBr_2$	0·508	0·500	0·482	0·461	0·435	0·402	a
$Sr(NO_3)_2$	0·511	0·507	0·498	0·487	0·473	—	
$BaCl_2$	0·511	0·507	0·497	—	—	—	
$BaBr_2$	0·508	0·501	0·485	0·466	0·443	—	
BaI_2	0·507	0·499	0·481	0·459	0·432	0·398	a
$AgNO_3$	0·510	0·505	0·494	0·480	0·464	0·442	
$CuSO_4$	0·510	0·505	—	—	—	—	d
$ZnSO_4$	0·510	0·505	0·494	0·481	—	—	b
$ZnCl_2$	0·506	0·496	0·474	0·448	0·418	—	a
$Pb(NO_3)_2$	0·511	0·507	0·498	0·487	—	—	d
$Co(NO_3)_2$	0·509	0·502	0·486	0·468	0·447	—	
$Al_2(SO_4)_3$	0·507	0·498	0·477	—	—	—	
$Th(NO_3)_4$	0·508	0·501	0·484	—	—	—	
NH_4Cl	0·501	0·487	0·457	—	—	—	e
NH_3	0·487	0·460	0·416	0·383	—	—	f

† (a) The values of λ for high concentrations were obtained by extrapolation.
(b) Inconsistent results.
(c) Braun gives the following values of λ (kcal/m.hr.°C, $t = 29$°C) for high concentrations:

	50%	60%	70%	80%	90%	95%
H_2SO_4	0·416	0·392	0·371	0·354	0·318	0·289
H_3PO_4	0·427	0·410	0·399	0·392	0·385	—

(d) Data after Jaeger, uncertain.
(e) Data after various authors, inconsistent and uncertain.
(f) Recalculated data of Braun and Lees.

The simplest method is to use the additivity rule:

$$\lambda_m = \lambda_1 m_1 + \lambda_2 m_2 \tag{6.26}$$

where λ_m is the thermal conductivity of the mixture,
λ_1 and λ_2 are the thermal conductivities of pure components 1 and 2,
and m_1 and m_2 are the mass fractions of components 1 and 2 ($m_1+m_2 = 1$).

It may be seen from Table 6.17 that in many cases the deviations from this additivity rule are usually small for the above class of mixture. According to Tsederberg[372] this is applicable when the molecules of both components of the mixture are non-polar, or when the molecules of one component are non-polar and the permanent dipole moment of the molecule of the other component is lower than 1·55 D. In these cases deviations from the rule do not exceed 4·4%.

Cecil and Munch[19] have investigated the thermal conductivities of binary and ternary mixtures of water, methanol, carbon tetrachloride, ethyl bromide and chloroform with organic liquids composed of comparatively large molecules (such as adipates, phosphates, phthalates, tricresyl phosphate). These authors find that Eq. (6.26) is approximately valid when applied to mixtures of this type.

If the molecules of one of the components of the mixture have dipole moments in excess of 1·55 D, Eq. (6.26) becomes less reliable, errors in the predicted conductivities being ≈ 5·7–12%. If the molecules of both components of the liquid mixture have large dipole moments, the additivity rule is of little use since deviations may then be as large as 28%.

Filippov and Novoselova[100] have suggested an equation for estimating the thermal conductivities of binary liquid mixtures, which is based on the observation that the conductivity of a mixture usually has a value between these for the pure components, being rather lower than that calculated from the additive rule (6.26) (cf. Table 6.17 from which it may be seen that Δ is usually positive). The above authors propose an equation of the following form:

$$\lambda_m = \lambda_2 m_2 + \lambda_1(1-m_2) - 0{\cdot}72(\lambda_2-\lambda_1)m_2(1-m_2) \tag{6.27}$$

where λ_m is the thermal conductivity of the mixture,
λ_1 and λ_2 are the thermal conductivities of pure components 1 and 2 of the mixture ($\lambda_2 > \lambda_1$),
and m_2 is the mass fraction of the component with the higher conductivity (λ_2).

The authors initially developed this equation as a means of predicting thermal conductivities for mixtures of non-polar compounds and they checked it against data for 13 binary systems of this type over the entire concentration range at the temperatures 15°C and 30°C. Tsederberg[372] subsequently pointed out that Eq. (6.27) may be used even when the components of the mixture are strongly polar. For example the equation predicts the conductivities of mixtures of ethanol and water with an error not exceeding 5%.

A third method for predicting the thermal conductivities of liquid mixtures has been suggested by Bates, Hazzard and Palmer,[19] who have shown that the data for the water–ethanol and water–methanol systems are adequately represented by the equation

$$\lambda_m \sinh\varphi = \lambda_1 \sinh(m_1\varphi) + \lambda_2 \sinh(m_2\varphi) \tag{6.28}$$

where λ_m is the thermal conductivity of the mixture of components 1 and 2,
λ_1 and λ_2 are the thermal conductivities of pure components 1 and 2,
m_1 and m_2 are the mass fractions of components 1 and 2 in the mixture ($m_1+m_2 = 1$),
and φ is a constant characteristic of the given pair of components of the mixture (φ depends only slightly on temperature).

The function sinh is defined by the relationship

$$\sinh\varphi = 0{\cdot}5(e^{\varphi} - e^{-\varphi})$$

and its value is readily available in standard tables. At a temperature of 20°C, the values of the constant φ are as follows:

for the water–ethanol system $\varphi = 0{\cdot}94$
for the water–methanol system $\varphi = 0{\cdot}90$

In order to evaluate the constant φ for a given pair of compounds, the thermal conductivity for at least one mixture must be known.

Following Bondi,[28] Riedel recommends the following equation for estimating the thermal conductivities of mixtures:

$$\lambda_m = x_1\lambda_1 + x_2\lambda_2 - fbx_1x_2 \tag{6.29}$$

where λ_m is the thermal conductivity of the mixture of components 1 and 2,
λ_1 and λ_2 are the thermal conductivities of pure components 1 and 2,
and x_1 and x_2 are the mole fractions of components 1 and 2.
The function f is given by:

$$f = \left(\frac{E_1^\circ}{M_1}\right)^{1/2} - \left(\frac{E_2^\circ}{M_2}\right)^{1/2}$$

M_1 and M_2 are the molecular weights of components 1 and 2,
E_1° and E_2° are the standard energies of vaporization, calculated by summing contributions listed by Bondi (Table 6.14).

The constant b in Eq. (6.29) is equal to
$4{\cdot}5\times10^{-5}$ when E° is expressed in cal/mole or
$7{\cdot}0\times19^{-9}$ when E° is expressed in erg/mole.

Bondi recommends that if one of the components of the mixture is a compound with a large (flexible chain) molecule then volume fractions should be employed instead of the mole fractions x_1 and x_2 in Eq. (6.29). According to Bondi[28] Eq. (6.29) is more accurate than the Filippov equation (6.27).

Aqueous Solutions of Electrolytes

A method proposed by Riedel[310] is very satisfactory for predicting the thermal conductivities of dilute or moderately concentrated aqueous solutions of electrolytes. In this method the thermal conductivity of the solution is taken to be equal to the sum of the thermal conductivity λ_w of water and terms arising from the ions present in the solution, i.e.

$$\lambda_m = \lambda_w + \sum \alpha_i c_i \tag{6.30}$$

where c_i is the molar concentration of ion i,
α_i is the contribution of ion i as given by Table (6.19),
and λ_w is the thermal conductivity of water (0·515 kcal/m.hr.°C at 20°C.)

The ionic contributions α_i were deduced by Riedel from the conductivity data for salts listed in Table 6.18, on the arbitrary assumption that the contribution of the sodium ion α_{Na} was zero. Although the values of the ion contributions calculated in this way are not absolute values, they may be used in conjunction with Eq. (6.30), to give a correct value for the thermal conductivity λ_m. The contributions α of 36 of the more frequently encountered cations and anions are listed for a temperature of 20°C in Table 6.19.

Although very satisfactory for prediction purposes, Eq. (6.30) is not obeyed with sufficient accuracy for data correlation and Riedel recommends the use of directly determined values of thermal conductivities λ_m, as compiled in Table 6.18, when these are available. If the value of λ_m is required at a temperature other than 20°C it may be estimated by multiplying the value of $\lambda_{m,20°}$ by the respective coefficient f_T, interpolated from the values given in Table 6.20. The required value $\lambda_{m,T}$, of the conductivity at temperature T is then found from the equation

$$\lambda_{m,T} = \lambda_{m,20°} f_T \tag{6.31}$$

Although Riedel's method gives good results when applied to dilute or moderately concentrated solutions, it is not satisfactory when applied to very concentrated solutions. For example Osminin[267] finds that, for concentrated solutions of sodium and potassium hydroxides, the predicted conductivities are in error by as much as 35%.

TABLE 6.19. THERMAL CONDUCTIVITIES OF AQUEOUS IONIC SOLUTIONS ACCORDING TO EQ. 6.30. IONIC CONTRIBUTIONS, α_i AT A CONCENTRATION OF 1 MOLE/L (AT 20°C, WITH $\alpha_{Na^+} = 0$) (REF. 314)

Anion	α_i	X	Cation	α_i	X
OH^-	+0·0180	c	H^+	−0·0078	
F^-	+0·0018	b	Li^+	−0·0030	b
Cl^-	−0·0047				
Br^-	−0·0150		Na^+	0·0000	
I^-	−0·0236		K^+	−0·0065	
NO_2^-	−0·0040		NH_4^+	−0·0100	d
NO_3^-	−0·0060				
ClO_3^-	−0·0122	a	Mg^{2+}	−0·0080	
ClO_4^-	−0·0150	a	Ca^{2+}	−0·0005	
BrO_3^-	−0·0122	a	Sr^{2+}	−0·0034	
CO_3^{2-}	−0·0065	c	Ba^{2+}	−0·0066	
SiO_3^{2-}	−0·0080	a	Ag^+	−0·0090	a
SO_3^{2-}	−0·0020	a	Cu^{2+}	−0·0140	d
SO_4^{2-}	+0·0010		Zn^{2+}	−0·0140	d
$S_2O_3^{2-}$	−0·0070	a, c	Pb^{2+}	−0·0080	d
CrO_4^{2-}	+0·0010	a	Co^{2+}	−0·0100	a
$Cr_2O_7^{2-}$	−0·0137	a	Al^{3+}	−0·0280	a
PO_4^{3-}	+0·0180	a	Th^{4+}	−0·0375	a
$Fe(CN)_6^{4-}$	−0·0160	a			
CH_3COO^-	−0·0197	a			
$(COO)_2^{2-}$	−0·0030	a			

(a) The given value of α, estimated on the basis of measurements of the conductivity of one solution, is not very certain.
(b) Deviations possible in the region of high concentrations.
(c) May be applied to concentrations below 1 mole/l.
(d) Value not quite certain (not enough data).

TABLE 6.20. VALUES OF THE FACTOR f_T IN EQ. (6.31) FOR PREDICTING THE THERMAL CONDUCTIVITY OF AN ELECTROLYTE SOLUTION AT t°C FROM THAT AT 20°C

t, °C	f_T	t, °C	f_T	t, °C	f_T	t, °C	f_T
−40	0·838	0	0·946	40	1·048	80	1·119
−30	0·865	10	0·973	50	1·069	90	1·130
−20	0·892	20	1·000	60	1·088	100	1·137
−10	0·919	30	1·025	70	1·105	110	1·141

For predicting the thermal conductivities of concentrated electrolytic solutions Osminin[267] and Vargaftig[382] recommend the use of the following equation:

$$\lambda_e = \lambda_w \frac{c_{p,e}}{c_{p,w}} \left(\frac{\rho_e}{\rho_w}\right)^{4/3} \left(\frac{M_w}{M_e}\right)^{1/3} \qquad (6.32)$$

where λ_e is the thermal conductivity of the electrolyte solution,
λ_w is the thermal conductivity of water,
$c_{p,e}$ and $c_{p,w}$ are the heat capacities (at constant pressure) of the electrolyte solution and of water, respectively,

ρ_e is the density of the electrolyte solution,
ρ_w is the density of water,
M_w is the molecular weight of water,
M_e is the average "molecular weight" for the solution given by $M_e = x_w M_w + x_s M_s$, where x_w and x_s are the mole fractions of water and solute and M_w and M_s are the respective molecular weights.

The authors obtained this equation from the Vargaftig equation which is a modified form of Predvoditelev's equation (6.11) obtained from the latter by dividing the R.H.S by α. Applying the Vargaftig equation to the elc ctrolyte solution and to water respectively gives

$$\lambda_e = \frac{1}{\alpha_e} I c_{p,e} \rho_e^{4/3} M_e^{-1/3} \tag{6.33a}$$

and

$$\lambda_w = \frac{1}{\alpha_w} I c_{p,w} \rho_w^{4/3} M_w^{-1/3} \tag{6.33b}$$

Eq. (6.32) is obtained by dividing (6.33a) by (6.33b) if it is assumed that the factor α (which allows for molecular association) has the same value for the solution and for water. Equation (6.32) is suitable for estimating the thermal conductivities of concentrated solutions (e.g. of NaOH and KOH), for which the Riedel formula (6.30) fails.

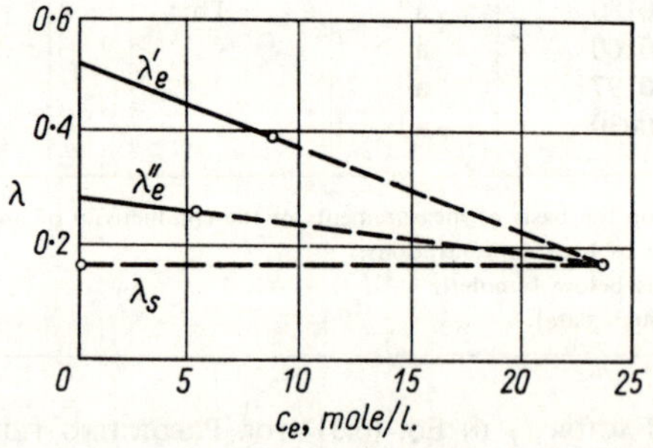

FIG. 6.12. Determination of λ_s, the thermal conductivity of the hypothetical supercooled salt, by extrapolating data for the conductivities λ'_e and λ''_e of solutions of this salt in various solvents (see Eq. (6.34)).

It is found empirically that the thermal conductivities of electrolytes in both aqueous and non-aqueous solvents vary in a roughly linear way with the molar concentration of the electrolyte c_e. Furthermore, if λ is plotted as a function of c_e for solutions of a given electrolyte in a series of solvents and the lines obtained are extrapolated to high values of c_e, these lines intersect at one point (Fig. 6.12). This point may be regarded as giving a fictitious value λ_s of the thermal conductivity of the supercooled salt. Eldarov[86] has used this finding as the basis of a method for predicting the thermal conductivities of solutions of electrolytes in aqueous or non-aqueous solvents. He determined λ_s graphically for a number of salts (cf. Fig. 6.12) and obtained the following values:

Salt	NaI	$ZnCl_2$	NaCl	KI
$\lambda_s, \dfrac{\text{kcal}}{\text{m.hr.°C}}$	0·110	0·150	0·130	0·075

If the values of λ_s and ρ_s are known for a given electrolyte, the thermal conductivity λ_e of a solution of that electrolyte in any solvent can be estimated from the formula

$$\lambda_e = \lambda_w + (\lambda_w - \lambda_s)\frac{C_e}{C_s} \tag{6.34}$$

where λ_w is the thermal conductivity of the pure solvent,
C_e is the molar concentration of the dissolved salt,
C_s $(= 1000\,\rho_s/M)$ is the concentration of fused supercooled salt,
ρ_s is the density of the fused supercooled salt,
and M is the molecular weight of the salt.

The practical utility of this equation is limited by a lack of tabulated values of λ_s and ρ_s. As Eldarov[85] points out Eq. (6.34) provides a method for predicting the thermal conductivities of non-aqueous electrolytic solutions for which the Riedel equation is not suitable.

Thermal Conductivities of Emulsions, Suspensions and Colloidal Solutions

This problem has not been sufficiently examined so far. Wang and Knudsen[391] find that experimental data for the conductivities of emulsions of mineral oils in water may be represented to within a few per cent by the additive law

$$\lambda_{em} = \lambda_w x_w + \lambda_{oil} x_{oil} \tag{6.35}$$

where λ_{em}, λ_w and λ_{oil} are the thermal conductivities of the emulsion, of water, and of oil, respectively
and x_w and x_{oil} are the volume fractions of water and oil in the mixture.

This formula is not satisfactory when applied to data for the emulsion of carbon tetrachloride in water, being then in error by as much as 85%. In this case rather better results may be obtained by taking the reciprocals of the conductivities to be additive quantities so that

$$\frac{1}{\lambda_{em}} = \frac{x_w}{\lambda_w} + \frac{x_{oil}}{\lambda_{oil}} \tag{6.36}$$

where the notation is the same as in the former equation. This equation adequately represents the data at 17·5°C, but at 27°C it may be in error by as much as 32%. In the light of these results it is clear that none of the equations presented above may be recommended for the estimation of the thermal conductivities of more than limited classes of emulsions. Even less information is available about the thermal conductivities of suspensions and colloidal solutions. The best that can be done when estimating the thermal conductivity of this type of mixture is to use the empirical rule that the thermal conductivity λ_m of an emulsion is usually about 10% lower than the thermal conductivity λ_l of the liquid phase in which the emulsion of macromolecules or colloidal particles is suspended[195] so that

$$\lambda_m = (0{\cdot}9)\,(\lambda_l) \tag{6.37}$$

This rule must be used with considerable caution.

EXAMPLE 6.3. Estimate the thermal conductivities λ_m of mixtures of ethyl formate and carbon tetrachloride of the following compositions at a temperature of 15°C (compositions are expressed in mass fractions):

	Ethyl formate	Carbon tetrachloride
(a)	$m_1 = 0{\cdot}3$	$m_2 = 0{\cdot}7$
(b)	$m_1 = 0{\cdot}5$	$m_2 = 0{\cdot}5$
(c)	$m_1 = 0{\cdot}7$	$m_2 = 0{\cdot}3$

Solution

The thermal conductivities of the pure components (found from Table 6.1) are as follows:

Ethyl formate $\lambda_1 = 380\times 10^{-6}$ cal/cm.sec.°C

Carbon tetrachloride $\lambda_2 = 250\times 10^{-6}$ cal/cm.sec.°C.

Application of the additivity rule (Eq. (6.25)) gives:

(a) $\lambda_m = 0{\cdot}3\times 380\times 10^{-6}+0{\cdot}7\times 250\times 10^{-6} = 289\times 10^{-6}$ cal/cm.sec.°C

(b) $\lambda_m = 0{\cdot}5\times 380\times 10^{-6}+0{\cdot}5\times 250\times 10^{-6} = 315\times 10^{-6}$ cal/cm.sec.°C

(c) $\lambda_m = 0{\cdot}7\times 380\times 10^{-6}+0{\cdot}3\times 250\times 10^{-6} = 341\times 10^{-6}$ cal/cm.sec.°C

Use of the Filippov and Novoselova equation (6.27) gives:

(a) $\lambda_m = 380\times 10^{-6}\times 0{\cdot}3+250\times 10^{-6}\times 0{\cdot}7-0{\cdot}72\times(380-250)\times 10^{-6}\times 0{\cdot}3\times 0{\cdot}7$
$= 269\times 10^{-6}$ cal/cm.sec.°C

(b) $\lambda_m = 380\times 10^{-6}\times 0{\cdot}5+250\times 10^{-6}\times 0{\cdot}5-0{\cdot}72\times(380-250)\times 10^{-6}\times 0{\cdot}5\times 0{\cdot}5$
$= 292\times 10^{-6}$ cal/cm.sec.°C

(c) $\lambda_m = 380\times 10^{-6}\times 0{\cdot}7+250\times 10^{-6}\times 0{\cdot}3-0{\cdot}72\times(380-250)\times 10^{-6}\times 0{\cdot}3\times 0{\cdot}7$
$= 321\times 10^{-6}$ cal/cm.sec.°C

An experimental value is available for case (b), viz.

$\lambda_m = 289\times 10^{-6}$ cal/cm.sec.°C

The error in the estimate given by the "additive" equation (6.23) is thus +9% while that given the formula of Filippov and Novoselova is +1·0%.

EXAMPLE 6.4. Use Bondi's equation (6.29) to calculate the thermal conductivity of an equimolar mixture of ethyl formate and carbon tetrachloride at a temperature of 15°C.

Solution

Pure component data

Ethyl formate: $M_1 = 74{\cdot}1$
$\lambda_1 = 380\times 10^{-6}$ cal/cm.sec.°C
$x_1 = 0{\cdot}5$

Estimation of E° (Table 6.14)

Contributions:

(C_2H_5)—		3130 cal/mole
—C=O	3430+200/3	3497 cal/mole
\O		
H	1970−1070	900 cal/mole
	$E_1^\circ =$	7527 cal/mole

Carbon tetrachloride: $M_2 = 153{\cdot}8$
$\lambda_2 = 250\times 10^{-6}$ cal/cm.sec.°C
$x_2 = 0{\cdot}5$
$E_2^\circ = 7850$ cal/mole†

The conductivity of the mixture is given by

$$\lambda_m = 380\times 10^{-6}\times 0{\cdot}5+250\times 10^{-6}\times 0{\cdot}5-\left[\left(\frac{7527}{74{\cdot}1}\right)^{1/2}-\left(\frac{7850}{153{\cdot}8}\right)^{1/2}\right]\times 4{\cdot}5\times 10^{-5}\times 0{\cdot}5\times 0{\cdot}5$$
$$= 283\times 10^{-6} \text{ cal/cm.sec.°C}$$

EXAMPLE 6.5. Estimate the thermal conductivity λ_m of a liquid mixture containing 45% by weight of oxygen and 55% of nitrogen at −200°C.

Available Data

(i) The thermal conductivities of the pure liquefied compounds are virtually identical so
$\lambda_1 = \lambda_2 = 0{\cdot}179$ kcal/m.hr.°C
(subscript 1 refers to oxygen)

(ii) The thermal conductivity λ_m' of a mixture containing 15% by weight of oxygen + 85% of nitrogen is given by
$\lambda_m' = 0{\cdot}172$ kcal/m.hr.°C

Note. As seen from the above conductivity data, the curve representing λ_m as a function of composition must show a minimum. In the case under consideration $\lambda_1 = \lambda_2$ with the result that the equation of Filippov and Novoselova reduces to the normal additive relationship (Eq. (6.25)). It cannot therefore be rigidly accurate when applied to this system and it could appear preferable to use Eq. (6.28).

† Estimated from the heat of vaporization.

Solution

The constant φ in Eq. (6.28) may be determined from the known conductivity λ'_m of the 15 wt. % mixture. On substituting $m'_1 = 0{\cdot}15$, $m'_2 = 0{\cdot}85$, $\lambda_m = 0{\cdot}172$ in Eq. (6.28) we obtain:

$$0{\cdot}172 \sinh\varphi = 0{\cdot}179 \sinh(0{\cdot}15\varphi) + 0{\cdot}179 \sinh(0{\cdot}85\varphi)$$

$$\text{whence } \frac{0{\cdot}172}{0{\cdot}179} = \frac{\sinh(m_1\varphi)+\sinh(m_2\varphi)}{\sinh\varphi} = 0{\cdot}960$$

This equation may be solved for ϕ by trial and error.

a. Taking $\varphi = 1$, we obtain

$$0{\cdot}15\,\varphi = 0{\cdot}15 \qquad \sinh 0{\cdot}15 = 0{\cdot}1506$$
$$0{\cdot}85\,\varphi = 0{\cdot}85 \qquad \sinh 0{\cdot}85 = 0{\cdot}9561$$
$$\sinh 1{\cdot}00 = 1{\cdot}1752$$
$$\frac{0{\cdot}1506+0{\cdot}9561}{1{\cdot}1752} = 0{\cdot}943 < 0{\cdot}960$$

b. Taking $\varphi = 0{\cdot}80$, we obtain

$$0{\cdot}15\,\varphi = 0{\cdot}12 \qquad \sinh 0{\cdot}12 = 0{\cdot}1203$$
$$0{\cdot}85\,\varphi = 0{\cdot}68 \qquad \sinh 0{\cdot}68 = 0{\cdot}7336$$
$$\frac{0{\cdot}1203+0{\cdot}7336}{1{\cdot}1752} = 0{\cdot}962 > 0{\cdot}960$$

By interpolating we find $\varphi = 0{\cdot}81$

λ_m for the desired mixture may now be calculated from Eq. (6.28), by setting

$$\varphi = 0{\cdot}81, \qquad m_1 = 0{\cdot}45, \qquad m_2 = 0{\cdot}55$$

whence

$$\lambda_m \sinh 0{\cdot}81 = 0{\cdot}179 \sinh(0{\cdot}45\times 0{\cdot}81) + 0{\cdot}179 \sinh(0{\cdot}55\times 0{\cdot}81)$$

Now

$$\sinh(0{\cdot}81) = 0{\cdot}9015$$
$$\sinh(0{\cdot}45\times 0{\cdot}81) = \sinh(0{\cdot}355) = 0{\cdot}3625$$
$$\sinh(0{\cdot}55\times 0{\cdot}81) = \sinh(0{\cdot}446) = 0{\cdot}4600$$

$$\text{so } \lambda_m(\text{calculated}) = \frac{0{\cdot}179}{0{\cdot}9015}\times(0{\cdot}3625+0{\cdot}4600) = 0{\cdot}164 \text{ kcal/m.hr.°C}$$

The experimentally determined value (Fig. 6.6) is 0·166 kcal/m.hr.°C so the predicted value is in error by 1·2%.

EXAMPLE 6.6 (Ref. 310). Estimate the thermal conductivity λ of an aqueous solution containing 30% by weight of aluminium nitrate at a temperature of 80°C.

Solution

The molecular weight of aluminium nitrate (M) is 213·0 g/mole and the density ρ of the solution at 20°C is 1·280 g/cm³. The concentration c of aluminium nitrate in the solution is thus given by

$$c = \frac{1280\times 30}{100\times 213{\cdot}0} = 1{\cdot}80 \text{ moles per litre}$$

The following conductivity contributions may be obtained from Table 6.19:

for Al^{3+} $\quad \alpha_1 = -0{\cdot}0280$

for NO_3^- $\quad \alpha_2 = -0{\cdot}0060$

The thermal conductivity of the solution at 20°C is then given by Eq. (6.30) to be $\lambda_{m,20} = 0{\cdot}515 - 1{\cdot}80\times \times(0{\cdot}280+3\times 0{\cdot}0060) = 0{\cdot}432$ kcal/m.hr.°C. The conductivity at 80°C may now be estimated from Eq. (6.31) and Table 6.20. According to this Table the value of the multiplier f_T is 1·119 at 80°C. Hence $\lambda_{m,80} = 0{\cdot}432\times 1{\cdot}119 = 0{\cdot}483$ kcal/m.hr.°C.

EXAMPLE 6.7. Use the Vargaftig equation (6.32) to estimate the thermal conductivity λ_e of an aqueous solution containing 32·0% by weight of sodium nitrate at a temperature of +20°C.

Solution

The following data are available from standard tabulations:

Heat capacity of the solution	$c_{p,e} = 0{\cdot}769$ cal/g.°C
Heat capacity of water	$c_{p,w} = 0{\cdot}999$ cal/g.°C
Density of the solution	$\rho_e = 1{\cdot}243$ g/cm³
Density of water	$\rho_w = 0{\cdot}998$ g/cm³
Molecular weight of sodium nitrate	$M_s = 85$

Molecular weight of water $M_w = 18$
Thermal conductivity of water $\lambda_w = 0{\cdot}515$ kcal/m.hr.°C
100 g of solution contains
32 g of $NaNO_3$ i.e. $32/85 = 0{\cdot}386$ moles
and 68 g of H_2O i.e. $68/18 = 3{\cdot}780$ moles

The mean molecular weight M_e of the solution is thus given by

$$M_e = \frac{0{\cdot}386\times 32+3{\cdot}780\times 18}{0{\cdot}386+3{\cdot}780} = 19{\cdot}3$$

On substituting the above values in Eq. (6.30) it is found that

$$\lambda_e = 0{\cdot}515\times\frac{0{\cdot}769}{0{\cdot}999}\times\frac{1{\cdot}243^{4/3}}{0{\cdot}998}\times\frac{18^{1/3}}{24{\cdot}1} = 0{\cdot}516 \text{ kcal/m.hr.°C}$$

The value of λ_e obtained by interpolation from the experimental data listed in Table 6.18 is 0·494 kcal/m.hr.°C. The error in the predicted value is thus −4·4%.

c. Selection of the Best Method for Estimating Mixture Thermal Conductivities

When deciding which of the available methods is the most appropriate, consideration should be given to the nature of the substances which constitute the mixture. If the mixture consists of non-polar molecules, the additive law (6.26) is usually sufficiently accurate, though slightly better accuracy may be obtained by using the equation of Filippov and Novoselova (Eq. (6.27)), or of Bondi and Riedel (Eq. (6.29)): these equations are substantially better than (6.26) if one of the components is slightly polar.

In the case of liquid mixtures containing one or several kinds of strongly polar molecules, the additive law is insufficiently accurate to be of use. Either of Eqs. (6.27) or (6.28) may then be used, though the latter equation is somewhat inconvenient to use and involves prior knowledge of the thermal conductivity of at least one mixture of the system under consideration.

Owing to the scarcity of experimental data it is difficult at present to gauge the reliability of the methods discussed above though some estimates of the magnitudes of the errors which may be incurred have been made by Tsederberg,[372]. These have been given in section 5b. All the methods given for estimating the thermal conductivities of aqueous electrolyte solutions are fairly accurate. Both Riedel's method based on summing ionic contributions (Eq. (6.30)) and Vargaftig's equation (6.32) are accurate to within a few per cent. The Vargaftig equation is more universal, but its use involves prior knowledge of certain physicochemical properties of the solution (such as c_p and ρ).

Methods for estimating the thermal conductivities of heterogenous liquid systems (such as suspensions, colloidal solutions, and emulsions) are not yet sufficiently developed. In doubtful cases it is often assumed (Kern[195]) that, for design purposes, the thermal conductivity of a heterogeneous mixture may be taken to be equal to the thermal conductivity of the pure liquid dispersing phase multiplied by 0·9 (Eq. (6.37)).

CHAPTER 7

DIFFUSION IN GASES

1. BASIC CONCEPTS AND NOMENCLATURE

The coefficient of diffusion, like the coefficients of viscosity and heat conduction is a transport property. This property is of importance in the design of many types of separation equipment since it is one of the factors which governs the rate at which molecules of a given species can leave one phase and penetrate into the bulk of a second phase in contact with the first. This phenomenon takes place, for example, in distillation and gas absorption towers and in liquid/liquid extraction units.

A distinction should be drawn between cases in which mass transfer takes place in one direction only and cases in which two different kinds of molecules diffuse in opposite directions. In the first of these cases *unidirectional diffusion* is said to occur, while the second phenomenon is known as *counterdiffusion.* If in the latter case the numbers of molecules diffusing in opposite directions are equal, the process is called *equimolal counterdiffusion.*

The number of moles of type A diffusing per unit time through unit cross-sectional area perpendicular to the direction of diffusion is proportional to the negative of the concentration gradient, $-(\mathrm{d}c_A/\mathrm{d}x)$, of the diffusing compound, where c_A is the concentration of the diffusing compound A and x is the distance measured along the direction of diffusion. The gradient has a negative sign because the concentration c_A decreases in the direction of diffusion whereas the distance x increases in this direction. If there is no bulk flow of the mixture as a whole, the net rate of flow of component A is equal to the rate of diffusion of that component. Now consider the general case when the gas mixture may be undergoing bulk flow. Let N_A and N_B be the net molar rates of flow of A and B in the direction x through unit cross section area in a plane normal to that direction. In unit time the net transfer of component A per unit area in the x direction is given by

$$N_A = -D_{AB}\frac{\mathrm{d}c_A}{\mathrm{d}x} + (N_A+N_B)\left(\frac{C_A}{C_A+C_B}\right) \tag{7.1}$$

In this expression, the first term on the right hand side takes account of transport of A by diffusion which results from the concentration gradient, $\mathrm{d}c_A/\mathrm{d}x$ of this component: the proportionality constant D_{AB} is termed the *diffusivity* or *coefficient of diffusion* of A in B. This quantity is usually expressed in $\mathrm{cm^2/sec}$ or $\mathrm{m^2/sec}$. The last term in the above equation takes account of the transport of A due to *bulk motion* of the gas mixture; this term is the product of the total molal flow rate (N_A+N_B) and the mole fraction of A in the mixture. N_A and N_B are related to average resultant velocities of transfer w_A and w_B of A and B by the equations

$$\begin{aligned} N_A &= w_A C_A \\ N_B &= w_B C_B \end{aligned} \tag{7.2}$$

Equations (7.1) and (7.2) may be combined and rearranged to give

$$-(dc_A/dx) = \left(\frac{1}{D_{AB}}\right)\frac{C_A C_B(w_A - w_B)}{(C_A + C_B)} = Kc_A c_B(w_A - w_B) \tag{7.3}$$

where

$$K = \frac{1}{D_{AB}(C_A + C_B)} \tag{7.4}$$

The coefficient K is a characteristic of the diffusing component A and the medium B through which it is diffusing. In the case of equimolal counterdiffusion of A and B

$$N_A = -N_B \tag{7.5}$$

In this case therefore the *bulk flow* term in equation (7.1) is zero and the equation reduces to

$$N_A = -D_{AB}\frac{dc_A}{dx} \tag{7.6}$$

or

$$dN_A = -D_{AB} S \frac{dc_A}{dx} d\tau \tag{7.7}$$

where S is the cross-section area through which diffusion is taking place,
and τ is the time.

Equation (7.7) is a formulation of the first law of Fick (1855). In the case of unsteady-state diffusion

$$\frac{\partial c_A}{\partial \tau} = D\frac{\partial^2 c_A}{\partial x^2} \tag{7.8}$$

This is a formulation of Fick's second law.

In the case of unidirectional steady state diffusion (when only compound A diffuses) $w_B = 0$, and Eq. (7.3) reduces to

$$-\frac{dc_A}{dx} = Kc_A c_B w_A \tag{7.9}$$

Since $N_A = w_A c_B$ (Eq. (7.2))

$$-\frac{dc_A}{dx} = KN_A c_B \tag{7.10}$$

and, by Eq. (7.4), we get the equation of unidirectional diffusion

$$N_A = -D\frac{(c_A + c_B)}{c_B}\frac{dc_A}{dx_B} \tag{7.11}$$

This equation may be transformed to give

$$N_A = \frac{D(c_A + c_B)}{xc_{Bm}}(c'_A - c''_A) \tag{7.12}$$

where

$$c_{Bm} = \frac{c''_B - c'_B}{\ln(c''_B/c'_B)} \tag{7.13}$$

c'_B and c''_B are the concentrations of the stagnant (nondiffusing) component at two points at a distance x apart from each other along the direction of diffusion $c''_B - c'_B = c'_A - c''_A$ (because $c_A + c_B$ = constant).

Replacing the concentrations c by partial pressures, we obtain

$$N_A = \frac{DP}{RTxp_{Bm}}(p'_A - p''_A) \tag{7.14}$$

This formula is another form of the equation of unidirectional steady-state diffusion.

Self-diffusion consists in the intermixing by natural molecular motion of the molecules of a substance into other molecules of the same substance. The self-diffusion process is characterized by the self-diffusion coefficient $D_{1,1}$ (for compound 1 in 1). A review of the literature on gaseous diffusion up to 1961 has been given by Liley (cf. Ref. 5).

2. SOME RELATIONSHIPS BASED ON THE KINETIC THEORY OF GASES

The elementary kinetic theory of gases leads to the following equation relating the coefficient of self-diffusion $D_{1,1}$ to the mean molecular speed $\overline{w}$ and the mean free path $\overline{l}$ of the molecules (Ch. 3, section 1):

$$D_{1,1} = \tfrac{1}{3}\overline{w}\overline{l} \tag{7.15}$$

When mixtures of several kinds of molecules 1, 2, 3 ... are considered, the simple equations, (3.8) and (3.9), are unsuitable for calculating the mean free paths of the molecules. According to Maxwell, the kinetic theory of gases leads to the following expression for the mean free path of molecules of type 1 in a gas mixture.

$$\overline{l}_1 = \frac{1}{\pi(n_1\sigma_1^2\sqrt{2} + n_2\sigma_{1,2}^2\sqrt{1+m_1/m_2} + n_3\sigma_{1,3}^2\sqrt{1+m_1/m_3} + \ldots)} \tag{7.16}$$

where $\sigma_{1,2} = 0{\cdot}5(\sigma_1+\sigma_2)$, $\sigma_{1,3} = 0{\cdot}5(\sigma_1+\sigma_3)$ ...,
$\sigma_1, \sigma_2, \sigma_3$ ... are the diameters of molecules 1, 2, 3 ...,
m_1, m_2, m_3 ... are the masses of molecules 1, 2, 3 ...,
n_1, n_2, n_3 ... are the numbers of molecules 1, 2, 3 ... in 1 cm^3,
and $\overline{l}_1$ is the mean free path for molecules of type 1 in the mixture of molecules 1, 2, 3 ... .

According to the above theory, the diffusion coefficient $D_{1,2}$ for a mixture of gases 1 and 2 is given by

$$D_{1,2} = \tfrac{1}{3}\overline{l}\overline{w} \tag{7.17}$$

In practice, this simplified equation is not obeyed very closely and it is convenient to define correction factors φ by the equations[182]

$$D_{1,2} = \tfrac{1}{2}\overline{l}\,\overline{w}\varphi_{1,2} \quad \text{and} \quad D_{1,1} = \tfrac{1}{2}\overline{l}\,\overline{w}\varphi_{1,1} \tag{7.18}$$

A similar factor appears in the analogous equations for the viscosity and thermal conductivity of gases (Eqs. (3.16) and (5.3)).

Dividing Eq. (7.18) by Eq. (3.16) we obtain

$$D_{1,1} = f_1 \frac{\mu}{\rho} \qquad (1 \cdot 0 < f_1 < 0 \cdot 5) \tag{7.19}$$

where $f_1 = \varphi_{1,1}/\varphi$ (on the average, $f_1 = 1 \cdot 32$),
μ is the viscosity,
and ρ is the gas density.

An analogous relationship between the viscosity and diffusivity $D_{1,2}$ is obtained from the simplified kinetic theory of gases

$$D_{1,2} = \frac{f_2}{3}(x_1 \bar{l}_1 \bar{w}_1 + w_2 \bar{l}_2 \bar{w}_2) \tag{7.20}$$

$$D_{1,2} = f_2 \left(x_1 \frac{\mu_1}{\rho_1} + x_2 \frac{\mu_2}{\rho_2} \right) \tag{7.21}$$

where f_2 is a correction factor,
x_1, x_2 are the mole fractions of the constituents of the mixture,
$\bar{l}_1$, $\bar{l}_2$ are the mean paths of molecules of types 1 and 2,
$\bar{w}_1$, $\bar{w}_2$ are the mean speeds of molecules of types 1 and 2,
μ_1, μ_2 are the viscosities of gases 1 and 2,
and ρ_1 and ρ_2 are the partial densities of these gases.

Although the above simplified equations are not accurately obeyed in practice, they do provide an interpretation of some of the broad aspects of diffusivity behaviour. For example, since the mean free paths of the molecules decrease with increase of pressure it follows from the above equations that the coefficients of diffusion $D_{1,1}$ and $D_{1,2}$ should be inversely proportional to pressure. It also follows from these equations that, at constant pressure, the coefficients should be proportional to the quantity $T^{3/2}$. This follows since the mean molecular speed $\bar{w}$ increases proportionally to $T^{1/2}$ (Eqs. (3.6) and (3.7)) while the mean free path $\bar{l}$ of the molecules is directly proportional to the absolute temperature T (being proportional to volume which is itself proportional to the absolute temperature). Maxwell, Clausius, Boltzmann, Sutherland, Chapman and others have shown that for a gas consisting of rigid elastic spheres, the diffusivity $D_{1,2}$ is given by:

$$D_{1,2} = \frac{3}{8(n_1+n_2)\sigma_{1,2}^2} \left(\frac{kT(m_1+m_2)}{2\pi m_1 m_2} \right)^{\frac{1}{2}} \tag{7.22}$$

(the notation is the same as in Eqs. (7.16) and (7.18)). Equation (7.22) may be transformed to give

$$D_{1,2} = \frac{bT^{3/2}}{p\sigma_{1,2}^2} \left(\frac{M_1+M_2}{M_1 M_2} \right)^{\frac{1}{2}} \tag{7.23}$$

where M_1 and M_2 are the molecular weights of constituents 1, 2 of the mixture,
$D_{1,2}$ is the coefficient of diffusion (diffusivity) for a binary system,
b is a constant,
p is the pressure
and T is the absolute temperature.

Sutherland has modified this equation, and introduced a factor which takes into account the effect of the forces of intermolecular attraction upon the mean free paths of the molecules. Sutherland's equation is

$$D_{1,2}=\frac{bT^{3/2}}{p\sigma_{1,2}^2}\left(\frac{M_1+M_2}{M_1M_2}\right)^{\frac{1}{2}}\frac{1}{1+C_{1,2}/T} \qquad (7.24)$$

where $C_{1,2}$ is Sutherland's constant for the mixture of compounds 1 and 2.

The equations presented above form the basis for most of the semi-empirical methods for estimating diffusivities. These methods will be discussed later.

According to the kinetic theory of gases the viscosities and diffusivities of gases are interrelated by equations of the form of (3.82). Using this equation it is possible to calculate the coefficient of diffusion $D_{1,2}$, if the viscosities of the pure compounds and of the mixture are known. This method may be particularly useful if values of $D_{1,2}$ are required either at very low or at very high temperatures, for which it is much easier to measure the viscosity than the diffusivity. Weissman and Mason[398] have carried out such calculations and have shown that fairly accurate results may be obtained in this way: however, the calculation procedure is somewhat laborious.

3. EXPERIMENTAL DATA FOR GASEOUS DIFFUSION COEFFICIENTS

Illustrative values of the coefficient $D_{1,1}$ for the process of self-diffusion of para-hydrogen and hydrogen are listed below:[182]

T, °K	273	85	20
$D_{1,1}$, cm²/sec	1·26	0·172	0·00816

Further values of the coefficient of self-diffusion $D_{1,1}$ for various gases, rounded off according to Andrussov (cf. Ref. 5) are compiled in Table 7.1.

TABLE 7.1. SELF-DIFFUSION COEFFICIENTS FOR VARIOUS GASES AT 0°C AND A PRESSURE OF ONE ATMOSPHERE AS GIVEN BY ANDRUSSOV

Gas	$D_{1,1}$	Gas	$D_{1,1}$	Gas	$D_{1,1}$
H_2	1·6040	C_2H_6	0·0900	CH_2F_2	0·0808
He	1·3856	O_2	0·1920	FCl	0·0876
D_2	0·8434	CH_3OH	0·1090	COS	0·0643
CH_4	0·1882	CH_3F	0·1107	SO_2	0·0701
NH_3	0·1930	H_2S	0·1135	Cl_2	0·0581
H_2O	0·2763	HCl	0·1210	CS_2	0·0462
Ne	0·4388	F_2	0·1626	Kr	0·0760
C_2H_2	0·1269	Ar	0·1569	CF_4	0·0508
CO	0·1647	CO_2	0·1062	Xe	0·0476
N_2	0·1549	N_2O	0·0888	CCl_4	0·0221
C_2H_4	0·1080	C_3H_8	0·0575		
NO	0·1549	CH_3Cl	0·0689		

The relationship between the values of the diffusion coefficients in gas mixtures and those of self-diffusion has been examined by Carman and Stein.[53]

Values of the diffusivities $D_{1,2}$ for a number of gaseous inorganic and organic systems are listed in Tables 7.2 and 7.3, respectively. It may be seen from Table 7.3 that gaseous

TABLE 7.2. COEFFICIENTS OF DIFFUSION FOR VARIOUS INORGANIC SYSTEMS (REFS. 11, 182, 402)

Diffusing gas	Medium into which the gas diffuses	p, mm Hg	t, °C	Diffusion coefficient D, cm²/sec
Ammonia	Air	760	0	0·198
Argon	Helium	751	15	0·703
Bromine	Argon	760	0	0·0861
Bromine	Carbon dioxide	760	0	0·0863
Bromine	Hydrogen	760	0	0·563
Bromine	Hydrogen chloride	760	0	0·0869
Bromine	Methane	760	0	0·156
Carbon dioxide	Hydrogen	749	18	0·606
Carbon monoxide	Hydrogen	760	0	0·622
Carbon monoxide	Oxygen	760	0	0·185
Hydrogen	Ammonia	760	25	0·849
Hydrogen	Helium	760	25	1·64
Hydrogen	Hydrogen chloride	760	21	0·795
Hydrogen	Nitrogen	755	12·5	0·739
Hydrogen	Nitrogen	760	15	0·743
Hydrogen	Nitrous oxide	760	0	0·535
Hydrogen	Oxygen	753	14	0·778
Hydrogen	Sulphur dioxide	760	0	0·483
Iodine	Air	760	20	0·081
Iodine	Nitrogen	10	19·4	5·70
Mercury	Hydrogen	760	0	0·53
Mercury	Nitrogen	3	19·4	32·3
Nitrogen	Carbon dioxide	760	15	0·158
Nitrogen	Carbon monoxide	760	15	0·211
Nitrous oxide	Carbon dioxide	760	15	0·107
Oxygen	Ammonia	760	20	0·253
Sodium	Nitrogen	760	15	20·4
Water	Air	760	0	0·219
Water	Carbon dioxide	760	0	0·138
Water	Hydrogen	760	0	0·747

diffusion coefficients increase with temperature; this is at least qualitatively in accord with the behaviour predicted by the simple kinetic theory of gases. Klibanova *et al.*[202] have found that the coefficients of diffusion for water vapour and carbon dioxide increase about fifteen fold when the temperature is raised from 0 to 1200°C (Table 7.4).

TABLE 7.3. COEFFICIENTS OF DIFFUSION FOR VARIOUS ORGANIC SYSTEMS AT A PRESSURE OF ONE ATMOSPHERE (REFS. 11, 182, 402)

Diffusing compound	t, °C	Diffusion coefficient D, cm²/sec		
		hydrogen	air	carbon dioxide
Acetic acid	0	0·4040	0·1061	0·0713
Acetic acid	98·5	0·7481	0·1965	0·1321
Amyl iso-butyrate	0	0·1694	0·0423	0·0308
Amyl iso-butyrate	97·7	0·3182	0·0786	0·0564
Amyl propionate	0	0·1891	0·0466	0·0341
Benzene	0	0·294	0·0751	0·0527
Benzene	45	0·3993	0·1011	0·0715
Butyl alcohol	0	0·2716	0·0681	0·0476
Butyl alcohol	99·05	0·5045	0·1265	0·0848
Butyl iso-butyrate	0	0·1850	0·0474	0·0332
Butyl iso-butyrate	97·9	0·3415	0·0876	0·0612
Butyric acid	0	0·2325	0·0605	0·0424
Butyric acid	98·6	0·4905	0·1263	0·0884
iso-Butyl acetate	0	0·2312	0·0592	0·0419
iso-Butyl acetate	97·9	0·4155	0·1055	0·0745
iso-Butyl butyrate	0	0·1850	0·0474	0·0332
iso-Butyl butyrate	97·9	0·3415	0·0876	0·0612
iso-Butyl propionate	0	0·2029	0·0529	0·0366
iso-Butyl valerate	0	0·1724	0·0426	0·0305
iso-Butyl valerate	97·8	0·3177	0·0782	0·0568
Ethyl acetate	0	0·2727	0·0709	0·0487
Ethyl acetate	46·1	0·3729	0·0970	0·0666
Ethyl alcohol	0	0·378	0·1016	0·0685
Ethyl alcohol	66·9	0·543	0·1475	0·1026
Ethyl butyrate	0	0·2239	0·0574	0·0407
Ethyl iso-butyrate	0	0·2237	0·0552	0·0400
Ethyl iso-butyrate	96·1	0·4267	0·1121	0·0784
Ethyl ether	0	0·296	0·0775	0·0552
Ethyl ether	19·9	0·341	0·0893	0·0636
Ethyl formate	0	0·3357	0·0852	0·0572
Ethyl formate	46·2	0·4383	0·1108	0·0751
Ethyl propionate	0	0·2373	0·0631	0·0450
Ethyl propionate	90·3	0·4019	0·1092	0·0806
Ethyl valerate	0	0·2050	0·0505	0·0366
Ethyl valerate	97·6	0·3784	0·0932	0·676
Formic acid	0	0·5131	0·1315	0·0879
Formic acid	84·9	0·8830	0·2244	0·1519
Hexyl alcohol	0	0·1998	0·0499	0·0351
Hexyl alcohol	99·0	0·3712	0·0927	0·0651
Methyl acetate	0	0·3277	0·0840	0·0577

Continued overleaf

TABLE 7.3 (*cont.*)

Diffusing compound	t, °C	Diffusion coefficient D, cm^2/sec		
		hydrogen	air	carbon dioxide
Methyl acetate	46·2	0·4531	0·1126	0·0760
Methyl alcohol	0	0·5001	0·1325	0·0880
Methyl alcohol	49·6	0·6738	0·1809	0·1234
Methyl butyrate	0	0·2422	0·0641	0·0439
Methyl butyrate	92·1	0·4308	0·1139	0·0809
Methyl iso-butyrate	0	0·2568	0·0642	0·0450
Methyl iso-butyrate	66·65	0·3913	0·0991	0·0696
Methyl propionate	0	0·2945	0·0745	0·0529
Methyl propionate	66·8	0·4564	0·1146	0·0820
Propionic acid	0	0·3333	0·0847	0·0595
Propionic acid	98·85	0·6182	0·1570	0·1104
Propyl alcohol	0	0·3153	0·0803	0·0577
Propyl alcohol	83·5	0·5434	0·1379	0·0976
Propyl butyrate	0	0·2059	0·0523	0·0364
Propyl butyrate	97·9	0·3801	0·0965	0·0673
Propyl iso-butyrate	0	0·2120	0·0549	0·0388
Propyl iso-butyrate	97·1	0·3897	0·0991	0·0714
Propyl formate	0	0·2811	0·0714	0·0490
Propyl formate	66·8	0·4234	0·1065	0·0738
Propyl propionate	0	0·2121	0·0554	0·0396
Propyl propionate	96·5	0·3864	0·1010	0·0721
Propyl valerate	0	0·1891	0·0466	0·0341
Propyl valerate	97·6	0·3490	0·0859	0·0629
iso-Valeric acid	0	0·2118	0·0555	0·0375
iso-Valeric acid	98·05	0·3934	0·1031	0·0696

System	t, °C	D, cm^2/sec	System	t, °C	D, cm^2/sec
Air–diphenyl	218	0·160	H_2–CCl_4	23	0·345
Air–nitrobenzene	25	0·0868	H_2–C_2H_6	0	0·439
Air–n-octane	25	0·0602	H_2–C_2H_4	20	0·483
CO_2–CH_4	0	0·153	H_2–CH_4	0	0·625
CO_2–C_3H_8	25	0·0863	H_2–C_3H_8	27	0·450
CO–C_2H_4	0	0·151	N_2–C_2H_4	27	0·187
C_6H_6–He	25	0·384	C_3H_8–C_2H_6	20	0·085
			C_2H_6–CH_4	20	0·163

TABLE 7.4. COEFFICIENTS OF DIFFUSION FOR THE SYSTEMS WATER VAPOUR–AIR AND CARBON DIOXIDE–AIR AS FUNCTIONS OF TEMPERATURE

Gases	Water vapour–air									
T, K°	373	573	773	923	1073	1093	1273	1373	1463	1493
$D_{1,2}$	0·353	0·849	1·556	2·19	2·626	2·808	3·253	3·94	4·48	4·49
Gases	Carbon dioxide–air									
T, °K	293	400	600	800	1000	1300	1500			
$D_{1,2}$	0·151	0·273	0·555	0·915	1·32	1·97	2·45			

Experiments also confirm the second conclusion resulting from the kinetic theory of gases namely that, at moderate pressures, the diffusion coefficient should be inversely proportional to pressure. This is illustrated in Table 7.5 which is taken from the work of Loschmidt. This table shows the product Dp of the diffusivity D and the pressure p for the systems carbon dioxide–air, carbon dioxide–hydrogen and hydrogen–oxygen as a function of pressure at subatmospheric pressures down to about 300 mm Hg. It is seen that for a given pair of gases this product is approximately constant. This relationship does not, however, hold at high pressures and high densities. For example, at a temperature of 42°C, the product $D\rho$, of the self-diffusivity and density, is about one-half as large for CO_2 gas at a density ρ of 0·7 g/cm^3 as it is at moderate pressures.

TABLE 7.5. DIFFUSION COEFFICIENTS FOR THE SYSTEMS CARBON DIOXIDE–AIR, CARBON DIOXIDE–HYDROGEN AND HYDROGEN–OXYGEN AS FUNCTIONS OF PRESSURE

Compound	$D_{1,2}$, cm^2/sec	t, °C	p, mm Hg	$Dp/760$
Carbon dioxide–air	0·1653	17·6	751	0·163
Carbon dioxide–air	0·3376	15·2	364	0·162
Carbon dioxide–air	0·4139	15·7	309	0·164
Carbon dioxide–hydrogen	0·6142	12·8	757	0·612
Carbon dioxide–hydrogen	0·9184	15·4	510	0·616
Hydrogen–oxygen	0·8012	11·4	748	0·790
Hydrogen–oxygen	1·1718	15·8	512	0·791

Krichevskii, Khazanova and Linshitz[204] have investigated the diffusion of gases in the vicinity of the critical point and find that under these conditions the diffusivity becomes very small (at least 10^3 times smaller than at atmospheric pressure).

At high pressures, the diffusion coefficient decreases with increase of pressure more rapidly than would result from the inverse proportionality law. The following values of the ratio of self-diffusivity for nitrogen at 50°C and pressure p, $(D_{1,1})_p$, to the value of this coefficient under similar conditions and a pressure of 1 atm $(D_{1,1})_1$, have been reported:[182]

p, atm	1	15·37	104·5	320·4	541·7	742·1	965·8
$(D_{1,1})_p/(D_{1,1})_1$	1	0·064	0·0084	0·0026	0·0016	0·0013	0·0011

These figures indicate that the relationship pD = constant does not hold very accurately at high pressures.

High values of the diffusion coefficient are found for systems containing a gas with a low molecular weight (hydrogen, helium); this is in agreement with the predictions of the kinetic theory of gases (Eqs. (7.22), (7.23) and (7.24)).

In the above-mentioned binary systems the diffusing component was present in small amount only. If the composition of the phase into which this component diffuses is allowed to change sufficiently, the value of the diffusion coefficient may change noticeably, i.e. the diffusion coefficient is composition dependent. For example the table below shows some data obtained by Deutsch[182] for mixtures of hydrogen (first component, subscript 1) and carbon dioxide (second component, subscript 2) and helium and argon:

Mole ratio	$D_{1,2}$	Mole ratio	$D_{1,2}$
H_2–CO_2 = 3/1	0·594	He–Ar = 2·65/1	0·678
1/1	0·605	1·66/1	0·696
1/3	0·633	1/1	0·706
		0·311/1	0·731

These data show that for these two systems the effect of composition is not very large, not exceeding about 8%.

The magnitude of this effect is rather larger in the nitrogen–carbon dioxide system as illustrated in Table 7.6. The data shown in this table were obtained by Krichevskii, Khazanova, Lesnevskaya and Polyakova[205] and cover an extensive composition range for pressures up to about 75 atmospheres.

4. METHODS FOR ESTIMATING THE COEFFICIENTS OF DIFFUSION OF GASES

a. Methods Based on the Kinetic Theory of Gases Extended to Allow for the Intermolecular Forces

In order to apply these methods it is necessary to know the values of force constants σ and ε in the Lennard-Jones equation (1.29).

The methods of calculation are discussed in more detail by Bird, Hirschfelder and Curtiss.[25] For the case of two non-polar gases 1 and 2, these authors give the approximate relationship

$$D_{1,2} = \frac{0{\cdot}0026280}{p\sigma_{1,2}^2 \Omega_{1,2}^{(1,1)*}} \sqrt{\frac{T^3(M_1+M_2)}{2M_1 M_2}} \tag{7.25}$$

TABLE 7.6. THE COEFFICIENT OF DIFFUSION OF NITROGEN INTO A MIXTURE OF NITROGEN AND CARBON DIOXIDE UNDER ELEVATED PRESSURES ($D_{N_2} \times 10^3$ cm²/sec)

Temperature and pressure	Mole fraction of nitrogen in the mixture			
	0·25	0·30	0·45	0·60
Temperature 25°C				
6·0 atm	22·2	22·7	23·0	23·5
13·8	8·45	9·20	9·7	10·4
24·0	4·55	5·20	5·7	6·35
47·0	1·83	2·30	2·65	3·13
58·6	1·07	1·45	1·80	2·20
Temperature 28·15°C				
6·0 atm	27·5	28·4	29·4	30·5
13·8	11·7	12·1	12·5	13·0
20·0	7·30	7·55	8·0	8·4
24·0	5·40	5·70	5·93	6·17
47·0	2·00	2·30	2·60	3·00
58·6	1·30	1·65	1·93	2·25
65·8	1·00	1·23	1·50	1·77
Temperature 31·5°C				
24·0 atm	4·97	5·03	6·10	—
47·0	2·03	2·43	2·83	—
58·6	1·65	2·00	2·37	—
70·0	0·90	1·05	1·20	—
74·0	0·33	0·43	0·53	—

where $D_{1,2}$ is the diffusivity, cm²/sec,
p is the pressure, atm,
T is the absolute temperature, °K,
M_1 and M_2 are the molecular weights of gases 1 and 2,
$\sigma_{1,2}$ is a "force constant" calculated from Eq. (1.39), Å,
$\Omega_{1,2}^{(1,1)*}$ is a function, the value of which is taken from Table 3.1 for $T^* = kT/\varepsilon_{1,2}$,
and $\varepsilon_{1,2}$ is a "force constant" calculated from Eq. (1.39) (($\varepsilon_{1,2}/k$) is expressed in °K).
Equation (7.25) may also be employed if one of the gases is polar and the other non-polar if appropriate force constants are used. Kuong[211] has constructed a nomograph (Fig. 7.1) for the quick solution of Eq. (7.25). The use of this nomograph is illustrated in Example 7.4.

Wilke and Lee[402] have shown that the assumption of a constant numerical coefficient in Eq. (7.25) is not legitimate. According to these authors the value of this coefficient varies with the molecular weight of the diffusing gas, the equation for the diffusion coefficient taking the form

$$D_{1,2} = \frac{BT^{3/2}}{p\sigma_{1,2}^2 \Omega_{1,2}^{(1,1)*}} \left(\frac{M_1+M_2}{M_1 M_2}\right)^{\frac{1}{2}} \tag{7.26}$$

The coefficient B is a function of the quantities M_1 and M_2, being given by

$$B = 0{\cdot}00214 - 0{\cdot}000492 \left(\frac{M_1+M_2}{M_1 M_2}\right)^{\frac{1}{2}} \tag{7.27}$$

The introduction of the variable reduces the average error in $D_{1,2}$ by approximately one half.

Values of the function $\Omega_{1,2}^{(1,1)*}$ may most conveniently be obtained by interpolation from Table 3.1. Alternatively an equation due to Ning Hsing Chen[257] may be used:

$$\Omega^{(1,1)*} = 1{\cdot}075\left(\frac{kT}{\varepsilon}\right)^{-0{\cdot}1615} - 2\left(\frac{10kT}{\varepsilon}\right)^{-0{\cdot}741\log(10kT/\varepsilon)} \tag{7.28}$$

where k is Boltzmann's constant,
and ε is the Lennard-Jones force constant.
This equation gives highly accurate values for Ω (the average error is as low as 0·078%).

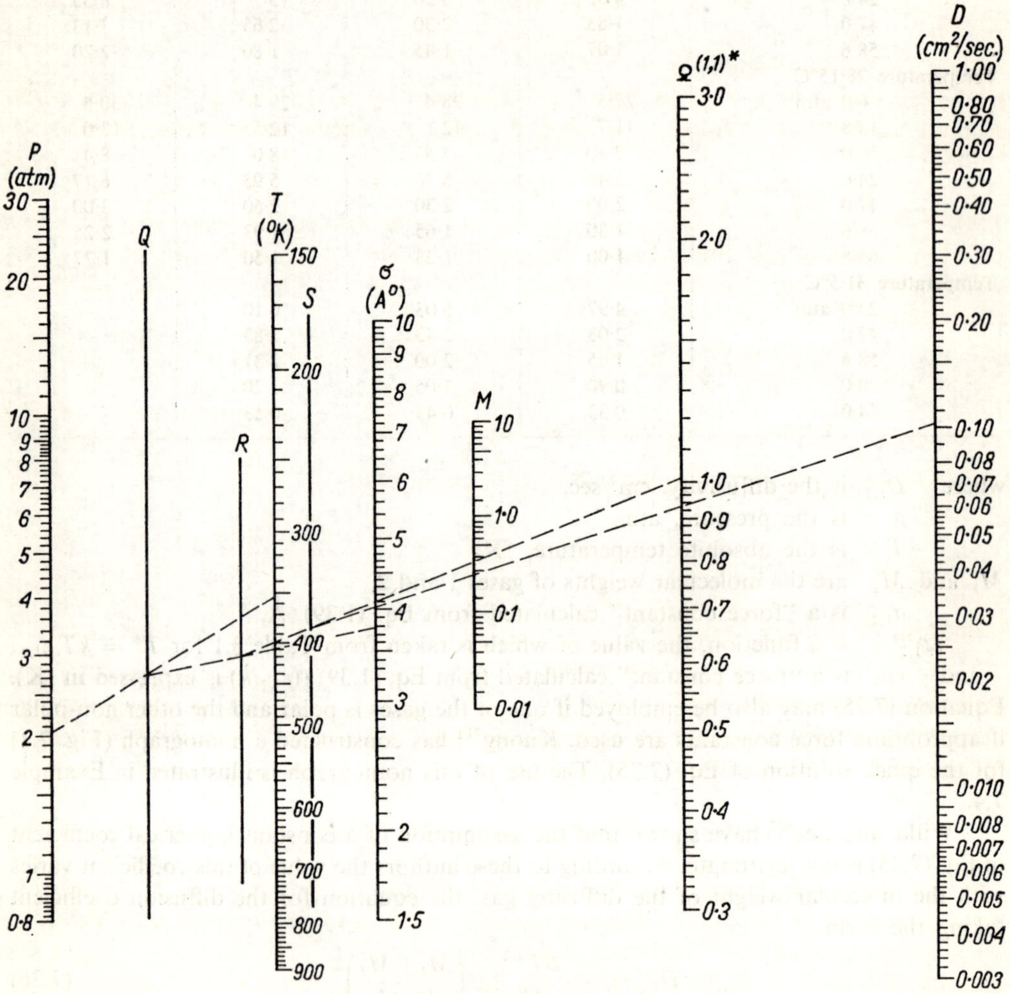

FIG. 7.1. Nomograph for estimating gaseous diffusivities from Hirschfelder's equation (7.25).

Ning Hsing Chen and Othmer,[258] have combined Eqs. (7.25) and (7.28) with the following expressions for the force constants:

$$\varepsilon/k = 1{\cdot}276T_c^{0{\cdot}9061} \tag{7.29}$$

$$\sigma = 0{\cdot}5894V_c^{0{\cdot}4006} \tag{7.30}$$

(These authors consider these expressions to be more accurate than Eqs. (1.33) and (1.34)). As a result they obtain the relationship

$$D_{1,2} = \frac{0{\cdot}43\left(\frac{T}{100}\right)^{1{\cdot}81}\left(\frac{1}{M_1}+\frac{1}{M_2}\right)^{0{\cdot}5}}{p\left(\frac{T_{c1}T_{c2}}{10{,}000}\right)^{0{\cdot}1406}\left[\left(\frac{V_{c1}}{100}\right)^{0{\cdot}4}+\left(\frac{V_{c2}}{100}\right)^{0{\cdot}4}\right]^2} \tag{7.31}$$

where T is the absolute temperature, °K,
T_c is the critical temperature, °K,
V_c is the critical volume, cm³/mole,
and p is the pressure, atm.

In the opinion of these authors, Eq. (7.31) is more accurate than Eq. (7.25). A still more accurate value for the diffusion coefficient D', may be obtained by using the formula

$$D' = Df_D \tag{7.32}$$

The correction factor f_D is taken from Table 3.3 for the appropriate value of $T^* = kT/\varepsilon$.

In the case of self-diffusion, i.e. when the molecules of some gas diffuse into a gas with the same or almost the same physical properties (for example, the diffusion of the molecules of ortho- into para-hydrogen), Eq. (7.25) becomes more simple, since then $M_1 = M_2 = M$, $\sigma_1 \approx \sigma_2$, $\varepsilon_1 \approx \varepsilon_2$ and the diffusivity $D_{1,1}$ is given by

$$D_{1,1} \approx \frac{0{\cdot}0026280}{p\sigma^2\Omega^{(1,1)*}}\sqrt{\frac{T^3}{M}} \tag{7.33}$$

In this case Eq. (3.82) leads to the simple relationship:

$$D_{1,1} = \frac{6}{5}\mu\bar{A}^*/\rho \tag{7.34}$$

where ρ is the gas density, g/cm³,
and μ is the gas viscosity.

$\bar{A}^*$ is given in Table 3.3 as a function of $T^* = kT/\varepsilon$.

It should be noted that the quantity $\bar{A}^*$ varies only very slightly with temperature, its value being always close to 1·1. It results therefrom that the Schmidt number should be independent of temperature, i.e. that

$$Sc = \frac{\mu}{\rho D} \approx 0{\cdot}75 \tag{7.35}$$

The use of the correction factor f_D when estimating the coefficient of diffusion is usually superfluous, since the correction which it introduces rarely exceeds a few per cent.

Ibrahim and Kuloor[165] find that, in accordance with Eq. (7.35) the value of the modified Schmidt number, $Sc' = \mu V_m/D$ for the diffusion of various gases through air varies linearly with the molecular weight M of the diffusing gas. This fact led them to suggest the use of the equation:

$$D_{1,2} = \frac{100\mu V_m}{0{\cdot}122M - 1{\cdot}3} \tag{7.36}$$

where $D_{1,2}$ is the coefficient of diffusion in the system gas (1)–air (2),
μ is the viscosity of gas (1),

V_m is the molal volume of gas (1),

and M is the molecular weight of gas (1).

In the case of polar gases, a correction factor should be employed. The mean deviation of the calculated values from the experimental ones is about 7%.

Variation of Diffusion Coefficients with Temperature

According to the kinetic theory of gases in its simplest form the diffusion coefficients of gases should vary proportionally to $T^{3/2}$, i.e.

$$D_{T_2} = D_{T_1}\left(\frac{T_2}{T_1}\right)^{\frac{3}{2}} \tag{7.37}$$

where D_{T_1} and D_{T_2} denote the values of diffusivity at temperatures T_1 and T_2. This relationship follows as a consequence of Eqs. (7.17) and (7.18), as shown in section 2. In practice Eq. (7.37) is found to be only approximately valid. A better correlation of diffusivity data may be obtained by empirically substituting the exponent $\frac{3}{2}$ by 1·8.

Sutherland recommends the use of an equation analogous to those already discussed for correlating viscosities and thermal conductivities as functions of temperature. Sutherland's equation is

$$D_{T_2} = D_{T_1}\left(\frac{T_2}{T_1}\right)^{\frac{5}{2}}\left(\frac{T_1+C_{AB}}{T_2+C_{AB}}\right) \tag{7.38}$$

where C_{AB} is Sutherland's constant for the pair of gases A and B. Sutherland's formula is obeyed much more accurately than Eq. (7.37) and it provides a good approximation to the actual dependence of D upon T. C_{AB} may be estimated using Arnold's method (Eqs. 7.46 and 7.47). Equations (7.25) and (7.32) may be used to predict the diffusion coefficient D_{T_2} for a pair of gases at temperature (T_1) if the diffusion coefficient D_{T_1} for the same pair of gases at temperature T_2 is known. The equation used is

$$D_{T_2} = D_{T_1}\left(\frac{T_2}{T_1}\right)^{\frac{3}{2}} \frac{\Omega_{T_1^*}^{(1,1)*}}{\Omega_{T_2}^{(1,1)*}} \frac{f_{DT_2}}{f_{DT_1}} \tag{7.39}$$

where $\Omega_{T_1}^{(1,1)*}$ and $\Omega_{T_2}^{(1,1)*}$ are taken from Table 3.1 for values of T^* corresponding to the temperatures T_1 and T_2 respectively,

and the correction factors f_{D,T_1} and f_{D,T_2} are taken from Table 3.3 for the appropriate values of T_1^* and T_2^*.

For many purposes it is sufficiently accurate to disregard the last term since the ratio f_{DT_2}/f_{DT_1} is very close to unity.

Variation of Diffusion Coefficients with Pressure

For a given temperature, the product of the diffusion coefficient D and the pressure p is approximately constant provided the pressure does not exceed several atmospheres. At higher pressures it is found that the product Dp decreases as the pressure is increased.

It has been shown that Enskog's theory[140] of dense gases (according to which a gas may be regarded as composed of rigid, spherical and perfectly elastic molecules) leads to the relationship

$$(Dp)_p = \frac{(Dp_0)^\circ z}{\gamma} \tag{7.40}$$

where $(Dp)_p$ is the product of the self diffusivity and pressure at pressure p,
$(Dp_0)^\circ$ is the same product at the same temperature (T) but at a moderate pressure p_0,
z is the compressibility factor of the gas ($z = pV/RT$),
and γ is a function of the molal volume of the gas and of the diameter of its molecules, as given by Enskog[88]

$$\gamma = \frac{yV}{b_0} \tag{7.41}$$

According to Enskog

$$y = \frac{pV}{RT} - 1 = \left(\frac{b_0}{V}\right) + 0{\cdot}6250\left(\frac{b_0}{V}\right)^2 + 0{\cdot}2869\left(\frac{b_0}{V}\right)^3 + 0{\cdot}115\left(\frac{b_0}{V}\right)^4 + \ldots \tag{7.42}$$

In this equation V is the molal volume of the gas at temperature T and pressure p and

$$b_0 = \frac{2}{3}\pi N\sigma^3 \tag{7.43}$$

where N is Avogadro's number,
and σ is the diameter of the molecules.
Values of y as a function of (b_0/V) are given in Table 7.7.

TABLE 7.7. DEPENDENCE OF y UPON (b_0/V), EQ. (7.42) (REF. 25)

b_0/V	y	b_0/V	y	b_0/V	y
0·3535	0·44	1·377	2·91	2·058	5·46
0·6250	0·91	1·527	3·43	2·160	5·99
0·8511	1·39	1·664	3·93	2·278	6·50
1·047	1·89	1·805	4·44	2·387	6·93
1·224	2·40	1·934	4·95	2·962	∞

If the compressibility factor z of the gas is known at the temperature T and pressure p, its molal volume V may be calculated and, provided the molecular diameter σ is known, b_0 may be calculated from Eq. (7.43).

Having determined b_0 and V, the value of y may be interpolated from the data given in Table 7.7, and, after calculating γ from Eq. (7.41), the ratio $(Dp)_p/(Dp_0)^\circ$, may be found from Eq. (7.40). This ratio characterizes the effect of pressure on the diffusion coefficient.

This calculation is only applicable when the pair of diffusing gases have similar molecular diameters σ. According to Reid and Sherwood,[306] Eq. (7.40) is only of very approximate validity. This is evident from Fig. 7.2, in which the curve shows the variation of $(Dp)_p/(Dp_0)^\circ$ with b_0/V as predicted by Eqs. (7.40) and (7.41). Pressure is indicated at the top of the figure.[140] The crosses and circles correspond to measurements of the self-diffusivity of carbon dioxide at 25°C and 42°C. The experimental points at a temperature of 25°C and a pressure exceeding 64 atm correspond to the liquid phase. For CO_2,

$$b = 98{\cdot}9 \text{ cm}^3/\text{mole}, \quad \sigma = 4{\cdot}28 \text{ Å}.$$

Durbin and Kobayashi[81] have measured the diffusivity in the systems: Kr–Ar, Kr–He, Kr–N_2 and in other systems at pressures up to 415 atm, and have compared their results with values predicted by Enskog's equations. They find considerable discrepancies between the estimated and measured values.

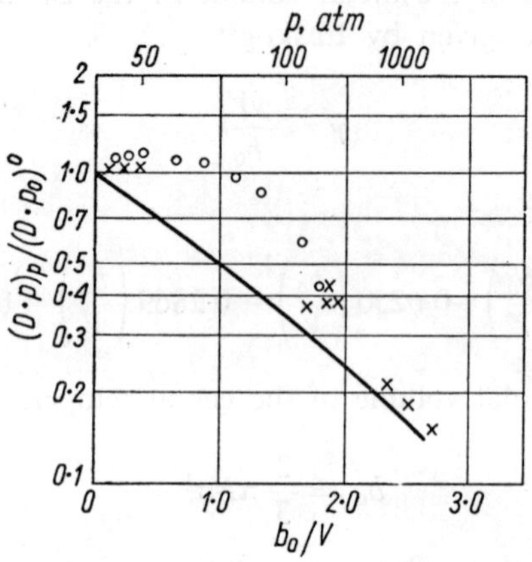

FIG. 7.2. Effect of pressure on gaseous diffusion coefficients. The full line corresponds to the theoretical Eqs. (7.40) and (7.41). Experimentally determined points for the self-diffusion of carbon dioxide are marked × (at 25°C) and o (at 42°C).

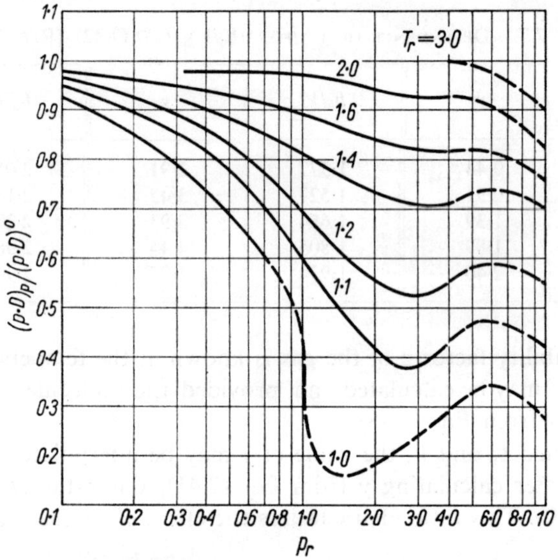

FIG. 7.3. The ratio $(pD)_p/(pD)^\circ$ as a function of reduced pressure and reduced temperature (after Slattery[331]). (D is the coefficient of self-diffusion and p the pressure).

Slattery[330] has suggested an alternative method of applying Enskog's theory. This method involves the use of a generalized plot showing $Dp/(Dp)^\circ$ as a function of reduced temperature and reduced pressure. This plot is, for the self-diffusion coefficient, $D_{1,1}$. Slattery's original plot has subsequently been extended[331,404] to cover a wider range of reduced pressures ($p_r = p/p_c$) and reduced temperatures ($T_r = T/T_c$). The extended plot is shown in Fig. (7.3). This plot enables the diffusivity at a pressure p and temperature T

to be predicted, provided the critical pressure and temperature are known (so that p_r and T_r may be calculated) and that the diffusivity D° at a moderate pressure p° and the same temperature is also known.

In the opinion of Slattery and Bird,[331] this method can be extended to predict binary diffusivities $D_{1,2}$ by using the pseudocritical pressure and temperature for the mixture in place of the critical temperature and pressure when evaluating the reduced parameters.

Gambill,[122] reviewing various methods for predicting diffusivities, stresses that the method recommended by Slattery and Bird has not yet been sufficiently checked. Reid and Sherwood[306] on the other hand, having checked the method in a number of cases, conclude that despite some deviations, the graph presented in Fig. 7.3 provides the best method of predicting diffusion coefficients at high pressures.

Burry and Koeller[51] have determined diffusivities in the system H_2–N_2 at pressures up to 1000 atm, and in the systems N_2–CH_4 and N_2–C_2H_6 at pressures up to 190 atm. The results obtained for the CH_4–C_2H_6 system are, on average, 31% higher than values predicted using Enskog and Thorne's method (Eq. (7.40)). For the N_2–H_2 system, this difference is 5·7%. Slattery and Bird's method (Fig. 7.3) gives much better results, the mean deviation for the first system (CH_4–C_2H_6) being one-sixth as large, viz., 5·3%.

b. Semi-empirical Equations for Estimating the Diffusion Coefficient

For the purpose of technological calculations, it is convenient to modify Eqs. (7.23) and (7.24), which are based on the kinetic theory of gases by replacing the average diameter $\sigma_{1,2}$ of the molecules by an approximately proportional quantity, i.e. the molal volume at the boiling point (cm³/mole). This volume, being an additive quantity, may if necessary be predicted from the molecular structure of the substance considered using the method of group contributions, as recommended by Le Bas (Table 1.19). Gilliland's equation and Arnold's equation are of this type.

Gilliland's Equation[127]

This is the simplest semi-empirical equation for predicting the diffusion coefficient $D_{1,2}$. Gilliland arrived at this equation by replacing the average diameter $\sigma_{1,2}$ of the molecules in Eq. (7.23) by the quantity $(V_{b1}^{1/3}+V_{b2}^{1/3})^3$ which should be proportional to $\sigma_{1,2}$. V_{b1} and V_{b2} are the molal volumes of substances 1 and 2 in the liquid state at their normal boiling temperatures. According to Gilliland's equation

$$D_{1,2} = \frac{0{\cdot}0043T^{3/2}}{p(V_{b1}^{1/3}+V_{b2}^{1/3})^2}\left(\frac{M_1+M_2}{M_1M_2}\right)^{\frac{1}{2}} \tag{7.44}$$

where $D_{1,2}$ is the kinematic coefficient of diffusion, cm²/sec,
p is the total pressure, atm
V_{b1} and V_{b2} are the molal volumes of the liquid compounds 1 and 2 at their normal boiling temperatures, calculated from the contributions given in Table 1.19, cm³/mole,
M_1 and M_2 are the molecular weights of components 1 and 2, g/mole,
and T is the absolute temperature, °K.

Some authors consider that a closer fit to the experimental data may be obtained if the power exponent 3/2 is replaced by a higher value, the magnitude of which lies between 1·8 and 2·2, depending on the compound involved. At temperatures close to room temperature, Gilliland's equation gives fairly good results; at elevated temperatures the deviations between the measured and calculated values may become large. The advantage of this equation is its simplicity and easy applicability.

Arnold's Equation

Arnold[13] has modified Sutherland's equation (7.24) to a form which is more suitable for technological calculations by replacing the average value of the diameters $\sigma_{1,2}$ of the molecules of the diffusing gases by the average value of the molal volumes of these compounds at their normal boiling temperatures, a suitably modified value of the constant coefficient then being used. This author also suggests a method of calculating Sutherland's constant. According to Arnold's equation:

$$D_{2,1} = D_{1,2} = \frac{0{\cdot}00837T^{5/2}}{p(V_{b1}^{1/3}+V_{b2}^{1/3})^2(T+C_{1,2})}\left(\frac{M_1+M_2}{M_1M_2}\right)^{\frac{1}{2}} \tag{7.45}$$

The notation and units of the quantities appearing in this equation are the same as in Gilliland's equation (7.44). The formula recommended for calculating the value of Sutherland's constant $C_{1,2}$ is

$$C_{1,2} = F\sqrt{C_1C_2} \tag{7.46}$$

According to Vogel[386]

$$C_1 = 1{\cdot}47T_{b1} \qquad C_2 = 1{\cdot}47T_{b2} \tag{7.47}$$

where T_{b1} and T_{b2} denote the normal boiling temperatures of compounds 1 and 2. The value of the coefficient F is stated to depend as follows on the ratio of the molal volumes at the normal boiling point (V_{b2}/V_{b1}) of the diffusing compounds (component 2 is taken to be the component with the higher molar volume):

V_{b2}/V_{b1}	1	2	3	4	6	8	10
F	1·00	0·98	0·953	0·920	0·875	0·838	0·805

Arnold's equation was proposed earlier than that of Gilliland. It is more accurate (in particular the effect of temperature is better allowed for), but slightly more tedious in application and requires a knowledge of the normal boiling points of the diffusing gases.

Slattery's Equation

Slattery[330] has proposed a semi-empirical equation for calculating diffusivities as a function of a reduced temperature T_r given by

$$T_r = \frac{T}{(T_{c1}T_{c2})^{\frac{1}{2}}} \tag{7.48}$$

Slattery's equation is based on the presumption that the diffusion coefficient for any two gaseous compounds 1 and 2 may be represented as a function of T_r by an equation of the form

$$(pD_{1,2})_r = aT_r^b \tag{7.49}$$

where $(pD_{1,2})_r$ is a "reduced" value of $(pD_{1,2})$ and a and b are constants. Transformation of the equations of the classical kinetic theory of gases leads, after some empirical simplifications, to the relationship

$$D_{1,2} = \frac{aT_r^b(p_{c1}p_{c2})^{1/3}(T_{c1}T_{c2})^{5/12}}{p}\left(\frac{M_1+M_2}{M_1M_2}\right)^{\frac{1}{2}} \tag{7.50}$$

where $D_{1,2}$ is the diffusion coefficient for gases 1 and 2, cm^2/sec,
T_r is the "reduced" temperature given by Eq. (7.49),
T_{c1} and T_{c2} are the critical temperatures of compounds 1 and 2, °K,
p_{c1} and p_{c2} are the critical pressures of compounds 1 and 2, atm,
M_1 and M_2 are the molecular weights of compounds 1 and 2, g/mole,
p is the pressure, atm,
and a, b are constants.

The values of the constants a and b are as follows:

	a	b
Diffusion of two non-polar gases or self-diffusion	$2{\cdot}74\times10^{-4}$	1·823
Diffusion in a system of water vapour and a non-polar gas	$3{\cdot}63\times10^{-4}$	2·334

Slattery's equation is slightly less accurate than that of Arnold.

Andrussov's Equations

If the molecular weights M_1 and M_2 of the gases taking part in the diffusion process differ greatly, the above equations are found to be unsatisfactory. Andrussov (cf. Ref. 5) recommends the use of the following equation for estimating $D_{1,2}$ at 0°C and a pressure of p atm:

$$D_{1,2} = \frac{17{\cdot}2(1+\sqrt{M_1+M_2})}{p(V_1^{1/3}+V_2^{1/3})^2M_1^{1/2}M_2^{1/2}} \quad (\text{cm}^2/\text{sec}) \tag{7.51}$$

where V_1 and V_2 are "mean molecular volumes" of gases 1 and 2 as tabulated by the author. According to Andrussov, the self-diffusion coefficient $D_{1,1}$ at temperature T and pressure p may be calculated from the equation:

$$D_{1,1} = \frac{A_0(1+\sqrt{2M_1})}{pV_1^{2/3}M_1}\left(\frac{T}{T_0}\right)^{n_D} \quad (\text{cm}^2/\text{sec}) \tag{7.52}$$

The exponent n_D in this equation is given by $n_D = n_\mu+1$, where n_μ is the power exponent in Andrussov's viscosity equations ((3.38) and (3.39)). At 0°C and atmospheric pressure, the constant $A_0 = 4{\cdot}3$.

Equation (7.52) leads to an alternative route for estimating $D_{1,2}$. After calculating the values of the self-diffusivities $D_{1,1}$ and $D_{2,2}$ we can find the value of the coefficient of diffusion $D_{1,2}$ by using the following equation

$$D_{1,2} = \frac{1+0{\cdot}5\delta^2}{\dfrac{\beta_1}{(1+\beta_2\delta)D_{2,2}}+\dfrac{\beta_2}{(1+\beta_1\delta)D_{1,1}}} \tag{7.53}$$

where $\delta = \dfrac{V_1^{1/3}-V_2^{1/2}}{V_1^{1/3}+V_2^{1/3}}$, $\quad\beta_1 = \dfrac{\sqrt{M_1}}{\sqrt{M_1}+\sqrt{M_2}}$, $\quad\beta_2 = 1-\beta_1$

Kuvshinskii's Equation

According to Kuvshinskii,[215] if $D_{1,1}$ and $D_{2,2}$ are known, $D_{1,2}$ may be calculated from the simple equation

$$D_{1,2} \approx \sqrt{D_{1,1}D_{2,2}} \tag{7.54}$$

This equation gives satisfactory results, provided the numerical values of $D_{1,1}$ and $D_{2,2}$ are similar; for example, for the CO–O_2 or He–H_2 systems, the error of estimation does not exceed 5%. In cases where $D_{1,1}$ differs appreciably from $D_{2,2}$ (for example in the system CO_2–H_2), the error in the calculated value of $D_{1,2}$ may be as high as 25%.

Othmer and Hung Tsung Chen's Equation

Othmer and Hung Tsung Chen[272] considered the similarity of the temperature dependence of diffusivity and viscosity, and transformed Eq. (7.23), introducing in place of the average diameter of the molecules of the components a quantity $(V_{c1}^{0\cdot4}+V_{c2}^{0\cdot4})$ derived from the critical volumes.

After a series of transformations, these authors arrive at the equation

$$D_{1,2}p = 2{\cdot}52\times10^{7}\mu^{2\cdot74}\left[\frac{\left(\dfrac{1}{M_1}+\dfrac{1}{M_2}\right)^{0\cdot5}}{(V_{c1}^{0\cdot4}+V_{c2}^{0\cdot4})^2}\right]^{1\cdot23} \tag{7.55}$$

where μ is the viscosity of air (cP) at the temperature of diffusion,
V_{c1} and V_{c2} are the molal critical volumes of diffusing gases 1 and 2,
and M_1 and M_2 are the molecular weights of gases 1 and 2.

The nomograph shown in Fig. 7.4 is based on this equation and it provides a rapid method for solving it.

The use of this nomograph is illustrated in the small inset at the bottom right hand corner of the figure. The numbers on this inset indicate the sequence in which the points on the nomograph are located. The use of the nomograph is also illustrated in Example 7.1 given below. This is the most accurate of the empirical methods for predicting mixture diffusivities, and may be applied to mixtures containing polar gases.

EXAMPLE 7.1. Estimate the diffusion coefficient $D_{1,2}$ (0°C, 1 atm) for the H_2O–H_2 system at 0°C and atmospheric pressure using (a) Andrussov's equations, (b) Kuvshinskii's equation, and (c) the nomograph of Othmer and Chen.

Available data:[5]

Gas	M_i	$V_i^{1/3}$	D_{ii}
H_2	2·02	2·00	1·6040
H_2O	18·02	2·46	0·2763

The experimental value of $D_{1,2}$ is 0·747 cm²/sec.

Solution

a. Application of the Andrussov's equations

(i) Applying Eq. (7.51)

$$D_{1,2} = \frac{17{\cdot}2(1+\sqrt{2{\cdot}02+18{\cdot}02})}{1\times(2{\cdot}00+2{\cdot}46)^2\times\sqrt{2{\cdot}02\times18{\cdot}02}} = 0{\cdot}786 \text{ cm}^2\text{/sec}$$

The error in the estimated value is +5·2%.

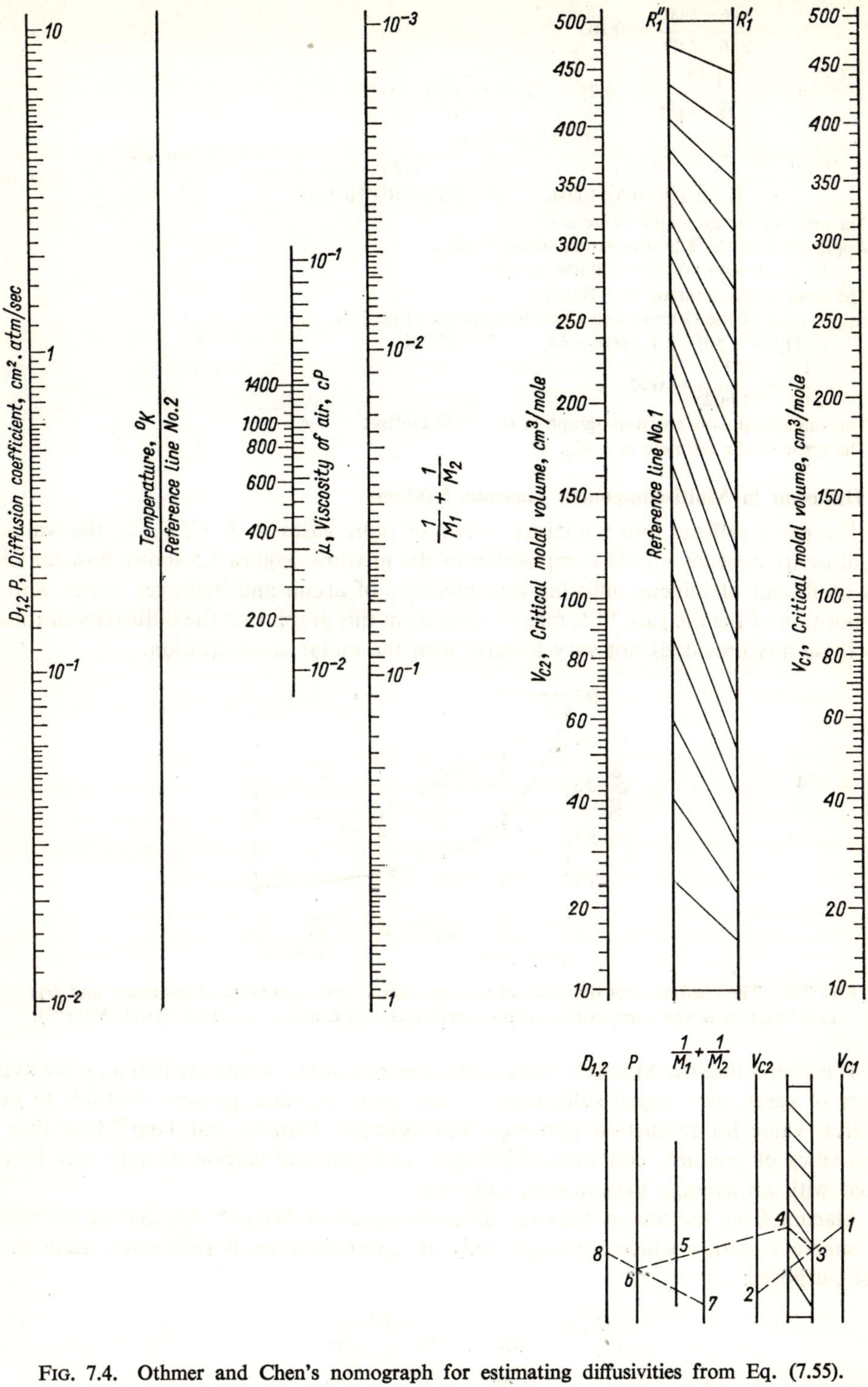

FIG. 7.4. Othmer and Chen's nomograph for estimating diffusivities from Eq. (7.55).

(ii) Applying Eq. (7.53)

$$\delta = \frac{2\cdot46-2\cdot00}{2\cdot46+2\cdot00} = 0\cdot103$$

$$\beta_1 = \frac{\sqrt{18}}{\sqrt{18}+\sqrt{2}} = 0\cdot75 \qquad \beta_2 = 1-0\cdot75 = 0\cdot25$$

$$D_{1,2} = \frac{1+0\cdot5\times(0\cdot103)^2}{\dfrac{0\cdot75}{[1+0\cdot25(0\cdot103)]\,1\cdot6040}+\dfrac{0\cdot25}{[1+0\cdot75(0\cdot103)]\,0\cdot2763}} = 0\cdot776 \text{ cm}^2/\text{sec}$$

The error in the estimated value is $+3\cdot9\%$.

b. Application of the Kuvshinskii equation (7.54)

$D_{1,2} = 1\cdot6040\times0\cdot2763 = 0\cdot666$ cm²/sec

The error in the estimate is $-10\cdot8\%$.

c. Application of the Othmer and Chen nomograph (Fig. 7.4)

$V_c(H_2O) = 56, \quad V_c(H_2) = 65, \quad T = 273°K$

$$\frac{1}{2\cdot02}+\frac{1}{18\cdot02} = 0\cdot55$$

The value read from the nomograph is $D = 0\cdot78$ cm²/sec.

The error in the estimate is $+4\%$.

c. Diffusion in Multi-component Gaseous Systems

If a gas A diffuses into a mixture of two or more gases $A, B, C, D, \ldots$, the value of its diffusivity depends on the composition of the mixture. Figure 7.5 shows how the diffusion coefficient of toluene diffusing into mixtures of argon and hydrogen varies with the composition of the mixture.[401] It may be seen from this graph that the diffusivity of toluene into these mixtures does not vary linearly with the molar concentration.

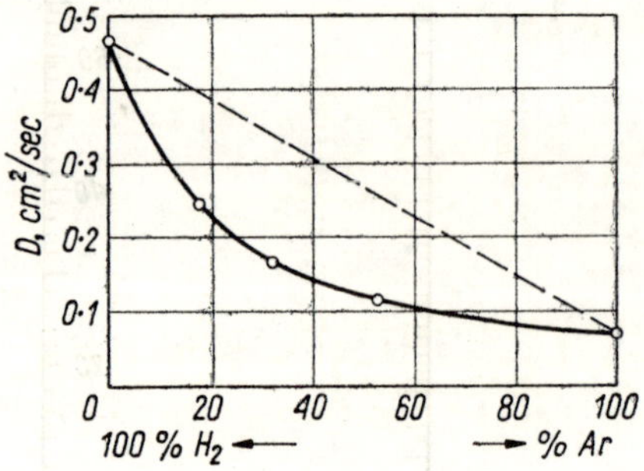

FIG. 7.5. The coefficient of diffusion of toluene vapour into a mixture of hydrogen and argon as a function of the composition of the mixture ($t = 0°C$ and $p = 1$ atm) (after Wilke[401]).

It is found that the Maxwell–Stefan diffusion equations, which are based on the kinetic theory of gases, are obeyed sufficiently closely by many ideal gaseous mixtures to be of practical value for prediction purposes. For example, Duncan and Toor[80] find that the diffusivities of mixtures containing hydrogen, nitrogen and carbon dioxide may be estimated with an average deviation of only 0·4%.

Starting from the Stefan–Maxwell diffusion equation, Wilke[401] has derived the following simple equation which although only an approximation is sufficiently accurate for most purposes:

$$D'_A = \frac{1-y_A}{\dfrac{y_B}{D_{AB}}+\dfrac{y_C}{D_{AC}}+\dfrac{y_D}{D_{AD}}+\ldots} \tag{7.56}$$

In this equation D'_A is the diffusion coefficient of component A into a mixture of components $B, C, D \ldots$,

y_A, y_B, y_C are the mole fractions of components $A, B, C \ldots$ in the mixture,

and D_{AB}, D_{AC} and D_{AD} are the diffusion coefficients for the binary systems AB, AC and AD, respectively.

Wilke and Fairbanks[401] find that Eq. (7.56) is obeyed with a relatively high accuracy. In Fig. 7.5 the curve corresponding to Eq. (7.56) (full line) passes almost exactly through the experimental points.

d. Examples of the Prediction of Gaseous Diffusion Coefficients

EXAMPLE 7.2. Estimate the diffusion coefficient $D_{1,2}$ for the binary system benzene–carbon dioxide at a temperature of 45°C ($T = 318{\cdot}2$°K) and atmospheric pressure.

Solution

The following physico-chemical data are first obtained from standard tabulations

(i) Molecular weights

Benzene	$M_1 = 78{\cdot}11$
Carbon dioxide	$M_2 = 44{\cdot}01$

(ii) Contributions for calculating the molal volume at the boiling temperature are taken from Table 1.19

C Carbon	14·8 cm³/mole
H Hydrogen	3·7 cm³/mole
Six-membered ring	15 cm³/mole
CO_2 Carbon dioxide	34·0 cm³/mole

(iii) Boiling temperatures

Benzene	$T_{b1} = 353{\cdot}3$°K
Carbon dioxide	$T_{b2} = 194{\cdot}7$°K

(iv) Critical constants (Table 1.31)

Benzene	$T_{c1} = 561{\cdot}7$°K,	$p_c = 47{\cdot}7$ atm,
Carbon dioxide	$T_{c2} = 304{\cdot}3$°K,	$p_c = 73{\cdot}0$ atm

The following molal volumes may now be calculated:

Benzene $V_{b1} = 6\times 14{\cdot}8 + 6\times 3{\cdot}7 - 15 = 96$ cm³/mole

CO_2 $V_{b2} = 34{\cdot}0$ cm³/mole

a. Gilliland's method

$$D_{1,2} = \frac{0{\cdot}0043\times(318{\cdot}2)^{3/2}}{1\times(96^{1/3}+34^{1/3})^2}\times\left(\frac{78{\cdot}11+44{\cdot}01}{78{\cdot}11\times 44{\cdot}01}\right)^{1/2} = 0{\cdot}0748 \text{ cm}^2/\text{sec}$$

b. Arnold's method

$V_{b1}/V_{b2} = 96/34 = 2{\cdot}82$

The value of F is now interpolated from the data taken from the respective table

$F = 0{\cdot}96$

$C_{1,2} = 1{\cdot}47\times 0{\cdot}96\times\sqrt{353{\cdot}3\times 194{\cdot}7}$

$C_{1,2} = 368$

$$D_{1,2} = \frac{0{\cdot}00837\times(318{\cdot}2)^{5/2}}{1\times(96^{1/3}+34^{1/3})^2(318{\cdot}2+368)}\left(\frac{78{\cdot}11+44{\cdot}01}{78{\cdot}11\times 44{\cdot}01}\right)^{1/2} = 0{\cdot}0678 \text{ cm}^2/\text{sec}$$

c. Slattery's method

$$T_r = \frac{318{\cdot}2}{\sqrt{561{\cdot}7\times 304{\cdot}3}} = 0{\cdot}770$$

$$D_{1,2} = \frac{2{\cdot}74\times 10^{-4}\times 0{\cdot}77^{1{\cdot}023}\times 47{\cdot}7^{1/3}\times 73^{1/3}\times 561{\cdot}7^{5/12}\times 304{\cdot}3^{5/12}}{1}\times$$

$$\times\left(\frac{78{\cdot}11+44{\cdot}01}{78{\cdot}11\times 44{\cdot}01}\right)^{1/2} = 0{\cdot}0664 \text{ cm}^2/\text{sec}$$

Comparison of the results

According to Table 7.3, the diffusion coefficient in the benzene–carbon dioxide system is

$D_{1,2} = 0{\cdot}0715\ \text{cm}^2/\text{sec}$ (45°C, 1 atm)

	Method	$D_{1,2}$	Error
(a)	Gilliland	0·0748	+4·6%
(b)	Arnold	0·0678	−5·2%
(c)	Slattery	0·0664	−6·9%

EXAMPLE 7.3. Estimate by means of the method of Hirschfelder, Bird and Curtiss, the value of the kinematic diffusion coefficient $D_{1,2}$ in the nitrogen–carbon monoxide system at 15°C and atmospheric pressure.

Solution

Force constants (Table 1.16)

Nitrogen $\varepsilon_1/k = 91{\cdot}5°\text{K}$ $\sigma_1 = 3{\cdot}681\ \text{Å}$

Carbon monoxide $\varepsilon_2/k = 110{\cdot}3°\text{K}$ $\sigma_2 = 3{\cdot}590\ \text{Å}$

Molecular weights

Nitrogen $M_1 = 28$

Carbon monoxide $M_2 = 28$

$$\frac{\varepsilon_{1,2}}{k} = \sqrt{\frac{\varepsilon_1\varepsilon_2}{k^2}} = \sqrt{91{\cdot}5\times 110{\cdot}3} = 100{\cdot}4°\text{K}$$

$$T^* = \frac{kT}{\varepsilon_{1,2}} = \frac{273+15}{100{\cdot}4} = 2{\cdot}87$$

Interpolation from Table 3.1 gives for $T^* = 2{\cdot}87$, $\Omega^{(1,1)*} = 0{\cdot}960$

Correction factor (Table 3.3) $f_D = 1{\cdot}0035$

$$\sigma_{1,2} = \frac{3{\cdot}681+3{\cdot}390}{2} = 3{\cdot}635$$

$$D_{1,2} = \frac{0{\cdot}002628\times 288^{3/2}}{1\times(3{\cdot}635)^2\times 0{\cdot}960}\left(\frac{28+28}{2\times 28\times 28}\right)^{1/2}\times 1{\cdot}0035 = 0{\cdot}197\ \text{cm}^2/\text{sec}$$

The value determined experimentally is 0·211 cm²/sec.

The error in the estimate is thus −6·6%.

EXAMPLE 7.4 (Ref. 211). Estimate by means of Kuong's nomograph (Fig. 7.1) and by the method of Hirschfelder, Bird and Curtiss (Eq. (7.25)), the diffusivity in the carbon dioxide (1)–carbon monoxide (2) system at a temperature of 80°C and a pressure of 2 atm.

Solution

From Table 1.16

Carbon dioxide $\sigma_1 = 3{\cdot}996\ \text{Å}$, $\varepsilon_1/k = 190°\text{K}$

Carbon monoxide $\sigma_2 = 3{\cdot}590\ \text{Å}$, $\varepsilon_2/k = 110°\text{K}$

Molecular weights $M_1 = 44$, $M_2 = 28$

Estimation of the force constants for the mixture:

$\sigma_{1,2} = (3{\cdot}996+3{\cdot}590)/2 = 3{\cdot}79\ \text{Å}$

$\varepsilon_{1,2}/k = [(190)(110)]^{1/2} = 144°\text{K}$

Estimation of $T^* = kT/\varepsilon_{1,2}$, $T = 273+80 = 353°\text{K}$

$T^* = 353/144 = 2{\cdot}45$

For $T^* = 2{\cdot}45$, Table 3.1 gives (after interpolation)

$\Omega^{(1,1)*} = 1{\cdot}007$

$$\left(\frac{M_1+M_2}{2M_1M_2}\right)^{1/2} = \left(\frac{44+28}{2\times 44\times 28}\right)^{1/2} = 0{\cdot}242$$

Operations on the nomograph:

(1) Connection of $P = 2$ with $T = 353$ determines the point on the Q-axis.
(2) Connection of the point on the Q-axis with $\sigma = 3{\cdot}79$ determines the point on the R-axis.
(3) Connection of the point on the R-axis with $\Omega = 1{\cdot}007$ determines the point on the S-axis.
(4) Connection of the point on the S-axis with $M = 0{\cdot}242$ determines $D = 0{\cdot}103\ \text{cm}^2/\text{sec}$.
(Using Eq. (7.25) we also obtain $D = 0{\cdot}103$).

EXAMPLE 7.5. Estimate the value of diffusivity $D_{1,2}$ for the carbon dioxide–air system at 1300°K and atmospheric pressure. Make use of the value of $D_{1,2} = 0{\cdot}151$ cm²/sec (20°C, 1 atm) given in Table 7.4. Employ Eq. (7.39) and disregard the correction (f_{T2}/f_{T1}).

Solution

Force constants

Air $\varepsilon_1/k = 97\ 0$

Carbon dioxide $\varepsilon_2/k = 190$

$$\frac{\varepsilon_{1,2}}{k} = 135{\cdot}4$$

Table 3.1 gives for

$$T_1^* = \frac{kT_1}{\varepsilon_{1,2}} = \frac{293}{135{\cdot}4} = 2{\cdot}162, \qquad \Omega_{T_1}^{(1,1)*} = 1{\cdot}047$$

and for $$T_2^* = \frac{kT_2}{\varepsilon_{1,2}} = \frac{1300}{135{\cdot}4} = 9{\cdot}58, \qquad \Omega_{T_2}^{(1,1)*} = 0{\cdot}7491$$

Hence, $$D_{1300} = 0{\cdot}151\left(\frac{1300}{293}\right)^{3/2} \times \frac{1{\cdot}047}{0{\cdot}7491} = 1{\cdot}96 \text{ cm}^2/\text{sec}.$$

According to Table 7.4, the experimental value of D is 1·97 cm²/sec.

The error in the estimate is thus −0·5%.

EXAMPLE 7.6. Estimate the diffusion coefficient $D_{1,2}$ in the carbon monoxide–nitrogen system at a temperature of 50°C and a pressure of 105 atm. According to Table 7.4 the value of the coefficient at 15°C and a pressure of 1 atm is $D_{15}^{\circ} = 0{\cdot}211$ cm²/sec.

Critical constants (Table 1.31):

Nitrogen $T_{c1} = 126$°K, $p_{c1} = 33{\cdot}5$ atm

Carbon monoxide $T_{c2} = 134{\cdot}5$°K, $p_{c2} = 34{\cdot}6$ atm.

Solution

Pseudocritical parameters ($x_1 = x_2 = 0{\cdot}5$) are

$$T_{pc} = 0{\cdot}5\times 126 + 0{\cdot}5\times 134{\cdot}5 = 130{\cdot}2\text{°K}$$

$$p_{pc} = 0{\cdot}5\times 33{\cdot}5 + 0{\cdot}5\times 34{\cdot}6 = 34{\cdot}0 \text{ atm}$$

Diffusion coefficient at 50°C, 1 atm (Eq. (7.37))

$$D_{50}^{\circ} = D_{15}^{\circ}\left(\frac{T_2}{T_1}\right)^{3/2} \qquad D_{50}^{\circ} = 0{\cdot}211\left(\frac{273+50}{273+15}\right)^{3/2} = 0{\cdot}236 \text{ cm}^2/\text{sec}$$

Reduced parameters

$$T_r = \frac{323}{130{\cdot}2} = 2{\cdot}48, \qquad p_r = \frac{105}{34} = 3{\cdot}08$$

According to Fig. 7.3

$$(pD)_p/(pD)^{\circ} = 0{\cdot}95$$

$$\frac{105D_p}{1D^{\circ}} = 0{\cdot}95, \qquad D_p = \frac{0{\cdot}95\times 0{\cdot}236}{105} = 0{\cdot}00214 \text{ cm}^2/\text{sec}$$

Note. This estimated value cannot be compared with an experimental value since one is not available. However, the ratio D_p/D° (= 0·00214/0·236 = 0·009) is very close to that for the self-diffusivity of nitrogen under a pressure of 104·5 atm for which $D_p/D^{\circ} = 0{\cdot}0084$; this fact suggests that the estimated result may be close to the actual value.

EXAMPLE 7.7. Estimate the diffusion coefficient D for the vapour of ethyl ether diffusing into a mixture containing 15 vol. % of ethyl ether, 60 vol. % of carbon dioxide and 25 vol. % of hydrogen at a temperature of 0°C and atmospheric pressure.

Solution

The following diffusivities for the constituent binary systems are taken from Table 7.3:

Ether–carbon dioxide $D_{1,2} = 0{\cdot}0552$ cm²/sec

Ether–hydrogen $D_{1,3} = 0{\cdot}296$ cm²/sec

Assuming that the ideal-gas law may be applied to the mixture

$$y_1 = 0{\cdot}15, \quad y_2 = 0{\cdot}60, \quad y_3 = 0{\cdot}25$$

so

$$D_1' = \frac{1-0{\cdot}15}{\dfrac{0{\cdot}60}{0{\cdot}0552}+\dfrac{0{\cdot}25}{0{\cdot}296}} = \frac{0{\cdot}85}{11{\cdot}69} = 0{\cdot}0728 \text{ cm}^2/\text{sec}.$$

5. CHOICE OF THE BEST METHOD FOR ESTIMATING GASEOUS DIFFUSION COEFFICIENTS

If the force constants σ and ε of the molecules of the diffusing gases are known, the most accurate method is to use Eq. (7.25) which was developed by Hirschfelder *et al.* According to Reid and Sherwood,[306] the average error in diffusivities estimated in this way is of the order of $\pm 6\%$. Wilke and Lee[402] estimate the average error to be $\pm 7\%$. This error is reduced to 3·9% when Eq. (7.26) (which contains a correction factor given by Eq. (7.27)) is used.

The average error in the estimated diffusivities increases considerably if the values of the force constants σ and ε are unknown, and have to be estimated from critical data or other physical properties such as the boiling temperature. In such cases diffusivities estimated from Eq. (7.25) are subject to an average error of 10–12%, while those estimated from Eqs. (7.26) and (7.27) are subject to an average error of 5–9%. The maximum error in diffusivities predicted from these equations is then of the order of 20–35%.

If no physico-chemical constants at all are known for the diffusing gases (excluding of course the structural formulae of the molecules), the only equation which may be applied is the semi-empirical equation (7.44) of Gilliland. The quantities appearing in this equation must then be calculated by the method of group contributions. The average error in diffusivities obtained from Gilliland's equation is variously estimated to be 16% (Ref. 306) and up to 20% (Ref. 402) (the maximum error is 46·8%). When the normal boiling points of the diffusing compounds are known, Arnold's equation (7·45) should be used. This equation is only insignificantly inferior in accuracy to Hirschfelder's equation (if the values of σ and ε are unknown), but is not so troublesome in use. The average error in estimates obtained from Arnold's equation is about 10%.[306] Wilke and Lee estimate the average error to be 8·4%, and the maximum error to be 20·5%.

Results of the same accuracy may be obtained using Slattery's method (Eq. (7.50)). The use of this equation, however, requires knowledge of the critical constants of the diffusing compounds. The authors estimate the average error to be 7%, while Reid and Sherwood estimate it to be 10%.

Chen and Othmer[258] have calculated by various methods the diffusivities of 71 pairs of gases at temperatures close to room temperature. The deviations of the calculated values from those determined experimentally are listed in Table 7.8.

It may be seen from this Table that when the values of the critical volumes V_{c1} and V_{c2} are known, good results are obtained using the nomograph of Othmer and Chen.

If the diffusion coefficient is known at one temperature and it is necessary to predict the value of this coefficient at another temperature, the calculation may be performed most accurately by using Eq. (7.39). If the range of temperatures considered is not large (say only a degree or two), sufficiently accurate results may be obtained by employing Eq. (7.37).

TABLE 7.8. COMPARISON OF THE ACCURACY OF SEVERAL METHODS FOR PREDICTING GASEOUS DIFFUSIVITIES (AFTER REFS. 258, 272)

Method	Equation	Average absolute deviation %	Maximum absolute deviation %	Average absolute deviation for the system air–water (293–1493°K)
Arnold	(7·45), (7·46)	16·01	90·5	18·4
Gilliland	(7·44)	16·80	60·2	39·3
Hirschfelder, Bird, Spotz (σ, ε/k estimated)	(7·25) (1·33), (1·34)	11·23	30·1	23·5
Hirschfelder, Bird, Spotz (σ, ε/k viscosity)	(7·25)	6·12	17·9	5·64
Slattery	(7·50)	9·90	137	82·1
Chen, Othmer	(7·31)	5·64	25·2	8·9
Othmer, Chen (nomogram)	(7·55) Fig. 7.4	5·75	—	7·6

Equation (7.38) containing Sutherland's constants is slightly inferior in accuracy to Eq. (7.39), but when temperature differences do not exceed a few hundred degrees it is sufficiently accurate for technological calculations.

If the diffusivity is known at a "moderate" pressure (i.e. in the range in which $pD =$ $=$ constant), and it is necessary to predict the value of this coefficient at a high pressure and the same temperature, use may be made of the graph presented in Fig. 7.3. Unfortunately, this method is liable to be very inaccurate, though it is the only one so far proposed which is suitable for use in technological calculations.

When estimating diffusion coefficients in multi-component mixtures it is best to apply Wilke's equation (7.56), which yields quite satisfactory results.

6. SCHMIDT NUMBER

Just as the Prandtl number characterizes those physical properties of a fluid which are of importance in the processes of heat transfer, so the *Schmidt number*, Sc, is characteristic of the physical properties of two components 1 and 2, which are of importance in a mass transfer process involving these components. If gas 1 diffuses into gas 2, then

$$\mathrm{Sc}_{1,2} = \mu_2/\rho_2 D_{1,2} \tag{7.57}$$

where μ_2 is the viscosity of gas 2,
ρ_2 is the density of gas 2,
and $D_{1,2}$ is the diffusion coefficient of gases 1 and 2.

In some publications $\mathrm{Sc}_{1,2}$ is referred to as the *diffusional Prandtl number* with the symbol Pr_d instead of the Schmidt number.

According to the simple kinetic theory of gases, Sc = Pr = 1. However, in the more exact theory, the value of the Schmidt number is found to be equal to 0·83 (whereas the Prandtl number for various gases takes values ranging from 0·65 to 0·9).

The value of the Schmidt number is usually calculated by finding the values of ρ, μ and D, and substituting the values in Eq. (7.57). This value depends, however, on the composition of the gas mixture.

At "moderate" pressures (i.e. in the range in which viscosity does not depend on pressure and in which pD = constant), μ and D may be eliminated by introducing the relationship of Hirschfelder *et al.* (Eq. (3.22)) for μ, and Eq. (7.25) for D into Eq. (7.57). Assuming the validity of the perfect-gas law, the density ρ_2 is given by

$$\rho_2 = \frac{M_2 p}{82{\cdot}06T} \tag{7.58}$$

Substitution of Eqs. (7.58), (3.22) and (7.25) into Eq. (7.57) leads to the equation

$$\mathrm{Sc}_{1,2} = 1{\cdot}18\left(\frac{\sigma_{1,2}}{\sigma_2}\right)^2\left(\frac{\Omega_{1,2}^{(1,1)*}}{\Omega_2^{(2,2)*}}\right)\left(\frac{M_1}{M_1+M_2}\right)^{0{\cdot}5} \tag{7.59}$$

where σ_1 and ε_1 are the Lennard-Jones force constants for molecules of type 1,
σ_2 and ε_2 are the Lennard-Jones force constants for molecules of type 2,
$\sigma_{1,2}$ and $\varepsilon_{1,2}$ are force constants describing the interactions between molecules of type 1 and 2. These may be estimated using Eq. (1.39),
$\Omega_{1,2}^{(1,1)*}$ is a function of $T^* = kT/\varepsilon_{1,2}$ and is taken from Table 3.1,
$\Omega_2^{(2,2)*}$ is a function of $T^* = kT/\varepsilon_2$ and is taken from Table 3.1,
and M_1 and M_2 are the molecular weights of gases 1 and 2.

The application of Eq. (7.59) is illustrated by the following example.

EXAMPLE 7.8. Estimate the Schmidt number for the diffusion of benzene into air at $T = 318$°K and $p = 1$ atm. (Molecular weights: M_1 (benzene) = 78, M_2 (air) = 29.)

Solution

Force constants: (Table 1.16)

Benzene	$\sigma_1 = 5{\cdot}270$ Å	$\varepsilon_1/k = 440°$
Air	$\sigma_2 = 3{\cdot}617$ Å	$\varepsilon_2/k = 97°$

$$\sigma_{1,2} = \frac{5{\cdot}270+3{\cdot}617}{2} = 4{\cdot}443 \text{ Å}$$

$$kT/\varepsilon_{1,2} = \frac{318}{\sqrt{440\times 97}} = 1{\cdot}542$$

$$kT/\varepsilon_2 = \frac{318}{97} = 3{\cdot}276$$

Interpolation from Table 3.1 gives

for $kT/\varepsilon_{1,2} = 1{\cdot}542$, $\Omega_{1,2}^{(1,1)*} = 1{\cdot}188$
and for $kT/\varepsilon_2 = 3{\cdot}276$, $\Omega_2^{(2,2)*} = 1{\cdot}016$

Substituting the above values in Eq. (7.59)

$$\mathrm{Sc}_{1,2} = 1{\cdot}18\left(\frac{4{\cdot}443}{3{\cdot}617}\right)^2\times\left(\frac{1{\cdot}188}{1{\cdot}016}\right)\times\left(\frac{78}{78+29}\right)^{0{\cdot}5} = 1{\cdot}78$$

This may be compared with the "experimental" value of the Schmidt number, as calculated from the following data:

The dynamic viscosity of air (0°C, 1 atm) $\mu_0^\circ = 1708\times 10^{-5}$ cP,
Sutherland's constant $C = 112$.
The viscosity at 318°K, 1 atm:

$$\mu^\circ = 0{\cdot}01708\left(\frac{273+112}{318+112}\right)\left(\frac{318}{273}\right)^{3/2} = 0{\cdot}0192 \text{ cP}$$

$$\mu_2^\circ = 0{\cdot}0192 \text{ cP}$$

The density of air at 0°C, 1 atm $\rho_0 = 0{\cdot}001293$ g/cm³

The density of air at 318°K, 1 atm $\rho_2 = \dfrac{0{\cdot}001293 \times 273}{318} = 0{\cdot}00111$ g/cm³.

The diffusion coefficient of benzene into air (at 318°K, 1 atm):

$D_{1,2} = 0{\cdot}1011$ cm²/sec.

Substituting the numerical data in Eq. (7.57) gives

$$\mathrm{Sc}_{1,2} = \frac{0{\cdot}000192}{0{\cdot}00111 \times 0{\cdot}1011} = 1{\cdot}72$$

The error in the estimated value obtained from Eq. (7.59), is thus 3·5%.

CHAPTER 8

DIFFUSION IN LIQUIDS

1. EXPERIMENTAL DATA

Diffusion processes in liquids are many times slower than in gases, the diffusion coefficient D (defined as for gases by Fick's law (Eq. 7.8) being 10^4 to 10^5 times smaller. For this reason liquid diffusivities are often expressed not in cm^2/sec, but in $cm^2/24$ hr. (1 $cm^2/sec = 86{,}400$ $cm^2/24$ hr).

There are several groups of liquid systems for which a knowledge of diffusion rates may be of practical importance. For example, solutions of non-electrolytes and of electrolytes, fused salts and molten metals. Diffusion phenomena in systems containing dissolved ionized electrolytes should strictly be considered separately since in this case it is the ions which diffuse and not the molecules. Illustrative values of the diffusion coefficients for all the above types of system are given in this section.

It is found experimentally that binary liquid diffusivities depend on concentration as well as on temperature so the concentration c has been listed alongside the temperature t in the diffusivity tables in this section.

Table 8.1 lists values of the diffusion coefficients of non-electrolytes in aqueous solutions. Table 8.2 gives diffusivities for a number of compounds dissolved in methanol, Table 8.3 is for compounds dissolved in benzene, and Table 8.4 for compounds dissolved in other solvents. The data listed in these tables are mainly from Wilke[399] but are supplemented with data taken from other sources,[11,182,306] and refer mainly to organic compounds.

Representative values of diffusivities in fused salts are listed in Table 8.5 while some values for amalgams are given in Table 8.6. Table 8.7 lists diffusion coefficients for some molten metals. The large values of these coefficients at high temperatures in fused salts and molten metal systems should be noted.

The values of the diffusion coefficients in aqueous solutions of some electrolytes are listed in Table 8.8. The concentrations in this table are expressed as normalities, i.e. as the number of gram-equivalents of the compounds in 1 l. of the solution. The diffusion coefficients D are expressed in $cm^2/24$ hr.

Comparison of the data in Table 8.8 shows irregularities in the dependence of the coefficient D upon concentration. In the case of electrolyte solutions, the value of D sometimes increases with increase in concentration, and sometimes decreases (in the case of non-electrolytes it always decreases, see below).

The effect of concentration on the diffusion coefficients of some non-electrolytes dissolved in water and in ethanol is illustrated both in Table 8.1 and in Fig. 8.1. Figure 8.1 shows the diffusivities D of some carbohydrates in aqueous solutions as functions of concentration.

TABLE 8.1. DIFFUSION COEFFICIENTS FOR AQUEOUS SOLUTIONS OF SOME NON-ELECTROLYTES (REFS. 216, 399)

Solute	t, °C	c, mole/l.	$D\times10^5$, cm²/sec
Acetamide	20	0	1·05
Acetic acid	12·5	0·01	0·91
Acetonitrile	15	0	1·26
Alloxan	20	0	0·67
Allyl alcohol	15	0	0·90
Ammonia	15	1	1·78
iso-Amyl alcohol	15	0	0·69
Arabinose	20	0	0·70
Bromine	12	0·1	0·90
n-Butyl alcohol	15	0	0·77
Caffeine	10	0	0·42
Carbon dioxide	18	0	1·71
Chlorine	12	0·1	1·40
Ethanol	15	0	1·00
Glucose	15	0	0·52
Glycerol	10	0·125	0·63
Hydrogen	25	0	3·36
Hydrogen sulphide	15	0	1·43
Hydroquinone	20	0	0·78
Iodine	25	0·05	1·25
Lactose	20	0	0·46
Maltose	20	0	0·426
Maltose	10	0	0·328
Mannitol	20	0	0·605
Methanol	15	0	1·28
Nicotine	20	0	0·53
Nitrogen	22	0	2·02
Nitrous oxide	16	0	1·54
Oxygen	25	0	2·60
Pentaerythritol	20	0	0·69
Propanol	15	0	0·87
Pyrogallol	15	0	0·56
Raffinose	20	0·05	0·36
Resorcinol	20	0	0·77
Saccharose	12	0	0·36
Tartaric acid	15	0	0·61
Urea	12	0	0·99
Urethane	15	0	0·87

TABLE 8.2. DIFFUSION COEFFICIENTS IN METHANOL FOR A SOLUTE CONCENTRATION OF 1% AND A TEMPERATURE OF 15°C (AFTER WILKE[399])

Solute	$D\times10^5$, cm²/sec	Solute	$D\times10^5$, cm²/sec
Acetanilide	1·50	Bromobenzene	1·75
Acetonitrile	2·64	Bromonitrobenzene	1·43
Acetyl diphenylamine	0·98	Bromophenol	1·34
Allyl alcohol	1·80	Bromopropionic acid	1·35
iso-Amyl alcohol	1·34	3-Bromopropylene	2·22
Aniline	1·49	Carbon tetrachloride	1·70
Benzaldehyde	1·66	Chloral hydrate	1·16
Bromoaniline	1·41	Chloroacetic acid	1·52

Continued overleaf

TABLE 8.2 (*cont.*)

Solute	$D \times 10^5$, cm²/sec	Solute	$D \times 10^5$, cm²/sec
Chloroaniline	1·34	Hydroquinone	1·25
Chlorobenzoic acid	1·29	Iodobenzene	1·65
Chlorobromomethane	2·50	Iodoform	1·33
Chloroform	2·07	Iodopropionic acid	1·36
1-Chlorohydrin	1·30	3-Iodopropylene	1·72
Chloronitrobenzene	1·68	Lactic acid	1·36
Chlorophenol	1·32	*m*-Nitrobenzaldehyde	1·24
Dibenzylamine	0·86	Nitrophenol	1·38
p-Dibromobenzene	1·55	Phenetole	1·83
Dibromonaphthalene	1·33	Phenol	1·40
Dichloroacetic acid	1·36	Phenolphthalein	0·78
1,1-Dichlorohydrin	1·36	Phenyl acetate	1·62
Dichloronaphthalene	1·52	Phthalic acid	1·30
Diiodomethane	1·68	Propionic acid	1·62
m-Dinitrobenzene	1·56	Pyridine	1·58
Dinitronaphthalene	1·32	Salol	1·29
2,4-Dinitrophenol	1·40	Tribromoacetic acid	1·23
Ethyl bromide	2·40	2,4,6-Tribromophenol	1·12
Ethyl iodide	2·16	Trichloroacetic acid	1·45
Ethyl nitrate	2·20	2,4,6-Trichlorophenol	1·21
Ethylene bromide	1·95	2,4,6-Trinitrophenol	1·32
Ethylene iodide	1·56	Urethane	1·41
Furfural	1·70	Vanillin	1·00

TABLE 8.3. DIFFUSION COEFFICIENTS IN BENZENE FOR A SOLUTE CONCENTRATION OF 1% AND A TEMPERATURE OF 15°C (REFS. 11, 399)

Solute	$D \times 10^5$, cm²/sec	Solute	$D \times 10^5$, cm²/sec
Acetyl diphenylamine	0·90	Ethyl ether	2·21
iso-Amyl alcohol	1·48	Ethylene bromide	1·97
Benzaldehyde	1·73	Ethylene iodide	1·40
Benzoic acid	1·36	Hexachlorobenzene (0%, 8°C)	1·02
Bromine (0·1 molar, 12°C)	2·00	Iodine (0·05 molar, 20°C)	1·95
Bromoaniline	1·41	Iodobenzene (0%, 8°C)	1·35
Bromobenzene	1·86	Iodoform	1·38
Bromoform (18°C)	1·62	Methyl salicylate	1·56
α-Bromonaphthalene	1·30	Naphthalene (0%, 8°C)	1·19
Bromonitrobenzene	1·33	Nitrobenzene	1·84
Bromophenol	1·34	α-Nitronaphthalene	1·39
Chloroacetic acid	1·48	Phenanthrene (0%, 8°C)	0·95
Chloroaniline	1·56	Phenol (5°C)	1·27
Chlorobenzene	1·42	Phthalic acid	1·37
Chloroform	2·11	Propanol	1·60
α-Chloronaphthalene	1·20	Quinone	1·68
Chloronitrobenzene	1·70	Salicylaldehyde	1·78
Chlorophenol	1·42	1,2,4,5-Tetrachlorobenzene (0%, 8°C)	
p-Dibromobenzene	1·37		1·24
Dibromonaphthalene	1·25	1,2,3-Trichlorobenzene (0%, 8°C)	
p-Dichlorobenzene	1·90		1·34
Dichloronaphthalene	1·40	1,2,3-Trichloropropane	1·72
m-Dinitrobenzene	1·54	2,4,6-Trinitrophenol	1·39
Dinitronaphthalene	1·23	Trinitrotoluene	1·39

TABLE 8.4. DIFFUSION COEFFICIENTS IN VARIOUS SOLVENTS (REFS. 11, 175, 399)

Solvent	Solute	Concentration mole/l.	t, °C	$D \times 10^5$, cm²/sec
Acetic acid	iodine	0·1	20	1·03
Acetone	bromoform	0	20	2·73
iso-Amyl alcohol	phenol	0	19	0·20
Anisole	iodine	0·1	20	1·13
Bromoform	acetone	0·25	20	0·272
Bromoform	amyl alcohol	1	20	0·516
Bromoform	benzene	1	20	1·69
Bromoform	ethyl alcohol	0·5	20	0·976
Bromoform	ethyl ether	0·25	20	0·342
Bromoform	methanol	0·5	20	1·930
Bromoform	propyl alcohol	1	20	0·770
Carbon disulphide	bromine	0·1	16	3·60
Carbon disulphide	phenol	0	19	3·40
Carbon tetrachloride	iodine	0·1	20	1·36
Chloroform	iodine	0·05	10	1·93
Chloroform	phenol	0	10	1·60
Cyclohexane	bromobenzene	0	7	1·14
Decalin	α-bromonaphthalene	0	7	0·34
Decalin	bromobenzene	0	7	0·47
Dibenzylether	α-bromonaphthalene	0	7	0·149
Ethyl acetate	iodine	0·1	20	2·15
Ethyl alcohol	acetal	1	20	1·13
Ethyl alcohol	acetamide	0	20	0·67
Ethyl alcohol	acetoin	0	20	0·556
Ethyl alcohol	aconitine	0·04	20	0·27
Ethyl alcohol	allyl alcohol	1	20	0·95
Ethyl alcohol	amyl alcohol	1	20	0·76
Ethyl alcohol	azobenzene	0·2	20	0·740
Ethyl alcohol	bromoform	0·5	20	0·976
Ethyl alcohol	bromonaphthalene	0·5	20	0·740
Ethyl alcohol	brucine	0·062	20	0·28
Ethyl alcohol	camphor	0·5	20	0·694
Ethyl alcohol	chloral	1	20	0·614
Ethyl alcohol	chloroform	1	20	1·25
Ethyl alcohol	glycerol	0	20	0·532
Ethyl alcohol	hydroquinone	0	20	0·488
Ethyl alcohol	iodine	0·1	8	0·90
Ethyl alcohol	iodobenzene	0·25	20	0·970
Ethyl alcohol	phenol	0·1	20	0·80
Ethyl alcohol	pyridine	0·5	20	1·11
Ethyl alcohol	resorcinol	0	20	0·451
Ethyl alcohol	saligenin	0	20	0·608
Ethyl alcohol	stearic acid	0·2	20	0·59
Ethyl ether	bromobenzene	0	7	3·50
Ethyl ether	bromoform	0	17	3·30
Ethyl ether	nitrobenzene	0	8	3·24
Ethyl ether	phenol	0	19	3·60
Ethylbenzene	bromobenzene	0	7	1·44
Heptane	iodine	0·1	20	2·67
Hexane	bromobenzene	0	7	2·60
Tetrachloroethane	tetrabromoethane	0·06	50	0·94
Tetralin	bromobenzene	0	7	0·48

Continued overleaf

TABLE 8.4 (*cont.*)

Solvent	Solute	Concentration mole/l.	*t*, °C	$D \times 10^5$, cm²/sec
Tetralin	α-bromonaphthalene	0	7	0·36
Toluene	bromobenzene	0	7	1·59
Toluene	iodine	0·1	20	1·95
m-Xylene	bromobenzene	0	7	1·51
m-Xylene	iodine	0·1	20	1·68

TABLE 8.5. DIFFUSIVITIES IN FUSED SALTS (REFS. 11, 178)

Diffusing compound	Diffusion into	*t*, °C	$D \times 10^5$, cm²/sec	Diffusing compound	Diffusion into	*t*, °C	$D \times 10^5$, cm²/sec
$AgNO_3$	KNO_3	360	4·56	KBr	KNO_3	360	3·01
$AgNO_3$	KNO_3	390	4·86	KI	KNO_3	360	2·96
$AgNO_3$	$NaNO_3$	330	4·57	$PbCl_2$	KCl, LiCl	530	2·06
$AgNO_3$	$NaNO_3$	360	5·06	$PbCl_2$	KCl, LiCl	720	4·40
AgBr	KBr	780	4·92	$NaNO_3$	KNO_3	360	5·22
AgI	KI	720	4·63	$Sr(NO_3)_2$	KNO_3	360	2·81
AgI	KI	780	5·10	$Sr(NO_3)_2$	$NaNO_3$	345	4·17
AgCl	KCl, LiCl	480	4·61	TlBr	KBr	770	4·28
AgCl	KCl, LiCl	740	6·64	TlCl	KCl, LiCl	520	3·10
$Ba(NO_3)_2$	KNO_3	370	2·06	TlI	KI	720	3·14
$Ba(NO_3)_2$	$NaNO_3$	360	3·71	$TlNO_3$	KNO_3	345	3·17
				$TlNO_3$	$NaNO_3$	360	4·31

TABLE 8.6. DIFFUSIVITIES IN AMALGAMS (AFTER JOLIET[178])

Metal	*t*, °C	$D \times 10^5$, cm²/sec	Metal	*t*, °C	$D \times 10^5$, cm²/sec
Li	25	0·93	Zn	25	2·4
Na	25	0·86	Cd	25	2·0
K	25	0·71	Sn	25	2·1
Rb	7·3	0·53	Bi	25	1·5
Cs	25	0·64	Pb	25	2·1
Ag	16	1·11	Ba	7·8	0·60
Au	25	0·73	Sr	9·4	0·54
Tl	25	1·18	Ca	10·2	0·62

TABLE 8.7. DIFFUSIVITIES IN MOLTEN METALS (AFTER JOLIET[178])

Metals	*t*, °C	$D \times 10^5$, cm²/sec	Metals	*t*, °C	$D \times 10^5$, cm²/sec
Mg into Al	700	7·54	Rh into Pb	500	3·52
Au Bi	500	5·22	Pt Pb	490	1·95
Si Fe	1480	2·4	Ag Sn	500	4·8
Si Fe	1560	10·8	Au Sn	500	5·37
Au Pb	500	3·7	Pb Sn	500	3·68

TABLE 8.8. DIFFUSIVITIES OF SOME ELECTROLYTES IN AQUEOUS SOLUTIONS (AFTER D'ANS[11])

Compound	Concentration g-equiv./l.	*t*, °C	*D*, cm²/24 hr	Compound	Concentration g-equiv./l.	*t*, °C	*D*, cm²/24 hr
$CaCl_2$	0·29	9	0·68	KNO_3	0·05	18	1·25
	0·37	9	0·94		0·80	18	1·10
	1·5	9	0·72		2·5	18	1·01
$CdSO_4$	0	16·8	0·304	KOH	0·1	13·5	1·72
$CoCl_2$	0·0062	18	0·600		0·9	13·5	1·86
$CuCl_2$	1·5	10	0·43		3·9	13·5	2·43
$CuSO_4$	0·10	17	0·39	K_2SO_4	0·02	19·6	1·01
	0·50	17	0·29		0·28	19·6	0·86
	1·95	17	0·23		0·95	19·6	0·79
HCl	0·43	0	1·39	LiBr	2·3	10	0·80
	2·0	0	1·52	LiCl	0·01	9	0·757
	2·0	11	2·12		4·2	9	0·724
	5·0	0	1·86	$MgSO_4$	0·5	15·5	0·461
	6·5	11	2·67		3·0	15·5	0·509
	8	0	2·31		4·5	15·5	0·627
	9	11·5	2·93	NH_3	0·686	4	1·06
HNO_3	0·84	5·5	1·50		3·55	4·5	1·06
	3·0	6	1·54	NaBr	2·9	10	0·86
	20	9	1·94	Na_2CO_3	2·4	10	0·39
H_2O_2	0·011	20	0·76	NaCl	0·05	18	1·09
H_2SO_4	0·005	18	1·30		2·00	18	1·11
	0·35	18	1·32		5·0	18	1·28
	0·85	18	1·34	NaI	1	10	0·80
	2·85	18	1·60	$NaNO_3$	0·6	13	0·90
	4·85	18	1·90		6	13	0·77
	9·85	18	2·36	NaOH	0·1	12	1·11
KBr	1	10	1·13		3·9	12	0·985
K_2CO_3	3	10	0·60	Na_2SO_4	1·4	10	0·66
KCl	0·02	25	1·685	$Ni(NO_3)_2$	0·068	18	0·802
	0·5	25	1·573	$ZnSO_4$	0·55	19·5	0·36
KI	0·01	10	1·117		2·95	19·5	0·33
	5·5	10	1·185				

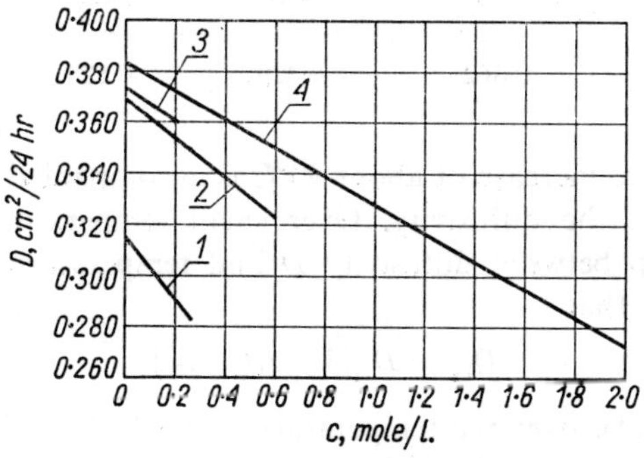

FIG. 8.1. The diffusivities *D* of aqueous solutions of some carbohydrates as functions of solute concentration (after Jellinek[175]); *1*—raffinose, *2*—maltose, *3*—lactose, *4*—saccharose ($t = 20°C$).

Effect of Concentration on the Diffusion Coefficients of some Non-electrolytes Dissolved in Water and in Ethanol. D is expressed in $cm^2/24$ hr
($t = 20°C$)

Solute	Solvent	Concentration, mole/l.						
		10	5	2	1	0·5	0·25	0
Acetamide	water	0·685	0·795	0·860	0·890	0·898	0·900	0·910
Resorcinol	water	—	—	0·524	0·594	0·623	0·652	0·663
Glycerol	ethanol		0·234†	0·304	0·353	0·422	0·431	0·460

† Concentration 4 mole/l.

As it is shown in Table 8.1 and Fig. 1, the diffusion coefficient usually falls in such systems as the concentration of the solute increases.

In the case of electrolyte solutions it frequently occurs that the diffusivity first decreases rapidly and then starts to increase slowly as the concentration is increased. This behaviour is illustrated for aqueous solutions of sodium chloride and potassium chloride at 18·5°C in Fig. 8.2.

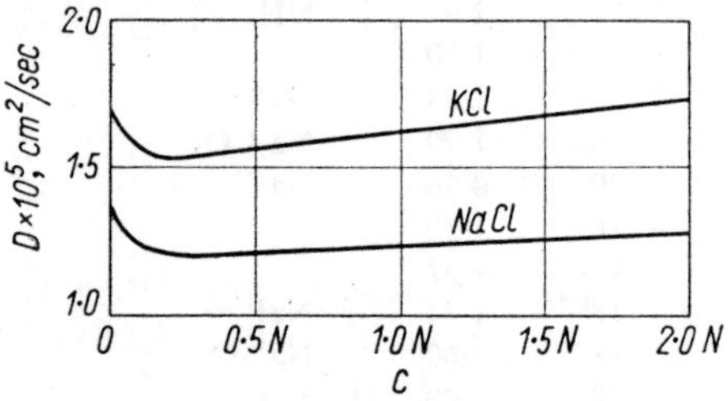

FIG. 8.2. Diffusivity D as a function of concentration for aqueous solutions of NaCl and KCl (after Gordon[133]).

A rise of temperature speeds up the process of diffusion, producing a corresponding rise in the diffusion coefficient. Cohen and Bruins[62] have obtained the following data for the variation of the diffusivity D with temperature t for the diffusion of symmetrical tetrabromoethane into a very dilute solution of this compound in symmetrical tetrachloroethane as solvent:

t, °C	0·44	7·70	15·00	25·00	35·61	51·10
D, $cm^2/24$ hr	0·303	0·362	0·429	0·528	0·640	0·825

In this case, a temperature increase of about 50°C is accompanied by more than a $2\frac{1}{2}$-fold increase in the value of the diffusivity. Over small temperature ranges (e.g. of a few degrees), the relationship between diffusivity D and temperature t may be taken to be approximately linear so that

$$D_{t_2} = D_{t_1}[1+\alpha(t_2-t_1)] \tag{8.1}$$

For many compounds, over the temperature range 10°C to 20°C, it is found that α lies between 0·03 to 0·04, the value being well approximated by the equation

$$\alpha = 0{\cdot}026-0{\cdot}021 \log D \tag{8.2}$$

Both relationships are of an empirical character.

Comparison of the diffusion coefficients listed in Tables 8.1–8.8 shows that the coefficients for different substances diffusing in a given solvent vary widely between each other. Similarly large differences are found when comparison is made of the diffusion coefficients for a given solute in various different solvents. Miller,[250] who investigated the diffusion of iodine in various solvents at a temperature of 20°C, found the following values of D (cm^2/sec):

Heptane	$D = 2{\cdot}386$	Benzene	$D = 1{\cdot}670$
Carbon disulphide	$D = 2{\cdot}697$	Carbon tetrachloride	$D = 1{\cdot}177$
Bromobenzene	$D = 1{\cdot}038$	Methanol	$D = 1{\cdot}572$
Ethyl acetate	$D = 1{\cdot}859$		

The effect of pressure on diffusion in liquids becomes perceptible only at high pressures, above a hundred atmospheres say. By way of illustration some values for the self-diffusivities of mercury and carbon disulphide are listed below:[140]

a. *Mercury* (30°C)

p, atm	1	1887	3676	5883	8097
D_p/D_0	1·000	0·957	0·908	0·877	0·828

b. *Carbon disulphide* (0°C)

p, atm	1	500	1000	2000	4000	6000	10,000
D_p/D_0	1·00	0·92	0·86	0·72	0·56	0·40	0·07

In this table D_p and D_0 represent the diffusivities at pressure p and at 1 atm, respectively.

As a further illustration of the effect of pressure on diffusivities in liquid systems we may refer to the investigations of Cohen and Bruins[62] on the diffusion of cadmium into mercury. At 20°C they find that at

$$p = 1 \text{ atm}, \qquad D = 1{\cdot}520\times10^{-5}\ \text{cm}^2/\text{sec},$$
$$\text{and at } p = 1500 \text{ atm}, \qquad D = 1{\cdot}466\times10^{-5}\ \text{cm}^2/\text{sec}$$

Thus even an increase of 1500 atmospheres in the pressure produced a decrease in the diffusion coefficient of only 5%.

The effect of pressure on liquid diffusivities may usually be neglected for the purpose of technological calculations.

2. THEORETICAL ESTIMATION OF LIQUID DIFFUSION COEFFICIENTS

Owing to the fact that no satisfactory theory of the liquid state has so far been developed, it is impossible to provide any exact picture of the diffusion phenomenon in liquids. In consequence no rigorous theoretical methods for estimating diffusion coefficients for liquids have yet been developed.

Early authors used the Van der Waals theory of condensed gases as their starting point. Some contemporary treatments are based on the quasi-crystalline theory of liquid structure, though the application of this theory to the interpretation of kinetic phenomena (momentum, energy and mass transfer) in liquids entails difficulties,[182] and the results of these investigations are not yet suitable for use in technological calculations.

Following some earlier work by Sutherland,[355] Einstein applied Stokes' law to the process of diffusion and obtained the following relationship:

$$\frac{D\mu}{kT} = \frac{1}{6\pi r}\,\frac{1+(3\mu/\beta r)}{1+(2\mu/\beta r)} \tag{8.3}$$

where D is the diffusion coefficient,
μ is the dynamic viscosity,
k is Boltzmann's constant,
r is the radius of the diffusing molecule,
and β is the molecular friction coefficient: this quantity increases with the size of the diffusing molecule.

In the extreme cases where $\beta = 0$ and $\beta = \infty$ Eq. (8.3) reduces to

$$\frac{D\mu}{kT} = \frac{1}{4\pi r} \qquad \text{if} \qquad \beta = 0 \tag{8.4}$$

or

$$\frac{D\mu}{kT} = \frac{1}{6\pi r} \qquad \text{if} \qquad \beta = \infty \tag{8.5}$$

Experiments indicate that, in fact, if the diffusing molecule is very large, so that β is large, Eq. (8.5) is approximately applicable. If this diameter is of the same order of magnitude as the diameter of the solvent molecules, Eq. (8.4) is applicable. It sometimes happens, when the diffusing molecule is considerably smaller than the molecule of the solvent (e.g. in the case of the diffusion of water in glycerol), that the numerical factor in the denominator on the right-hand side of the equation is less than four.

If the approximation is made of calculating the molecule's diameter, $\sigma = 2r$, from the molal volume V of the diffusing compound using the relation

$$\sigma = (V/N)^{1/3} \tag{8.6}$$

(where N denotes Avogadro's number), then for the process of self-diffusion equation (8.4) reduces to

$$\frac{D\mu}{kT} = \frac{1}{2\pi}\left(\frac{N}{V}\right)^{1/3} \tag{8.7}$$

As Li and Chang[222] have established, the latter relationship holds approximately in the majority of cases investigated by them (the deviations, however, may be as high as $\pm 40\%$).

In the case of the diffusion of large molecules at a temperature of 293°K, Eqs. (8.3) and (8.5) take the form

$$D = \frac{1{\cdot}84}{\mu r} \tag{8.8}$$

where D is expressed in $cm^2/24$ hr, r in Å, and μ in cP. Diffusivities calculated from Eq. (8.8) are for the most part in approximate agreement with the measured values. For the diffusion of molecules with molecular weight $M > 1000$ in water, Polson[290] has found that the relationship

$$D = 2{\cdot}74 \times 10^{-5} M^{-1/3} \tag{8.9}$$

is valid. This formula is a further simplification of Eq. (8.3).

According to the transition state theory (or "absolute rate theory") developed by Eyring,[130] the diffusivity D may be calculated from the equation

$$D = \left(\frac{V}{N}\right)^{2/3} \left(\frac{kT}{h}\right) \mathrm{e}^{\frac{\Delta S}{R}} \mathrm{e}^{-\frac{\Delta H}{RT}} \tag{8.10}$$

where V is the molal volume of the solvent,
T is the absolute temperature,
ΔS is the entropy of activation of the diffusion process,
ΔH is the heat of activation of the diffusion process,
N is Avogadro's number,
k is Boltzmann's constant,
h is Planck's constant,
and R is the gas constant.

In this theory it is presumed that the mechanism of activation in the diffusion process is the same as in the process of internal friction. According to the mechanism accepted by Eyring, the molecules of a diffusing compound penetrate into empty spaces, or "holes", between the molecules of the solvent. For the formation of a "hole" a quantum of energy, i.e. the activation energy, is required. The size of this quantum depends on the nature of the solvent and on the size of the space formed. Diffusion of large molecules requires large empty spaces and, since large empty spaces occur less readily than small ones, large molecules diffuse more slowly. According to this theory, the activation energy for the formation of an empty space (or "hole") is proportional to the latent heat of vaporization of the solvent.

The relationship between the coefficient D of self-diffusion and viscosity μ is, according to the absolute rate theory, given by the equation

$$\frac{D\mu}{kT} = \frac{\lambda_1}{\lambda_2 \lambda_3} \tag{8.11}$$

where λ_1 is the distance between two adjacent liquid layers,
λ_2 is the distance between two adjacent molecules in the direction of diffusion,
and λ_3 is the distance between two adjacent molecules in a direction perpendicular to the direction of diffusion.

In practice, the diffusion equations based on the "hole" theory are not very accurate when applied to systems of practical interest. Furthermore the necessary data which would characterize the system are not usually available. They are not, therefore, of much value for technological calculations. However, in the most simple cases good agreement between calculation and experiment has been obtained (Eyring *et al.* (cf. Ref. 5)).

From statistical mechanical considerations Kamal and Canjar[185,186] have derived a formula by means of which they were able to calculate the values of diffusivities in several score of cases, with an average deviation from the experimental values of about 13%. According to these authors, the diffusion coefficient D is equal to the product of two functions F and H. The function F depends on the properties of the solvent, whereas H depends on the properties of the solute:

$$D = F_{\text{solvent}} \times H_{\text{solute}} \tag{8.12}$$
$$F_{\text{solvent}} = AB \tag{8.13}$$
$$H_{\text{solute}} = RT/(24L - 39RT) \tag{8.14}$$

where $A = [1 \cdot 6(v_0/v) + 2 \cdot 56(v_0/v)^2 G](v/v_0)^{2/3}$
$G = 1 + 2 \cdot 5(v_0/v) + 4 \cdot 5864(v_0/v)^2$
$B = 6 \cdot 19 \times 10^{-5}(T/M)^{1/2} V_0^{1/3}$
D is the binary diffusion coefficient, cm^2/sec,
L is the heat of vaporization of the solute at temperature T, °K,
M is the molecular weight,
R is the gas constant,
V_0 is the molal volume,
v is the total volume per molecule,
and v_0 is the occupied volume per molecule.

The ratio v_0/v may be calculated by trial-and-error from the formula of Collins and Navidi,[63] provided that the acoustic velocity u_s is known:

$$u_s = \frac{1 - \frac{2}{3}(v_0/v)^{\frac{1}{3}}}{1 - (v_0/v)^{\frac{1}{3}}} \left[\frac{C_p RT/M}{C_p[1 - \frac{2}{3}(v_0/v)]^{\frac{1}{3}} - R} \right]^{\frac{1}{2}}$$

where C_p is the heat capacity at constant pressure,
and $R = 8 \cdot 314 \times 10^7$ g.cm²/(sec)².mole.°K.

The above equations are suitable for calculating the diffusivity D when the solvent and the solute are liquids, but they cannot be used if either a solid or a gas is dissolved. The authors of this method have compiled the values of F and H for a number of compounds (Tables 8.9 and 8.10).

TABLE 8.9. VALUES OF F_{solvent} (EQ. (8.12.)) (AFTER KAMAL[186])

Solvent	°C	A	$F \times 10^3$	Solvent	°C	A	$F \times 10^3$
C_6H_6	15	11·189	5·925	$C_6H_5CH_3$	25	11·575	6·109
	25	10·763	5·807	CH_3OH	15	10·583	6·745
CCl_4	15	11·131		C_2H_5OH	25	6·285	3·831
	25	10·714	4·239	H_2O	15	6·252	4·050
C_6H_5Cl	25	12·115	5·676		25	6·295	4·151
C_6H_5Br	25	12·677	5·094				

Kamal and Canjar have made use of Darken's equation[70] to find the dependence of diffusivity upon concentration:

$$D = (N_1 D_2^\circ + N_2 D_1^\circ) \left[1 + N_2 \left(\frac{\partial \ln \gamma_2}{\partial N_2} \right) \right]$$

where D_i° is the coefficient of diffusion of component i in infinite dilution into a solution with the same composition as that of the solution under consideration (D_i° is determined, for example, by means of the isotope method),
N_i is the mole fraction of component i,
and γ_i is the activity coefficient of component i.

TABLE 8.10. VALUES OF H_{solute} (EQ. (8.12)) (AFTER KAMAL[186])

Solute	°C	$H\times10^3$	Solute	°C	$H\times10^3$
Acetic acid	15	2·312	Ethyl ether	15	3·833
Acetone	15	3·546	Ethylene bromide	15	2·913
Allyl alcohol	15	2·328	Formic acid	6	4·159
Aniline	15	2·178	Furfural	15	2·178
Benzaldehyde	15	2·364	Isoamyl alcohol	15	2·282
Benzene	25	3·460	Methanol	15	2·861
Bromobenzene	15	2·403	Methyl iodide	7·5	3·912
n-Butyl alcohol	15	2·049	Nitrobenzene	15	2·322
Carbon tetrachloride	25	3·576	Phenol	5	1·889
Chloral hydrate	15	2·033	Propanol	15	2·243
Chlorobenzene	15	2·591	Propionic acid	15	3·008
Chloroform	15	3·328	Pyridine	20	2·723
Ethanol	15	2·456	Toluene	25	3·037
Ethyl bromide	15	4·025	Water	25	2·591

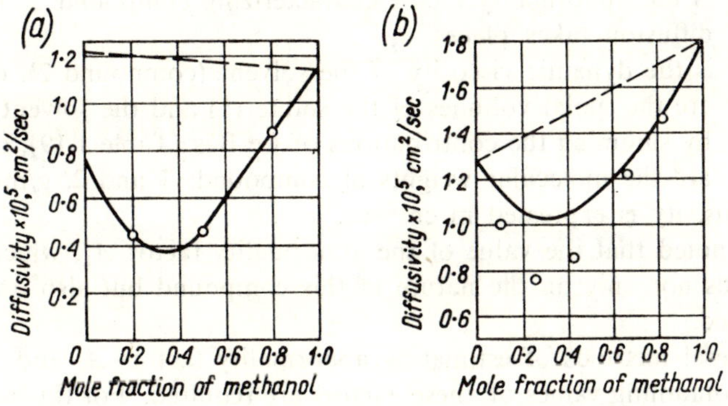

FIG. 8.3. Effect of concentration on diffusivity; (a) ethanol–water system, (b) methanol–water system calculated data—solid line, ideal calculated data—dashed line, o experimental points.

In the case of ideal solutions the term

$$\left[1+N_2\left(\frac{\partial \ln \gamma_2}{\partial N_2}\right)\right]$$

is unity, but for non-ideal solutions the value of this term must be calculated. The authors obtain excellent agreement between the calculated and experimental values for the ethanol–water system (Fig. 8.3); for the methanol–water system (Fig. 8.3b) the agreement is not so good, though the character of both curves is the same.

The self-diffusion coefficients in liquid mixtures were also the object of investigations undertaken by Carman, Stein and Miller.[53,251]

3. SEMI-EMPIRICAL METHODS FOR ESTIMATING LIQUID DIFFUSION COEFFICIENTS

a. Arnold's Method

Arnold[14] has applied his equation (7.45) (which was developed for gaseous diffusivities) to the estimation of diffusion coefficients in liquid systems. He finds, however, that this equation cannot be used directly for liquids without the introduction of factors characterizing both the solute and the solvent in which the diffusion is taking place. These factors have been referred to as *abnormality factors.* The molal volumes of the compounds appear in the equation as quantities proportional to the diameters of the molecules taking part in the process of diffusion. In contrast to the equation for gases (7.45), Arnold's equation relating to liquids possesses no term containing Sutherland's constant $C_{1,2}$. The value of the diffusion coefficient D_1 (cm²/sec) at 20°C, when solute 1 is diffusing into solvent 2, may be calculated, according to Arnold, on the basis of the equation

$$D_1 = \frac{0{\cdot}0100}{A_1 A_2 \mu_2^{1/2}(V_1^{1/3}+V_2^{1/3})^2}\left(\frac{M_1+M_2}{M_1 M_2}\right)^{\frac{1}{2}} \tag{8.15}$$

where A_1 is the abnormality factor characterizing compound 1 (the diffusing compound),

A_2 is the abnormality factor characterizing compound 2 (into which the diffusion takes place),

μ_2 is the dynamic viscosity of the solvent (compound 2), cP,

V_1 and V_2 are the molal volumes of the solute (1) and the solvent (2), estimated by summing the contributions of Le Bas (Table 1.19), cm³/mole,

and M_1 and M_2 are the molecular weights of compounds 1 and 2, g/mole.

The resulting diffusivity is expressed in cm²/sec.

It should be noted that the value of the abnormality factor A_1, which characterizes the solute, depends not only on the nature of this compound but also on the nature of the solvent involved.

Since no general methods of estimating abnormality factors A_1 and A_2 are known, extensive tables containing values of these factors are required. For this reason the practical utility of Arnold's method is, perhaps, questionable.

TABLE 8.11. VALUES OF THE ABNORMALITY FACTOR A_1 IN ARNOLD'S EQUATION (8.15) FOR SOLUTES (AFTER ARNOLD[14])

Solute	Solvent		
	Benzene	Methanol	Water
Acetamide	—	1·54	1·15
Acetic acid	1·38	1·53	1·27
Amyl alcohol	1·29	1·29	1·16
Chlorophenol	1·26	1·26	—
Formic acid	1·29	1·53	1·32
Glycerol	—	1·65	1·26
Phenol	1·32	1·35	1·21
Pyridine	—	1·26	1·60

TABLE 8.12. VALUES OF THE ABNORMALITY FACTOR A_2 IN ARNOLD'S EQUATION (8.12) FOR SOLVENTS (AFTER ARNOLD[14])

Solvent	A_2	Solvent	A_2
Acetic acid	0·86	Ethyl benzoate	1·0
Acetone	1·15	Ethyl ether	0·90
Amyl alcohol	0·97	Heptane	0·66
Benzene	1·0	Methanol	2·0
Carbon disulphide	1·0	Nitrobenzene	1·35
Carbon tetrachloride	0·94	Propanol	1·36
Chloroform	1·0	Toluene	1·0
Ethyl acetate	1·06	Water	4·70
Ethyl alcohol	2·0	*m*-Xylene	0·97

Tables 8.11 and 8.12 list some values of abnormality factors. For non-associated compounds and for dissolved gases the value of the abnormality factor A_1 is unity.

b. Wilke's Method

Working from Eyring's assumptions about diffusion (section 3) and introducing the Stokes–Einstein equation (8.7), Wilke[399] concludes that a quantity F, called by him the *diffusion factor*, may be useful for calculating values of the diffusivity D (cm²/sec) in liquids. F is defined by the equation

$$F = \frac{T}{D\mu} \tag{8.16}$$

where F the diffusion factor is expressed in °K.sec/cm².cP,
T is the absolute temperature, °K,
and μ is the dynamic viscosity of the solvent, cP.

For some of the pairs of components considered, the diffusion factor F was found to be almost completely independent of temperature. F is a function of the volume V_b of the solute at its normal boiling point (i.e. of a quantity which is approximately proportional to the volume of the diffusing compound) and depends also on the nature of the solvent.

Wilke has calculated values of the diffusion factor F for many systems in which the solute neither reacts with the solvent nor dissociates into ions. He finds that when F is plotted as a function of log V_b the points corresponding to different solutes in the same solvent lie on a common curve. By considering a range of solvents he obtained a series of such curves, each characterized by a definite value of the parameter ϕ, known under the name of *solvent factor*. This is the ratio of the diffusivities of compounds with the same molal volumes V_b in the solvent considered and in water. For water $\phi = 1{\cdot}0$; in general ϕ is small for solvents with a large molal volume and vice versa. For benzene $\phi = 0{\cdot}70$ and for methanol $\phi = 0{\cdot}82$. ϕ may readily be calculated for other solvents provided one diffusivity in that solvent is known; otherwise it is usually a fairly good approximation to take the value of ϕ to be 0·90. These values of ϕ are approximately valid for diffusing compounds with a molal volume $V_b < 150$. If the molal volume is larger, the value of ϕ approaches unity (when $V_b > 1000$).

Wilke[399] has presented the relationship between the diffusion factor F and molal volume V_b graphically in the form of a bunch of curves corresponding to varying values of the solvent factor ϕ. In order to find the molal volume V_b of a solute, the table of contributions given by Le Bas may be used (Table 1.19). To apply Wilke's method, it is necessary to know the viscosity μ of the solvent.

The method of calculation is as follows: first, the molal volume V_b of the compound is estimated, then the value of the diffusion factor F for this volume is read from Fig. 8.4.

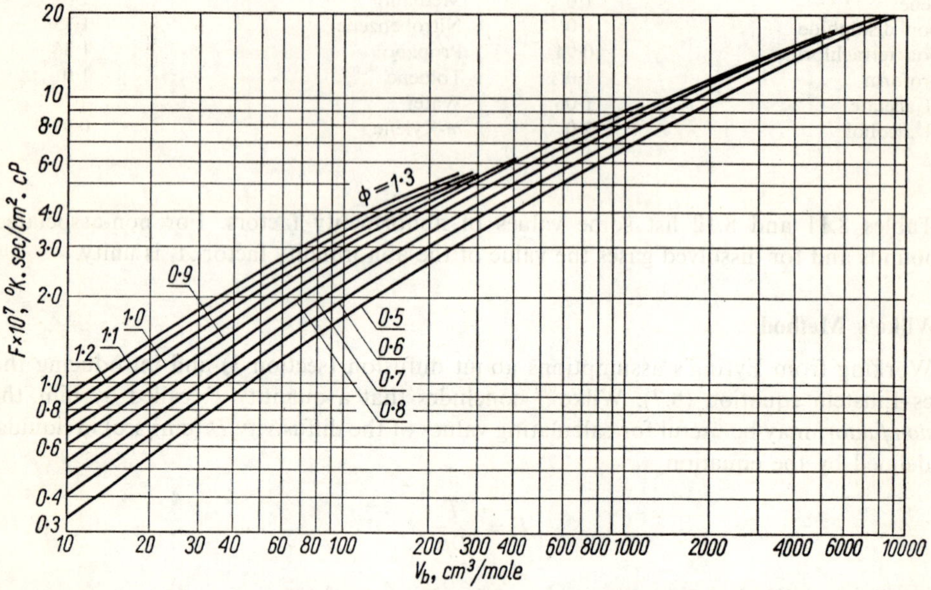

FIG. 8.4. Graph for determining the diffusion factor F in Eq. (8.16) (after Wilke[399]).

If the diffusion process occurs in water, the correct curve in the graph will be that for $\phi = 1{\cdot}0$; for other solvents we assume, according to what has been said above, the following curves: for benzene $\phi = 0{\cdot}7$, for methanol $\phi = 0{\cdot}82$, for other solvents $\phi \approx 0{\cdot}9$. Knowing the value of F from the graph as well as the viscosity μ and temperature T, we can calculate the required value of the diffusivity D by means of Eq. (8.16).

In the case of diffusion in water the average error of estimation is of the order of $\pm 6{\cdot}3\%$, in methanol $\pm 11{\cdot}5\%$, in benzene $\pm 12{\cdot}5\%$.[399]

Wilke and Chang[403] have presented an empirical relationship between the coefficient of diffusion in liquids $D_{1,2}$ (cm²/sec), the molecular weight of the solvent M (g/mole), the absolute temperature T (°K), the dynamic viscosity μ (cP), and the molal volume V (cm³/mole) of the solute:

$$D_{1,2} = \frac{7{\cdot}4\times 10^{-8}(xM)^{\frac{1}{2}}\,T}{\mu V^{0{\cdot}6}} \tag{8.17}$$

where x is a coefficient, which allows for association of the solvent. For non-associated compounds, e.g. ether, benzene and heptane, $x = 1{\cdot}0$. For water $x = 2{\cdot}6$, for methanol $x = 1{\cdot}9$, for ethanol $x = 1{\cdot}5$. For calculations, use is made of Fig. 8.5. A nomograph for solving Eq. (8.17) is presented in Fig. 8.6. The use of this nomograph is illustrated

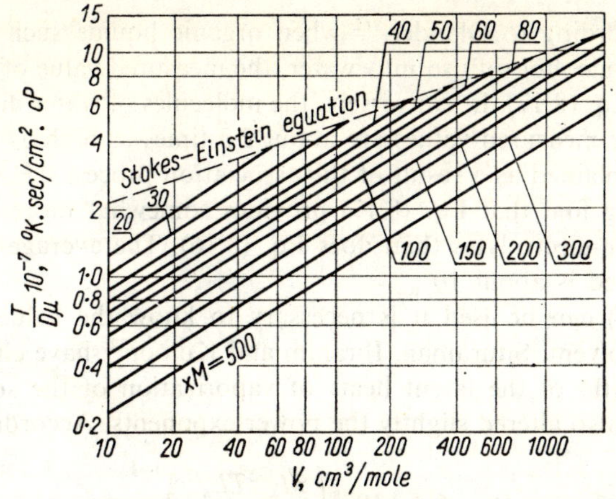

FIG. 8.5. Variation of $(T/D\mu)$ with molal volume V of solute (after Wilke and Chang[403]).

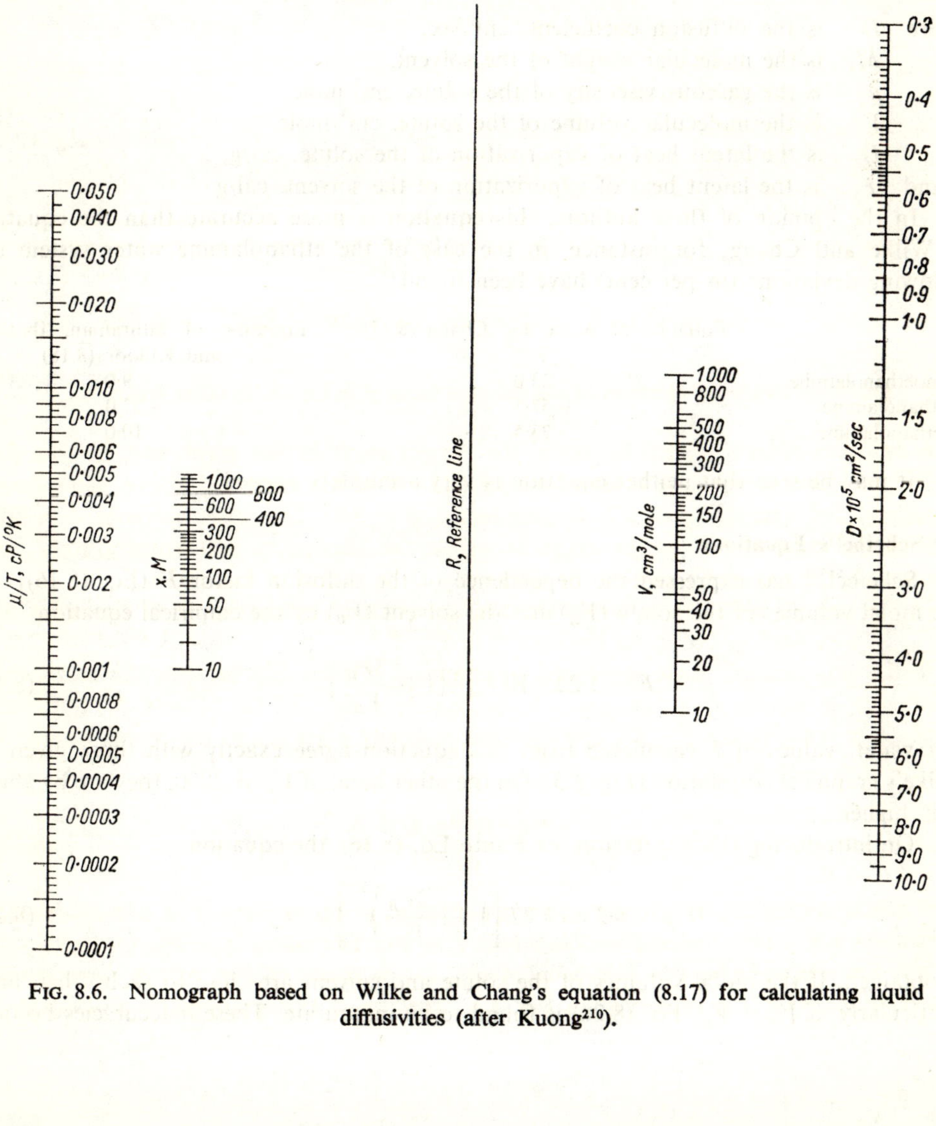

FIG. 8.6. Nomograph based on Wilke and Chang's equation (8.17) for calculating liquid diffusivities (after Kuong[210]).

in Example 8.6. According to Olander,[262] when organic liquids such as acetone, aniline, furfural, or iso-butyl alcohol diffuse into water, the measured value of $D_{1,2}$ is very close to that calculated from Eq. (8.17). If, conversely, the molecules of water diffuse into the above mentioned liquids, the measured values are about 2·3 times as high as the estimated ones, which fact may be explained as a result of the association process $4H_2O \rightleftarrows (H_2O)_4$.

Wilke and Chang find that Eq. (8.17) provides somewhat more accurate values for the diffusivities of non-electrolytes than does Eq. (8.16). The average error of estimation when Eq. (8.17) is used is about 10%.

Before Eq. (8.17) can be used it is necessary to know the value of the association coefficient x for the solvent. Situraman, Ibrahim and Kuloor[328] have eliminated this factor by introducing the ratio of the latent heats of vaporization of the solvent, L_s, and the solute, L. They have also altered slightly the power exponents. According to these authors

$$D = 5{\cdot}4\times 10^{-8}\left(\frac{M_s^{1/2}TL_s^{1/3}}{\mu V^{1/2}L^{0{\cdot}3}}\right)^{0{\cdot}93} \tag{8.18}$$

where D is the diffusion coefficient, cm²/sec,
M_s is the molecular weight of the solvent,
μ is the gaseous viscosity of the solute, cm³/mole,
V is the molecular volume of the solute, cm³/mole,
L is the latent heat of vaporization of the solute, cal/g,
and L_s is the latent heat of vaporization of the solvent, cal/g.

In the opinion of these authors, this equation is more accurate than the equation of Wilke and Chang; for instance, in the case of the ethanolamine–water system the following deviations (in per cent) have been found:

	Equation of Wilke and Chang (8.17)	Equation of Situraman, Ibrahim and Kuloor (8.18)
Monoethanolamine	+23·0	−9·0
Diethanolamine	+57·5	+27·0
Triethanolamine	+23·5	+10·0

It may be seen that neither equation is very accurately obeyed.

c. Scheibel's Equations

Scheibel[321] has expressed the dependence of the diffusion factor F (Eq. (8.16)) on the molal volumes of the solute (V_A) and the solvent (V_B) by the empirical equation

$$F = 1{\cdot}22\times 10^7 V_A^{1/3}\left(1+\frac{3V_B}{V_A}\right)^{-2/3} \tag{8.19}$$

For water, values of F calculated from this equation agree exactly with those given by Wilke's graphical correlation (Fig. 8.5). On the other hand, if $V_A \approx 3000$, they are by about 10% higher.

On introducing this expression for F into Eq. (8.16), the equation

$$D_{AB} = 8{\cdot}2\times 10^{-8}T\left[1+\left(\frac{3V_B}{V_A}\right)^{2/3}\right]\mu^{-1}V_a^{-1/3} \tag{8.20}$$

is obtained. If the molal volumes of the solute and solvent are close to each other, and, particularly, if $V_A < V_B$, Eq. (8.20) is found to be inaccurate. These inaccuracies become marked with water as solvent when $V_A \leqslant V_B$, with benzene when $V_A \leqslant 2V_B$, with methanol when $V_A \leqslant 1{\cdot}5V_B$ and with other solvents when $V_A \leqslant 2{\cdot}5V_B$. The following equations are then recommended:

Solvent: water

$$D = 25{\cdot}2\times 10^{-8}T\mu^{-1}V_A^{-1/3} \tag{8.21}$$

Solvent: benzene

$$D = 18{\cdot}9\times 10^{-8}T\mu^{-1}V_A^{-1/3} \tag{8.22}$$

Other solvents

$$D = 17{\cdot}5\times 10^{-8}T\mu^{-1}V_A^{-1/3} \tag{8.23}$$

Reid and Sherwood[306] have found Scheibel's equations to yield slightly more accurate results than Wilke's method (section 3b); with water, methanol and benzene as solvents, the average errors of estimation are 9, 12, and 10%, respectively.

d. Ibrahim and Kuloor's Equations

The equations of Wilke, Wilke and Chang, and of Scheibel, discussed above, all contain a factor which characterizes the solvent (ϕ in Wilke's method, x in Eq. (8.17)). According to Ibrahim and Kuloor[163,164] the values of these factors may be estimated from the surface tension of the solvent, thus enabling them to be eliminated from the diffusivity equations. The authors develop their correlation in terms of the modified Schmidt number

$$\text{Sc}' = \frac{\mu V}{D} \tag{8.24}$$

where μ is the viscosity of the solute in the gaseous state,
V is the molal volume of the solute,
and D is the diffusivity in the liquid phase.

They find that Sc′ varies linearly with the molecular weight M of the solute so that

$$\text{Sc}' = aM+b \tag{8.25}$$

The values of the coefficients a and b are related to the nature of the solvent. For diffusion in aqueous solutions (Fig. 8.7) $a = 13$ and $b = 100$ so that

$$\text{Sc}' = 13M-100 \quad \text{and} \quad D = \frac{\mu V}{13M-100} \tag{8.26}$$

For diffusion in benzene

$$\text{Sc}' = 4{\cdot}5M, \qquad D = \frac{\mu V}{4{\cdot}5M} \tag{8.27}$$

For diffusion in methanol

$$\text{Sc}' = 4M-75, \qquad D = \frac{\mu V}{4M-75} \tag{8.28}$$

Ibrahim and Kuloor find that the coefficient a in Eq. (8.25) is approximately proportional to surface tension σ of the solvent so that, for two different solvents (1 and 2)

$$a_1/a_2 = \sigma_1/\sigma_2 \tag{8.29}$$

The values of the surface tension at 20°C are:

for water	$\sigma_w = 72{\cdot}5$ dynes/cm
for benzene	$\sigma_b = 29{\cdot}0$ dynes/cm
for methanol	$\sigma_m = 23{\cdot}0$ dynes/cm

For water as solvent $a = a_w = 13$. It follows then from Eq. (8.29) that for benzene as solvent a should be given by

$$a = a_b = a_w \frac{\sigma_b}{\sigma_w} = 13 \times \frac{29{\cdot}0}{72{\cdot}5} = 5{\cdot}2$$

(according to Eq. (8.27) $a_b = 4{\cdot}5$),
and for methanol

$$a_m = a_w \frac{\sigma_m}{\sigma_w} = 13 \times \frac{23{\cdot}0}{72{\cdot}5} = 4{\cdot}1$$

(according to Eq. (8.28) $a_m = 4$).

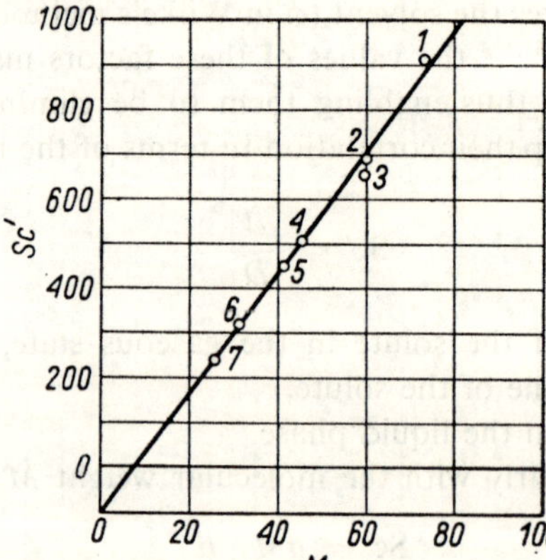

FIG. 8.7. Relationship between the modified Schmidt number $Sc' = \mu V/D$ and the solute molecular weight M for diffusion in water (after Kuloor[163]); *1*—butanol, *2*—acetic acid, *3*—propanol, *4*—ethanol, *5*—propylene, *6*—methanol, *7*—acetylene.

In order to determine the constant b for a given solvent one has to know the diffusivity of at least one substance in this solvent. As seen above $b_w = -100$, $b_b = 0$ and $b_m = -75$.

For predicting the diffusivities of inorganic gases such as O_2, H_2S, Cl_2, CO, NH_3 and NO in water, Ibrahim and Kuloor recommend the equation

$$D = \frac{\mu V}{6{\cdot}5M + 30} \tag{8.30}$$

The average deviations of the measured diffusivities from those calculated from Eqs. (8.26)–(8.30) are about 10%. Owing to their simplicity, these equations are very useful for obtaining a rapid estimate of the value of a liquid diffusivity.

e. Othmer and Thakar's Method

The semi-empirical equations for estimating liquid diffusivities considered above, are based on simple considerations. Othmer and Thakar[271] have developed a method for predicting liquid diffusivities which utilizes the similarity which exists between the physico-chemical properties of the substance considered and those of a suitable reference compound (Chapter 1, section 7). These authors develop their correlation on the basis of Eyring's equation which may be written in the form

$$D = Ke^{-E_d/RT} \tag{8.31}$$

where K is a constant,
and E_d is the activation energy for diffusion: the magnitude of E_d will depend on the soluted solvent.

On taking logarithms and differentiating, this equation gives, for a given solute and solvent,

$$\mathrm{d}(\log D) = \frac{\mathrm{d}T}{RT^2} E_d \tag{8.32}$$

A similar equation may be obtained for the viscosity, namely

$$\mathrm{d}(\log \mu) = \frac{\mathrm{d}T}{RT^2} E_v \tag{8.33}$$

where μ is the dynamic viscosity,
and E_v is the activation energy for viscous flow.

Dividing Eq. (8.32) by Eq. (8.33) and integrating

$$\log D = -\frac{E_d}{E_v} \log \mu + C' \tag{8.34}$$

where C' is the integration constant for the given solute and solvent. It is found experimentally that the ratio (E_d/E_v) usually lies between $-1{\cdot}07$ and $-1{\cdot}15$, so that it is reasonable to take as an average value $E_d/E_v = -1{\cdot}1$.

The authors first applied Eq. (8.34) to aqueous solutions. The viscosity of water, μ_w°, at 20·06°C is 1 cP, so at this temperature Eq. (8.34) reduces to $\log D_w^\circ = C$. It follows therefore that the integration constant in Eq. (8.34) for the diffusion of a given solute in an aqueous solution is equal to the logarithm of the diffusivity of this solute in water at 20·06°C. Othmer and Thakar find that this coefficient may be predicted with an average absolute error of about 5% by means of the empirical equation

$$D_w^\circ \times 10^5 = \frac{14{\cdot}0}{V_m^{0{\cdot}6}} \quad (\mathrm{cm^2/sec}) \tag{8.35}$$

where V_m is the molal volume of the solute, calculated on the basis of Le Bas' table of group contributions (Table 1.19). On substituting $\log D_w^\circ$ (obtained by transforming Eq. (8.35)) for C' in Eq. (8.34) the following equation is obtained (this equation is applicable over a range of temperatures since C' is independent of temperature):

$$\log D_w = -1{\cdot}1 \log \mu_w + \log \frac{14 \times 10^{-5}}{V^{0{\cdot}6}}$$

This equation may be rewritten in the form

$$D_w = \frac{14 \times 10^{-5}}{\mu_w^{1{\cdot}1} V_m^{0{\cdot}6}} \tag{8.36}$$

where D_w is the diffusion coefficient of the solute in a dilute aqueous solution at temperature t,
and μ_w is the viscosity of water at temperature t.

This equation enables diffusivities in dilute aqueous solutions to be predicted with reasonable accuracy. According to Reid and Sherwood,[306] the average absolute error in the estimated diffusivities is only 8%.

More recently Othmer and Thaker have developed an equation (8.40) for predicting the diffusivities of non-aqueous solutions. This may be derived by taking the temperature dependence of the diffusivity in the non-aqueous solvent q to be given by Eq. (8.32) while that of the viscosity of water is given by Eq. (8.33). On dividing the second expression by the first and integrating, a relationship analogous to Eq. (8.34) is obtained, namely

$$\log D_q = -\left(\frac{E_{dq}}{E_{vw}}\right)\log \mu_w + C'' \tag{8.37}$$

where E_{dq} is the activation energy for diffusion in the non-aqueous solution,
E_{vw} is the activation energy for viscous flow in water,
D_q is the diffusivity in the non-aqueous solution,
and C'' is the integration constant for the given solute and solvent.

Using the empirical result that $E_{dw}/E_{vw} = -1{\cdot}1$, and assuming that $E_{dq}/E_{dw} = L_q/L_w$ (where L_q and L_w denote the heats of vaporization of the solvent and of water), Eq. (8.37) reduces to

$$\log D_q = -1{\cdot}1\frac{L_q}{L_w}\log \mu_w + k \tag{8.38}$$

In order to calculate the value of the integration constant, the authors assume that the diffusivities and viscosities at 20·06°C are inversely related so that

$$D_q^\circ/D_w^\circ = \mu_w^\circ/\mu_q^\circ \tag{8.39}$$

where $\mu_w^\circ = 1$. On substituting $\log D_q^\circ$, as given by Eqs. (8.39) and (8.35), for k in Eq. (8.38), the following formula is finally obtained:

$$D_q = \frac{14\times 10^{-5}}{V_m^{0\cdot 6}\mu_q^\circ \mu_w^{1\cdot 1(L_q/L_w)}} \tag{8.40}$$

where D_q is the diffusion coefficient in solvent q at temperature T, cm²/sec,
V_m is the molal volume of the solute, cm³/mole,
μ_q° is the viscosity of the solvent q at 20·06°C, cP,
μ_w is the viscosity of water at temperature T, cP,
L_q is the molal heat of vaporization of solvent q at temperature T, cal/mole,
and L_w is the molal heat of vaporization of water at temperature T, cal/mole.

Unfortunately diffusivities predicted from this equation are not very accurate. According to Reid and Sherwood[306] the average absolute error in the case of methanol solutions is 12% while for benzene solutions it is 16%.

Othmer and Thakar have constructed a nomograph, which makes it possible to solve Eqs. (8.36) and (8.38) very quickly. This nomograph is presented in Fig. 8.8.

For a non-aqueous solvent, the nomograph is used as follows: The point corresponding to the temperature t is first located on the left-hand axis and a straight line is then drawn from this point through the point L_q/L_w on the short oblique axis to intersect with the reference axis 0′ (this axis has no scale). Next, a straight line is drawn from the point thus determined on the reference axis 0′ through the point corresponding to the viscosity μ_q of the solvent on the first axis to the right of the axis 0′ to intersect the second reference axis 0″ (this axis carries a temperature scale since in the case of aqueous solutions it is used for the water temperature t_w). Finally a third line is drawn from the point previously located on the axis 0″ through the point corresponding to the molal volume V_m of the solute (on the next axis to the right) to intersect the extreme right-hand axis. This intersection point gives the diffusivity D_q (cm²/sec).

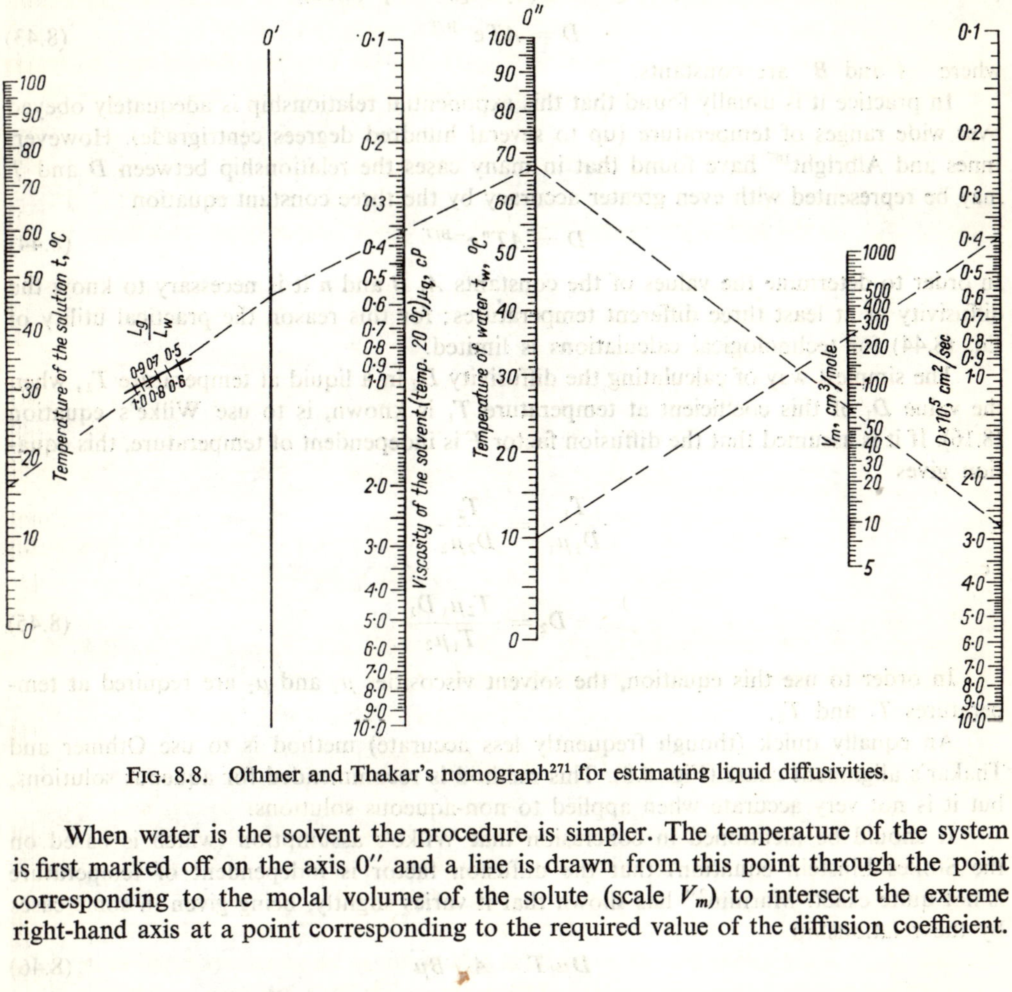

FIG. 8.8. Othmer and Thakar's nomograph[271] for estimating liquid diffusivities.

When water is the solvent the procedure is simpler. The temperature of the system is first marked off on the axis 0″ and a line is drawn from this point through the point corresponding to the molal volume of the solute (scale V_m) to intersect the extreme right-hand axis at a point corresponding to the required value of the diffusion coefficient.

4. EFFECT OF TEMPERATURE ON LIQUID DIFFUSION COEFFICIENTS

Changes in diffusivity occurring over very small temperature ranges (say of ten degrees or less) can be predicted by assuming a linear relationship to exist between this coefficient and temperature and using an empirical relationship to estimate the temperature coefficient (Eqs. (8.1) and (8.2)). At $t_1 = 20°C$, Eq. (8.1) takes the form

$$D_{t_2} = D_{20}[1+\alpha(t_2-20)] \qquad (8.41)$$

where D_{20} denotes the diffusion coefficient at 20°C. The value of the coefficient α may be calculated by means of the empirical formula

$$\alpha = 0{\cdot}02\mu^{1/2}\rho^{1/3} \tag{8.42}$$

where μ is the dinamic viscosity of the solvent at 20°C, cP,
and ρ is the density of the solvent, g/cm^3.

Theoretical considerations suggest that the relationship between diffusivity and temperature should have the following form (Eyring,[130] Eq. (8.10)):

$$D = ATe^{-B/T} \tag{8.43}$$

where A and B are constants.

In practice it is usually found that this exponential relationship is adequately obeyed over wide ranges of temperature (up to several hundred degrees centrigrade). However, Innes and Albright[167] have found that in many cases the relationship between D and T may be represented with even greater accuracy by the three constant equation

$$D = AT^n e^{-B/T} \tag{8.44}$$

In order to determine the values of the constants A, B and n it is necessary to know the diffusivity at at least three different temperatures; for this reason the practical utility of Eq. (8.44) for technological calculations is limited.

The simplest way of calculating the diffusivity D_2 in a liquid at temperature T_2, when the value D_1 of this coefficient at temperature T_1 is known, is to use Wilke's equation (8.16). If it is assumed that the diffusion factor F is independent of temperature, this equation gives

$$\frac{T_1}{D_1\mu_1} = \frac{T_2}{D_2\mu_2}$$

i.e.

$$D_2 = \frac{T_2\mu_1 D_1}{T_1\mu_2} \tag{8.45}$$

In order to use this equation, the solvent viscosities μ_1 and μ_2 are required at temperatures T_1 and T_2.

An equally quick (though frequently less accurate) method is to use Othmer and Thakar's alignment chart (Fig. 8.8). This method is recommended for aqueous solutions, but it is not very accurate when applied to non-aqueous solutions.

It should be mentioned in conclusion that Wilke's assumption (which is based on the Stokes–Einstein equation) that the diffusion factor is independent of temperature is not quite exact. Marinin[235] has shown that it varies slightly, being given in some cases by the relationship

$$D\mu/T = A + B\mu \tag{8.46}$$

where A and B are constants, which are characteristic of the given system.

5. EFFECT OF SOLUTE CONCENTRATION ON LIQUID DIFFUSIVITIES

The diffusivities of the systems considered in Fig. 8.1 vary linearly with the concentration c of the solute. This simple behaviour is by no means always encountered; the curves representing the effect of c upon D often exhibit maxima as well as minima.

For example, Fig. 8.9 shows diffusivity as a function of composition for the system chloroform–ethyl ether. (The effect of composition on the viscosity μ and on the product $D.\mu$ is also shown.) It may be seen that the diffusivity curve shows a pronounced maximum.

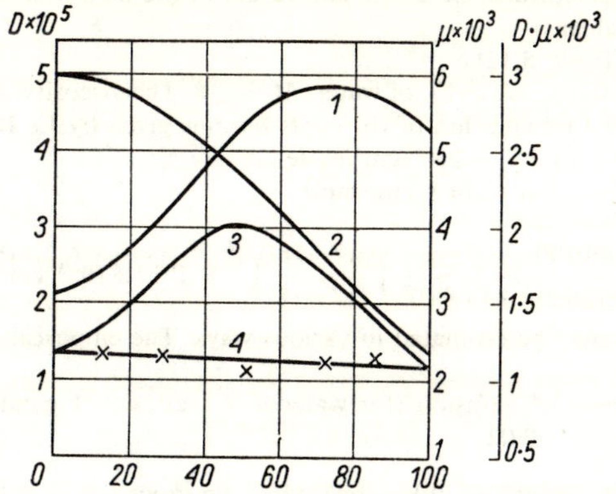

FIG. 8.9. Effect of liquid composition in the system chloroform-ethyl ether on: *1*—the diffusion coefficient D, *2*—the dynamic viscosity μ, *3*—the product $D.\mu$, *4*—the ratio $(D.\mu)/(d\ln a/d\ln c)$ (after Jost[182]).

Fick's law is based on the simplified assumption that the diffusion rate should be proportional to the concentration gradient of the solute. A more exact analysis indicates that the rate of diffusion may be controlled by gradients in the chemical potentials of the components in the system rather than by concentration differences. The activity of the diffusing solute should, therefore, be introduced into the diffusion equations in place of the concentration. The following relationship is then obtained:

$$D = D_0 \frac{d\ln a}{d\ln c} = D_0 \left(1 + \frac{d\ln \gamma}{d\ln c}\right) \tag{8.47}$$

where D is the diffusivity in a solution with concentration c,
D_0 is the diffusivity in a very dilute solution,
a is the activity of the solute referred to concentration,
c is the concentration of the solute,
and γ is the activity coefficient.

This equation has been checked against experimental data by Powell, Roseveare and Eyring.[295,296] It worked well for the systems chloroform–ether, chloroform–acetone and n-propanol–water, but in other cases it would appear that the thermodynamic correction term is less successful.

The effect of concentration changes on diffusivities in solutions of non-electrolytes have been analysed by Kincaid, Eyring and Stearn.[200] It is found that in near-ideal systems the product $D.\mu$ varies approximately linearly with the concentration of the solute. This finding is sometimes used to facilitate interpolations and extrapolations in technological calculations, though it is not applicable to non-ideal systems as may be seen from Fig. (8.9). It is also sometimes found that the ratio $(D.\mu)/(d\ln a/d\ln c)$ varies linearly with the composition of the mixture (as in curve 4 in Fig. 8.9). In other cases, however, this rule is involved and it may be observed that in general the changes in diffusivities arising from changes in composition do not allow for using the simple methods.

Examples of Calculating the Diffusivity in Liquid Solutions of Non-ionized Compounds

EXAMPLE 8.1a. Estimate the diffusivity of ammonia in water at 50°C using Arnold's method (this example is taken from Ramm[303]).

Solution

The calculation is first performed for 20°C. The values of the abnormality factors are:

$A_1 = 1$ (dissolved gas)

$A_2 = 4{\cdot}7$ (water, Table 8.12)

The molecular weight of ammonia, $M_1 = 17$, of water, $M_2 = 18$. The viscosity of water at 20°C, $\mu_W = 1$ cP. The molal volumes (calculated by addition of the contributions given by Le Bas, Table (1.19))

NH_3: $V_1 = 15{\cdot}6+3\times 3{\cdot}7 = 26{\cdot}7$ cm³/mole

H_2O: $V_2 = 2\times 3{\cdot}7+7{\cdot}4 = 14{\cdot}8$ cm³/mole

At 20°C

$$D_{20} = \frac{0{\cdot}0100}{1\times 4{\cdot}7\times 1^{1/2}(26{\cdot}7^{1/3}+14{\cdot}8^{1/3})^2}\left(\frac{17+18}{17\times 18}\right)^{1/2} = 2{\cdot}17\times 10^{-5}\ \text{cm}^2/\text{sec}$$

The diffusivity at 50°C may be estimated in various ways. The empirical equations (8.41) and (8.42) will first be used:

the temperature coefficient of diffusion (for water $\mu = 1$ cP, $\rho = 1$ g/cm³) is

$\alpha = 0{\cdot}02\times 1^{1/2}\times 1^{-1/3} = 0{\cdot}02$

so

$$D_{50} = 2{\cdot}17\times 10^{-5}[1+0{\cdot}02(50-20)] = 3{\cdot}47\times 10^{-5}\ \text{cm}^2/\text{sec}$$

It should be noted that this method for estimating the effect of temperature upon the diffusivity may be inaccurate particularly since in the case considered the temperature range is as large as 30°. In this case it is better to use Wilke's equation (8.45) and to assume that the diffusion factor F does not vary with temperature so that

$$D_2 = \frac{T_2\mu_1 D_1}{\mu_2 T_1}$$

Viscosity of water at 20°C $\mu_1 = 1{\cdot}005$ cP

at 50°C $\mu_2 = 0{\cdot}5494$ cP

Hence

$$D_{50} = \frac{(273+50)\times 1{\cdot}005\times 2{\cdot}17\times 10^{-5}}{(273+20)\times 0{\cdot}5494} = 4{\cdot}36\times 10^{-5}\ \text{cm}^2/\text{sec}$$

It may be seen that the two methods for calculating D_{50} from D_{20} lead to noticeably different values for D_{50}. This difference would be reduced somewhat if the values for ρ and μ used in Eqs. 8.41 and 8.42 were taken for an intermediate temperature between 20°C and 50°C rather than for 20°C.

EXAMPLE 8.1b. Estimate the diffusivity of ammonia in water at 15°C (the experimental value, taken from Table 18.1, is $D_{15} = 1{\cdot}78\times 10^{-5}$ cm²/sec).

Using the value for D_{20} obtained in example 8.1a, D_{15} is given by Eq. (8.42) to be

$$D_{15} = 2{\cdot}17\times 10^{-5}[1+0{\cdot}02(15-20)] = 1{\cdot}95\times 10^{-5}\ \text{cm}^2/\text{sec}$$

The error in this estimated value is $+9{\cdot}5\%$.

EXAMPLE 8.2. Use Wilke's method to estimate the diffusivity of butyl alcohol in a very dilute aqueous solution at 15°C (288°K).

Solution

The molal volume of the solute is first calculated using the group contributions given in Table 1.19 whence

$$V = 14{\cdot}8\times 4+3{\cdot}7\times 10+12 = 108{\cdot}2\ \text{cm}^3/\text{mole}$$

The value of the diffusion factor read from Fig. 8.3 for $V = 108{\cdot}2$ and $\phi = 1{\cdot}0$ (solvent–water) is

$F = 3{\cdot}1\times 10^7$ °K.sec/cm².cP

Viscosity of water at 15°C

$\mu = 1{\cdot}1406$ cP

According to Eq. (8.16), D is given by

$$D = \frac{T}{\mu F} = \frac{288}{1{\cdot}1406\times 3{\cdot}1\times 10^7} = 0{\cdot}81\times 10^{-5}\ \text{cm}^2/\text{sec}$$

The value determined experimentally (Table 8.1)

$D = 0{\cdot}77\times 10^{-5}$ cm²/sec

The error of estimation is $+5{\cdot}2\%$.

EXAMPLE 8.3. Calculate the diffusivity D_{70} of butyl alcohol in a dilute aqueous solution at 70°C (343°K), using the value for 15°C calculated in Example 8.2. The viscosity of water at 70°C is: $\mu_{70} = 0{\cdot}406$ cP.

Solution

Using Eq. (8.16) and assuming F to be independent of temperature we obtain

$$F = \frac{T}{\mu_1 D_1} = \frac{T}{\mu_2 D_2} \quad \text{and} \quad D_2 = \frac{T_2 \mu_1 D_1}{T_1 \mu_2}$$

so

$$D_{70} = \frac{343 \times 1{\cdot}141 \times 0{\cdot}81 \times 10^{-5}}{288 \times 0{\cdot}406} = 2{\cdot}71 \times 10^{-5} \text{ cm}^2/\text{sec}$$

EXAMPLE 8.4. Solve the previous example using Kamal and Canjar's method.

Solution

The value of F for water at 15°C is found from Table 8.9 to be $F = 4{\cdot}050 \times 10^{-3}$.
H for n-butanol at 15°C is found from Table 8.10 to be $H = 2{\cdot}049 \times 10^{-3}$.
Substitution of these values in Eq. (8.12) gives

$D = FH = 4{\cdot}050 \times 10^{-3} \times 2{\cdot}049 \times 10^{-3}$

i.e. $D = 8{\cdot}3 \times 10^{-6} = 0{\cdot}83 \times 10^{-5}$ cm²/sec

The error of estimation is +8%.

EXAMPLE 8.5. Use Wilke and Chang's method to estimate the value of the diffusivity D of bromoform in benzene at 18°C (291°K).

Solution

For benzene, the coefficient $x = 1$. The viscosity of benzene at 18°C is $\mu = 0{\cdot}669$ cP and the molecular weight of benzene $M = 78$.

The molal volume of bromoform, $CHBr_3$ (according to Table 1.19) is

$V = 14{\cdot}8 + 3{\cdot}7 + 27{\cdot}0 \times 3 = 99{\cdot}5$ cm³/mole

According to Eq. (8.17), the diffusivity D is thus

$$D = \frac{7{\cdot}4 \times 10^{-3} \times 78^{0{\cdot}5} \times 291}{0{\cdot}669 \times 99{\cdot}5^{0{\cdot}6}} = 1{\cdot}79 \times 10^{-5} \text{ cm}^2/\text{sec}$$

The value found experimentally (Table 8.3) is

$D = 1{\cdot}62 \times 10^{-5}$ cm²/sec

EXAMPLE 8.6 (Ref. 210). Estimate by means of Kuong's nomograph (Fig. 8.6) the value of the diffusivity of mannitol $C_6H_{14}O_6$ in a dilute aqueous solution at 70°C, assuming the viscosity of this solution to be equal to that of water, $\mu = 0{\cdot}4061$ cP.

Solution

The molal volume of the solute is estimated by summing the contributions given in Table 1.29 whence

$V_b = 14{\cdot}8 \times 6 + 3{\cdot}7 \times 14 + 7{\cdot}4 \times 6 = 185{\cdot}0$ cm³/mole

For water the association parameter is $x = 2{\cdot}6$, hence $xM = 2{\cdot}6 \times 18 = 46{\cdot}8$

$\mu/T = 0{\cdot}4061/(273+70) = 0{\cdot}001185$

A line is now constructed on the nomograph in Fig. 8.6 passing from the point 0·001185 on the μ/T scale, through the point 46·8 on the scale xM, and extended to intersect the reference line R. A second line is now constructed passing from this point on the axis R, through the point 185 on the V_b-axis and extended to intersect the diffusivity axis at a point corresponding to $D = 1{\cdot}86 \times 10^{-5}$ cm²/sec.

The value found experimentally is $D = 1{\cdot}56 \times 10^{-5}$.

The error of estimation thus is 19%.

EXAMPLE 8.7. Use Scheibel's formula to estimate the value of the diffusivity D for mannitol ($CH_2OH(CHOH)_4$) in a very dilute aqueous solution at 20°C.

Solution

Since the solute molecules are large, Eq. (8.20) should be employed.

The molal volume of mannitol, estimated by summing the contributions given in Table 1.19, is

$V_1 = 14{\cdot}8 \times 6 + 3{\cdot}7 \times 14 + 7{\cdot}4 \times 6 = 185{\cdot}0$ cm³/mole

The molal volume of water is

$$V_2 = 3{\cdot}7\times2+7{\cdot}4 = 14{\cdot}8 \text{ cm}^3/\text{mole}$$

so

$$D_{1,2} = \frac{8{\cdot}2\times10^{-8}\times293\left[1+\left(\frac{3\times14{\cdot}8}{185}\right)^{2/3}\right]}{1{\cdot}185^{1/3}} = 0{\cdot}585\times10^{-5} \text{ cm}^2/\text{sec}$$

The value of D determined experimentally is

$$D = 0{\cdot}605\times10^{-5} \text{ cm}^2/\text{sec.}$$

The error of estimation is thus 3·5%.

EXAMPLE 8.8 (Ref. 271). Use Othmer and Thakar's method to estimate the value of the diffusivity D for caffeine in water at 10°C (the molal volume of caffeine is $V_m = 232$ cm³/mole, the viscosity of water at 10°C is $\mu_w = 1{\cdot}3077$ cP).

Solution

a. Using equation (8.36)

$$D = \frac{14\times10^{-5}}{1{\cdot}3077^{1{\cdot}1}\times232^{0{\cdot}6}} = 0{\cdot}397\times10^{-5} \text{ cm}^2/\text{sec}$$

b. Using the nomograph presented in Fig. 8.8. Knowing the temperature and the molar volume of the solute, the straight line *1* may be constructed, whence $D = 0{\cdot}40\times10^{-5}$ cm²/sec. The experimental value is $0{\cdot}41\times10^{-5}$ cm²/sec.

EXAMPLE 8.9 (Ref. 276). Use Othmer and Thakar's method to estimate the diffusivity D for iodine in carbon disulphide at 16°C.

Solution

The ratio of the heats of vaporization, L_q/L_w, is 0·59, the viscosity of carbon disulphide at 20°C, μ_q, is 0·365 cP, the viscosity of water at 16°C, μ_w, is 1·11 cP, the molal volume of iodine is $V_m = 37{\cdot}0\times2 = 74{\cdot}0$ cm³/mole (Table 1.19).

a. Substitution in Eq. (8.40) gives

$$D_q = \frac{14\times10^{-5}}{74^{0{\cdot}6}\times0{\cdot}365\times1{\cdot}1^{1{\cdot}1\times0{\cdot}59}} = 2{\cdot}71\times10^{-5} \text{ cm}^2/\text{sec}$$

b. Solution by means of the nomograph, Fig. 8.8. Using the above values for t, μ_w and V_m the dotted lines shown in Fig. 8.8 are drawn successively from the left to the right-hand side of the figure, giving

$$D_q = 2{\cdot}8\times10^{-5} \text{ cm}^2/\text{sec}$$

The value determined experimentally is $D_q = 2{\cdot}97\times10^{-5}$ cm²/sec.

EXAMPLE 8.10. Use Ibrahim and Kuloor's equation to estimate the diffusivity of n-propanol in an aqueous solution at 15°C.

Solution

The viscosity of n-propanol vapour at 0°C is given by Table 3.4 to be $\mu_0 = 682\times10^{-5}$ cP, and the Sutherland constant is $C = 515{\cdot}6$.

The viscosity of n-propanol vapour at 15°C is thus estimated to be

$$\mu = 682\times10^{-5}\left(\frac{273{\cdot}2+515{\cdot}6}{288{\cdot}2+515{\cdot}6}\right)\times\left(\frac{288{\cdot}2}{273{\cdot}2}\right)^{3/2} = 702\times10^{-5} \text{ cP} = 7{\cdot}02\times10^{-5} \text{ P}$$

The molal volume of n-propanol, estimated by summing the contributions, in Table 1.19, is found to be

$$V = 14{\cdot}8\times3+3{\cdot}7\times8+12 = 86 \text{ cm}^3/\text{mole}$$

The molecular weight, M, of n-propanol is 60·1, so the estimated value of D is

$$D = \frac{\mu V}{13M-100} = \frac{7{\cdot}02\times10^{-5}\times86}{13\times60{\cdot}1-100} = 0{\cdot}90\times10^{-5} \text{ cm}^2/\text{sec}$$

The measured value is $0{\cdot}87\times10^{-5}$ cm²/sec, so the error of estimation is $-3{\cdot}3\%$.

6. SELECTION OF THE BEST METHOD FOR ESTIMATING THE DIFFUSIVITIES OF NON-IONIZED COMPOUNDS IN LIQUIDS

The accuracy of the available methods appears to be approximately the same; fo these methods only that of Kamal and Canjar[185,186] (Eq. (8.12)) has any very substantial theoretical basis. This method is quick to apply and gives good results, provided the liquids forming the system are listed in Tables 8.9 and 8.10.

The other methods discussed above are more empirical in character.

The equations of Wilke (8.16), Wilke and Chang (8.17), Scheibel (8.19)–(8.23) and of Ibrahim and Kullor (8.26)–(8.28) involve the use of a factor characteristic of the solvent (e.g. Wilke's factor ϕ, the association factor x, etc.). If this factor is not known these methods cannot be used. This difficulty may be avoided by using Wilke and Chang's equation, as modified by Sitaraman, Ibrahim and Kuloor (Eq. (8.18)) or by calculating the value of the coefficient a in Eq. (8.25) from the ratio of surface tensions (Eq. (8.29)).

Reid and Sherwood[306] have compared a substantial number of experimental diffusivities with values predicted by various methods. As a result of this comparison they recommend the use of Wilke and Chang's method (Eq. (8.17)). The use of the nomograph given in Fig. 8.6 greatly expedites calculations by this method.

The method of Othmer and Thakar, a nomograph for which is given in Fig. 8.8, is recommended in the case of aqueous solutions; however, it is not very satisfactory for non-aqueous solvents.

Diffusivities estimated by the above empirical methods may be subject to considerable error. It is advisable, therefore, to use several methods and to compare the results.

7. ESTIMATION OF DIFFUSION COEFFICIENTS IN ELECTROLYTE SOLUTIONS

Diffusion of a dissolved ionized electrolyte involves the diffusion of both cations and anions. This phenomenon was first treated quantitatively by Nernst (1888); his considerations refer to weak electrolytes, which according to the old theory of Arrhenius, are almost completely ionized in very dilute solutions. According to this simplified theory, the ionic mobilities do not depend under such conditions on the concentration of the electrolyte.

By the term ion mobility we understand the velocity u (cm/sec) of an ion in an electric field under a potential gradient of 1 V/cm.

This mobility is usually expressed in $cm^2.sec^{-1}.V^{-1}$.

The equivalent conductances of the cation, l_C, and of the anion, l_A, depend on the mobility being given by

$$l_C = Fu_C \quad \text{and} \quad l_A = Fu_A \tag{8.48}$$

where F is Faraday's constant, equal to 96,500 of international coulombs per gram-equivalent.

The equivalent conductances l_C° and l_A° of the ions in an infinitely dilute solution are referred to as the limiting equivalent conductances.

These quantities are expressed in $ohm^{-1}.cm^2.(gram\text{-}equivalent)^{-1}$.

According to Nernst, in the case of univalent ions,

$$\frac{1}{D^\circ} = \left(\frac{1}{u_C} + \frac{1}{u_A}\right)\frac{F}{2RT} \tag{8.49}$$

or

$$\frac{1}{D^\circ} = \left(\frac{1}{l_C} + \frac{1}{l_A}\right)\frac{F^2}{2RT} \tag{8.50}$$

where D° is the diffusivity in a very dilute solution, cm²/sec,
R is the gas constant equal to 8·316 Joule/°K.mole,
T is the absolute temperature, °K,
l_C and l_A are the limiting equivalent conductances of cation and anion, $ohm^{-1}.cm^2.(gram\text{-}equivalent)^{-1}$,
and F is Faraday's constant.

TABLE 8.13. EQUIVALENT IONIC CONDUCTANCES AT INFINITE DILUTION FOR CATIONS IN AQUEOUS SOLUTIONS (AFTER D'ANS[11]) $l_C^\circ(ohm^{-1}.cm^2.(gram\text{-}equivalent)^{-1})$

Cation	Temperature, °C			
	0	18	25	100
$H^+(H_2O)$	225	315	349·7	637
Li^+	19·1	33·4	38·68	120
Na^+	25·85	43·5	50·10	150
K^+	40·3	64·6	73·5	200
Rb^+	43·5	64·3	76·4	
Cs^+	44	68	76·8	200
NH_4^+	40·3	64	73·7	184·3
$\frac{1}{2}Be^{2+}$			45	
$\frac{1}{2}Mg^{2+}$	28·5	45	53·06	170
$\frac{1}{2}Ca^{2+}$	30·8	51	59·5	187
$\frac{1}{2}Sr^{2+}$	31	51	59·5	
$\frac{1}{2}Ba^{2+}$	33·6	55	63·7	200
$\frac{1}{2}Ra^{2+}$	33	56·5	66·8	
$\frac{1}{3}Al^{3+}$	29		63	
$\frac{1}{3}Sc^{3+}$			64·7	
$\frac{1}{3}La^{3+}$	35·0	59·2	69·7	220
$\frac{1}{3}Ce^{3+}$			67	
$\frac{1}{3}Pr^{3+}$			65·4	
$\frac{1}{3}Nd^{3+}$			64·3	
$\frac{1}{3}Sm^{3+}$			65·8	
$\frac{1}{3}Cr^{3+}$			67·0	
$\frac{1}{2}Mn^{2+}$	27	44·5	53·5	
$\frac{1}{2}Fe^{2+}$	28	44·5	53·5	
$\frac{1}{3}Fe^{3+}$			68	
$\frac{1}{2}CO^{2+}$	28	45	54	
$\frac{1}{2}Ni^{3+}$	28	45	54	
$\frac{1}{2}Cu^{2+}$	28	45·3	56	
Ag^+	33	54·36	61·9	180
$\frac{1}{2}Zn^{2+}$	28	45·0	53·5	
$\frac{1}{2}Cd^{2+}$	28	45·1	54	
Tl^+	43·3	66	74·9	
$\frac{1}{2}Pb^{2+}$	37·5	60·5	70	

TABLE 8.14. EQUIVALENT IONIC CONDUCTANCES AT INFINITE DILUTION FOR ANIONS IN AQUEOUS SOLUTIONS (AFTER D'ANS[11]), l_A° (ohm^{-1}.cm^2. (gram-equivalent)$^{-1}$)

Anion	Temperature, °C			
	0	18	25	100
OH^-	105	174	200	446
F^-	—	46·6	55·4	—
Cl^-	41·4	65·5	76·32	207
ClO_2^-			52	
ClO_3^-	36	55·0	64	172
ClO_4^-	37·3	59·1	68	179
Br^-	43·1	67·6	78·4	—
BrO_3^-	31	49·0	56	155
I^-	42·0	66·5	76·9	—
IO_3^-	21	33·9	41·0	127
IO_4^-	—	49	55·6	—
SH^-	40	57	65	—
SO_3H^-	27	—	50	—
$\frac{1}{2}SO_3^{2-}$			72	
$\frac{1}{2}SO_4^{2-}$	41	68·3	79·8	256
$\frac{1}{2}S_2O_3^{2}$			85	
$\frac{1}{2}S_2O_4^{2-}$	34		66·5	
$\frac{1}{2}S_2O_6^{2-}$			93	
$\frac{1}{2}S_2O_8^{2-}$			86	
SCN^-	41·7	56·6	66·5	
$\frac{1}{2}SeO_4^{2-}$		65	75·7	
N_3^-			69·5	
NO_2^-	44	59	72	
NO_3^-	40·2	61·7	71·42	189
NCO^-		54·8	64·6	
$PO_4H_2^-$		28	36	
$\frac{1}{2}PO_4H^{2-}$			57	
$AsO_4H_2^-$			34	
$SbO_4H_2^-$			31	
CO_3H^-	—	—	44·5	—
$\frac{1}{2}CO_3^{2-}$	36	60·5	72	—
CN^-			78	
$\frac{1}{2}CrO_4^{2-}$	42	72	85	—
$\frac{1}{2}MoO_4^{2-}$			74·5	
$\frac{1}{2}WO_4^{2-}$	35	59	69·4	
MnO_4^-	36	53	62·8	
ReO_4^-		46·5		
$HCOO^-$		47		
CH_3COO^-	20	34	41	130
$\frac{1}{2}(C_2O_4)^{2-}$		63		
$\frac{1}{2}(C_4H_4O_6)^{2-}$		55		

If the ions are not univalent, and the valence of the cation is z_C and of the anion z_A, the coefficient 2 in the denominator of the above equation is replaced by the expression $(1/z_C+1/z_A)$. Values of the equivalent ionic conductances l_C° and l_A° at infinite dilution are listed in Tables 8.13 and 8.14.

In practice Nernst's equation is found to give good results for diffusivities in very dilute solutions. For example, for potassium chloride, the mobilities of the ions are

$$u_{K+} = 6{\cdot}75 \times 10^{-4} \text{ cm}^2/\text{sec.V}$$

and

$$u_{Cl-} = 6{\cdot}85 \times 10^{-4} \text{ cm}^2/\text{sec.V}$$

On substituting these values in Eq. (8.49), it is found that, at 18°C

$$D^\circ = 1{\cdot}70 \times 10^{-5} \text{ cm}^2/\text{sec}$$

The experimental value of D° is $1{\cdot}69 \times 10^{-5}$ at this temperature.

Nernst's equation is also suitable for estimating the effect of temperature on diffusion coefficients. If Tables 8.13 and 8.14 do not contain values of l_C and l_A for the temperature T, at which the diffusivity D_T° is required, Reid and Sherwood[306] recommend that the diffusivity D_{25}° at 25°C be first calculated and that this value should then be "corrected" for temperature using the equation

$$D_T^\circ = D_{25}^\circ \frac{T}{334\mu_T} \tag{8.51}$$

where μ_T denotes the viscosity of water at temperature T°K, cP.

Diffusion coefficients in electrolytic systems are found to depend on concentration (Fig. 8.2). There is no easy method for predicting concentration effects theoretically, and empirical or semi-empirical methods must at present be used for technological calculations.

For concentrations up to 2 N, use may be made of a semi-empirical formula put forward by Gordon[133]

$$D = D^\circ \frac{V}{x_1 V_1} \frac{\mu_1}{\mu} \left(1 + m \frac{\partial \ln \gamma}{\partial m}\right) \tag{8.52}$$

where D is the diffusion coefficient of the electrolyte in a m molar solution, cm²/sec,
D° is the diffusion coefficient of the electrolyte at infinite dilution, cm²/sec,
V is the volume of solution, cm³,
V_1 is the partial molal volume of solvent, cm³/mole,
x_1 are the moles of the solvent in V cm³ of solution,
m is the molality of the solution, moles/1000 g solvent,
μ is the dynamic viscosity of the solution, P,
μ_1 is the dynamic viscosity of the solvent, P,
and γ is the mean molar activity coefficient (as taken from tables).

The equation enables diffusivities to be estimated with fair accuracy, though lack of some of the necessary data (in particular of V_1 and γ) may make the use of Gordon's formula rather difficult. This type of difficulty is even more pronounced in the case of certain more accurate equations, which have been developed. A review of such equations is given by Reid and Sherwood.[306]

EXAMPLE 8.11. Estimate the diffusivity D of lithium chloride in a very dilute aqueous solution at 25°C ($T = 298$°K).

Solution

According to Eq. (8.50)

$$D^\circ = \frac{2RT}{\left(\frac{1}{l_C} + \frac{1}{l_A}\right) F^2}$$

From Table 8.13 for Li^+ $\quad l_C = 38{\cdot}68$
From Table 8.14 for Cl^- $\quad l_A = 76{\cdot}32$

Hence

$$D_{25}^{\circ} = \frac{2 \times 8{\cdot}316 \times 298}{\left(\frac{1}{38{\cdot}68} + \frac{1}{76{\cdot}32}\right) \times 96500^2} = 1{\cdot}37 \times 10^{-5} \text{ cm}^2/\text{sec}$$

An experimental value for the diffusivity is given in Table 8.8 for a 0·01 N solution, though this is for a temperature of 9°C. From this table

$$D_9 = 0{\cdot}757 \text{ cm}^2/24 \text{ hr}$$
$$= 0{\cdot}876 \times 10^{-5} \text{ cm}^2/\text{sec}.$$

According to Eq. (8.51)

$$D_9 = D_{25} \times \frac{(273 \times 9)}{334\,\mu_9}$$

The viscosity of water at 9°C, μ_9, is 1·346 cP, hence

$$D_9 = 1{\cdot}37 \times 10^{-5} \times \frac{282}{334 \times 1{\cdot}346} = 0{\cdot}858 \times 10^{-5} \text{ cm}^2/\text{sec}$$

The difference between the calculated and observed values is 2·05%.

EXAMPLE 8.12. (This is an extension of an example given by Reid and Sherwood.[306]). Use Gordon's equation (8.52) to estimate the value of the diffusivity D of sodium hydroxide in a 2 N aqueous solution at 25°C.

Solution

The following physico-chemical data are first taken from standard tabulations:

(i) Densities of aqueous solutions of NaOH at 15°C (after D'Ans[11])

Mass of 1 l., g	1010	1020	1030	1040	1050	1060	1070	1080	1090	1100
NaOH, %	0·95	1·83	2·72	3·61	4·51	5·41	6·31	7·21	8·10	9·00
NaOH, g/l.	9·57	18·69	27·97	37·51	47·30	57·29	67·47	77·81	88·28	99·00

(ii) Viscosities of aqueous solutions of NaOH (cP) at 25°C (Refs. 11, 184)

Concentration of the solution, moles/1000 g water	0·125	0·25	0·5	1·0	2·0
Viscosity, cP	0·903	0·944	0·991	1·105	1·42

(iii) Viscosity of water at 25°C $\mu_1 = 0{\cdot}8937$ cP.

(iv) Mean activity coefficients of NaOH in an aqueous solution at 25°C (Ref. 184)

Concentration (moles/1000 g water)	0·05	0·1	0·5	1·0	2·0	3·0
Mean γ	0·818	0·766	0·690	0·678	0·709	0·784

By interpolating the above density data it is found that, for a 2 N solution containing 80 g NaOH/1000 cm³:

the mass of 1 l. of the solution is equal to 1082·3 g, and that this volume of solution contains 1002·3 g of water.

If the volume V of the solution is 1000 cm³, then

$$x_1 = \frac{1002{\cdot}3}{18} = 55{\cdot}6 \text{ moles } H_2O/1000 \text{ cm}^3 \text{ solution.}$$

Examination of the density data shows that the number of water moles per 1000 cm³ of solvent is almost constant for all concentrations given in the table, and amounts to about 55·5.

It follows that the quantities V/x_1 and V_1 in Eq. (8.52) are almost equal, thus enabling the equation to be simplified.

Since the mass of 1 l. of the molal solution (1000+80 = 1080 g) is almost the same as the value calculated for the 2 N solution (1082·3 g) the normality may in this case be taken to be equal to the molality without appreciable error.

The variation of the activity coefficient γ with molality of the solution is shown in Fig. 8.10. The gradient $(d\gamma/dm)$ at $m = 2$ is found by drawing a tangent to the curve on this figure and is given by $\tan\alpha = (d\gamma/dm) = \frac{0{\cdot}059}{1{\cdot}25} = 0{\cdot}0475$. Now $d\ln\gamma = d\gamma/\gamma$ and since, when $m = 2$, $\gamma = 0{\cdot}709$

$$\frac{m}{\gamma}\frac{d\gamma}{dm} = \frac{2}{0{\cdot}709}\times 0{\cdot}0475 = 0{\cdot}135$$

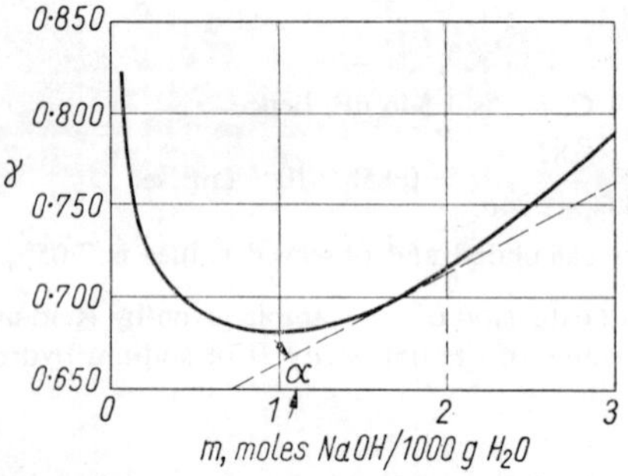

FIG. 8.10. Mean ionic activity coefficient for solution of NaOH in water indicating graphical determination of the quantity $d\gamma/dm = \tan\alpha$ (γ is the activity coefficient in the aqueous solution and m is the molality of the solution).

D_{25}° may be estimated by substituting the values

$$l_{NA} = 50{\cdot}10$$

and $l_{OH} = 200$

from tables (8.13) and (8.14) into Eq. (8.50) whence

$$D_{25}^{\circ} = \frac{2\times 8{\cdot}316\times 298}{\left(\frac{1}{50{\cdot}1}+\frac{1}{200}\right)\times(96{,}500)^2} = 2{\cdot}14\times 10^{-5}\ \text{cm}^2/\text{sec}$$

D_{25} is now estimated by substituting the appropriate quantities into Gordon's formula (8.52) (making the simplification of setting $V/x_1 \approx V_1$), whence

$$D_{25} = 2{\cdot}14\times 10^{-5}\times\frac{0{\cdot}894}{1{\cdot}42}\times(1+2\times 0{\cdot}135) = 1{\cdot}71\times 10^{-5}\ \text{cm}^2/\text{sec}$$

8. DIFFUSION IN SOLUTIONS CONTAINING SEVERAL DISSOLVED ELECTROLYTES

Because of the mutual interaction of different ions in an aqueous solution containing several dissolved electrolytes, the diffusion process does not proceed according to the simple scheme used by Nernst to derive Eq. (8.50).

Attempts at applying this equation to the phenomena of diffusion in solutions containing several kinds of ions have led Vinograd and McBain[385] to the formulation of relationships which, owing to the lack of the necessary data and also their low accuracy, are of little use in technological work.

REFERENCES

1. ABAS-ZADE, A. K., *Dokl. Akad. Nauk S.S.S.R.* **68**, 665 (1949).
2. ALTENBURG, K., *Chemische Technik*, **7**, 383 (1955).
3. ALTENBURG, K., *Z. phys. Chem.* **216**, 126, 139, 146 (1961); **217**, 71 (1961).
4. ALTENBURG, K., *Z. Elektrochem.* **65**, 801, 805 (1962); *Chem. Ing. Techn.* **34**, 13 (1962).
5. American Society of Mechanical Engineers, *Progress in International Research on Thermodynamic and Transport Properties, Second Symposium on Thermophysical Properties*, Jan. 24–26, 1962. Academic Press, New York.
6. ANDRADE, E. N., *Phil. Mag.* **17**, 497, 698 (1934).
7. ANDRUSSOV, L., *Z. Elektrochem.* **60**, 412 (1956).
8. ANDRUSSOV, L., *Z. Elektrochem.* **62**, 608 (1958).
9. ANDRUSSOV, L., *Chem. Proc. Eng.* **43** (2), 65 (1962).
10. AMDUR, I., *Amer. Inst. Chem. Eng. J.* **8**, 521 (1962).
11. D'ANS, J., LAX, E., *Taschenbuch für Chemiker und Physiker*. Berlin 1943.
12. ARNOLD, J. H., *Chem. Phys.* **1**, 170 (1933).
13. ARNOLD, J. H., *Ind. Eng. Chem.* **22**, 1091 (1930).
14. ARNOLD, J. H., *J. Amer. Chem. Soc.* **52**, 3937 (1930).
15. ARRHENIUS, S. A., *Z. phys. Chem.* **1**, 285 (1887).
16. AUERBACH, N., *Experientia*, **4**, 473 (1948) (after Partington[282]).
17. BACHINSKII, A. I., *Vremennik obshchestva im. Ledentseva*, prilozh. No. 3 (1913); BATSCHINSKII A., *Phys. Z.* **13**, 1157 (1912); *Z. phys. Chem.* **84**, 643 (1913).
18. BAIKUNTH NATH MISRA, YATENDRA PAL VARSHNI, *J. Chem. Eng. Data*, **6**, 194 (1961).
19. BATES, O. K., HAZZARD, G., PALMER, G., *Ind. Eng. Chem. Anal. Ed.* **10**, 314 (1938); BARRAT, T.: *Int. Crit. Tables*, **5**, 227 (1929); CECIL, O. B., MUNCH, R. H., *Ind. Eng. Chem.* **48**, 437 (1956).
20. BATSANOV, S. S., *Zhurn. Neorg. Khim.* **3**, 241 (1958).
21. BEARMAN, R. J., VAIDHYANATHAN, V. S., *J. Chem. Phys.* **34**, 264 (1961).
22. BELYAEV, A. I., ZHEMCHUZHINA, E. A., FIRSANOVA, L. A.: *Fizicheskaya khimiya rasplavlennykh solei* (*Physical Chemistry of Fused Salts*). Moscow 1957.
23. BENDIRSKII, V. A., NIKITIN, N. S., TAT'EVSKII, V. M., *Zhurn. Fiz. Khim.* **36** (1), 63 (1962).
24. BERGMANN, L.: *Der Ultraschall und seine Anwendung in Wissenschaft und Technik*. 6th Ed. Zurich 1954.
25. BIRD, R. B., HIRSCHFELDER, J. O., CURTISS, C. F., *Trans. Amer. Soc. Mech. Eng.* **76**, 1011 (1954); *Molecular Theory of Gases and Liquids*, New York 1954.
26. BOCK, H., *Ann. Physik.* (6), **8**, 134 (1950).
27. BONDI, A., SIMKIN, D. J., *Amer. Inst. Chem. Eng. J.* **6**, 191 (1960); **6**, No. 4, 3D (1960).
28. BONDI, A., *A.I.Ch.E.J.* **8**, 610 (1962).
29. BONDI, A., *Ind. Eng. Chem. Fundamentals* **2** (2), 95 (1963).
30. BORN, M., GREEN, H. S., *Proc. Roy. Soc.* **A 188**, 10 (1946).
31. BORN, M., GREEN, H. S., *Proc. Roy. Soc.* (London) **A 190**, 455 (1947).
32. BOWDEN, S. T., *J. Chem. Phys.* **23**, 2454 (1955).
33. BOSWORTH, R. C. L.: *Heat Transfer Phenomena*. Sydney—New York.
34. BREBACH, W. J., THODOS, G., *Ind. Eng. Chem.* **50**, 1095 (1958).
35. BRETSZNAJDER, S.: *Zagadnienia projektowania procesów przemysłu chemicznego* (*Problems of the Chemical Process Design*). Warszawa 1956.
36. BRETSZNAJDER, S.: *Własności gazów i cieczy* (*Properties of Gases and Liquids*). Warszawa 1962.
37. BRIDGMAN, P. W., *Proc. Amer. Acad.* **57**, 75 (1922); **59**, 139, 154 (1923); *Amer. J. Sci.* **7**, 81 (1924).
38. BRIDGMAN, P. W., *The Physics of High Pressure*. London 1949.
39. BRILLOUIN, L., *J. d. Phys.* **3**, 326, 362 (1922).
40. BRILLOUIN, L., *Compt. Rend.* **159**, 27 (1914).
41. BROENSTED, J. N., KOEFOED, J., *Kgl. Danske Videnskab. Selskab. Mat. fys. Medd.* **22**, No. 17, 1 (1946).

42. Brokaw, R. S., *Ind. Eng. Chem.* **47**, 2398 (1955).
43. Brokaw, R. S., *Journ. Chem. Phys.* **29**, 391 (1958).
44. Brokaw, R. S., *Journ. Chem. Phys.* **32**, 1005 (1960).
45. Brokaw, R. S., *Journ. Chem. Phys.* **35**, 1569 (1961).
46. Bromley, L. A., Wilke, C. R., *Ind. Eng. Chem.* **43**, 1641 (1951).
47. Bromley, L. A.: *Thermal Conductivity of Gases at Moderate Pressures.* University of California Radiation Laboratory Rept. UCRL-1852 (after Gambill[115] and Reid and Sherwood[306]).
48. Buehler, C. A., *J. Phys. Chem.* **42**, 1209 (1938).
49. Buddenberg, J. W., Wilke, C. R., *Ind. Eng. Chem.* **41**, 1345 (1949).
50. Buddenberg, J. W., Wilke, C. R., *J. Phys. Colloid Chem.* **55**, 1491 (1951).
51. Burry, V. J., Jr., Koeller, R. C., *Amer. Inst. Chem. Eng. J.* **6**, 274 (1960).
52. Butler, J. N., Brokaw, R. S., *J. Chem. Phys.* **26**, 1636 (1957).
53. Carman, P. C., Stein, L. H., *Trans. Faraday Soc.* **52**, 619 (1956); Miller, L., Carman, P. C., *Trans. Faraday Soc.* **57**, 2143 (1962).
54. Carr, N. L., *J. of Petrol. Technol.* (X. 1959), p. 47.
55. Carr, N. L., Parent, J. D., Peck, R. E., *Chem. Eng. Progress Symp. Ser.* **51**, No. 16, 91 (1955).
56. Carr, Wolczynsky, *J. Amer. Chem. Soc.* **56**, 2541 (1934).
57. Chapman, S., *Phil. Trans.* **217**, 115 (1916); *Phil. Mag.* **48**, 593 (1924).
58. Chapman, S., *Phil. Trans.* **A 211**, 459 (1912).
59. Chapman, S., Cowling, T. G.: *The Mathematical Theory of Non-Uniform Gases.* Cambridge 1952.
60. Cheung, H., Bromley, L. A., Wilke, C. R., *Amer. Inst. Chem. Eng. J.* **8**, 221 (1962).
61. Chirkin, V. S.: *Teplofizicheskie svoistva materialov* (*Thermophysical Properties of Materials*). Moscow 1959.
62. Cohen, E., Bruins, R., *Z. phys. Chem.* **103**, 432 (1923); **109**, 422 (1924).
63. Collins, F. C., Navidi, M. H., *J. Chem. Phys.* **22**, 1254 (1954) (after Kamal and Canjar[186]).
64. Comings, E. W., Egly, R. S., *Ind. Eng. Chem.*, **32**, 714 (1940); Comings, E. W., Mayland, B. J., *Chem. Met. Eng.* **52** (3) 115 (1945); Comings, E. W., Mayland, B. J., Egly, R. S.: The Viscosity of Gases at High Pressures. *Univ. Illinois, Eng. Expt. Sta. Bull. Ser.* 354 *Urbana Ill.* 1944–1945 (last work after Reid and Sherwood[306]).
65. Comings, E. W., Mayland, B. J., *Chem. Met. Eng.* **52**, No. 3, 115 (1945).
66. Comings, E. W., Nathan, M. F., *Ind. Eng. Chem.* **39**, 964 (1947); Leng, D. E., Comings, E. W., *Ind. Eng. Chem.* **49**, 2042 (1957).
67. Cornelissen, J., Waterman, H. L., *Chem. Eng. Sci.* **4**, 238 (1955).
68. Cox, W. M., Wolfenden, J. H., *Proc. Roy. Soc.* (London) **A 145**, 475 (1934); Kaminsky, M., *Z. Naturforsch.* **12a**, 424 (1957).
69. Damasius, G., Thodos, G., *Ind. Eng. Chem. Fundamentals* **2**, No. 1, 73 (1963).
70. Darken, L. S., *Trans. Amer. Inst. Mining Engrs* (*Metals Division*), **175**, 184 (1948) (after Kamal and Canjar[185]).
71. Davis, D. S., *Ind. Eng. Chem.* **33**, 1537 (1941).
72. Davis, D. S., *Ind. Eng. Chem.* **34**, 1231 (1942).
73. Davis, D. S.: *Chemical Engineering Nomographs.* New York 1944.
74. Davis, D. S., *Brit. Chem. Eng.* **3**, 210 (1958).
75. Davis, D. S., *Brit. Chem. Eng.* **4**, 478 (1959).
76. Deich, A. Ya., *Zhurn. Fiz. Khim.* **36**, 656, 1777 (1962).
77. Denbigh, K. G., *Trans. Faraday Soc.* **36**, 936 (1940).
78. Denbigh, K. G., *J. Soc. Chem. Ind.* **64**, 61 (1946).
79. Dovzhenko, F. P., *Zhurn. Fiz. Khim.* **31**, 2517 (1957).
80. Duncan, J. B., Toor, H. L., *Amer. Inst. Chem. Eng. J.* **8**, No. 1, 38 (1962).
81. Durbin, L., Riki Kobayashi, *J. Chem. Phys.* **37**, 1643 (1962).
82. Efremov, Yu. V., Golubev, I. F., *Zhurn. Fiz. Khim.* **36**, 1222 (1962).
83. Einstein, A., *Ann. d. Physik* (4), **19**, 289 (1906).
84. Eilers, H., *Kolloid Z.* **97**, 313 (1941).
85. Eldarov, F. G., *Zhurn. Fiz. Khim.* **32**, 2443 (1958); **36**, 1414 (1962).
86. Eldarov, F. G., *Zhurn. Fiz. Khim.* **34**, 1205 (1960).
87. Ember, G., Ferron, J. R., Wohl, K., *J. Chem. Phys.* **37**, 891 (1962).
88. Enskog, D., *Svensk. Vet. Akad. Handl.* **63**, No. 4 (1922); *Arkiv Mat. Astron. Fys.* **16**, No. 16 (1922); **21 A**, No. 13 (1928) (after Partington[282]).
89. Eötvös, R., *Ann. Phys.* **27**, 448 (1886).

90. EUCKEN, A., *Phys. Z.* **12**, 1101 (1911); **14**, 324 (1913).
91. EUCKEN, A., JAKOB, M.: *Der Chemie-Ingenieur*. Leipzig 1933.
92. EUCKEN, A.: *Lehrbuch der chemischen Physik*. Leipzig 1944.
93. EWING, C. T., WALKER, B. E., GRAND, J. A., MILLER, R. R., *Chem. Eng. Progress Symp. Ser.* **53**, No. 20, 19 (1957).
94. EYRING, H., *J. Chem. Phys.* **4**, 283 (1936).
95. FALKENHAGEN, H., *Physik Z.* **32**, 745 (1931); FALKENHAGEN, H., VERNON, E. L., *Physik Z.* **33**, 140 (1932); FALKENHAGEN, H., DOLE, M., *Physik Z.* **30**, 611 (1929).
96. FALKOVSKII, V. B., *Zhurn. Fiz. Khim.* **18**, 1639 (1948).
97. FARCAS, A. (Ed.): *Physical Chemistry of the Hydrocarbons*. Vol. II. New York 1953.
98. FAUST, O., *Z. phys. Chem.* **79**, 97 (1912).
99. FILIPPOV, L. P., *Dissertatsia Mosk. Gos. Univ.* 1951.
100. FILIPPOV, L. P., NOVOSELOVA, N. S., *Vestnik Moskov. Gos. Univ.* **10**, No. 3, 37 (1955).
101. FILIPPOV, L. P., *Zhurn. Fiz. Khim.* **31**, 582, 1999 (1957); *Vestnik Moskov. Gos. Univ.* No. 1, 111 (1956); No. 2, 95 (1956).
102. FILIPPOV, L. P., *Zhurn. Fiz. Khim.* **37**, 201 (1963).
103. FILIPPOVA, G. P., ISHKIN, I. P., *Inzhenerno-Fizicheskii Zhurn.* **4** (10), 30 (1961).
104. FLYNN, L. W., THODOS, G., *J. Chem. Eng.* Data **6**, 457 (1961).
105. FORMAN, J. Ch., THODOS, G., *Amer. Inst. Chem. Eng. J.* **4**, 356 (1958); **6**, 206 (1960).
106. FOURIER, A., *Compt. Rend.* **221**, 234 (1945).
107. FRANCK, E. U., *Chem. Ing. Techn.* **25**, 238 (1953).
108. FRANCK, E. U., JOST, W., *Z. Elektrochem.* **62**, 1054 (1958).
109. FRIEND, J. N., HARGREAVES, W. D., *Phil. Mag.* **34**, 643, 810 (1943); **35**, 136, 631 (1944); **36**, 731 (1945).
110. GALINKER, V. S., TIAGAI, V. A., FENERLI, G. N., *Zhurn. Fiz. Khim.* **36**, 2638 (1962).
111. GAMBILL, W. R., *Chem. Engng.* **65**, No. 8, 146 (Apr. 1958); **65**, No. 9, 143 (May, 1958).
112. GAMBILL, W. R., *Chem. Engng.* **65**, No. 19, 169 (1958).
113. GAMBILL, W. R., *Chem. Engng.* **65**, No. 21, 157 (1958).
114. GAMBILL, W. R., *Chem. Engng.* **65**, No. 23, 157 (1958).
115. GAMBILL, W. R., *Chem. Engng.* **65**, No. 17, 121 (1958).
116. GAMBILL, W. R., *Chem. Engng.* **66**, No. 12, 181 (1959).
117. GAMBILL, W. R., *Chem. Engng.* **66**, No. 14, 157 (1959).
118. GAMBILL, W. R., *Chem. Engng.* **66**, No. 1, 127 (1959).
119. GAMBILL, W. R., *Chem. Engng.* **66**, No. 3, 123 (1959).
120. GAMBILL, W. R., *Chem. Engng.* **66**, No. 5, 151 (1959).
121. GAMBILL, W. R., *Chem. Engng.* **66**, No. 16, 129 (1959).
122. GAMBILL, W. R., *Chem. Engng.* **66**, No. 25, 169 (1959).
123. GAMSON, B. W., *Chem. Eng. Progress* **45**, 154 (1949).
124. GARDNER, K. A., *Trans. Amer. Soc. Mech. Eng.* **65**, 404 (1943).
125. GERSH, S. I.: *Glubokoe okhlazhdene* (*Low-temperature Refrigeration*). Moscow 1947, 1949.
126. GIELESSEN, J., SCHMATZ, W., *Z. phys. Chem. N.F.* **27**, 157 (1961).
127. GILLILAND, E. R., *Ind. Eng. Chem.* **26**, 681 (1934).
128. GIRIFALCO, L. A., *J. Chem. Phys.* **23**, 2446 (1955).
129. GLASER, F., GEBHARDT, F., *Chem. Ing. Techn.* **31**, 743 (1959).
130. GLASSTONE, S., LAIDLER, K. J., EYRING, H.: *The Theory of Rate Processes*. New York 1941.
131. GOLUBEV, I. F.: *Viazkost' gazov i gazovykh smesei* (*Viscosity of Gases and Gaseous Mixtures*). Moscow 1959.
132. GONIKBERG, M. G., *Zhurn. Fiz. Khim.* **21**, 7 (1947) (after Filippova and Ishkin[103]).
133. GORDON, A. R., *J. Chem. Phys.* **5**, 522 (1937).
134. GREEN, H. S.: *The Molecular Theory of Fluids*. Amsterdam 1952.
135. GRIEVES, R. B., THODOS, G., *Amer. Inst. Chem. Eng. J.* **8**, 550 (1962).
136. GRUNBERG, L., NISSAN, A. H., *Nature*, **164**, 799 (1949).
137. GRUNBERG, L., NISSAN, A. H., *Ind. Eng. Chem.* **42**, 886 (1950).
138. GRUSS, H., SCHMICK, H., *Wiss. Veröff. Siemens-Konzern* **7**, 202 (1928) (after Lehmann[217]).
139. GURNEY, R. W.: *Ionic Processes in Solution*. McGraw-Hill. London 1953.
140. HAMANN, S. D.: *Physico-Chemical Effects of Pressure*. London 1957.
141. HAMMICK, D. L., ANDREW, L. W., *J. Chem. Soc.* 754 (1929).
142. HARKINS, W. D., ROBERTS, *Phys. Rev.* **15**, 137 (1920); *J. Am. Chem. Soc.* **44**, 653 (1922).

143. Hatschek, E., *Trans. Faraday Soc.* **9**, 80 (1913).
144. Hatschek, E., *Kolloid Z.* **27**, 163 (1920).
145. Hatschek, E.: *The Viscosity of Liquids*. London 1928.
146. Helfand, E., Rice, S. A., *J. Chem. Phys.* **32**, 1642 (1960).
147. Henning, F.: *Wärmetechnische Richtwerte*. Berlin 1938.
148. Henning, F., Zipperer, L., *Gas- u. Wasserfach*, **79**, 69 (1936).
149. Herzog, R., *Ind. Eng. Chem.* **36**, 997 (1944).
150. Higgins, T. H. S., Raw, C. J. G., *J. Chem. Phys.* **27**, 623 (1957).
151. Hind, R. K., McLaughlin, E., Ubbelohde, U. R., *Trans. Faraday Soc.* **56**, 328 (1960).
152. Hiroshi Shimotake, Thodos, G., *Amer. Inst. Chem. Eng. J.* **9** (1), 68 (1963); *J. Chem. Eng. Data* **8** (1), 88 (1963).
153. Hirschfelder, J. O., Bird, R. B., Spotz, E. L., *J. Chem. Phys.* **16**, 968 (1948).
154. Hirschfelder, J. O., Bird, R. B., Spotz, E. L., *Chem. Revs.* **44**, 205 (1949).
155. Hirschfelder, J. O.: *Sixth Symposium on Combustion*, p. 351. Reinhold. New York 1957; *J. Chem. Phys.* **26**, 274, 282 (1957).
156. Hirschfelder, J. O., Buehler, R. J., McGee, H. A., Jr., Sutton, J. R., *Ind. Eng. Chem.* **50**, 375, 386 (1958).
157. Hirschfelder, J. O., Taylor, M. H., Kihara, T., Rutherford, R., *The Physics of Fluids* **4**, 663 (1961).
158. Hilsenrath, J., Touloukian, I. S., *Trans. ASME* **76**, 967 (1954).
159. Hobler, T.: *Ruch ciepła i wymienniki* (*Heat Transfer and Exchangers*). Warszawa 1959.
160. Hougen, O. A., Watson, K. M., Ragatz, R. A.: *Chemical Process Principles*. Vol. I, 2nd Ed., 1954, Vol. II, 2nd Ed. 1959. New York.
161. Hudson, G. H., McCoubrey, J. C., Ubbelohde, A. R., *Trans. Faraday Soc.* **56**, 1144 (1960).
162. Hurwic, J.: (a) Dielectric constant and dipole moment; (b) Dipole moments (in Polish in the collective work *Chemia i technika*), Vol. IV. Warszawa 1949.
163. Ibrahim, S. H., Kuloor, N. R., *Brit. Chem. Eng.* **5** (1), 795 (1960).
164. Ibrahim, S. H., Kuloor, N. R., *Brit. Chem. Eng.* **6** (11), 781 (1961).
165. Ibrahim, S. H., Kuloor, N. R., *Brit. Chem. Eng.* **6**, 862 (1961).
166. Ianson, N. Ya., *Zhurn. Fiz. Khim.* **35**, 2108 (1961).
167. Innes, K. K., Albright, L. F., *Ind. Eng. Chem.* **49**, 1793 (1957).
168. Irving, J. P., Smith, J. M., *Amer. Inst. Chem. Eng. J.* **7** (1), 91 (1961).
169. Jackson, W. M., *J. Phys. Chem.* **60**, 789 (1956).
170. Jäger, G., *Wien Ber.* **105**, 15 (1896).
171. Janz, G. J.: *Estimation of Thermodynamic Properties of Organic Compounds*. New York 1958.
172. Jeans, J. A.: *Dynamical Theory of Gases*. 1921.
173. Jeferson, T. B., Witzell, O. W., Sibbitt, W. L., *Ind. Eng. Chem.* **50**, 1589 (1958).
174. Jellinek, K.: *Lehrbuch der physikalischen Chemie*. Vol. I. Stuttgart 1928.
175. Jellinek, K.: *Lehrbuch der physikalischen Chemie*. Vol. II. Stuttgart 1937.
176. Joffe, J., *Ind. Eng. Chem.* **39**, 837 (1947); *Chem. Eng. Progr.* **45**, 160 (1949).
177. Johnson, A. I., Huang, C. J., *Chem. Eng.* **61**, No. 2, 205 (1954).
178. Joliet, J. F.: *C.R.* **234**, 428 (1950); **236**, 1259 (1953).
179. Jones, G., Dole, M., *J. Amer. Chem. Soc.* **51**, 2950 (1929).
180. Joshi, K. M., Saxena, S. C., *Physica*, **27**, 329 (1961).
181. Jossi, J. A., Stiel, L. I., Thodos, G., *Amer. Inst. Chem. Eng. J.* **8**, 59 (1962).
182. Jost, W.: *Diffusion in Solids, Liquids, Gases*. New York 1952.
183. Junk, W. A., Comings, E. W., *Chem. Eng. Progr.* **49**, 263 (1953).
184. *Kalendarz Chemiczny* (*Handbook of Chemistry*—collective work). Warszawa 1954–1955.
185. Kamal, M. R., Canjar, L. N., *Amer. Inst. Chem. Eng. J.* **8**, 329 (1962).
186. Kamal, M. R., Canjar, L. N., *Chem. Engng.* **69**, No. 25, 159 (1962).
187. Kaminsky, M., *Z. physik. Chem.* Neue Folge **12**, 206 (1957).
188. Kaminsky, M., *Disc. Faraday Soc.* 171 (1957).
189. Kammerlingh-Onnes, H., *Comm. Phys. Lab. Leiden*, No. 71, 25, 74 (1901); 1188 (1910); *Konink. Akad. Wetens* 273 (1912); Keesom, W. H., *Comm. Phys. Lab. Leiden*. Suppl. 24a, 23 (1912).
190. Kapustinskii, A. F., Ruzavin, I. I., *Zhurn. Fiz. Khim.* **29**, 2222 (1955).
191. Karapet'yants, M. H., Chen-Huang Iuc, *Khim. Prom.* **1963**, No. 3, 192.
192. Katayama, M., *Sci. Rept., Tohoku Imp. Univ.* **4**, 373 (1916); *J. Chem. Soc.* 110, II, 219 (1916).
193. Kay, W. B., *Ind. Eng. Chem.* **28**, 1014 (1936).

194. Kendall, J., Monroe, K. P., *J. Amer. Chem. Soc.* **39**, 1802 (1917).
195. Kern, D. Q.: *Progress Heat Transfer*. New York 1950.
196. Ketelaar, J. A. A.: *Chemical Constitution*. New York 1953.
197. Keyes, F. G., *Trans. Amer. Soc. Mech. Eng.* **74**, 1303 (1952).
198. Kharbanda, O. P., *Ind. Chemist.*, **31**, 187 (1955).
199. Kincaid, J. H., Eyring, H., *J. Chem. Phys.* **6**, 620 (1938).
200. Kincaid, J. G., Eyring, H., Stearn, H. E., *Chem. Revs.* **28**, 301 (1941).
201. Kirilov, I. I., *Dokl. Akad. Nauk S.S.S.R.* **114**, No. 1, 106 (1957).
202. Klibanova, T. M., Pomerantsev, V. V., Frank-Kamenetskii, D. A. *Zhurn. Tekhn. Fiz.* **12**, 14 (1942).
203. Kopylov, N. I., *Inzhenerno-fizicheskii Zhurn.* **3**, No. 11, 97 (1960).
204. Krichevskii, I. R., Khazanova, M. E., Linshitz, L. R., *Inzhenerno-fizicheskii Zhurn.* **3**, No. 10, 117 (1960).
205. Krichevskii, I. R., Khazanova, M. E., Lesnevskaya, L. S., Polyakova, Z. A., *Khim. Prom.* **1962**, No. 2, 105; *Int. Chem. Engng.* **2**, 438 (1962).
206. Krieger, F. J., *Rand Corporation Report RM* 646 (1951), after *Thermodynamic and Transport ProPerties of Gases, Liquids and Solids*, p. 43 (Symposium ASME, Febr. 23–26, 1959).
207. Krieve, W. F., Mason, D. M., *Amer. Inst. Chem. Eng. J.* **7**, 277 (1961).
208. Kundt, A., *Annalen d.Ph.u.Ch.N.F.* **12**, 538 (1881); Slovinski, E. J., Jr., Gates, E. E., Waring, Ch. E., *J. phys. Chem.* **61**, 808 (1957); Hocott, C. R., *Trans. Amer. Inst. Min. Met. Engr.* **132**, 184 (1939); Jones, D. T. *ibid.* **118**, 245 (1931); Hough, E. W., Rzasa, H. J., Wood, B. B., *J. Petrol, Technol.* **3**, 57 (1951); Hauser, E. A., Michaels, A. S., *J. Phys. Colloid. Chem.* **52**, 1157 (1948); *J. Phys. Chem.* **55**, 408 (1951).
209. Kunitz, M., *J. Gen. Physiol.* **9**, 715 (1926).
210. Kuong, J. F., *Chem. Engng.* **68**, No. 12, 258 (1961).
211. Kuong, J. F., *Petroleum Refiner.* **41** (3), 175 (1962).
212. Kuong, J. F., *Brit. Chem. Eng.* **8**, 356 (1963).
213. Kurgaev, E. F., *Dokl. Akad. Nauk S.S.S.R.* **132**, 392 (1960).
214. Kurtz, S. S., Sankin, A., *Ind. Eng. Chem.* **46**, 2186 (1954).
215. Kuvshinskii, E. V., *Zhurn. Tekhn. Fiz.* **18**, 67 (1948).
216. Landolt-Börnstein: *Physikalisch-Chemische Tabellen.*
217. Lehmann, H., *Chem. Techn.* **9**, 530 (1957).
218. Leland, Th. W., Jr., Mueller, W. H., *Ind. Eng. Chem.* **51**, 597 (1959).
219. Lenoir, J. M., Comings, E. W., *Chem. Eng. Progr.* **47**, 223 (1951).
220. Lenoir, J. M., Junk, W. A., Comings, E. W., *Chem. Eng. Progr.* **49**, 542 (1953).
221. Lewis, W. K., Squires, L., *Oil and Gas J.* p. 92, 15.XI.1934.
222. Li, J. C. M., Chang, P., *J. Chem. Phys.* **23**, 518 (1955).
223. Licht, W., Stechert, D. G., *J. Phys. Chem.* **48**, 23 (1944).
224. Liley, P. E.: *Thermodynamic and Transport Properties of Gases, Liquids and Solids*, p. 45. Symposium ASME, Febr. 23–26 (1959).
225. Lima, F. W., *J. Phys. Chem.* **56**, 1052 (1952).
226. Lindsay, A. L., Bromley, L. A., *Ind. Eng. Chem.* **42**, 1508 (1950).
227. Litvinov, N. D., *Zhurn. Fiz. Khim.* **35**, 1745 (1961).
228. Loschmidt, J., *Wien. Ber.* **2**, 468 (1870).
229. Lutskii, A. E., *Zhurn. Fiz. Khim.* **29**, 983 (1955).
230. Lutskii, A. E., *Zhurn. Fiz. Khim.* **29**, 1162 (1955).
231. Lutskii, A. E., *Zhurn. Fiz. Khim.* **29**, 983 (1955); **30**, 1547 (1956).
232. Lutskii, A. E., *Zhurn. Fiz. Khim.* **31**, 377 (1957).
233. Lutskii, A. E., Obukhova, E. M., *Zhurn. Fiz. Khim.* **31**, 1693 (1957).
234. Manley, R. H. J., Mason, S. B., *J. Colloid Sci.* **7**, 354 (1952).
235. Marinin, V. A., *Zhurn. Fiz. Khim.* **30**, 129 (1956).
236. Markov, V. K., *Zhurn. Fiz. Khim.* **27**, 919 (1953).
237. Mason, E. A., *J. Chem. Phys.* **28**, 1000 (1958).
238. Mason, E. A., Saxena, S. C., *Physics of Fluids*, **1**, 361 (1958).
239. Mason, E. A., von Ubisch, H., *Physics of Fluids*, **3**, 355 (1960).
240. Maxwell, J. C., *Phil. Trans.* **156**, 249 (1860).
241. Maxwell, J. C., *Collected Works*, Vol. II. Cambridge 1890.
242. Mayinger, F., *Int. J. Heat and Mass Transfer*, **5**, 807 (1962).

243. McAllister, R. A., *Amer. Inst. Chem. Eng. J.* **6**, 427 (1960).
244. McLeod, D. B., *Trans. Faraday Soc.* **19**, 38 (1923).
245. McGowan, J. C., *Chem. and Ind.*, May 31 (1952).
246. Meissner, H. P., Michaels, A. S., *Ind. Eng. Chem.* **41**, 2782 (1949).
247. Meissner, H. P., *Chem. Eng. Progress* **45**, 149 (1949).
248. Michaels, A. S., Alexander, R. S., Becker, C. L., *Ind. Eng. Chem.* **42**, 2332 (1950).
249. Mikheev, M. A.: *Osnovy teploperedachi* (*Principles of Heat Transfer*). Moscow 1949.
250. Miller, C. C., *Proc. Roy. Soc.* (London) A **106**, 724 (1924).
251. Miller, L., Carman, P. C., *Trans. Faraday Soc.* **55**, 1831 (1959); Carman, P. C., Miller, L., *Trans Faraday Soc.* **55**, 1838 (1959).
252. Miller, D. G., *Ind. Eng. Chem. Fundamentals*, **2**, No. 1, 78 (1963).
253. Misic, D., Thodos, G., *Amer. Inst. Chem. Eng. J.* **7**, 264 (1961).
254. Monchick, L., Mason, E. A., *J. Chem. Phys.* **35**, 1676 (1961).
255. Neogy, R. K., Chosh, B. N., *J. Indian Chem. Soc.* **30**, No. 2, 113 (1953).
256. Newton, R. H., *Ind. Eng. Chem.* **27**, 302 (1935).
257. Ning Hsing Chen, *Ind. Eng. Chem.* **51**, 1494 (1959).
258. Ning Hsing Chen, Othmer, D. F., *J. Chem. Eng. Data*, **7** (1), 37 (1962).
259. Nokay, R., *Chem. Engng.* **66**, No. 4, 147 (1959).
260. Novikov, I. I., Vukalovich, M. P.: *Uravnenie sostoyaniya real'nykh gazov* (*Equation of State of Real Gases*). Moscow 1948.
261. Novikov, I. I., *Atomnaya Energiya*, **2**, 468 (1957).
262. Olander, D. R., *Amer. Inst. Chem. Eng. Journ.* **7** (1), 175 (1961).
263. O'Neal, C. Jr., Brokaw, R. S., *The Physics of Fluids*, **5** (5), 567 (1962).
264. Onnes, H. K., *Commun. Phys. Lab. Univ. Leiden.* No. 12 (1894).
265. Onsager, L., Fuoss, R. M., *J. Phys. Chem.* **36**, 2689 (1932).
266. Osipov, O. A., *Dokl. Akad. Nauk S.S.S.R.* **102**, 1171 (1955).
267. Osminin, I. P., *Vestnik Mosk. Gosud. Univ.* **12**, No. 2, 117 (1957).
268. Othmer, D. F., Conwell, J. W., *Ind. Eng. Chem.* **37**, 1112 (1945).
269. Othmer, D. F., Josefowitz, S., *Ind. Eng. Chem.* **38**, 111 (1946).
270. Othmer, D. F., *Chimie et Industrie*, **59**, 552 (1948).
271. Othmer, D. F., Thakar, M. S., *Ind. Eng. Chem.* **45**, 589 (1953).
272. Othmer, D. F., Hung Tsung Chen, *Ind. Eng. Chem. Des. Dev.* **1** (4), 249 (1962).
273. Owens, E. J., Thodos, G., *Amer. Inst. Chem. Eng. J.* **3**, 454 (1957).
274. Owens, E. J., Thodos, G., *Amer. Inst. Chem. Eng. J.* **6**, 676 (1960).
275. Palenko, A. M.: *Teploemkost' prirodnogo gaza* (*Heat Capacity of Natural Gas*). Kiev 1956.
276. Palmer, G., *Ind. Eng. Chem.* **40**, 89 (1948).
277. Panchenkov, G. M., *Dokl. Akad. Nauk S.S.S.R.* **59**, 329 (1945); **63**, 701 (1948); **51**, 361, 453 (1946); *Zhurn. Fiz. Khim.* **20**, 811, 1011 (1946); **21**, 187 (1947).
278. Panchenkov, G. M.: *Teoriya vyazkosti zhidkostei* (*Theory of Viscosity of Liquids*). Moscow 1947.
279. Panchenkov, G. M., *Zhurn. Fiz. Khim.* **24**, 1390 (1950).
280. Panchenkov, G. M., *Zhurn. Fiz. Khim.* **34**, 1883 (1960).
281. Parisot, P. E., Johnson, E. F., *J. Chem. Eng. Data*, **6**, 263 (1961).
282. Partington, J.: *An Advanced Treatise on Physical Chemistry*. London 1949.
283. Pavlov, K. F., *Zhurn. Khim. Prom.* **6** (1936).
284. Pavlov, K. F., Romankov, P. G., Noskov, A. A.: *Primery i zadachi po kursu protsessov i apparatov khimicheskoi promyshlennosti* (*Examples and Problems on Chemical Engineering*). Leningrad 1959.
285. Perry, J. H.: *Chemical Engineers' Handbook*. 2nd Ed. New York 1941.
286. Pilař, A.: *Lineárni závislosti v chemicko-inženýrských výpočtech* (*Linear Relationships in Chemical Engineering Calculations*). Praha 1957.
287. Pitzer, K., *J. Chem. Phys.* **7**, 583 (1939); de Boer J., *Physica*, **14**, 139 (1948); *Reports on Progress in Physics*, **12**, 305 (1949).
288. Pitzer, K. S., Lippman, D. Z., Curl, R. F., Huggins, C. M., Petersen, D. E., *J. Amer. Chem. Soc.* **77**, 3433 (1955).
289. Plank, R., *Chemie-Ing.-Technik*, **31**, 576 (1959).
290. Polson, A., *J. Phys. Colloid Chem.* **54**, 649 (1950).
291. Poltorak, O. M., *Zhurn. Fiz. Khim.* **31**, 1949 (1957).
292. Popovich, S., Pop, M., *Zhurn. Fiz. Khim.* **35** (1), 48 (1961).
293. Pospekhov, D. A., *Zhurn. Prikl. Khim.* **26**, 644 (1953).

294. POSPEKHOV, D. A., *Zhurn. Fiz. Khim.* **30**, 228 (1956).
295. POWELL, R. E., ROSEVEARE, W. E., EYRING, H., *Ind. Eng. Chem.* **33**, 430 (1941).
296. POWELL, R. E., ROSEVEARE, W. E., EYRING, H., *J. Appl. Phys.* **12**, 669 (1941).
297. PRATAP, R., NARSIMHAN, G., *Brit. Chem. Engng.* **7**, 757 (1962).
298. PRAUSNITZ, J. M., GUNN, R. D., *Amer. Inst. Chem. Eng. J.* **4**, 430, 494 (1958).
299. PREDVODITELEV, A. S., *Zhurn. Eksp. Tekhn. Fiz.* **4**, No. 8, 1 (1934); **8**, No. 3 (1938); *Izv. VTI* No. 9–10 (1946); *Zhurn. Fiz. Khim.* **22**, 339 (1948); *Izv. Akad. Nauk S.S.S.R. OTN* No. 4 (1948); *Vestnik Mosk. Gosud. Univ.* No. 3 (1949).
300. PREDVODITELEV, A. S.: *Collection of Works Devoted to Lazarev, P.P.*, pp. 84–112, 1956.
301. PRIGOGINE, I.: *Molecular Theory of Solutions.* Interscience. New York 1957; PRIGOGINE, I., BELLEMANS, A., NAAR-COLIN, C., *J. Chem. Phys.* **26**, 751 (1957).
302. PRINS, J., *Z. Phys.* **56**, 617 (1929); MARK, H., *Z. Phys.* **54**, 505 (1929); KRATKY, O., *Phys. Z.* **34**, 482 (1933).
303. RAMM, V. M.: *Absorptsionnye protsessy v khimicheskoi promyshlennosti* (*Absorption Processes in Chemical Industry*). Moscow 1951.
304. RAMSAY, W., SHIELDS, J., *Phil. Trans.* **184**, 647 (1893).
305. RAO, R. V. G., SUBBA RAO, V. V., *Z. phys. Chem.* **221**, 357 (1963).
306. REID, R. C., SHERWOOD, T. K.: *The Properties of Gases and Liquids.* New York 1958, 2nd Ed. 1966.
307. REIK, H. G., *Z. Elektrochem.* **59**, 35, 126 (1955).
308. RICHARDS, TH. W., MATHEWS, J. H., *Z. phys. Chem.* **61**, 449 (1908); RICHARDS, TH. W., PALITZSCH, *J. Amer. Chem. Soc.* **41**, 58 (1919).
309. RIEDEL, L., *Z. Elektrochem.* **53**, 222 (1949); *Chem. Ing. Techn.* **24**, 353 (1952); **26**, 259 (1954); **28**, 419 (1956).
310. RIEDEL, L., *Chem. Ing. Techn.* **23**, 59 (1951).
311. RIEDEL, L., *Chem. Ing. Techn.* **23**, 321 (1951).
312. RIEDEL, L., *Chem. Ing. Techn.* **26**, 83, 679 (1954); **27**, 209, 475 (1955); **28**, 557 (1956).
313. RUBIN, F., *Chem. Eng. Progr.* **47**, 541 (1951).
314. SAKIADIS, B. C., COATES, J., *Amer. Inst. Chem. Eng. J.* **1**, 275 (1955); A Literature Survey of Ultrasonic Velocities in Liquids and Solution, *Louisiana State Univer. Eng. Expt. Sta. Bull.* 46. Baton Rouge Louisiana 1954.
315. SAXENA, S. C., NARAYANAN, T. K. S., *Ind. Eng. Chem. Fundamentals*, **1** (3), 191 (1962).
316. SCHÄFER, K., FOZ GAZULLA, O. R., *Z. Phys. Kol. Chem.* **B52**, 299 (1942).
317. SCHAEFER, CH. A., THODOS, G., *Ind. Eng. Chem.* **50**, 1585 (1958).
318. SCHAEFER, CH. A., THODOS, G., *Amer. Inst. Chem. Eng. J.* **5**, 367 (1959).
319. SCHEFFY, J. W., JOHNSON, E. F., *J. Chem. Eng. Data*, **6**, 245 (1961).
320. SCHEIBEL, E. G., *Ind. Eng. Chem.* **46**, 1574, 2007 (1954).
321. SCHEIBEL, E. G., *Ind. Eng. Chem.* **46**, 2007 (1954).
322. SCHMIDT, C., *Gas u. Wasserfach*, **25**, 92 (1942).
323. SEIHUN CHANG, TAIKYUE REE, EYRING, H., MATZNER, I.: *Progress in International Research on Thermodynamic and Transport Properties.* Second Symposium on Thermophysical Properties ASME, Jan. 24–26, 1962.
324. SEINER, J. A., *Chem. Engng.* **65**, No. 19, 176 (1958).
325. SHIROKOV, M. F.: *Viazkost' zhidkostei. Soveshchanie po viazkosti zhidkostei* (*Viscosity of Liquids. Conference on the Viscosity of Liquids*). Akad. Nauk S.S.S.R., Vol. II, p. 35, 1944.
326. SHTRAUF, E. A.: *Molekularnaya fizika* (*Molecular Physics*). Moscow 1949.
327. SHUKLA, R. P., BHATNAGAR, R. P., *J. Phys. Chem.* **59**, 989 (1955); **60**, 809 (1956).
328. SITURAMAN, R., IBRAHIM, S. H., KULOOR, N. R. (after *Chem. Eng. Sci.* **17**, 1087 (1962)).
329. SLATER, J. C., KIRKWOOD, J. G., *Phys. Rev.* **37**, 682 (1931).
330. SLATTERY, J. C.: *Diffusion Coefficients and the Principles of Corresponding States.* M. S. Thesis. University of Wisconsin, 1955 (after Wilke[404]).
331. SLATTERY, J. C., BIRD, R. B., *Amer. Inst. Chem. Eng. J.* **4**, 137 (1958).
332. SLAVYANSKII, V. T., *Zhurn. Fiz. Khim.* **27**, 1776 (1953); SHIMANSKII, N. I., RAVIKOVICH, S. D., *ibid.* **29**, 48 (1955).
333. SMITH, J. F., *Ind. Eng. Chem.* **22**, 1246 (1930); **23**, 416 (1931); *Trans. Amer. Soc. Mech. Engrs.* **58**, 719 (1936).
334. SMITH, R. L., WATSON, K. M., *Ind. Eng. Chem.* **29**, 1408 (1937); KURATA, F., KATZ, D. L., *Trans. Amer. Inst. Chem. Eng.* **38**, 995 (1942); DAVIS, P. C., BERTUZZI, A. F., GORE, T. L., KURATA, F., *J. Petrol. Technol.* **6**, 37 (1954); GRIEVES, R. B., THODOS, G., *Amer. Inst. Chem. Eng. J.* **6**, 561 (1960); ETTER, D. O., KAY, W. B., *J. Chem. Eng. Data* **6**, 409 (1961).

335. Smith, A. S., Brown, G. G., *Ind. Eng. Chem.* **35**, 705 (1943).
336. Smith, W. T., Greenbaum, S., Rutledge, G. P., *J. Phys. Chem.* **58**, 443 (1954).
337. Souders, M., *J. Amer. Chem. Soc.* **60**, 154 (1938).
338. Souders, M., Matthews, C. S., Hurd, C. O., *Ind. Eng. Chem.* **41**, 1037, 1048 (1949).
339. Srinivasan, M. K., Prasad, B., *Phil. Mag.* **33**, 258 (1942).
340. Srivastava, B. N., Barna, A. K., *J. Chem. Phys.* **32**, 427 (1962).
341. Stefan, J., *Wien. Ber.* **65** (2), 361 (1872).
342. Stefan, J., *Wien. Ann.* **42**, 269 (1891).
343. Steinhardt, R. G., Jr., *J. Phys. Chem.* **64**, 170 (1960).
344. Steinour, H. H., *Ind. Eng. Chem.* **36**, 618, 840, 901 (1944).
345. Stewart, G., *Rev. Mod. Phys.*, **2**, 116 (1930); **4**, 1 (1932); *Phys. Rev.* **32**, 153 (1927); **33**, 889 (1929); **35**, 291, 296, 726 (1930); **37**, 9 (1931); **38**, 931 (1931).
346. Stiel, L. I., Thodos, G., *Amer. Inst. Chem. Eng. J.* **7**, 611 (1961); **8**, 229 (1962).
347. Stiel, L. I., Thodos, G., *J. Chem. Eng. Data*, **7**, 234 (1962).
348. Stiel, L. I., Thodos, G,. *Amer. Inst. Chem. Eng. J.* **8**, 527 (1962).
349. Stolarov, E. A., Ipatev, V. V., Teodorovich, V. P., *Zhurn. Fiz. Khim.* **24**, 166 (1950).
350. Stolarov, E. A., *Zhurn. Fiz. Khim.* **24**, 279 (1950); Stolarov, E. A., Ipatev, V. V., Teodorovich, V. P., *Zhurn. Fiz. Khim.* **24**, 166 (1950).
351. Su, G. J., Huang, P. H., Chang, Y. M., *J. Amer. Chem. Soc.* **68**, 1403 (1946).
352. Sugden, S., *J. Chem. Soc.* **125**, 1177 (1924).
353. Sutherland, W., *Phil. Mag.* **36**, 507 (1893).
354. Sutherland, W., *Phil. Mag.* **40**, 421 (1895).
355. Sutherland, W., *Phil. Mag.* **9**, 781 (1905).
356. Syrkin, I. K., Dyatkina, M. E.: *Khimicheskaya sviaz' i stroyene molekul* (*Chemical Bond and Molecular Structure*). Moscow 1946.
357. Tamura, M., Kurata, M., *Bull. Chem. Soc. Japan*, **25**, 32 (1952).
358. Tarzimanov, A. A., *Inzhenerno-Fizicheskii Zhurnal*, **4** (9), 86 (1961).
359. Tat'evskii, V. M.: *Khimicheskoye stroyenie uglevodorodov* (*Chemical Structure of Hydrocarbons*). Moscow 1952.
360. Tat'evskii, V. M., *Zhurn. Fiz. Khim.* **35**, 2090 (1961); Tat'evskii, V. M., Bendirskii, V. A., *Zhurn. Obshch. Khim.* **28**, 1733 (1958).
361. Tat'evskii, V. M., Iarovoi, S. S., *Zhurn. Fiz. Khim.* **35**, 2409 (1961).
362. Taylor, G. J., *Proc. Roy. Soc.* (London) **138A**, 41 (1932).
363. Teplo- i massoperenos (Heat and Mass Transfer). Papers from the 1st Congress 5–9.VI.1961, Minsk, S.S.S.R. Amirkhanov, A. I., Adamov, A. P., Levina, L. N., Vol. I, p. 105. 1962.
364. Thiesen, M., *Verh. Dtsch. phys. Ges.* **4**, 348 (1902).
365. Thodos, G., *Amer. Inst. Chem. Eng. J.* **1**, 165, 168 (1955); **2**, 508 (1956); **3**, 428 (1957); Forman, J. Ch., Thodos, G., *ibid.* **4**, 356 (1958).
366. Thomas, L. H., *J. Chem. Soc.* **1946**, 573.
367. Thorpe, T. E., Rodger, J. W., *Phil. Trans.* **185**, 397 (1895); **189**, 90 (1897).
368. Tomassi, W.: *Jony i atomy jako składniki materii* (*Ions and Atoms as Constituents of Matter*). Warszawa 1952.
369. Toporov, A. P., Brodskaya, G. A., *Zhurn. Fiz. Khim.* **34**, 1879 (1960).
370. Tripathi, R. C., *J. Indian Chem. Soc.* **18**, 411 (1941).
371. Tsederberg, N. V., *Teploenergetika*, **3**, No. 7, 7 (1956).
372. Tsederberg, N. V., *Teploenergetika*, **3**, No. 9, 42 (1956).
373. Tsederberg, N. V., *Teploenergetika*, **4**, No. 1, 45 (1957).
374. Ulibin, S. A., *Teploenergetika*, **9** (8), 93 (1962).
375. Usmanov, A. G., Bakirov, N. U., *Zhurn. Fiz. Khim.* **36**, 591 (1962).
376. Uyehara, O. A., Watson, K. M., *Natl. Petroleum News*, Tech. Sec. **36**, R. 764 (1944) (after Hougen, O. A., Watson, K. M.: *Chemical Process Principles*, Vol. III, 870 (1947)).
377. Van der Waals, J. D., *Z. phys. Chem.* **13**, 657 (1894).
378. Van der Wyk, A. J. A., *Nature*, **138**, 845 (1936).
379. Vand, V. J., *J. Phys. Colloid Chem.* **52**, 277, 300 (1948).
380. Vargaftig, N. B., *Izv. Vsesoyuznogo Teplotekhn. Instituta*, No. 8 (1948).
381. Vargaftig, N. B., *Izv. Vsesoyuznogo Teplotekhn. Instituta*, No. 7 (1951).
382. Vargaftig, N. B., Osminin, I. P., *Teploenergetika*, **3**, No. 7, 11 (1956).
383. Vargaftig, N. B., Tarzimanov, A. A., *Teploenergetika*, **8**, No. 6, 5 (1961).

384. Vargaftig, N. B., Zaitseva, L. S., *Inzhenerno-Fizicheskii Zhurn.* **6**, No. 5, 3 (1963).
385. Vinograd, J. R., McBain, J. W., *J. Amer. Chem. Soc.* **63**, 2008 (1941).
386. Vogel, H., *Ann. Physik*, **43**, 1235 (1914).
387. Vogel, A., *J. Chem. Soc.* 1833 (1948).
388. Volyak, L. D., *Teploenergetika*, **5**, No. 7, 33 (1958).
389. Walden, P., *Z. Elektrochem.* **14**, 712 (1908); *Z. phys. Chemie*, **65**, 129, 257 (1908–9); **66**, 385 (1909); *Z. anorg. Chem.* **15**, 7, 41 (1926).
390. Walker, R. E., Westenberg, A. A., *J. Chem. Phys.* **32**, 436 (1962).
391. Wang, R. H., Knudsen, J. G., *Ind. Eng. Chem.* **50**, 1667 (1958).
392. Waring, C. E., Becher, P., *J. Chem. Phys.* **15**, 488 (1947).
393. Watson, K. M., *Ind. Eng. Chem.* **35**, 398 (1943).
394. Wassiliewa, A., *Physik Z.* **5**, 737 (1904).
395. Weber, H. F., Wiedemanns A., *Ann. phys. Chem.* **10**, 103, 304, 472 (1880); **48**, 173 (1893).
396. Weinaug, C. F., Katz, D. L., *Ind. Eng. Chem.* **35**, 239 (1942); Reno, G. J., Katz, D. L., *Ind. Eng. Chem.* **35**, 1091 (1942).
397. Weissberger, A. (Editor): *Physical Methods of Organic Chemistry*. Vol. I, Part II. C.P. Smyth. New York 1949.
398. Weissman, S., Mason, E. A., *J. Chem. Phys.* **37**, 1289 (1962).
399. Wilke, C. R., *Chem. Eng. Progr.* **45**, 218 (1949).
400. Wilke, C. R., *Journ. Chem. Phys.* **18**, 517 (1950).
401. Wilke, C. R., *Chem. Eng. Progr.* **46**, 95 (1950); Fairbanks, D. F., Wilke, C. R., *Ind. Eng. Chem.* **42**, 471 (1950).
402. Wilke, C. R., Lee, C. Y., *Ind. Eng. Chem.* **47**, 1253 (1955).
403. Wilke, C. R., Chang, P., *Amer. Inst. Chem. Eng. J.* **1**, 264 (1955).
404. Wilke, C. R., Prausnitz, J. M., *Ind. Eng. Chem.* **49**, 577 (1957).
405. Wohl, K., *Zts. phys. Chem.* **B 14**, 36 (1931).
406. Zawidzki, J., *Z. phys. Chem.* **39**, 129 (1909).
407. Zdanovskii, A. B., *Zhurn. Fiz. Khim.* **29**, 209 (1955).
408. Zdanovskii, A. B., *Zhurn. Fiz. Khim.* **34**, 1380 (1960); **36**, 657 (1962).

INDEX

OTHER TITLES IN THE SERIES IN CHEMICAL ENGINEERING

Vol. 1. WILKINSON: Non-Newtonian Fluids
Vol. 2. JAMRACK: Rare Metal Extraction by Chemical Engineering Techniques
Vol. 3. NAGIEV: The Theory of Recycle Processes in Chemical Engineering
Vol. 4. BRADLEY: The Hydrocyclone
Vol. 5. ŠTĚRBÁČEK and TAUSK: Mixing in the Chemical Industry
Vol. 6. HOBLER: Mass Transfer and Absorbers
Vol. 7. ROSENBROCK and STOREY: Computational Techniques for Chemical Engineers
Vol. 8. STRAUSS: Industrial Gas Cleaning
Vol. 9. KING: Phase Equilibrium in Mixtures
Vol. 10. BROWN and RICHARDS: Principles of Powder Mechanics